Birkhäuser

Modeling and Simulation in Science, Engineering and Technology

More information about this series at http://www.springer.com/series/4960

Luca D'Acci
Editor

The Mathematics of Urban Morphology

Foreword by Michael Batty

Birkhäuser

Editor
Luca D'Acci
Interuniversity Department of Regional
and Urban Studies and Planning
Politecnico di Torino
Torino, Italy

Erasmus School of Social and Behavioural
Sciences
Erasmus Universiteit Rotterdam
Rotterdam, The Netherlands

ISSN 2164-3679 ISSN 2164-3725 (electronic)
Modeling and Simulation in Science, Engineering and Technology
ISBN 978-3-030-12380-2 ISBN 978-3-030-12381-9 (eBook)
https://doi.org/10.1007/978-3-030-12381-9

Library of Congress Control Number: 2019930642

Mathematics Subject Classification (2010): 91D10, 91B72

This book is published under the imprint Birkhäuser, www.birkhauser-science.com by the registered company Springer Nature Switzerland AG
The registered company address is: Gewerbestrasse 11, 6330 Cham, Switzerland

Foreword: The Morphology of Cities

I once took the ICE train from Frankfurt to Weimar to give a talk about cellular automata modelling to architectural students at the original Bauhaus. In Weimar, I not only visited the buildings where Walter Gropius set up his school that fostered the modern movement, encapsulated in Louis Sullivan's hallowed phrase 'form follows function', but I also discovered that it was in Weimar where Goethe spent most of his life, writing of many things but in particular about 'form', exploring the way plants grew and metamorphized. In 1817, there he wrote: "Form is a thing in motion, in the process of becoming, of passing away. The study of form is the study of transformation. The study of metamorphosis is the key to all the signs of nature." (J. W. von Goethe, *Zur Morphologie*, Stuttgart, Germany: J. G. Cotta Publishers, 1817, 201).

Form follows function has been the dominant mantra not only for modern architecture throughout most of the twentieth century, but for urbanism too and most historical studies of city form are structured according to the notion that cities develop around markets where there is competition to trade goods and where nearness to these points of exchange is more valued than locations further away. Thus, cities are denser nearer their centres and their form thus represents a balance between centralizing forces that concentrate on activities and decentralizing forces that attempt to capture more space, thus diffusing activities as far from their centres as possible. This, then, is the economic trade-off that determines the form of the city with respect to the juxtaposition of all its key activities.

Traditionally, the form of the city is one whose descriptions are both verbal and visual. In his introductory essay which follows next, the editor Luca D'Acci takes several quotes from Italo Calvino's *Invisible Cities* (Harcourt, Brace, and Jovanovich, 1978), which is a speculation on ideal cities but expressed entirely verbally. In contrast, many works focus on the visual structure of cities, often as maps and plans, but also linking to the third dimension and betraying a strong architectural and aesthetic interest that has been the dominant force in the descriptive study of urban morphology (for example see Spiro Kostoff's books *The City Shaped*, Thames and Hudson, 1993 and *The City Assembled*, Thames and Hudson, 1999). Exploring urban morphology using more abstract and mathematical methods, however, apart from occasional forays into the geometry of cities in the nineteenth century, did not begin until the late 1950s when M.R.G. Conzen first began to study the geographic structure of small towns at the level of plots and streets. In the 1970s, various approaches coming from the formal study of building layouts, as for example, in work from the Centre for Land Use and Built Form Studies (see Lionel March and Philip Steadman's wonderful statement of this new movement in *The Geometry of Environment*, RIBA Press, 1971), provided another twist to these perspectives. Thinking about the geometric structure of cities as it was reflected in the design process also generated another theme in the serious development of urban morphology. Christopher Alexander's contributions, particularly his book *Notes on the Synthesis of Form* (Harvard, 1964) strongly define this tradition.

The mathematical approach, however, did not really make itself felt until about 30 years ago and this followed in the wake of complexity theory. In the 1950s and 1960s, the systems approach became fashionable and although this was entirely consistent with formal and computational approaches to cities and their design, the city itself resisted any formal theories other than those that were developed within location theory and urban economics. The idea of form was thus implicit in these largely aggregate spatial theories. The problem was partly that the systems approach focused on thinking of cities as machines, to be fashioned and planned from the top-down, whereas more generally in science from that time, the idea that systems evolved from the bottom up was beginning to make itself felt. Essentially cities (and many other kinds of social system—the economy, for example) came to be thought of as more like organisms than machines. Biology thus began to replace physics as the dominant metaphor and the notion that cities largely evolve rather than being planned in any sense, took hold. In other words, complexity theory emerged and it was this that introduced the new thinking about urban morphology of which this book is a statement.

Much of what has developed since then which can be badged rather loosely under the label of mathematical morphology, is illustrated in the collection of articles gathered together here. The book is largely structured around the main key approaches. The properties of cities with respect to scale and size are introduced in diverse ways and new geometries based on fractals constitute the first part of the book, where form and function using ideas from statistical physics are presented. Transformations of form in cities, of course, are key to complex systems and the processes of generating forms using ideas from cellular automata follow these first chapters on fractals. The focus then changes to networks. There are a number of chapters on elementary methods of looking at cities as network structures as in space syntax, where there is a slow transition to ideas that are emerging in network science. The emphasis here, however, is on streets as networks rather than on clusters and one of the slowly emerging themes in this whole domain involves how networks and clusters might be reconciled—in particular through ideas such as percolation theory.

The wider context of complexity theory as a frame of reference for studies of urban morphology is then presented and this is followed by a series of studies which do not fit into the classic frame of fractals, cellular automata or statistical physics. These are then followed by commentaries, vignettes almost, some of which recall historical developments, others of which are more speculative summaries of the field, defining further directions for research. In fact, although the urban morphological work of M.R.G. Conzen in the 1960s was the first of the formal approaches to be developed in this broad domain, it is in this last section that the Conzenian tradition is best explained, in fact by his son and others. In fact, the younger Conzen's own contribution to this book presented in the last part focuses on how we might integrate the many approaches developed here with his father's tradition and those of his colleagues such as Whitehand. However, the major part of the book deals with different approaches but the last part introduces speculations on the future of the field and on ways in which these ideas can be connected to developments in computing, transport and related features of the smart city that are energizing the field. The collection of essays here is quite rounded and although there are areas of morphology that are not covered in detail, all the pointers to the wider field mean that the diligent reader can easily navigate the broad domain to which the contributions here ascribe.

In the first part which is devoted to fractals, the key focus is on how self-similar structures that reflect the hierarchy of central places and of streets—that is of locations and interactions, nodes and networks—sort themselves out through competition. The results of any competitive process result in the rich getting richer and the poor getting poorer. If a lower bound is put on poverty which might be a basic form of existence or survival even, then what happens is that the competitive process which is essentially random, generates power laws which relate the frequency of occurrence of an object with its size. All the papers in this section end up showing how the classic signatures of fractal morphology are power laws, often stated in the form of the counter-cumulative frequency distribution of city (and other) sizes, which is usually called Zipf's Law. This applies across the board, from the way streets are ordered to the correlations between clusters of self-similar development. Chen leads the way with a classic demonstration of correlation and fractals. Hsu and Zou illustrate rather nicely how power laws arise quite naturally from ideas about central places, while Bee and his colleagues show how Zipf, Gibrat and Pareto can all be related as power laws. Neillson and Gil do the same for streets also measuring organic growth, while Mengyuan and colleagues illustrate that for Brisbane, urban form is less fractal, meaning less circuitous than other city examples from around the world. This leads rather easily to a short second part where cellular automata models are introduced by Antoni and his colleagues, these generating fractal forms, while the section is concluded by a foray into the origins of cellular automata with respect to self-replicating forms which lie at the basis of computation, presented by Clarke.

The third part deals with networks, essentially street networks although the ideas might be applied to other forms. Volchenkov introduces the mathematics of networks and then there is a long chapter which summarizes space syntax by Mahbub that, in turn, is followed by a summary of applications of syntax by van Nes. Boeing then examines some very basic proprieties of networks, in particular the extent to which actual walking networks are more efficient than driving networks. He thus raises the thorny issue of how circuitous networks should be, thus implying that networks built out of straight-line segments are too simple to represent the properties of compactness and directedness in cities. In the fourth part, three contributions stand back a little from the technical expositions that dominate the book thus far, and these deal with complexity, per se. Goh and colleagues introduce emergence of complexity in urban morphology, focussing on generative models and these relate to those in the first part. Bellomo and Terna look at transformations and metamorphosis, following Goethe to an extent and invoking images from Escher. Jiang completes the section looking at Christopher Alexander's ideas of 'wholeness' with respect to entire city systems, where he focuses more on topological properties, but without losing ideas about hierarchy.

In the fifth part, a series of interesting but somewhat ad hoc set of ideas about mathematical morphology are introduced. Schirmer and Axhausen examine multiple scales directly and these have powerful links to earlier papers. In fact, most papers deal with development and clusters on one scale only and although many scales are dealt with, most of the contributions do not compare data or outcomes across scales. Raimbault then deals with interactions between networks and locations, while Huynh introduces percolation and point pattern analysis. Marshall and colleagues deal with compactness and Li and Ratti show how to extract morphologies from *Google Street View* in the form of street canyons and related volumes. Last but not least, spatial interaction models are introduced for the first time by Burger and colleagues and this serves to tie together key ideas involving locations, networks and clusters of development.

The book then moves to commentaries that provide a rather refreshing foil to the technical essays that dominate most of the book. Roger White first argues that in cellular modelling, which one might take as being morphological modelling, it is algorithms not mathematics that count. This is quite a controversial point of view but it reflects a number of key paradoxes and conflicts between different approaches which are also echoed in Trevor Barnes' ideas that we should begin to work hard on integrating mathematical models and urban social theory. Ron Johnston introduces his own experience with Conzenian urban morphology and its wider relation to social area analysis, and Peter Larkham takes a wide view of how one might think about integrating complexity ideas, fractals, cellular automata and networks into these traditional morphological approaches. Chris Rogers argues that infrastructure and engineering issues should be integrated with these ideas while Carola Hein and Tino Mager maintain that we need a deeper discussion of computation in the context of humanistic systems such as cities. Vitor Oliveira links agent-based models to traditional morphology and then Diane Davis and Andres Sevtsuk look at what might happen to streets when we move to an era of automated vehicles. Steffen Lehmann then argues that density is the key to the future city, thus echoing many of the contributions to this volume.

In this last part, Michael Conzen's commentary requires particular mention. Like Ron Johnston and Trevor Barnes, he augments his father's and his own approach which we have referred to here as a more descriptive geographical perspective on morphology, with a call for integrating the diversity of approaches presented here. His clear and balanced discussion of the way urban morphology has expanded its boundaries as a focus of study provides a useful synthesis of where the field is heading and thus provides a fitting point to the future. To reinforce his argument, he issues a challenge for the field: to develop the many alternative approaches on a single place, so a true comparative analysis might be developed. Many insights could flow from this. Such a project has been attempted before in comparing different land use transport models for the same place and there are many pitfalls. But considerable learning about how far each approach enriches our understanding of cities would be the result.

This book provides a useful perspective on the state of the art with respect to urban morphology in general and mathematics as tools and frames to disentangle the ideas that pervade arguments about form and function in particular. There is much to absorb in the pages that follow and there are many pointers to ways in which these ideas can be linked to related theories of cities, urban design and urban policy analysis as well as new movements such as the role of computation in cities and the idea of the smart city. Much food for thought. Read on, digest, enjoy.

London, UK

Michael Batty
The Bartlett School of Planning, CASA
University College London (UCL)

Contents

On Urban Morphology and Mathematics

Luca D'Acci

Abstract This chapter introduces the isolated and joined meaning of the terms Urban, Morphology and Mathematics. After a brief synthesis of the re-emerging new science of cities and urban morphology, it speculates about cities conceived between hard and soft sciences under the complexity perspective.

Cities are ultimately the results of human behaviour, whose understanding might be reductive if solely framed within a strictly 'mathematical' confine.[1] Human behaviour 'is so complex and influenced by such a wide range of factors that any claim to provide precise, deterministic prediction is unrealistic' (Inglehart 2018, p. 10).

But there is another side of the coin: quantification and mathematical modelling are means enabling us to discover partially predictable macro paths of our behaviours otherwise unreadable. Even if 'not deterministic [...,] some trajectories are more probable than others' (Inglehart 2018, p. 11): the mathematical language helps both in seeing these trajectories and in quantifying these probabilities.

1 Morphology, Urban and Mathematics

The expression 'mathematics of urban morphology', has three terms: mathematics, urban and morphology; whose individual meanings are the following:

'Morphology' is built from the suffix 'logy', from ancient Greek λογία, meaning 'the study of', and *morphé* meaning 'form', 'shape'. It intrinsically involves a

[1] Among many, we just recall as example William S. Jevons's ambition to mathematically formulate human behaviour as the ultimate in scientific soundness.

L. D'Acci (✉)
Politecnico di Torino, Castello del Valentino, Turin, Italy
e-mail: luca.dacci@polito.it

Erasmus Universiteit Rotterdam, Rotterdam, The Netherlands
e-mail: dacci@essb.eur.nl

L. D'Acci (ed.), *The Mathematics of Urban Morphology*,
Modeling and Simulation in Science, Engineering and Technology,
https://doi.org/10.1007/978-3-030-12381-9_1

quantitative component, as it is defined 'the scientific study of the structure and form of…'[2]; where 'scientific' evokes the *scientific method* introduced in the seventeenth century in the natural sciences 'consisting in systematic observation, measurement, and experiment […]'.[3]

'Urban' is related to the Latin word *urbs* used to indicate the physicality of the city, as an ensemble of built structures.[4] According to Pomponio,[5] who recalls the authority of Alfenus Varus, *urbs* comes from *urbum*, the handle of the plough, used in the Roman tradition to define the border of a new city, and *urbare* means the action of 'defining with the plough'. Varrone[6] prefers to associate the etymology of *urbs* to *orbis*; the *orbis urbis principium* of the Romans was the founding element of the edification of the city. The *orbis* constitutes the *principium urbis* and is the strip of land around the *urbs* which is generated by the plough (De Sanctis 2007).

'Mathematics', from Greek μάθημα máthēma, originally meant 'knowledge, study, learning'. Among many definitions,[7] is nowadays often defined as the science of number, quantity and space, as abstract concepts or as applied to other disciplines.[8]

The meanings arising when grouping together these terms—mathematics, urban and morphology—become:

'Urban morphology' is *the study of urban form*. It concerns the size, shape and physical structure of urban settlements. The American Planning Association sees urban morphology as the understanding of 'the spatial structure and character of an urban area by examining its patterns and the process of its development'[9]; Kropf (2017, p. 9) as 'concerned with the form and structure of cities, towns and villages, the way that they grow and change and their characteristics […]; and Batty as 'patterns of urban structure based on the way activities are ordered with respect to their locations'[10] stressing how the urban morphology, as [*physical*] patterns of urban structure, is linked with *activities*.

[2]https://dictionary.cambridge.org/dictionary/english/morphology (Cambridge University Press 2018).

[3]https://en.oxforddictionaries.com/definition/scientific_method (Oxford University Press 2018).

[4]The Latin used another term, *civitas*, to refer to the city as an ensemble of people, which said in Rousseau's words becomes: 'les maisons font la ville, mais […] le citoyens font la cite" (Rousseau 1762, reprint 2011, Livre 1, Chap. 6, note p. 57) (houses make the "ville" [urbs], but citizens make the "cite" [civitas]). While the Greeks used the word *polis* to indicate the city as the place where to exercise politics.

[5]"Urbs ab urbo appellata est: urbare est aratro definire. Et Varus ait urbum appellari curvaturam aratri, quod in urbe condenda adhiberi solet' Pomponio, *Digesta*. Libro 16, 239.

[6]Varrone, *De lingua Latina*. V, 143.

[7]A poetical one from Nicola Bellomo (Highly cited in mathematics, from 2014 by WEB of SCIENCE), is "Mathematics: a virtual bridge that starts from the phenomenon in search of the noumenon" (Nicola Bellomo personal communication, 3 May 2018).

[8]https://en.oxforddictionaries.com/definition/mathematics (Oxford University Press 2018).

[9]https://www.planning.org/tuesdaysatapa/2010/aug.htm.

[10]http://discovery.ucl.ac.uk/15183/1/15183.pdf.

'Mathematics of urban morphology' injects a double dose of 'quantitative' in the study of urban form: one from 'mathematics', one from the *scientific* study intrinsic in the term morphology we saw at the beginning of this introduction.

2 Urban Morphology

Urban morphology is a relatively young field which is becoming, as Whitehand (2016) underlines, more and more significant[11] and multidisciplinary[12] and its aim is to 'contribute to our understanding of the built environment as a complex physical object, a cultural artefact and quasi-natural phenomenon […]' (Kropf 2017 p. 9).

A better understanding of urban forms helps the availability of measures, models and analysis, which in turn helps to make effective previsions (that eventually lead to decisions) on urban form issues, which is among the most urgent needs and with the longest lasting effects, in the sustainable development arena of a world whose current urbanization and population growth rates are unique in history.

Rooted in the morphological tradition of European geographers interested in the urban layout forms in-between the nineteenth and twentieth centuries (human geographers such as Schlüter and Geisler, whose work Conzen later brought to the British school), urban morphological studies in urban design and architecture, began in Italy (Muratori and his student Caniggia) and France (Castex, Panerai, Depaule and later in the 1980s Borie, Micheloni, Pinon and Fortier) in the 1950s and 1960s as an attempt to explain the structural continuity of traditional cities, after the failure of the modern movement; while in North America still during the second half of the twentieth century urban layout studies focused on aesthetics or economics, but historians of architecture (Fitch, Garvan, Lewis, Mayer, Whitehill, Kennedy), made some early progress in the study of urban forms (Rashid 2017, pp. 23–25).

We might cluster urban morphology lines into four: the (mostly British) human geography field (influenced by Conzenian's description and classification approach, recently carried on by Whitehand and his Birmingham School[13]); the (mostly Italian) architecture/planning field (as the Caniggian school underlining urban structure's components); the utopian approach (in search for ideal cities, e.g. Le Corbusier,

[11] "[…] the number of indexed documents containing the term "urban morphology" increased from 26 in 1991–95 to 363 in 2011–15, more than doubling in successive 5-year periods from 1996–2000 to 2011–15 (Web of Science, All Databases, accessed 30 January 2016). This is almost certainly indicative of a significant increase in the actual amount of work undertaken […]' (Whitehand 2016 http://www.urbanform.org/online_public/2016_1_editorial.shtml).

[12] '[…] journals carrying documents that contain the term "urban morphology" […] In 1991–95 just 21 […] 2015 roughly triple' (Whitehand 2016 http://www.urbanform.org/online_public/2016_1_editorial.shtml).

[13] 'Founded in 1974, the Urban Morphology Research Group (UMRG), in the School of Geography, Earth and Environmental Sciences at the University of Birmingham, is the major centre in the United Kingdom for the study of the geographical aspects of urban form'. https://www.birmingham.ac.uk/research/activity/urban-morphology. See also Oliveira, V. (Ed) (2019) *J.W.R. Whitehand and the Historico-geographical Approach to Urban Morphology*. Springer.

Howard and Wright); and the anthropological perspective (Lynch, Jacobs, Alexander) whose interest was to observe 'what actually works in real cities. This sowed the seeds of a mathematical approach for quantifying urban morphology' (Xiao 2017, p. 42), where Alexander was among the first to attempt the introduction of formal mathematics to the subject of design.

3 Mathematical Treatment of Urban Forms

Since then, the mathematical treatment of the urban forms has become more and more sophisticated, and, generally speaking, mathematics has proved to be an efficient language to deal with the aim of the scientific tradition in urban studies (rooted in neoclassical economics and embracing urban and regional economics, urban and economic geography), which is to generate a theory of urban form which is scientific in nature: namely formal, deductive, and based on postulates of human behaviour.

The *scientisation* process[14] of the study of urban forms in the planning sector dates back at least to Ildefonso Cerda (1867), often regarded as the father or Urbanism, who promoted a rational urban method with the declared aim to define a kind of urban science, even if his top-down view is sometimes conceptually distant from the bottom-up emergence paradigm which mostly characterizes the new urban science of today framed within the recent science of complexity.[15]

This 'rational' attitude in planning urban forms, would travel even much further back into the past if we think about the strictly geometrical proportional rules used in Ancient China to design the size of ex novo cities and their growth; the Harappan civilization planned towns, the earliest known gridiron urban form planning (e.g. Harappa, ~2150 BC, Kalibangan, Lothal, Mohenjo-daro); the ancient Egyptian gridiron planning (e.g. Tel-el Amarna, ~1346 BC, and Kahun, 1853 BC) as the Hippodamian's of the Ancient Greeks (started with the rebuilding plan of Miletus, 479 BC); the *centuratio* of the Romans; or more recently the grid of European colonies, the Renaissance symmetries, order and idealistic rationality, and the Baroque lavish geometries (Morris 1994; Benevolo 1980).

However, the mathematical approach to urban forms of contemporary times is not much oriented toward the above examples—which we may better define as top-down planning of forms rather than the science of cities we intend today. The latter is more oriented towards the analytical (e.g. Geoffrey West's research) finding for universal laws, and in the phenomenological (e.g. Michael Batty's research) understanding of processes dealing with the emergence of various urban phenomena as forms, which

[14]Here used in the common sense, without the literal meaning of 'scientific approach' and without meaning the use of mathematics.

[15]One of the most known catch-phrase of complexity science is the antireductionism affirmation that 'the whole is more than the sum of its parts', ironically implicit in Hofstadter book: 'Reductionism is the most natural thing in the world to grasp. It's simply the belief that "a whole can be understood completely if you understand its parts, and the nature of their 'sum'". No on in her left brain could reject reductionism. (Hofstadter 1979).

are treated as a physical expression of our behaviours to be understood within the complexity science framework with the help of statistical physics, computation and formal models. Recent availability of disaggregated data, together with the betterment of computer capacity and mathematical tools, allow a proper introduction of the complexity paradigm in the study of cities, by, among many examples, multi-agent-based modelling and behavioural economic elements acting as theoretical/technical tools to simulate interactions whose positive/negative feedbacks induce emergent phenomena dynamically translated into urban forms. This enormous availability of data, presents a great opportunity for the science of cities, or, to use the words of Marina Alberti, a challenge to 'turn the unprecedented floods of data into new knowledge about urban systems and novel insights for their effective management' (Alberti 2017 p. 2).

If we enlarge the overview of the use of mathematics more generally to urban and regional modelling (which still influence urban forms), the modern history of these models began with the contribution of von Thünen (1826), the Prussian economist, agriculturalist and social theorist who wished to understand in more depth human society and economics thanks to mathematical formulation of socio-economic principles.

A century later, two theorists writing in the tradition of German location theory, the geographer Walter Christaller (1933) and the economist August Lösch (1940), proposed, independently, formal theories of the spatial locations of towns and cities which also indirectly relate to urban sizes, see also Isard (1965).

Mathematical language carried on being successfully used in urban phenomena with, to mention only a few, Alfred Weber (1929) in his optimal location models; Ira Lowry (1964) in his mathematical model for spatial organization of urban activities; William Alonso (1964) and Richard Muth (1969) in their urban economic reinterpretation of the agricultural von Thünen model, which in turn relates to the Ricardian rent theory (Ricardo 1821), more recently re-elaborated in much more sophisticated mathematics in Fujita (1989) and Fujita et al. (2001).

The modelling approach to cities, 'emphasizes the development of robust theoretical constructs [...] it seeks to fashion precise analytic representations of the world [...] creating formalized, empirically testable, descriptions of the social world [...] They may also be presented as mathematical formula [...] closely related to [...] physical sciences' (Koch and Latham 2017, p. 9).

To use the expression of John Landis, 'if there is a golden age of urban models, it is surely now' (Landis 2012, p. 323). The list of authors of mathematical formulations for models (static or dynamic, aggregated or disaggregated) in some way related to urban morphology to mention in order to provide a fairly complete picture is enormous, particularly in the last decades. We can, however, briefly describe three main groups within the research community adopting mathematical models for spatial structures, which started fragmenting by the 1980s: one group oriented itself towards inferential statistics rather than formal theory and models; another towards mathematical models rather than empirical; another abandoned both statistical and mathematical theoretical modelling preferring postmodern approaches (White et al. 2015, p. 53).

A fine way to express the importance of bridging theorists (close to the first group above mentioned) and empiricists (second group) is described from Melanie Mitchell in her eminently readable book: 'the more established the theory or principles, the more sceptical you have to be of any contradicting facts, and conversely the more convincing the contradicting facts are, the more sceptical you need to be of your supposed principles. This is the nature of science – an endless cycle of proud proposing and disdainful doubting' (Mitchell 2011, p. 295).

More critical to the scientific lack in urban studies is Geoffrey West, former President of the Santa Fe Institute, in the beginning of the chapter 'Toward a science of cities' of his book: 'Almost all theories of the city are largely qualitative, developed primarily from focused studies on specific cities or groups of cities supplemented by narratives, anecdotes, and intuition' (West 2017, p. 269).

To carry on later on the same book with the necessity to mix qualitative and quantitative approaches in the study of cities: 'much of what we experience in cities [...] is embodied in this quantitative framework. In this respect it should be viewed as complementary to traditional social science and economic theories whose character is typically more qualitative, more localized, more based on narrative, less analytic, and less mechanistic' (West 2017, p. 325).

As the Géographie Vautrin Lud prize[16] winner Michael Batty of the Centre for Advanced Spatial Analysis (CASA) tells us (2018), the idea of a science of cities is not new; what is new is the availability of data and technologies.

Using the words of Koonin and Holland (2014), the new opportunity is 'to collect and analyze data that will allow us to characterize and quantify the 'pulse of the city'. We are not alone in believing that a new science of cities is emerging'.

Batty bases this new urban science on three principles: relations, scale and prediction; Goodwin (2004) indicates science to identify quality in cities through analytic-cognitive tools; and Ball (2004) claims how patterns of movement and activity can provide a more vigorous approach to urban planning and he went so far as to define a 'Physics of Societies'.

Hillier believes in the enormous potential of scientific analysis of movement's patterns, and recommends that urban designers have to internalize this knowledge and, according to complexity science, to use the self-organizational behaviour of cities. He (Hillier 2004) noted that the making of cities is both fully an art and fully a science: 'The art of urban design, as I firmly believe it to be, does rest on the foundation of the science of space'.

Hanson's ideas (2004) follows Alexander's magnum opus, *The Nature of Order* which is based on organized complexity principles—recalling pioneers like Jacobs—arguing that order does not mean top-down design, but should emerge from the actions of countless individuals.

All the above quarrels are sympathetic to Batty's approach, and more generally, quoting Wolfram's opera (2002), to our *New Kind of Science*.

In one of the classics of urban science from Patrick Geddes written over a century ago, we read '[...] appear the methods of a Science of Cities – that our cities should

[16]See https://en.wikipedia.org/wiki/Vautrin_Lud_Prize.

be individually surveyed, scientifically compared: as their architecture long has been – cathedral with cathedral, style with style' (Geddes 1915).

More recently, also West, claims the urgency to approach urban studies scientifically, quantitatively, when he firmly says: 'we desperately need a serious scientific theory of cities' (West 2011).

As biology shows, despite life being the most complex and varied system in the universe, an astonishing simplicity is expressed by relationships among variables and the same universality was found in cities when numerous variables are plotted against city sizes (Batty 2013a, b, Bettencourt 2013); 'all the data shows it's the same, despite the fact that these cities have evolved independently. Something universal is going on. The universality is *us* – *we* are the city' (West 2011).

4 Cities Between Hard and Soft Sciences

Perhaps it is inside the 'us' the reason why most theories treat cities as a *social* system rather than spatial, or why we often see *urban studies* departments/institutes/groups under macro categories labelled *Social Science* (such as, among many, the University of Oxford, the University of Amsterdam and the University of Sheffield), *Arts and Humanities* (as the University of Cambridge), *Humanities* (as the University of Manchester), *Social and Political Sciences* (as the University of Glasgow), *Arts, Humanities and Social Sciences* (as Cardiff University[17]), or under multidisciplinary-interuniversity departments between *Hard* and *Soft* sciences (as Politecnico di Torino[18]).

The abovementioned statement of West ('The universality is *us* – *we* are the city') encloses the polyhedric nature of urban studies and urban forms, and is, in some senses, somewhat close to the Latin terms to define cities, *civitas*, or the Aristotle[19] conception of city, more radically rooted in the political nature of human beings,

[17]Although this university also categorizes urban studies under Physical Sciences and Engineering, inside the School of Architecture.

[18]DIST: the Interuniversity Department of Regional and Urban Studies and Planning between Politecnico of Torino and University of Torino, excitingly bringing in an effervescent team of 235 people (75 professors, 59 researchers, 66 visiting professors-researchers, and 34 technicians-administrators) Economists, Historians, Anthropologists, Engineers, Architects, Urbanists, Geographers, Naturalists, Computer Scientists, Sociologists, Mathematicians, Agronomist and Ecologists. It has been recently qualified as one of the best Italian Departments within the already prestigious list of Departments of Excellence in Research.

[19]'[...] man is by nature a political animal; and so even when men have no need of assistance from each other they none the less desire to live together. At the same time they are also brought together by common interest, so far as each achieves a share of the good life. The good life then is the chief aim of society, both collectively for all its members and individually' (Aristotle, *Politics*, III, 1278b). 'It is clear that all partnerships aim at some good, and that the partnership that is most authoritative of all and embraces all the others does so particularly, and aims at the most authoritative good of all. This is what is called the city or the political partnership' (1252a3). '[T]he city belongs among the things that exist by nature, and...man is by nature a political animal' (1252b30–1253a3).

whose common interest notion, *to koinon sympheron*, anticipates the Cicerone's *utilitas rei publicae*[20] and *utilitas communis*.[21] In this sense, the 'us' may be interpreted as an inextricable bundle of forces intrinsically (universally?) built in us, which, even if with significant variations due to culture, climate, geography and historical individual paths, acts in a perpetual search for an 'equilibrium' among often contradictory forces; one of these forces is the Aristotelian common interest.

As soon as we adventure on the learning of (whatever) forms, we discern that they are always due to the action of some sort of force. D'Arcy Thompson describes the form of an object as 'a "diagram of forces" in this sense, at least, that from it we can judge or deduce the forces that are acting or have acted upon it'; forms come from an 'interaction or balance of forces' (D'Arcy Thompson 1917, p. 11).

Forces shaping cities and their forms come from economics reasoning of individuals, firms, companies and governments, behaviours and instincts of humans, cultural processes, historic paths, geographic-environmental influences/constrictions, technology, utopia, politics, alliances, markets, flows, trades and communications, physical and socio-economics networks, private/public interests, striving for well-being (objective and subjective, personal and collective), … all acting contemporarily at different physical and temporal scales; all the above is 'us', a multiscale, multitemporal summation of our behaviours and decisions, conscious or not, rational or not, instinctive or not.

These forces constantly change their reciprocal 'equilibrium' and so the related urban forms, in a continuous dynamic evolution. This urban form metamorphosis is poetically expressed by Lefebvre 'la ville est un emploi du temps' (Lefebvre 2001, p. 224), and Baudelaire[22] 'Le vieux Paris n'est plus (la forme d'une ville change plus vite, hélas! que le cœur d'un mortel)', and mathematics is an efficient language for its understanding.

5 In Search of Universal Laws: From Calvino to Santa Fe

Modern urban scientists, supported from unprecedented amounts of data on almost every aspect of cities, their availability and mathematical–technical capacity of elaborations, are starting to systematically analyse and model cities, 'in the hope of uncovering underlying laws governing the dynamics and the evolution of these systems' (Barthelemy 2016, p. xii).

Statistical regularities systematically emerge despite immense variety among cities. We may expect two reasons for this universality: that systems with large numbers of components lead to collective behaviours with statistical regularities; and the existence of fundamental processes shared to all cities such as spatial organization of activities and residences, mobility, etc. … (Barthelemy 2016, p. 1).

[20]Cicerone M.T. (85 b.C. circa). *De inventione*. I, 40.

[21]Cicerone M.T. (44 b.C. circa). *De Officiis*. III.

[22]'Le Cygne' in the section 'Tableaux parisiens' of 'Les Fleurs du Mal'.

Of a different opinion is the architectural historian Berkeleyan professor Spiro Kostof: 'cities are too particular as phenomena – specific to moments in time and to the vicissitudes of site and culture – to be pinned down by absolute taxonomies' (Kostof 2014a, p. 8). Thought shared from West himself: 'It may be that the sort of quantitative "physics-inspired" theory of cities that I am advocating is simply not conceivable. Cities and the process of urbanization may be just "too complex" to be subjected to laws and rules that transcend their individuality in a useful way' (West 2017, p. 269), even if he is optimistically oriented towards a truly scientific approach to cities and its, at least up to a certain extent, universality principles: 'science at his best is the search for commonalities, regularities, principles, and universalities that transcend and underlie the structure and behaviour of any particular individual constituent, whether it be a quark, a galaxy, an electron, a cell, an airplane, a computer, a person, or a city. And it is at its very best when it can do that in a quantitative, mathematically computational, predictive framework [...]' (West 2017, p. 269).

In search of a unifying law, almost three decades ago, also the distinguished mathematician Mitchell Feigenbaum asked (and replied to) himself: 'Are there intrinsic geometries that describe various chaotic motions, that serve as a unifying way of viewing these disparate nonlinear problems, as kindred? I ask the question because I know the answer to be affirmative in certain broad circumstances. The moment this is accepted, then strongly nonlinear problems appear no longer as each one its own case, but rather coordinated and suitable for theorizing upon as their own abstract entity'; he later continues toward a both qualitative and quantitative universality of behaviours 'an even stronger notion than this generality of shared qualitative geometry is the notion of universality [...] this shared geometry is not only one of a qualitative similarity but also one of true quantitative identicality'[23] (Feigenbaum 1992, p. 4).

Also, the Nobel Prize Niels Bohr wrote about the use of mathematics with the help of unread phenomena: 'for more and more phenomena, their governing laws were wrung from Nature and their rules were recognized. Simultaneously, mathematics developed hand in hand with the natural sciences, and thus an understanding of the nature of a phenomenon soon came to also include the discovery of an appropriate mathematization of it' (Bohr 1992, p. 11). We should underline the other way around too: the progress and use of mathematics help the discovering of processes behind phenomena.

Mathematics is essential in this regard, for analysing data, founding scientific theories, creating quantified models, where 'quantified model' doesn't *necessarily* mean a model fitting to real data but 'a theoretically consistent model whose parameters are based on some mix of data and assumptions, so that realistic simulation exercises can be carried out' (Fujita et al. 2001, p. 347).

---

[23] He was referring to the universality (today known as the Feigenbaum constants) in the transition to chaos that several systems share, regardless of their *practical* nature, as long as are mathematizible in a certain way.

In a Santa Fe Institute lecture,[24] Melanie Mitchell interviews Luis Bettencourt, who replies: 'we need to create a statistical theory of what cities are. We have a very poor idea of their statistical character. For example, scaling laws only give you an idea of how a system behaves on the average, given its size, etc...., but when you look at a particular city it's never quite that number. What creates these deviations?'.

Bettencourt's reply and the quantitative approach among urban scientists towards universal laws, slightly remind me Kublai Kan's reply to Marco Polo in their imaginary dialogue in Calvino's romance: 'I have constructed a model of city in my mind from which deduct every possible city. It comprehends everything corresponding to the norm. Since the cities that exist diverge in varying degree from the norm, it is enough to predict the exceptions to the norm and calculate the most probable combinations'.

In a within city context, syntactic analysis of street networks of large numbers of cities highlighted another kind of universal law: 'spatially speaking, and at a deep enough level, cities seem to be the same kind of thing [...] there is at a deep enough level a *generic city*, that is, a structure that makes a city a city in the first place [...]' (Hillier 2016, p. 200).

Hillier's *generic city*, Bettencourt's 'statistical city', and Kublai's comprehensive city in the Calvino's romance, tease our thoughts about the goal of a science of cities, that, according to Barthelemy, 'will be reached when, considering a specific case, we can basically say what will happen and which ingredients it is necessary to introduce in a model in order to get more detailed information and predictions' (Barthelemy 2016, p. 3).

In Popper's idea, science is 'the art of systematic over-simplification – the art of discerning what we may with advantage omit' (Popper 1992), such as the exceptions we cannot predict from general laws.

Sceptical about quantitative universal meanings, if without being accompanied from proper qualitative historical/cultural analysis, is again the historian architect Kostof: 'city form is neutral until it is impressed with specific cultural intent. So there is no point in noticing the formal similarities [...] unless we can elaborate on the nature of the content that was to be housed within each, and the social premises of the designers', in fact we cannot be really able to read cities until we turn 'to the archives, the history books, the old maps – until we assemble all the evidence, some of it often contradictory, that will help explain how a particular downtown got the look it now has' (Kostof 2014b, pp. 10, 11).

Other historians, as the Professor of Ancient History Arjan Zuiderhoek, well show their consciousness of the importance of the understanding of some kind of universality in urban form elements; he writes in his recent book about the observation of 'striking similarities between cities across space and time, particularly in terms of layout and the general structure of urban landscapes' (Zuiderhoek 2017, p. 6) which, using the words of the archaeologist Monica Smith, suggests 'that the capacities for human interaction in concentrated locations [villages, towns, cities] are exercised

---

[24]Introduction to Complexity, 2018. Video-lecture 10.4. https://www.complexityexplorer.org/courses/89-introduction-to-complexity.

within a limited set of parameters' (Smith 2003, pp. 3–8), which may be seen in evolutionary terms by using the words of the anthropologist Glenn Storey: 'human nucleation behaviour into cities might be a form of group selection strategy that has proved eminently adaptable for humans and has fostered strong interspecific ties of cooperation' (Storey 2006, p. 23). Zuiderhoek continues promoting an interdisciplinary universal urbanism: 'Along such broad interdisciplinary lines, combining insights from human geography, ecology and evolutionary biology, we may eventually be able to arrive at some universal understanding of urbanism'; and, similarly to the Kublai's city model in the Calvino's romance, and the Bettencourt's ambition for a statistical theory of city, Zuiferhoek defends the importance that the understanding of such hidden urban spatial patterns revealing universal mechanisms at work has also for those historians and archaeologists qualitatively interested in specific urban cultures rather than quantitatively deduced macro-universal-laws: 'the broad comparative study of world urbanism does supply us with a rough cross-cultural template that can be used to sketch the outlines of a particular type of urbanism, in order to bring out, as sharply as possible, its cultural specificities' (Zuiderhoek 2017, p. 7).

There are signs for believing that these universal behaviours urban scientists are trying to uncover are linked with the self-organization property of complex systems as cities are.

6 Complex Self-organizing Systems Urban Approaches

Ilya Prigogine's original investigation on self-organization systems, in the 1950s and 1960s, focused on far-from-equilibrium chemical systems and the emergence of macro-scale spatial structures in chemical reactions.[25]

Prigogine's group and many others, extended their research about self-organization to other fields as biology and urban systems, continuing to focus on the spatial structure, the physical forms taken, or better saying, *emerging* from these systems.

The ultimate message is that, whatever complex system is being analysed—economic, social, urban, biological, chemical, physical—characteristic spatial structures emerge.

Three main schools posed the basements to complex self-organizing systems urban approach: Brussels school, Santa Fe school, and CASA school.

The Brussels school is associated with the research of Prigogine and investigates how *real* systems (whether chemical, physical, social …) behave (under energy input).

The Santa Fe school, from the Santa Fe Institute, focuses on the algorithmic logic behind complex systems; it is to some extent more related to *abstract* systems virtu-

[25]The classical example is the Belousov–Zhabotinsky reaction which takes place in a shallow dish where the concentration of chemicals is kept far from equilibrium, and visible patterns may appear, such as spirals, concentric circles, multi-armed spirals… (Nicolis and Prigogine 1989).

ally explored by computer modelling, seeking to capture universal, general ways to explain how self-organization and adaptation of complex systems happen. It emphasizes the *analytic-mathematical* traditions of physics toward underlying laws.

The CASA school is rooted in the work of Michael Batty whose *phenomenological* accent is linked with the social, physical and geographical sciences traditions, by passing through *abstract* simulations too. It lies in some sense in between the first two schools.

The language to understand and model these emergences of complex systems, regardless if governed by simple or complicated rules, is logical-mathematical.

All these systems 'come into existence by virtue of processes that create a spatial structuring of their constituent elements' (White et al. 2015, p. 4): for the urban systems, these processes are the 'us' of West which are studied by the related disciplines (sociology, demography, economics, psychology, organization theory, sociobiology, anthropology …).

Complexity and self-organizing systems theories' application to urban studies is 'an attempt to generalize and formalize the qualitative understandings developed within the framework of the humanities and social sciences' (White et al. 2015, p. 14).

7 Laplace's Demon in Cities

The aim of this scientific approach is to disclose the in-built complexity nature of the city and its morphology. Despite passing through formal methodologies and rigorous mathematical language, the scientific approach applied to cities has, somehow, the softness of social sciences and humanities in the necessary link with histories, contexts and (recalling the before mentioned Bettencourt and Calvino), especially, in their limited predictive power. In this sense, its quantitative nature should be used to scientifically extrapolate laws and formally write models, whose predictions should be *qualitatively*, rather than (as Laplace's demon would like) quantitatively, read, for the impossibility of predicting with certitude the output or the state that the system under investigation we will assume.

Within the inherent previsions incertitude of socio-economic systems as *civitas* (ensembles of people) and *urbs* (their physical manifestations, like urban forms ultimately are), mathematical language remains an essential tool to decipher their internal mechanisms.

It often helps better measurement and analyses, which, when they go together with more data availability and computational capacity, 'our ability to predict and understand and explain would be greater. […] With these advances, the thinking goes, it should be possible to get closer to Laplace's demon […]' (Feldman 2014, pp. 71, 72).

It is a positive feedback: 'As we gain knowledge, we create more sophisticated tools and these tools enable us to ask and answer new questions' (Wilensky and Rand 2015, p. 6).

8 A Science but not a Science?

Sir D'Arcy Thompson—a truly Leonardian man[26] whose Italian Renaissance polyhedric spirit would be so precious, in any epoch, for the highly multi-interdisciplinary nature of urban studies—starts his seminal book[27] *On Growth and Form*, stating that 'the criterion of truth science lay in its relation to mathematics'. D'Arcy was conveying the Kantian quote '*eine Wissenschaft, aber nicth Wissenschaft*' (referring to chemistry, Kant[28] wrote that 'it is a science but not a science') translating it into 'it is a science but not a Science'.

West (2017, p. 86), citing D'Arcy's interpretation of a Kant quote, reflects on the provocative (but for him still, with caution, valid in the spirit) argument that a science (West refers to biology) would become a Science when its principles can be mathematizable, without however misevaluating in any way sciences that per nature are (or were till now) predominantly qualitative.

If we recall what defines a science a science (systematic study of structures and behaviours through observation and experiment[29]), it is not a *condicio sine qua non* to have mathematical language explaining its structures to be classified 'science', even if it might be efficient and enlarge horizons in some occasions, or undoubtedly essential in others.

Among many authors promoting mathematical approach to urban studies were Brian Berry, who brought quantitative spatial analysis and scientific method into the study of cities, building a science of cities 'with the potential to elevate urban studies and geography to the status, influence, and certainty of physics and the other 'hard' sciences' (Wily 2017, p. 39); and Janet Anu-Lughoud, an earlier adopter of statistical model and computer-based data processing, she also likes to explore integrations of qualitative and quantitative methods in social research.

9 The Language of Mathematics

'Mathematics is a language, and an exceedingly beautiful one, and the applications of that language are vast and extensive' (Adam 2012, p. xiv). I would go even further in reminding us that mathematics, as music and arts (yet these two being culturally biased), is a *universal* language *among* humans and, perhaps more importantly, *between* humans and nature (there is a stimulating dispute if it is a language *of* humans, or *of* nature which humans happen to learn) that helps to parsimoniously

[26] 'An aristocrat of learning whose intellectual endowments are not likely ever again to be combined within one man'. That was his description by the Nobel Prize Sir Peter Medawar.

[27] Which the Nobel Prize Sir Peter Medawar described as 'the finest work of literature in all the annals of science'.

[28] Not the Kant of the *Critics of Judgment* but the one (later called by Stephen J. Gould 'physicalist') infatuated with Laplace, Euclid, Newton, of the *Critics of Pure Reason*.

[29] https://en.oxforddictionaries.com/definition/science (Oxford University Press 2018).

describe, understand and extend complicated phenomena, principles and reasoning that would have been impossible otherwise. It is both a support and an extension of our minds.

Steven Strogatz, the MIT's prized and Schurman Professor of Applied Mathematics at Cornell University, lyrically describes his being 'captivated by the mathematics of nature' and the hidden 'beautiful world that can be seen only through mathematics' in his discursive book (2003, p. 4), who his 'only' partially reminds the Pythagorean (everything is number[30]) and Galilean (mathematics is the language Nature wrote her book) views.

Similarly, the well-known mathematician and astronomer John Barrow from the University of Cambridge, is not hesitant in stating that the reason we were so good in releasing the mystery of the universe is because we discovered the language in which it was written, that, as three hundred years ago Galileo believed, is mathematics. According to Barrow, science itself exists because the natural world seems algorithmically squeezable. Mathematical formulas we call Natural Laws are efficient reductions of enormous sequences of data about changes of the state of the system (Barrow 1992, pp. 5, 93).

Mathematical equations, Marshall believed (1890, p. 10), are beneficial in 'helping a person to write down quickly, shortly and exactly, some of his thoughts'; thought shared from Samuelson in his first major opus (1947) indicate the role of equations in sharpening muddled thinking replacing it with methodical exactness.

The same happens when we deal with urban systems and their physical forms which are the ultimate result of a complex sequence of interconnected multiscale multitemporal socio-economic forces constrained by local contexts, driven inside historical distinctive paths.

History, not only mathematics, as Glaeser wisely starts one of his influential book, is essential for a complete urban study: 'understanding cities requires more than current statistics. Many urban areas are extraordinarily ancient and their structures reflect events far in the past. We also need to know history if we are to understand our cities' (Glaeser 2007, p. 2). And he carries on pointing for a gainful trade between social scientists working in anthropology or sociology, and economists.

10 The Art and Science of Cities[31]

The view of our world has regularly shifted between these two pendulum: scientific and humanistic. According to Snow (1959), one of the major barriers to solving our world's problems resides in the absence of dialogue between these 'two cultures'.

During the first half of the twentieth century, both sides were present such as the humanistic perspective of Lewis Mumford (1938) and the quantitative's of Christaller (1933), Lösch (1940), Reilly (1931) and others. During that period, the system theory

[30] Yet mathematics is not much interested in numbers on themselves but in their interrelations.

[31] Some parts from (D'Acci 2016).

approach was preeminent and, during the 50s, induced researchers to see systems as centrally ordered, and as a hierarchical sum of subsystems dominated by negative feedback.

Until the middle of the twentieth century, a standard theory of cities prevailed as an economic and transportation model, and based predominantly on the monocentric city. Ideas and models were built on statistical aggregations of units, exemplified in models based on macroeconomics, such as econometric, population and Keynesian models description.

In the 1950s the quantitative revolution criticized the scientific validity of the humanistic trend, which defined descriptive approaches.

In turn, in the early 1970s scholars adopting urban social theories, through Structuralist Marxist and phenomenological idealistic perspectives, in the qualitative revolution, criticized the positivistic-quantitative approach.

From that decade on also the underlying conceptual ideas changed: cities were observed as governed by positive feedback rather negatives, and from the bottom-up rather than top-down, giving space to the new complexity science.

According to Portugali et al. (2012), the urban complexity science can be seen as a second scientific culture of cities or as a junction between the scientific and humanistic culture.

Also Batty, even if pushes the pendulum to the quantitative side, would like to think that is humanistic too in some sense, according to his way of thinking about science; in the urban context you also need intuition: quantifications without intuition may be a black box.[32]

The quantitative-scientific approach finds universal rules aiming to see cities as part of the domain of nature and to be studied by the scientific method. The humanistic approach claims a difference between the human domain and the natural domain, so that studying cities and their phenomena as quantitative may lead to reductionism, and it finds soft hermeneutic methods more suitable.

This opposition may be only apparent, as it could be transformed into a profitable complementarity; namely referring to the scientific methods for what concerns urban phenomena that are objective and universal, and to the humanistic approach for phenomena that are not. We are also able, not always, to quantify qualitative phenomena and to qualitatively interpret quantified data.

Art is viewed as the opposing counterpart to science. Batty (2013a, b) explains terms of *science* and *art* as: 'by science, we mean an organized body of knowledge produced using commonly agreed tools and methods that can be reproduced over and over again by different individuals. [...] this is quite different from ideas and knowledge we consider art, since the production of art is individual and formed by an intuition that is personal'.

According to Croce and Read (Read 1949), for art we do not intend 'beauty'—which is a 'very fluctuating phenomenon, with manifestation in the course of history that are very uncertain and often baffling'—but *intuition*. 'As for a work of art, what we expect in a city is a personal element. Each city reveals original, unique

[32]Batty's personal conversation, 27 June 2018, University of Cambridge.

elements; each of them is *special* and, even more, is *differently* special for each of us' (D'Acci 2015).

The above call for wise 'intuition' claimed from Batty, and the systematic 'deviations' from the modelled trajectory augmented from Bettencourt, lies in what mathematical models and quantifications in the social and behavioural sciences—the ultimate foundation of cities—leave invisible. A concept well expressed in the words of the Director of the MIT System Dynamics Group, (Sterman 2002, p. 513): 'the most important assumptions of a model are not in the equations, but what's not in them; [...] not in the variables on the computer screen, but in the blank spaces around them'.

Science sees the many in the one; art the one in the many. However, this new science based on the complexity paradigm,[33] is a science that induces art, identified as personal uniqueness: each city emerges from unique contexts, from where the randomness of the micro-fluctuations, the unpredictability of the agent's behaviours' positive feedbacks and the contextual historic successions, generate although within universal macro patterns, unique scenarios; and each of them is personally read.

Bibliography

Adam J. A. (2012). *X and the city*. Princeton University Press.
Alberti M. (2017). Grand Challenge in Urban Science. *Frontiers in Built Environment*, Vol. 3.
Alonso W. (1964). *Location and Land Use*. Harvard University Press.
Aristotle. *Politics*. In: Aristotle. *Nicomachean Ethics*. Translation and historical introduction by Rowe C; philosophical introduction and commentary by Broadie S. Oxford University Press, 2002.
Ball P. (2004). *Order, Emergence and the Self-Made Tapestry*. In: The Prince's Foundation for the Built Environment, The Order of Nature. New Science, New Urbanism, New Architecture? Conference Proceedings, 20 Sept 2004.
Batty M. (2013a). *The new science of cities*. MIT press.
Batty M. (2013b). A Theory of City Size. *Science*, vol 340.
Barthelemy M. (2016). *The Structure and Dynamics of Cities*. Cambridge University Press.
Baudelaire C. (1857). *Les Fleurs du Mal*.
Bettencourt (2013). The Origins of Scaling in Cities. *Science*, vol 340.
Barrow J.D. (1992). *Perchè il mondo è matematico*. Editori Laterza (Traslation from Tortorella B).
Benevolo L. (1980). *The History of the City*. Scholar Press. London.
Bohr N. (1992). Introduction. Causality Principle, Deterministic Laws, and Chaos. In: Peitgen H., Jurgens H., Saupe D. (1992, new edition 2004). *Chaos and Fractals. New Frontiers of Sciences*. Springer.
Cerdà I. (1867). *Teoria General de la Urbanizacion*. Imprenta espanola, Madrid.
Christaller W. (1933). *Die zentralen Orte in Suddeutschland* (The Central Regions of Southern Germany). Jena.

[33] In a seminal paper about Science and Complexity, Weaver (1948) recognised how most of new humanity challenges (economic, social, political, ecological, biological) are organized complexity in their nature and 'require science to make a third great advance'. The latter started in the 1970s with Complexity Science which study how the behaviour of a system is shaped by the relationships among its parts.

Cicerone M.T. (85 b.C. circa). *De inventione*.
Cicerone M.T. (44 b.C. circa). *De Officiis*.
D'Acci L. (2016). Humanistic and scientific approaches to understanding cities. In: Inaugural Editorial of Urban Planning. *Urban Planning*, vol 1, no 1.
D'Acci L. (2015). Mathematize urbes by humanizing them. Cities as Isobenefit Landscapes: Psycho-Economical distances and Personal Isobenefit Lines. *Landscape and Urban Planning*, Volume 139, July 2015, pp. 63–81.
D'Arcy Thompson W. (1917). *On Growth and Form*. [Edition 2014, Edited by Bonner J.T., Cambridge University Press.]
De Sanctis G. (2007). Solco, muro, pomerio. Mélanges de l'École française de Rome – Antiquité, *MEFRA*, 119/2 – 2007, pp. 503–526.
Feigenbaum M.J. (1992). Foreword. In: Peitgen H., Jurgens H., Saupe D. (1992, new edition 2004). *Chaos and Fractals. New Frontiers of Sciences*. Springer.
Feldman D. (2014). *Chaos and Fractals*. Oxford University Press.
Fujita M. (1989). *Urban Economic Theory*. Cambridge University Press.
Fujita M., Krugman P., Venables A.J. (2001). The Spatial Economy. The MIT Press.
Geddes P. (1915). *Cities in Evolution*. Williams & Norgate.
Glaeser E.L. (2007). *Cities, Agglomeration and Spatial Equilibrium*. Oxford University Press.
Goodwin B. (2004). Toward a New Science of Qualities. In: *The Prince's Foundation for the Built Environment, The Order of Nature. New Science, New Urbanism, New Architecture?* Conference Proceedings, 20 September 2004.
Hanson B. (2004). *Christopher Alexander's New Paradigm*. In: The Prince's Foundation for the Built Environment, *The Order of Nature. New Science, New Urbanism, New Architecture?* Conference Proceedings, 20 September 2004.
Hillier B. (2016). What are cities for? And how does this relate to their spatial form? *The Journal of Space Syntax*, vol 6, Issue 2.
Hillier B. (2004). *The Art of Urban Design: Growth From the Science of Space*. In: The Prince's Foundation for the Built Environment, *The Order of Nature. New Science, New Urbanism, New Architecture?* Conference Proceedings, 20 September 2004.
Hofstadter D.R. (1979). *Godel, Escher, Bach: an Eternal Golden Braid*. Basic Books, New York.
Inglehart R.F. (2018). *Cultural Evolution*. Cambridge University Press.
Isard W. (1965). *Location and Space-Economy: A General Theory Relating to Industrial Location, Market Areas, Land Use, Trade, and Urban Structure*. MIT Press.
Kant I. (1781). *Critique of Pure Reason*.
Koch R., Latham A., (2017). *Key Thinkers on Cities*. Sage.
Koonin S.E., Holland M.J. (2014). *The value of big data for urban science*. In: *Privacy, Big Data and the Public Good*. Cambridge University Press.
Kostof S. (2014a). *The City Assembled*. Thames & Hudson. (First print 1992).
Kostof S. (2014b). *The City Shaped*. Thames & Hudson. (First print 1991).
Kropf K. (2017). *The Handbook of Urban Morphology*. Wiley.
Landis J.D. (2012). Modelling urban systems. In Weber R., Crane R. *The Oxford Handbook of Urban Planning*. Oxford University Press.
Lefebvre H. (2001). *Du Rural à l'urbain*. Anthropos. Paris.
Lösch A. (1940). *The Economics of Location*. Yale University Press.
Lowry I.S. (1964). *A Model of Metropolis*. Rand Corporation.
Marshall A. (1890). *Principles of Economics*. London.
Mitchell M. (2011). *Complexity. A guided tour*. Oxford University Press. Oxford.
Morris (1994). *History of Urban Form: Before the Industrial Revolutions*. Longman.
Mumford W. (1938). *The Culture of Cities*. Harcourt Brace & Company.
Muth R. (1969). *Cities and Housing*. University of Chicago Press.
Nicolis G., Prigogine I. (1989). *Exploring Complexity: An Introduction*. Freeman W.H., New York.
Oliveira, V. (Ed.) (2019). *J.W.R. Whitehand and the Historico-geographical Approach to Urban Morphology*. Springer.

Pomponio, *Digesta*. Libro 16.
Popper K. (1992).*The Open Universe*. Routledge, London.
Portugali, J., Meyer, H., Stolk, E., Tan, E. (Eds.) (2012). *Complexity Theories of Cities Have Come of Age*. Springer, Complexity Series.
Rashid M. (2017). *The Geometry of Urban Layouts*. Springer.
Read H. (1949). *The Meaning of Art*. Pelican Books.
Reilly W. J. (1931). *The law of retail gravitation*. W.J. Reilly Co, New York.
Ricardo D. (1821). *On the Principles of Political Economy and Taxation*. (Everyman Library, 1963, London).
Rousseau J-J. (1762). *Du contrat social*. Reprint, Paris, GF-Flammarion, 2011.
Samuelson P.A. (1947). *Foundations of Economic Analysis*. Harvard University Press.
Smith M.L. (2003). Introduction: the social construction of ancient cities. In: Smith M.L. (Ed.) *The Social Construction of Ancient Cities*. Washington DC & London.
Snow C.P. (1959). *The Two Cultures*. Cambridge University Press.
Sterman J.D. (2002). All models are wrong: reflections on becoming a systems scientist. *System Dynamic Review* 18: 4.
Storey G.R. (2006). Introduction: urban demography of the past. In Storey (Ed.) *Urbanism in the preindustrial world: cross-cultural approaches*. Tuscaloosa.
Strogatz S. (2003). SYNC. *How order emerges from chaos in the universe, nature, and daily life. Hachette*. New York.
Varrone, *De lingua Latina*. V.
von Thünen J. (1826). *Isolated State*. Wartenberg, C. (Trans) (Pergamon Press, 1966, New York).
Weaver W. (1948). Science and Complexity. *American Scientist* 36, p. 536.
Weber (1929). *Theory of the Location of Industries*. University of Chicago Press.
West G. (2011). *The surprising math of cities and corporations*. Ted talk 2011.
West G. (2017). *Scale*. Penguin Random House.
White R., Engelen G., Uljee I. (2015). *Modeling Cities and Regions as Complex Systems*. MIT Press.
Whitehand J.W.R. (2016). Editorial comment. *Urban Morphology*, vol 20, no 1.
Wilensky U, Rand W. (2015). *An introduction to agent-based modelling*. MIT Press.
Wily E. (2017). Brian J.L. Berry. In Koch R., Latham A., (2017). *Key Thinkers on Cities*. Sage.
Wolfram S. (2002). *A new kind of science*. Wolfram Media.
Xiao Y. (2017). *Urban Morphology and Housing Market*. Springer.
Zuiderhoek A. (2017). *The Ancient City*. Cambridge University Press.

Part I
Fractals

Fractal Dimension Analysis of Urban Morphology Based on Spatial Correlation Functions

Yanguang Chen

Abstract A number of mathematical models in urban studies are associated with spatial correlation functions. However, the theory and method of spatial correlation modeling have not been developed for urban morphology. Based on power-law urban density models, a density–density correlation function can be constructed for urban modeling. Using scaling analysis and spectral analysis, we can derive a set of fractal parameter equations, which can be employed to explore urban form and growth. The main results and findings are as follows: first, if urban density follows power law, the spatial correlation function and its energy spectral density will follow scaling law; second, the reasonable numerical ranges of fractal parameters can be derived by the ideas from multifractals; third, the spatial correlation modeling can be generalized to spatial autocorrelation and gravity models. As an example, the analytical process is applied to the city of Beijing in China to show how to use this method. A conclusion can be drawn that the scaling analysis, spectral analysis, and spatial correlation analysis can be integrated into a new framework to form the 3S analysis for urban morphology.

Keywords Fractals · Multifractals · Urban form · Scaling analysis · Spectral analysis · Spatial correlation analysis

1 Introduction

If a system has typical scales which can be represented by some characteristic length, it can be described with conventional mathematical methods based on calculus, linear algebra, or probability theory and statistics. However, urban growth and form represent a type of scaling phenomena, which bears no characteristic scale and cannot be effectively described by the traditional measures and mathematical theories. Fractal

Y. Chen (✉)
Department of Geography, College of Urban and Environmental Sciences, Peking University, Beijing 100871, China
e-mail: chenyg@pku.edu.cn

L. D'Acci (ed.), *The Mathematics of Urban Morphology*,
Modeling and Simulation in Science, Engineering and Technology,
https://doi.org/10.1007/978-3-030-12381-9_2

geometry provides a powerful tool for scale-free analysis, resulting in a number of interesting and revealing studies on urban systems (e.g., Ariza-Villaverde et al. 2013; Arlinghaus 1985; Arlinghaus and Arlinghaus 1989; Batty 1995; Batty and Longley 1994; Benguigui and Daoud 1991; De Keersmaecker et al. 2003; Frankhauser 1998, 2014; Thomas et al. 2010; Murcio et al. 2015). The fractal concept is on the basis of scaling symmetry, and symmetry suggests some invariance in transformation process. This suggests that self-similarity is equivalent to invariance under contraction or dilation (Mandelbrot 1989). In geometry, fractal property can be abstracted as power-law relations between linear sizes and the corresponding measurements. A city can be empirically treated as a random fractal system that possesses self-similar or self-affine structure (e.g., Batty 2005, 2008; Benguigui et al. 2000; Feng and Chen 2010; Frankhauser 1994; Thomas et al. 2007; Thomas and Frankhauser et al. 2008; White and Engelen 1993, 1994). In urban morphology, the spatial relation between the radius (a scale) from a city center and the corresponding urban density (a measurement) may follow the inverse power law (Batty and Longley 1994; Batty and Xie 1999; Frankhauser 1994; Makse et al. 1995). In many cases, a power law implies fractal or scaling. The typical inverse power function is Smeed's model on traffic network (Smeed 1963), which can be employed to describe urban density distribution and estimate the fractal dimension of urban form (Batty and Longley 1994).

However, the key process of fractal modeling in urban studies is not yet clear and remains to be further explored. Due to scale-free property, fractal phenomena cannot be described with traditional measures such as length, area, and density. It should be characterized by fractal dimension and the related scaling exponents. Fractal dimension can be defined from two angles of view. One is the logarithmic relation between the linear sizes of fractal copies and the corresponding entropy functions, and the other is the power-law relation between the linear sizes and the corresponding correlation functions. The inverse power law of urban density proved to be a special spatial correlation function (Takayasu 1990). Urban morphology and systems of cities have been modeled by using the ideas from spatial correlation (Chen and Jiang 2010; Makse et al. 1995, 1998). Spatial correlation is one of the most important ways of modeling both urban form and urban systems, and the correlation equations can be solved by scaling transform. In particular, a general spatial correlation function of cities can be constructed on the basis of urban density function (Chen 2011; Chen and Jiang 2010). If the correlation function is converted into energy spectrum by means of Fourier transform, the urban density will become spectral density (Chen 2008, 2013). Thus, spatial correlation analysis can be turned into spectral analysis and vice versa, and the two analytical processes are associated with scaling analysis (Chen 2009). A problem is how to integrate the scaling analysis, spectral analysis, and spatial correlation analysis into a logic framework to make a new process of spatial analysis.

This work is devoted to developing new methodological framework for urban analysis and revealing the theoretical relations between different fractal parameters of urban form. In previous studies, several problems have been solved (Chen 2008, 2010, 2013; Chen and Jiang 2010). First, a new analytical process is preliminarily developed for fractal cities. Second, a number of fractal parameter relations are theoretically derived. Third, the proper ranges of fractal parameters of urban morphology are partially revealed. This is an integrated study. The main new points are as below: First, the previous framework of models is improved according to new thinking. Second, the spatial correlation modeling is generalized to spatial interaction and spatial autocorrelation analysis. Third, a new concise 3S-based case study on urban form is presented to illustrate how to use the 3S analytical process. Fourth, based on the empirical analysis, the scaling break phenomenon of wave spectrums is revealed and associated with self-affine urban growth. The rest parts are organized as follows. In Sect. 2, scaling analysis, spectral analysis, and spatial correlation analysis will be combined into a new approach, which is based on the inverse power law on urban density. In Sect. 3, the national capital of China, Beijing, is taken as example to make an empirical analysis. In Sect. 4, the related questions are discussed, and the theoretical models are further developed. Finally, the discussion is concluded by summarizing the main points of this studies.

2 Theoretical Models

2.1 Basic Postulates

Urban morphology should be researched through proper concept of geographical space. If we examine a city by means of fractal methods based on a digital map or a remotely sensed image, the dimension of the embedding space of the city fractal is regarded as $d = 2$. In fact, a fractal city is always defined in a 2-dimension geographical space (Batty and Longley 1994; Frankhauser 1994). There are three common approaches to estimating the fractal dimension values of urban patterns, that is, box-counting methods (Benguigui et al. 2000; Chen 2012; Feng and Chen 2010; Shen 2002), area–radius scaling (Batty and Longley 1994; Chen 2010; Frankhauser 1998; White and Engelen 1993), and area–perimeter scaling (Batty and Longley 1994; Wang et al. 2005). Each method has its strong and weak points. If we want to research the spatial patterns of urban land use, the box-counting method is the best way of evaluating fractal dimension. However, if we want to explore the dynamic process of urban growth, the area–radius scaling is the best approach to estimating fractal dimension because this procedure is more consistent with the relation between urban core and periphery than other methods. In theory, the area–radius scaling is equivalent to and can be replaced by the density–radius scaling. The density–radius modeling is a good approach to research spatial correlation processes and patterns of city development. If we model 2-dimension spatial correlation through 1-dimension space, the density–radius relation is effective for scaling analysis (Fig. 1).

To explore the density–radius scaling relation of urban growth, the concepts of urban form and urban density should be clarified before the theoretical models are presented. Based on a 2-dimensional space, *urban form* can be defined as the spatial pattern of elements, which compose the city in terms of its networks, buildings, and

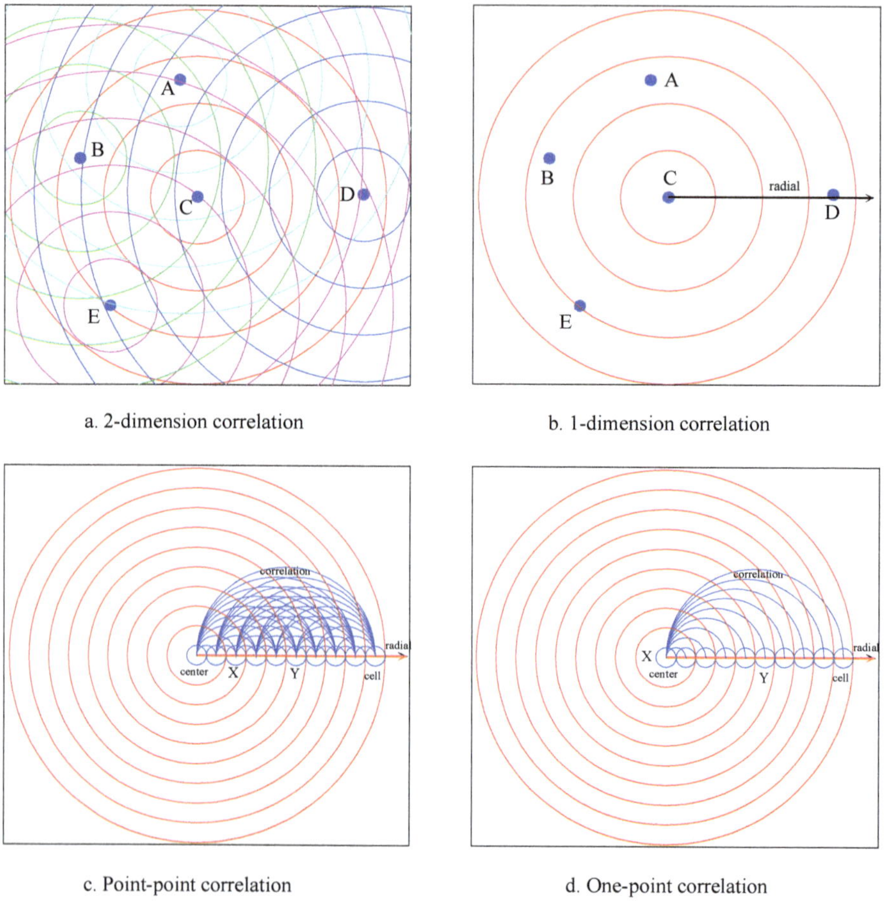

Fig. 1 The sketch maps of three types of spatial correlation processes (by Chen 2013). *Note* As schematic diagrams, only five urban elements are taken into account. The concentric circles are employed to compute average density of urban land use or traffic networks. The space between two immediate large circles forms a ring, the small circles falling between two circles represent cells along a radial line, and the arcs indicate spatial correlation between two cells. Subgraphs: **a** For the 2-dimension spatial correlation, we need five sets of concentric circles to determine spatial scaling; **b** For the 1-dimension spatial correlation, we need one set of concentric circles to compute average density; **c** The point–point correlation suggests density–density correlation; **d** The one-point correlation suggests central correlation

spaces (Batty and Longley 1994). Thus, *urban density* refers to the number of inhabitants, buildings, roads, streets, and so on, in given urbanized area. The urban density has different connotations and denotations for different spatial measurements. If we examine urban population distribution, the urban density implies urban population density; if we investigate the patterns of urban land uses, the urban density implies urban land use density; if we study the spatial structure of transport network, the urban density suggests the urban road density. Suppose that the fractal dimension of a city's spatial form will be evaluated by the remote sensing data for the purpose of exploring urban growth. The number of cells, $N(r)$, within the radius, r, from the city center can be used to measure the urbanized area. A cell is defined as the smallest image-forming unit or a pixel in a digital map. If the relation between measurement $N(r)$ and linear size r follow a power law such as $N(r) = N_0 r^D$, where D and N_0 are two parameters, then the urban morphology can be treated as a random fractal pattern, and the scaling exponent D is just the fractal dimension, which is termed *radial dimension* (Frankhauser 1998; Frankhauser and Sadler 1991). As an alternative way, the density–radius scaling can be employed to evaluate fractal dimension. In other words, we can examine the scaling relation between the radius r from a city center and the corresponding urban density $\rho(r)$. If the relation follow an inverse power law, and if the value of the scaling exponent falls between 0 and 1, then the urban growth can be regarded as a self-organized process. Self-organization of urban evolution results in a self-similar pattern with a fractional dimension value ranging from 0 to 2 (Chen 2010). The fractal morphology of cities can be described with a set of spatial correlation functions, which are equivalent to the corresponding entropy functions.

2.2 *Spatial Correlation Functions*

Scientific theoretical construction is often based on simple and clear prototype. First of all, for simplicity and without loss of generality, let us consider a monocentric city. For a fractal city with a growth core as the circle center, the area of concentric circles of radius r is $A(r) = \pi r^2$, where π denotes the circular constant. The marginal urban density $\rho(r)$ at distance r from the urban center can be expressed as (Batty and Longley 1994; Makse et al. 1995)

$$\rho(r) = \frac{\mathrm{d}N(r)}{\mathrm{d}A(r)} = \rho_1 r^{D_f - d} = \rho_1 r^{-a}, \tag{1}$$

where $\rho_1 = DN_0/(2\pi)$ refers to a proportionality coefficient, $a = d - D_f$ denotes the scaling exponent of density distribution ($a > 0$), d is the Euclidean dimension of the embedding space indicative of geographical space, and D_f is the *radial dimension* of urban morphology ($D_f < d$) (Frankhauser 1998). Equation (1) is identical in form to Smeed's model on traffic network density (Batty and Longley 1994; Smeed 1963). As there is no mathematical definition in Eq. (1) for the city center, we can specially define a central density $\rho(0) = \rho_0$ for the location $r = 0$. The ρ_0 value can be obtained by observation in empirical analyses. The urban density function is actually defined in a 1-dimension space based on the idea of statistical average, but it reflects the geographical information of a 2-dimension space (Chen 2010). In theory, the fractal dimension values of urban form (D_f) falls between 0 and 2, and the scaling exponent of urban density (a) varies from 0 to 1. In positive studies, sometimes we have $D_f > 2$, and thus the scaling exponent $a > 1$.

In terms of scaling notion, a fractal model is in fact based on a correlation function. A monofractal model is based on simple central correlation function, and a multifractal model is based on complex density–density correlation function. The former can be termed one-point correlation function, and the latter can be treated as point–point correlation function (Chen 2013). So, fractal analysis is usually related with correlation analysis. The global fractal dimension in multifractal theory is what is called generalized correlation dimension (Chen 2013; Chen and Jiang 2010; Grassberger and Procaccia 1983). Suppose that there exist two points on a radial from the city center, X and Y, on a digital map (Fig. 1c). Based on Eq. (1), the density–density correlation function can be expressed as

$$C(r) = \int_{-\infty}^{\infty} \rho(x)\rho(x + r)\mathrm{d}x = 2\rho_1^2 \int_0^{\infty} x^{D_f - d}(x + r)^{D_f - d}\mathrm{d}x, \tag{2}$$

where x denotes the distance of the first point (X) from the city center, and r represents the distance of the second point (Y) from the first point (X). Note that the central location ($x = 0$) is a discontinuity point in the density function. It is easy to demonstrate that the spatial correlation function follows the scaling law. The fractal essence is the scaling symmetry, which indicates the invariance under contraction or dilation transform (Mandelbrot 1989). For a function $f(x)$, if we contract or dilate the argument x by a constant scale factor λ, the function will not change in structure, but vary in size, and thus the result is $f(\lambda x) = \lambda^{\alpha} f(x)$, where α is a scaling exponent. In this case, the function conforms to the scaling law. Let $x = \xi y$, where ξ is a scale factor ($x \geq 0$, $y \geq 0$). A parameter relation can be derived by a scaling analysis as follows:

$$C(\xi r) = 2\rho_1^2 \int_0^{\infty} x^{D_f - d}(x + \xi r)^{D_f - d}\mathrm{d}x = \xi^{2H} C(r), \tag{3}$$

where $H = D_f - d + 1/2$, and H is the generalized Hurst exponent. Equation (3) is in fact a functional equation. In the theory of R/S analysis, the Hurst exponent is a scaling exponent associated with the autocorrelation coefficient of the time series (Hurst et al. 1965). The R/S analysis is normally termed “rescaled range analysis” (Feder 1988). This is a statistical method developed by Harold Edwin Hurst and his coworkers to analyze long-term continuous or regular records of natural phenomena (Hurst et al. 1965). It can also be employed to analyze the long orderly spatial series with proportional spacing. The rescaled range is a statistical measure of the variability of a time/space series based on two main measures/variables: one is the standard deviation (S), and the other, the range (R) of the data set, i.e., the difference between the highest and lowest values. From the slope of the logarithmic linear relation between the ratio of $R(\tau)$ to $S(\tau)$ and the time lag or spatial displacement τ, we can obtain a useful parameters, the Hurst exponent (H). Concretely speaking, for the increment series $\Delta x(i)$ of a space/time series $x(i)$, H is the scaling exponent of the ratio $R(\tau)/S(\tau)$ versus time lag or spatial displacement (τ) ($i = 1, 2, 3, \ldots; \tau = 1, 2, \ldots, i$). In other words, H is defined by the power function $R(\tau)/S(\tau) = (\tau/2)^H$ (Chen 2010; Feder 1988; Hurst et al. 1965). The exponent value falls in between 0 and 1, i.e., $0 \le H \le 1$. The value of $H = 1/2$ indicates a Brownian motion, while the values of $H \ne 1/2$ suggests the fractional Brownian motion (fBm) (Feder 1988; Mandelbrot 1983). Obviously, the solution to the above functional equation is

$$C(r) = C_1 r^{2(D_f - d)+1} = C_1 r^{2H} = C_1 r^b, \tag{4}$$

where C_1 refers to a proportionality coefficient related with the parameter ρ_1 in Eq. (1), and $b = 2H$ is a scaling exponent indicating fractal dimension.

A correlation dimension model can be derived from the spatial correlation function. Equation (4) represents a correlation function defined in a 1-dimension space, differing from the spatial correlation function defined in a 2-dimension space (Chen and Jiang 2010). For the latter, the scaling exponent is just the correlation dimension, but for the former, the scaling exponent is less than the correlation dimension. By dimensional analysis, we have

$$N(r) \propto rC(r) = r^{2-2(d-D_f)} = r^{2(D_f-1)} = r^{D_c}, \tag{5}$$

where $N(r)$ denotes the number of cells (pixels) within the radius of r from the center, and D_c is the density–density correlation dimension, or the *point-point correlation dimension* of urban density. Equation (5) gives a fractal dimension relation as below:

$$D_c = 2(D_f - 1) = b + 1, \tag{6}$$

which is important for our understanding fractal dimension. For the monocentric cities, the radial dimension can be directly calculated by the area–radius scaling (Batty and Longley 1994; Frankhauser 1998). However, for the polycentric cities, the area–radius scaling relations often break down and the radial dimension cannot be effectively estimated in a simple way. In this case, the spectral analysis is helpful due to the filter function of Fourier transform (Chen 2010).

A special case of the urban density–density correlation is the central correlation, which can be treated as "one-point correlation". Based on a 1-dimension space, the one-point correlation indicates the spatial correlation between a given point, e.g., a city's central point, and other points around the point, while the point–point correlation implies the spatial correlation between one point on a circle and another point on another circle around the center of a city (Chen 2013). Suppose that there is a radial line from the city center to the periphery. The density–density correlation implies the spatial relation between any two points on the radial line. If we fix one point to the center of city, we will have $x = 0$ (Fig. 1d). Thus, the integral in the point–point correlation function, Eq. (2), will vanish, and it is reduced to a special one-point correlation function as follows:

$$C_0(r) = \rho_0\rho_1 r^{D_f-d} = \rho_0\rho(r), \tag{7}$$

in which $C_0(r)$ denotes the one-point correlation measurement. This indicates that, if we fix one point in the central point of a city, the one-point correlation function is just proportional to the urban density function, and thus the correlation integral is proportional to the cell number within the radius of r from the center, $N(r)$. The radial dimension of cities is actually the one-point correlation dimension, that is, $D_0 = D_f$, where D_0 refers to the *one-point correlation dimension*. Therefore, $a = d - D_f = d - D_0$ indicates the scaling exponent of the one-point correlation. In this contexture, the one-point correlation represents central correlation, and the one-point correlation dimension suggests the spatial relation and interaction between a city's core and its periphery.

If a city takes on a polycentric pattern, it cannot be described by a simple monofractal model. In this case, the density–radius scaling should be substituted with wave-spectrum scaling of cities (Chen 2010, 2013). By means of Fourier transform, a correlation function can be converted into energy spectrum function and vice versa. The Fourier transform of the density–density correlation function also follows the scaling law. Based on Eq. (4), the scaling property of Fourier transform can be demonstrated as below:

$$S(\xi k) = \int_{-\infty}^{\infty} C(r)e^{-i2\pi\xi kr}\mathrm{d}r = \xi^{-2(D_f-1)}S(k), \tag{8}$$

where k denotes the wave number, which bears an analogy with the frequency in time series analysis. Thus, we have

$$S(k) \propto k^{-2(D_f-1)} = k^{-\beta}, \tag{9}$$

in which $\beta = 2(D_f - 1) = 2H + 1$ refers to the spectral exponent. Note that $2H = 2(D_f - d) + 1$ and $D_c = 2(D_f - 1) = b + 1$. According to Eq. (6), a parameter relation can be gotten as below:

$$\beta = D_c = 2(D_f - 1), \tag{10}$$

which suggests that the spectral exponent equals the density–density correlation dimension. Equation (10) has been empirically verified by Chen (2010). Comparing Eq. (5) with Eq. (9) yields the relation between the spectral density and the correlation function such as

$$N(r) \propto rC(r) \propto \frac{1}{S(k)}. \quad (11)$$

This implies that the pixel number of urban land use varies inversely as the spectral density.

For the one-point correlation, the cosine transformation relation between the correlation function and energy spectrum is not effective. The Fourier transform of the one-point correlation function can be expressed as

$$F(k) = \int_{-\infty}^{\infty} C_0(r)e^{-i2\pi kr}\mathrm{d}r = \rho_0\rho_1 \int_{-\infty}^{\infty} r^{D_0-d}e^{-i2\pi kr}\mathrm{d}r. \quad (12)$$

The equation proved to follow the scaling law under contraction or dilation, that is,

$$F(\xi k) = \rho_0\rho_1 \int_{-\infty}^{\infty} r^{D_0-d}e^{-i2\pi \xi kr}\mathrm{d}r = \xi^{-(D_0-d+1)}F(k). \quad (13)$$

This indicates the scaling relation of the energy spectrum as below

$$S_0(\xi k) = |F(\xi k)|^2 = \xi^{-2(D_0-d+1)}|F(k)|^2 = \xi^{-2(D_0-1)}S_0(k). \quad (14)$$

The solution to functional Eq. (14) is

$$S_0(k) \propto k^{-2(D_0-1)} = k^{-\beta_0}. \quad (15)$$

Apparently, the spectral exponent of the one-point correlation equals that of the point–point correlation, i.e., $\beta_0 = \beta$. To sum up, we have the following useful parameter relation:

$$\beta = 2(D_f - 1) = D_c = 2(D_0 - 1) = \beta_0, \quad (16)$$

which is equal to $b + 1$. Based on the parameter relationships shown in Eq. (16), more fractal dimension equations can be revealed for spatial analysis of urban morphology.

2.3 Fractal Parameter Equations

The fractal parameters based on wave-spectrum scaling are defined at the macro-level of urban morphology. macro-mathematical regularity is associated with micro-dynamic mechanism. It is necessary to investigate the spatial autocorrelation of urban growth at the micro-level. The macro-level is based on the urban density function, while the micro-level is based on the urban density increment function. An integral of Eq. (1) in the 2-dimension space yields the area–radius scaling relation (Batty and Longley 1994)

$$N(r) = N_1 r^{D_f}, \tag{17}$$

in which N_1 is a proportionality coefficient. From Eq. (17), an average density formula can be derived as follows (Batty and Longley 1994; Chen 2013; Longley et al. 1991):

$$\rho^*(r) = \frac{N(r)}{A(r)} = \frac{N_1}{\pi} r^{D_f - 2} \propto \frac{\mathrm{d}N(r)}{\mathrm{d}A(r)} = \rho(r), \tag{18}$$

where $\rho^*(r)$ refers to the average density within a radius of r from a city center, and $A(r) = \pi r^2$ represents the area within the circle of radius r. This indicates that the average density $\rho^*(r)$ is in proportion to the marginal density $\rho(r)$. Further, let us consider the variance of the density increment. Due to the symmetry of urban density function (from $-\infty$ to 0 then to ∞), the mean value of the density increment can be regarded as zero. So, the variance can be defined as

$$V(r) = \int_{-\infty}^{\infty} [\Delta\rho(x) - 0]^2 \mathrm{d}x = \int_{-\infty}^{\infty} [\rho(x + r) - \rho(x)]^2 \mathrm{d}x. \tag{19}$$

A scaling analysis of the above variance function yields

$$V(\xi r) = \rho_1^2 \int_0^{\infty} [(x + \xi r)^{D_f - 2} - x^{D_f - 2}]^2 \mathrm{d}x = \xi^{2H} V(r), \tag{20}$$

where $2H = 2(D_f - d) + 1$ has been given above. This suggests a scaling ratio such as $C(r)/V(r) = C(\xi r)/V(\xi r)$. The solution to functional Eq. (20) is

$$V(r) \propto r^{2H}. \tag{21}$$

If the spatial series of urban density changes is a white noise, the density increment $\Delta\rho(x) = \rho(x + r) - \rho(x)$ can be regarded as a Brownian motion, namely, random walk. Thus, we have $H = 1/2$, and this indicates $D_f = 2$. However, an urban density increment series is not a white noise. Because D_f value ranges from 0 to 2, the H value should vary from $-3/2$ to $1/2$. If so, the spatial process is always treated as an fBm process (Feder 1988; Peitgen et al. 2004). Of course, the H value comes between 0 and 1, and this will be clarified next.

Then, the self-affine fractal dimension can be derived by means of dimensional analysis. The method of dimensional analysis is useful in the theoretical studies of human geography (Haynes 1975; Haggett et al. 1977). Comparing Eq. (21) with Eq. (4) displays that the variance of density increment is in the proportion to the spatial correlation function, i.e., $V(r) \propto C(r)$. The square root of the variance is just the standard deviation

$$s(r) \propto r^H. \tag{22}$$

If the radial dimension D_f falls between 1.5 and 2, and the embedding space dimension is $d = 2$, the scaling exponent $a = d - D_f$ will come between 0 and 0.5, and thus the Hurst exponent $H = 1/2 - a$ will also fall between 0 and 0.5. According to the principle of dimensional consistency, a proportional relation appears as below:

$$\rho(r) \propto \frac{1}{s(r)} \propto r^{-H}. \tag{23}$$

Substituting Eq. (23) into Eq. (18) yields the following relation:

$$N(r) \propto \rho(r)A(r) \propto \frac{\pi r^2}{s(r)} \propto r^{2-H} = r^{D_s}. \tag{24}$$

Thus, another parameter equation can be obtained as follows:

$$D_s = 2 - H, \tag{25}$$

which denotes a special fractal dimension, termed *profile dimension* of urban morphology (Chen 2008). In literature, the parameter D_s is treated as the self-affine record dimension of the random walk (Feder 1988).

The above mathematical derivation and theoretical analysis result in a series of fractal parameter relations. These equations form a useful framework for the fractal study of urban systems. In terms of Eq. (3) or Eq. (20), if $d = 2$ as given, then we have

$$D_f = H + \frac{3}{2}, \tag{26}$$

which can also be derived from the dimensional analysis based on the point–point correlation function. The density function is in a proportion to r^{-a}, and squaring the function gives r^{-2a}. The integral of the squared density function is proportional to r^{1-2a}, and the second root of r^{1-2a} produces $r^{1/2-a}$. This indicates that $H = 1/2 - (d - D_f) = D_f - 3/2$ for $d = 2$. The result is the same as those proceeding from the scaling analysis of correlation function as well as variance function. Apparently, the scaling analysis and dimensional analysis reach the same goal by different routes.

A pair of useful fractal parameter relations can be further derived. First, substituting Eq. (25) into Eq. (26) yields an equation as follows:

$$D_f + D_s = \frac{7}{2}, \tag{27}$$

which gives the relation between the self-similar fractal dimension and self-affine fractal dimension of cities (Chen 2010). Second, integrating Eq. (16) into Eq. (26) or Eq. (27) yields another equation such as

$$\beta = 5 - 2D_s = 2H + 1, \tag{28}$$

which gives the relations between spectral exponent, Hurst exponent, and self-affine fractal dimension (Chen 2013; Feder 1988; Takayasu 1990). So far, we have had a set of fractal parameter relations, which comprises Eqs. (6), (16), (25), (26), (27), and (28).

The parameters and parameter equations can be integrated into a logic framework of fractal dimensions and relations. This framework represents a system of fractal parameters and scaling relations (Fig. 2). If the radial dimension D_f is calculated, then the one-point correlation dimension D_0, the point–point correlation dimension D_c, the wave spectral exponent β, the profile dimension D_s, the Hurst exponent H, and so on, can be estimated by the parameter equations (Table 1). Using this set of fractal parameters, we can make a systematic analysis of urban growth and form. To evaluate the radial dimension, we should select an urban center of concentric circles on a digital map. Given a center of circles, a radial dimension value is determined, and other fractal parameters will be in scale with the radial dimension in value. The valid ranges of different fractal parameters are as below: $0 \le D_f = D_0 \le 2, 0 \le D_c = \beta \le 2, 0 \le D_s \le 2, 0 \le \beta \le 3, 0 \le H \le 1$. In light of multifractal geometry, the one-point correlation dimension value must be greater than the point–point correlation dimension value, that is, $\beta = D_c \le D_f = D_0$. All these parameters are defined at two different levels: the parameters D_0, D_f, D_c, and β are at the macro-level, and D_s and H are at the micro-level. If the relation between macro and micro-levels of a city is overlooked, the acceptable domain of radial dimension will fall between 1 and 2. If $D_f > 2$ as given, then $\beta = D_c > D_f$; while if $D_f < 1$, then we will have $\beta = D_c < 0$. The two cases are illogic in multifractal theory. If we take the relation between macro and micro-levels into consideration, the feasible domain of the radial dimension will range from 1.5 to 2. When $D_f < 1.5$, we will have $D_s > 2$ and $H < 0$, and these values

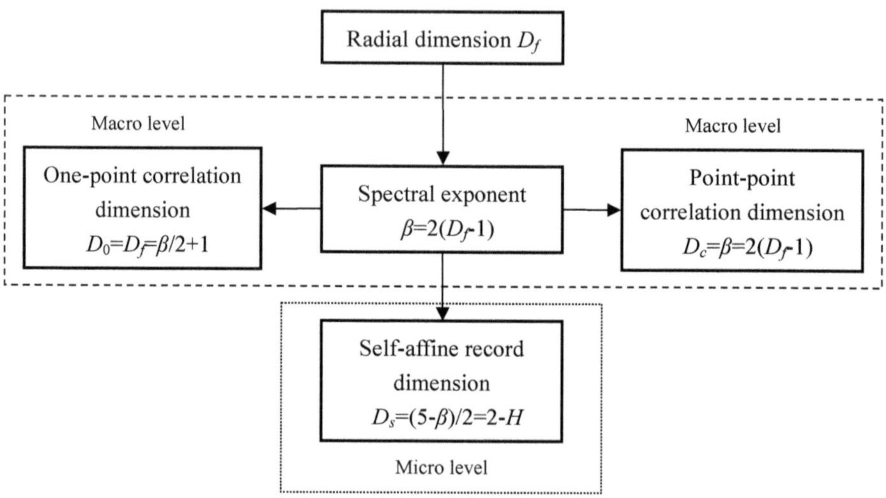

Fig. 2 A schematic diagram of the relationships between five fractal parameters for urban spatial analyses (by Chen 2013)

Table 1 The numerical relationships between radial dimension, correlation dimension, profile dimension, spectral exponent, and Hurst exponent

Radial dimension, one-point correlation dimension		Point–point correlation dimension, spectral exponent		Self-affine record dimension	Hurst exponent
D_f	D_0	D_c	β	D_s	H
0.5	**0.5**	−1	−1	3	−1
0.75	**0.75**	−0.5	−0.5	2.75	−0.75
1	**1**	**0**	**0**	2.5	−0.5
1.1	**1.1**	**0.2**	**0.2**	2.4	−0.4
1.2	**1.2**	**0.4**	**0.4**	2.3	−0.3
1.3	**1.3**	**0.6**	**0.6**	2.2	−0.2
1.4	**1.4**	**0.8**	**0.8**	2.1	−0.1
1.5	**1.5**	**1**	**1**	**2**	**0**
1.6	**1.6**	**1.2**	**1.2**	**1.9**	**0.1**
1.7	**1.7**	**1.4**	**1.4**	**1.8**	**0.2**
1.8	**1.8**	**1.6**	**1.6**	**1.7**	**0.3**
1.9	**1.9**	**1.8**	**1.8**	**1.6**	**0.4**
2	**2**	**2**	**2**	**1.5**	**0.5**
2.25	2.25	2.5	2.5	1.25	**0.75**
2.5	2.5	3	**3**	**1**	**1**

Note The proper numerical ranges of the fractal parameters are as follows: $0 < D_f, D_0, D_c < 2$; $0<\beta<3$; $0<H<1$; $-1<C_\Delta<1$. If a value exceeds the limiting sphere, the result will be meaningless. See Chen (2010, 2013)

are absurd. All in all, the reasonable range of both the radial dimension and profile dimension is from 1.5 and 2, that is, $1.5 \le D_f, D_s \le 2$.

2.4 New Analytical Framework for Urban Morphology

An analytical process of spatial correlation analysis based on scaling analysis and spectral analysis can be developed for urban studies. At macro-level, spatial correlation analysis of urban morphology can be made through Eqs. (1) and (4). Smeed's model, Eq. (1), is a one-point correlation function. However, its integral form, Eq. (4), is a point–point correlation function. What is more, the spatial autocorrelation coefficient defined in the 1-dimension space is associated with Hurst's exponent (Feder 1988), and can be expressed as

$$C_\Delta = 2^{2H-1} - 1 = 2^{2(d-D_s)-1} - 1, \tag{29}$$

Table 2 The autocorrelation coefficients at the micro-level and autocorrelation functions at the macro-level of cities

Radial dimension (D_f)	micro-level	macro-level	
	Autocorrelation coefficient (C_Δ)	One point spatial autocorrelation function [$C_0(r)$]	Point–point spatial autocorrelation function [$C(r)$]
0.5	−0.875	$r^{-1.5}$	$r^{-2.0}$
0.75	−0.823	$r^{-1.25}$	$r^{-1.5}$
1.0	−0.75	$r^{-1.0}$	$r^{-1.0}$
1.1	−0.713	$r^{-0.9}$	$r^{-0.8}$
1.2	−0.670	$r^{-0.8}$	$r^{-0.6}$
1.3	−0.621	$r^{-0.7}$	$r^{-0.4}$
1.4	−0.565	$r^{-0.6}$	$r^{-0.3}$
1.5	**−0.5**	$\mathbf{r^{-0.5}}$	$\mathbf{r^{0}}$ **= Constant**
1.6	**−0.426**	$\mathbf{r^{-0.4}}$	$\mathbf{r^{0.2}}$
1.7	**−0.340**	$\mathbf{r^{-0.3}}$	$\mathbf{r^{0.4}}$
1.8	**−0.242**	$\mathbf{r^{-0.2}}$	$\mathbf{r^{0.6}}$
1.9	**−0.129**	$\mathbf{r^{-0.1}}$	$\mathbf{r^{0.8}}$
2.0	**0**	$\mathbf{r^{0} = 1}$	$\mathbf{r^{1.0}}$
(2.25)	0.414	($r^{0.25}$)	($r^{1.5}$)
(2.5)	1	($r^{0.50}$)	($r^{2.0}$)

Note The bold numerals represent the proper scale of the fractal parameters of urban form. See Chen (2010, 2013)

in which C_Δ denotes a one-order autocorrelation coefficient, and $d = 2$ is the Euclidean dimension of the embedding space in which urban morphology is investigated. For a certain radial dimension value, we will get a corresponding autocorrelation coefficient or correlation function by means of fractal dimension equations (Table 2).

Using the concepts from multifractal, we can derive a reasonable interval of fractal dimension values of urban morphology. Theoretical correlation analysis shows that there are two special radial dimension values: $D_f = 1.5$ and $D_f = 2$. For $D_f = 1.5$, the density–density spatial correlation function will become a constant, that is, $C(0) = const$. Thus, the correlation function will be independent of the distance r. For $D_f < 1.5$, the density–density spatial correlation function will be inversely proportional to the distance r. This indicates that the spatial centripetal force for concentration growth surpasses the centrifugal force for urban deconcentration growth. In this instance, the major way of city development is inward space filling. For $D_f > 1.5$, the density–density spatial correlation function will be directly proportional to the distance r. This indicates the urban spatial centrifugal force overrides urban centripetal force. In this case, the main way of city development is outward spatial expansion. According to the ideas from multifractal theory, the density–density correlation dimension D_c must be less than the central correlation dimension D_f, that is $D_c \leq D_f = D_0$ (Chen 2011, 2013). However, if $D_f > 2$, we will have $D_c > D_f$. This relation violates the multifractal dimension principle. An inference is that, if $D_f < 1.5$, we will have a thin city growth (undergrowth); and if $D_f \geq 2$, we will have a fat city growth (overgrowth); if and only if the radial dimension falls between 1.5 and 2, urban growth will be consistent with a theoretical fractal growth at various scales.

As indicated above, the spectral exponent proved to be a density correlation dimension. Based on this result, the theoretical inferences from the spatial correlation analysis can be summarized as follows. If $\beta = D_c < 1$, we will have $D_f < 1.5$, thus the spatial correlation intensity is directly proportional to the distance r between two places. This suggests that city development is prone to filling in vacant space (e.g., spare land, open space) inwards. In this case, the focus of urban planning is the internal structure. If $\beta = D_c > 1$, we will have $D_f > 1.5$, and the spatial correlation intensity is inversely proportional to the distance r between two locations. This means that city development is inclined to growing outwards, and suburbs or even exurbs are gradually occupied by various buildings. In this instance, the focus of urban planning is external space. If $\beta = D_c = 1$, we will have $D_f = 1.5$, and the density–density correlation intensity has nothing to do with distance. Under these circumstances, urban evolution will fall into a self-organized critical state, in which the power-law distributions will emerge (Bak 1996). The self-organized criticality (SOC) is indeed a revealing concept in the theoretical research of cities (Batty and Xie 1999; Chen and Zhou 2006, 2008; Portugali 2000). The radial dimension $D_f = 2$ represents another special value of fractal dimension for urban morphology. If $D_f = 2$, we will have $C_0(0) = 1$, and this implies that the central correlation is independent of distance r. The interaction between city center and any urban place is the same as one another. This seems to be unaccountable in theory. The value $D_f = 2$ suggests that the density at one place equals that at another place. On the other hand, if $D_f = 2$, the autocorrelation coefficient $C_\Delta = 0$, and the fBm process will change to a random walk. The process of space filling within the urbanized area will stop due to the absence of vacant land.

In short, different fractal parameters are in fact related with one another. Each parameter value has its reasonable range (Table 1). The intersection of the proper ranges of the parameters D_0, D_c, D_s, H, and β suggests the feasible domain of the radial dimension D_f. Corresponding to the domain, all these fractal parameters make sense and accord with each other. If the fractal dimension values exceed the bounds ($D_f < 1.5$ or $D_f > 2$), the logic relations between different fractal parameters will be broken. In spite of the scaling nature of fractals, a fractal dimension is a measurement with characteristic length. If a city's fractal dimension value is too high (e.g., $D_f > 2$) or too low (e.g., $D_f < 1$), the possible problems of urban growth must be examined. Too low fractal dimension value indicates that the geographical space is not well developed for a city, while too high fractal dimension value implies that the geographical space may be overly filled. Both empirical and theoretical studies show that the advisable fractal dimension value of urban form is around $D_f = 1.7$ (Batty and Longley 1994; Chen 2010).

3 Case Study

3.1 Methodology

Based on the above-shown fractal parameter equations, the scaling analysis, spectral analysis, and spatial correlation analysis can be integrated into a logical framework to form a new analytical process for urban morphology. The new framework can be termed "3S analyses" method of urban geography (Fig. 3). The variables in the theoretical spatial correlation function are continuous ones. However, in practice, the spatial sampling is a discrete rather than a continuous process. So, the correlation functions must be discretized for empirical analyses. Consequently, spatial distance and displacement in the practical correlation functions are discrete variable and parameter. In fact, all quantitative analysis relates to three worlds: the *real world* (e.g., a city), the *mathematical world* (e.g., logical deduction), and *computational world* (e.g., algorithms and measurements) (Casti 1996). The theoretical derivations based on continuous variables are fulfilled in the mathematical world in the above section. The object of study, a Chinese city, is examined in the real world. The empirical analysis based on discrete variables will be made in the computational world in the following part.

A density–density correlation function can be defined for urban analysis either in the 1-dimension Euclidean space or in the 2-dimension Euclidean space. In other words, two approaches can be adopted to construct the point–point correlation functions: one is to construct the correlation in the 1-dimension space, and the other, to construct the correlation in the 2-dimension space. A practicable approach is to implement a 2-dimension correlation analysis through a 1-dimension space. It is indeed simpler to make a correlation analysis of cities in the 1-dimension geographical space than in the 2-dimension geographical space. The construction method and the related analytical process of spatial correlation function based on the 2-dimension space have been illustrated for systems of cities (Chen and Jiang 2010). This paper focuses on the density–density correlation function defined in a 1-dimension space. Now, let us consider a city as a system (a city's system rather than a system of cities), which comprises N geographical elements. The urban elements can be abstracted as "cells" or "pixels" on a digital map. If the density–density correlation function is defined in a 1-dimension space, we need a set of concentric circles for spatial measurements. In contrast, if the correlation function is constructed in a 2-dimension space, then we will need N sets of concentric circles for spatial data processing (Fig. 1).

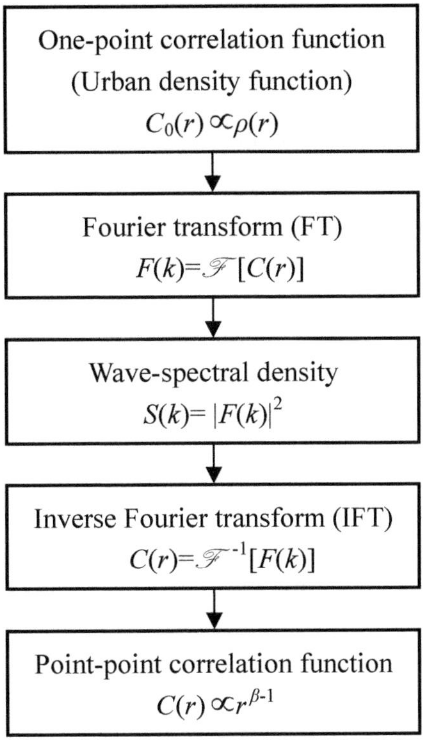

Fig. 3 A flow diagram of spatial correlation analysis based on scaling and spectral analyses (by Chen 2013)

The analytical process of the 1-dimension spatial correlation for 2-dimension geographical space can be divided into six steps. **Step 1**: Identify the center of mass (centroid) of an urban agglomeration. The center of a Western city is always inside the central business district (CBD). For Chinese cities, however, things are complicated. **Step 2**: Draw a set of concentric circles. The center of the circles should be the centroid of the urban cluster. The interval between the concentric circles is equal. That is to say, if we draw a radial line from the common center of these concentric circles to the periphery, the intersections of the circles and the radial are uniformly distributed over the straight line. The space between two circles can be treated as a "ring", and the "width" of the rings are the intervals, which may be very small. **Step 3**: Calculate the average density for each ring. This process is not only a spatial measurement but also a spatial mapping. As soon as the average density of urban element distribution between two circles is computed, the geographical information in the irregular 2-dimension space is mapped into the regular 1-dimension space. Then, the set of concentric circles can be converted into a continuous series of cells (Fig. 1). If the interval between two concentric circles is narrow enough, the size of these cells will be small enough. **Step 4**: Estimate the radial dimension by the radius–density scaling. As indicated above, we can evaluate the central correlation dimension by Eq. (1) in theory, but in empirical studies, it is Eq. (17) instead of

Eq. (1) that is suitable for determining the radial dimension. A problem is that the density series is too sensitive to stochastic perturbation in spatial measurements to give reliable results. **Step 5**: Compute the spectral exponent using the fast Fourier transform (FFT) and wave-spectrum scaling. In theory, the spectral analysis is based on Eq. (9), but in practice, the spectral analysis can be made by means of Eq. (15). As demonstrated above, the scaling exponent of wave-spectrum relation is actually the density–density correlation dimension. **Step 6**: Carry out spatial analysis for the urban morphology. Using the fractal parameter relations (Fig. 2), we can indirectly work out varied parameter values besides the spectral exponent and radial dimension. These parameters include the Hurst exponent and the profile dimension. With the help of the technology roadmap (Fig. 3), we can implement deeper spatial correlation analysis for urban form and growth.

3.2 Study Area, Datasets, and Results

The method of the 3S analysis and the related theory proposed above can be applied to the city of Beijing, the national capital of China. Beijing is a well-known megacity in the world, with an urban population of more than 16 million in 2010 (by the sixth census). The datasets were extracted from the remotely sensed images (Fig. 4). We have datasets of urban land use density for 11 years from 1984 to 2009. It has been shown that there are two approaches to estimating the radial dimension of urban morphology: one is the area–radius scaling based on Eq. (17), and the other is the energy spectrum scaling based on Eq. (9). The former is a direct approach, while the latter is an indirect approach. If urban growth is isotropic, the radius-area scaling relation will be well fitted to urban density data. Unfortunately, Beijing is of anisotropic growth, and the scaling relation cannot be directly fitted to the observational data. In this case, the energy spectrum scaling relation can be used as alternative approach.

The idea of spatial energy spectrum is defined in the mathematical world, based on continuous variable such as infinite spatial distance. However, in an empirical analysis for the real world, urbanized area is limited and variables are in discrete format. In this instance, the energy spectral density, $S(k)$, is substituted by the wave spectral density, $W(k)$, which is actually defined in the computational world. The wave spectrum is on the base of energy spectrum, and spectral density is as below:

$$W(k) = \frac{1}{N} S(k), \tag{30}$$

in which N refers to the number of data point of urban density indicative of the length of the sample path. For simplicity, the number is taken as $N = 64$ in this example. Thus, the energy spectrum relation is substituted by the wave-spectrum relation as follows:

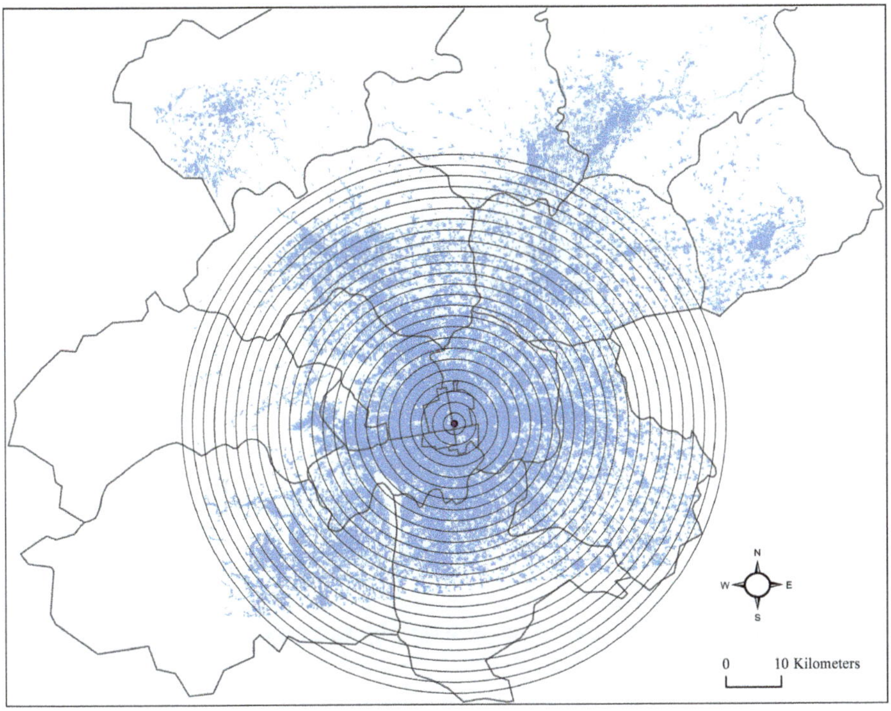

Fig. 4 A sketch map of wave spectral analysis for urban morphology of Beijing, the national capital of China (2009)

$$W(k) \propto k^{-\beta^*}, \tag{31}$$

where β^* denotes the predicted value of the wave spectral exponent. In light of Eq. (10), a fractal parameter can be derived from the wave-spectrum scaling relation such as

$$D_f^* = 1 + \frac{\beta^*}{2}, \tag{32}$$

which is termed "image dimension" of urban morphology (Chen 2013). A large number of mathematical experiments suggest an empirical relation between the radial dimension and its image dimension of a city as below

$$D_f = 1 + \frac{2}{5} D_f^*, \tag{33}$$

which is in fact an approximate relation (Chen 2010). The radial dimension reflects the space filling extent of the whole urban field, while the image dimension reflects the space filling degree of the central part of a city. Both the two parameters mirror the core-periphery relation of urban growth and form.

The procedure of data processing and spatial analysis of Beijing's morphology is as follows. **First, calculate the average urban density of Beijing in each year**. Based on remote sensing images, the center of Beijing city can be identified, and the urban land use density can be estimated by means of concentric circles. According to the requirement of the FFT algorithm, the number of circles is set as $N = 64$. Through ArcGIS technique, we can complete this computation. **Second, work out the wave spectral density**. FFT can be employed to turn Beijing's urban density into the spectral density. Today, it is easy to conduct this numerical transformation. Through Matlab or even MS Excel, we can fulfill this task of data conversion. **Third, fit the wave-spectrum scaling relation to the spectral density datasets**. The ordinary least squares (OLS) method can be utilized to make wave-spectrum analyses. For example, for Beijing in 1998, the wave-spectrum relation is $W(k) = 0.0001199k^{-1.7381}$. The goodness of fit is about $R^2 = 0.9770$, and the spectral exponent is estimated as $\beta^* = 1.7381$(Fig. 5). **Fourth, evaluate the radial dimension**. For the abovementioned example, according to Eq. (32), the image dimension is $D_f^* \approx 1.7381/2 + 1 \approx 1.8690$. Further, in light of Eq. (33), the radial dimension is $D_f \approx 1 + 2 * 1.8690/5 \approx 1.7476$. The rest may be done in the similar way (Table 3). **Fifth, estimate the other fractal parameters by the related formulae**. Using the fractal parameter equations presented in above section, we can figure out the spatial correlation dimension, D_c, the profile dimension, D_s, the Hurst exponent, H, and the autocorrelation coefficient, C_Δ. For the example abovementioned, we have $D_c = 1.7381$, $D_s = 1.6310$, $H = 0.3690$, and $C_\Delta = -0.1660$. The others may be treated by analogy (Table 4).

A discovery is that the wave-spectrum relations of Beijing's urban morphology do not follow scaling law globally. If we fit the power law relation between wave number and spectral density to the observational data, the data points cannot match the trend lines very well (Fig. 5). This suggests that the estimated values of spectral exponents are biased to some extent. In fact, the scattered points form two scaling ranges on double logarithmic wave-spectrum plots. The first scaling range falls between the wave numbers 1 and 9, and the second scaling range falls between the wave numbers 10 and 32 (Fig. 6). Thus, we have two spectral exponents for each year, and this suggests bi-fractal structure. Bi-fractals were found in the studies on both urban form and traffic networks (Benguigui and Daoud 1991; White and Engelen 1993, 1994). The essence of bi-fractals rests with self-affine growth. If urban growth is isotropic, no bi-fractal structure appears. However, if urban growth is anisotropic, bi-fractal structure will emerge. In short, anisotropic growth may lead to self-affine process, which in turn lead to bi-fractal pattern. The wave-spectrum relations imply that Beijing's urban growth bears self-affinity, which proceeds from anisotropic development in the last 30 years. In this case, all the fractal parameters can be estimated through bi-scaling ranges (Table 3).

Now, the spatiotemporal information of Beijing's city development can be revealed, and the main points are as below. **First**, the global radial dimension (D_f) falls between 1.3 and 2.2, while the local radial dimensions may be greater than 2

Table 3 The global and local fractal parameters of Beijing's urban form from 1984 to 2009

Year	Global parameters				Local parameters for core				Local parameters for periphery			
	β^*	R^2	D_s	D_f^*	β^*	R^2	D_s	D_f^*	β^*	R^2	D_s	D_f^*
1984	1.8472	0.9335	1.5764	1.9236	2.7643	0.9964	1.1178	2.3822	1.0628	0.9701	1.9686	1.5314
1988	1.6669	0.9059	1.6665	1.8335	2.6839	0.9968	1.1581	2.3419	0.8240	0.9837	2.0880	1.4120
1989	1.6160	0.9131	1.6920	1.8080	2.5673	0.9955	1.2164	2.2836	0.8664	0.9824	2.0668	1.4332
1991	2.1487	0.9029	1.4257	2.0743	3.1759	0.9861	0.9121	2.5879	0.6385	0.8468	2.1808	1.3192
1992	1.6437	0.9453	1.6782	1.8218	2.3580	0.9871	1.3210	2.1790	1.0337	0.9904	1.9832	1.5168
1994	1.5885	0.9334	1.7057	1.7943	2.3473	0.9849	1.3263	2.1737	0.9173	0.9849	2.0413	1.4587
1998	1.7381	0.9770	1.6310	1.8690	2.1872	0.9858	1.4064	2.0936	1.3463	0.9910	1.8269	1.6731
1999	1.4523	0.8353	1.7738	1.7262	2.6673	0.9869	1.1663	2.3337	0.4892	0.8914	2.2554	1.2446
2001	1.5111	0.8945	1.7444	1.7556	2.4900	0.9878	1.2550	2.2450	0.7538	0.9747	2.1231	1.3769
2006	1.5143	0.9075	1.7428	1.7572	2.4289	0.9930	1.2855	2.2145	0.7841	0.9807	2.1079	1.3921
2009	1.3926	0.8338	1.8037	1.6963	2.5882	0.9942	1.2059	2.2941	0.4199	0.8256	2.2900	1.2100

Table 4 The global fractal parameters and the derived scaling exponents of Beijing's urban form from 1984 to 2009

Year	β^*	D_s	D_f^*	D_f	D_c	H	C_Δ
1984	1.8472	1.5764	1.9236	1.7694	1.8472	0.4236	−0.1005
1988	1.6669	1.6665	1.8335	1.7334	1.6669	0.3335	−0.2062
1989	1.6160	1.6920	1.8080	1.7232	1.6160	0.3080	−0.2337
1991	2.1487	1.4257	2.0743	1.8297	2.1487	0.5743	0.1085
1992	1.6437	1.6782	1.8218	1.7287	1.6437	0.3218	−0.2188
1994	1.5885	1.7057	1.7943	1.7177	1.5885	0.2943	−0.2481
1998	1.7381	1.6310	1.8690	1.7476	1.7381	0.3690	−0.1660
1999	1.4523	1.7738	1.7262	1.6905	1.4523	0.2262	−0.3159
2001	1.5111	1.7444	1.7556	1.7022	1.5111	0.2556	−0.2874
2006	1.5143	1.7428	1.7572	1.7029	1.5143	0.2572	−0.2858
2009	1.3926	1.8037	1.6963	1.6785	1.3926	0.1963	−0.3436

and less than 1. This suggests that the peak area of urban density distribution is not near the city center. What is more, the average values of the local radial dimensions is close to the corresponding global radial dimension. This indicates that we can make wave spectral analysis using the approximate global spectral exponents. **Second**, the self-affine fractal dimension (D_s) went up and up, while the self-similar fractal dimension (D_f) went down and down over years. Generally speaking, the self-similar fractal dimension value of urban form goes up gradually due to urban space filling. The fractal dimension curve of urban growth is a logistic curve or quadratic logistic curve. However, for Beijing, the radial dimension change cannot be described with logistic function. In contrast, the self-affine profile dimension can be described with a quadratic logistic function. This indicates that it is the self-affine dimension rather than the self-similar dimension that can effectively reflect the space filling of Beijing because of anisotropic growth. **Third**, the Hurst exponent is less than 0.5, i.e., $H < 1/2$. This suggests that the spatial autocorrelation of Beijing urban growth is negative, and urban pattern is of heterogeneity. **Fourth**, the case in 1991 is exceptional. In this year, the radial dimension value is less than the correlation dimension values, that is, $D_f > D_c$, and the Hurst exponent is greater than 0.5. Of course, this may come from the quality of remote sensing image.

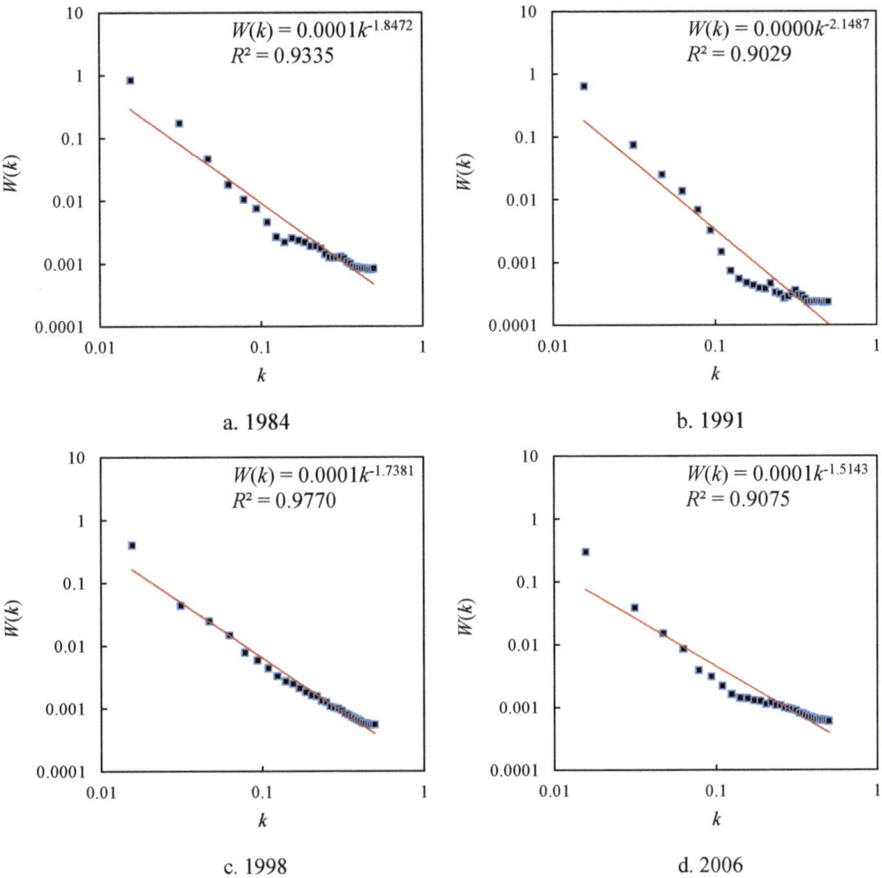

Fig. 5 The global wave-spectrum scaling relations of Beijing's urban morphology in four years (examples). *Note* The global scaling includes all the data points. For Beijing, the global wave-spectrum relation cannot be well fitted to the observational data. In this instance, the estimated spectral exponent values are biased

4 Questions and Discussion

4.1 Methodological Outline

A methodology of 3S analysis for cities has been presented, and the analytical process comprises scaling analysis, spectral analysis, and spatial correlation analysis. The key is the spatial correlation function. The spatial correlation is an important process in urban evolution. The scaling analysis and spectral analysis can be used to find the parameter solutions for a nonlinear correlation equation. The functions of the 3S method are as below: **First, the 3S analysis can be utilized to examine the spatial**

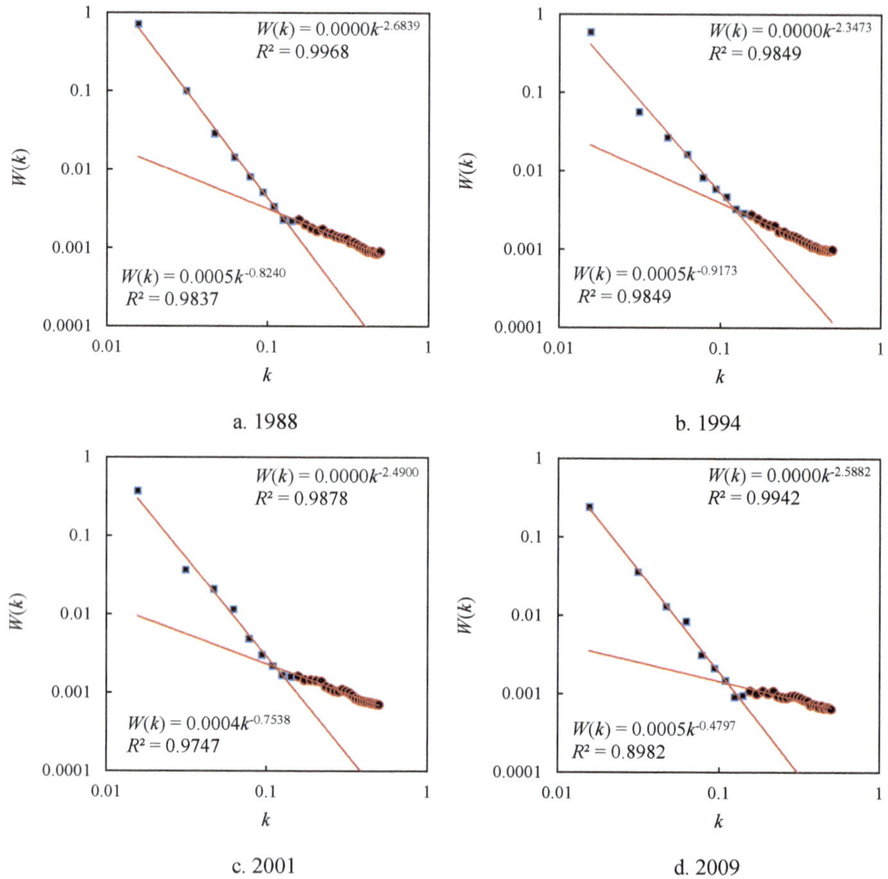

Fig. 6 The local wave-spectrum scaling relations of Beijing's urban morphology in four years (examples). *Note* The local scaling includes partial contiguous data points. On a log–log plot, the data points form two straight line segments, representing two scaling ranges. This suggests bi-fractals, which originated from self-affine process

interaction between urban components. The spatial correlation functions can be used to describe the patterns of local interactions of urban parts at the micro-level. Based on spatial correlation, spectral analysis can be used to explore the energy distribution of spatial interactions between different urban units. **Second, the 3S analysis can be used to investigate the relationships between different levels of a city's system**. The scaling analysis is important for geographers to understand the links between global and local levels. A scaling process involves various scales, ranging from the global level to the local level of urban structure. If a city evolves from a simple state into a complex critical state by self-organization, the power-law patterns will emerge, and the scaling analysis can be used to reveal the parameter relations in fractals, rank-size rule, and the law of allometric growth. The spectral analysis can be employed to examine the association of the spatial correlation (global level) with the autocorrelation (local level) of a city. **Third, the 3S analysis can be adopted to research the connection between structure of urban morphology and dynamics of urban growth**. Due to the absence of continuous sampling records within certain period, it is hard to model the spatial dynamics of urban evolution. Fortunately, the spatial structure of a city always reflects the dynamic information of city development. Combining correlation analysis and spectral analysis, we can bring to light the geographical information of urban spatial dynamics.

Spatial dimension is one of conundrums for mathematical modeling in scientific research. Reducing the spatial dimension makes it easy to model the correlation of urban elements. In this work, the correlation function is defined in the 1-dimension space, reflecting the geographical information in the 2-dimension space because the fractal dimension values fall between 1 and 2. All the fractal parameters based on the radial dimension display the 2-dimension spatial information of urban form and growth. If a correlation function is defined in the 2-dimension Euclidean space, it will cause three main problems as below: **First, mathematical modeling**. The spatial correlation cannot be directly constructed by the urban density function, Eq. (1), which is simple and fundamental in urban studies. **Second, spectral analysis**. It is difficult to make wave spectral analysis based on Fourier transform, which is helpful for us to reveal geographical spatial order and rules. **Third, computational effort**. The task quantity of numerical computation will be large if the pixels are too many. The common principle of choosing a model or method is the maximum ratio of "output" (explanation and prediction effects) to "input" (variable number, parameter number, computational effort). In many cases, the output–input ratio of fractal models do not decrease if we use the correlation function defined in the 1-dimension space to replace that defined in the 2-dimension space (Fig. 1).

The advantages and disadvantages of a method are relative, and sometimes an advantage becomes disadvantage under different conditions. One of the deficiencies of the spatial correlation analysis mainly rests with the model's mathematical structure. As indicated above, the urban density function based on monocentric cities is essentially defined in the 1-dimension Euclidean space. Although the fractal parameters based on the 1-dimension spatial correlation function reflect the 2-dimension spatial information, they cannot bring to light ALL the geographical processes and patterns of urban morphology. The shortcomings of the radial dimension and the related parameters are as below: First, these parameters cannot reflect the spatial correlation in all directions. The correlational direction is limited to the 1-dimension linear space of a city. Second, the parameters cannot effectively mirror the spatial morphology of heterogeneous cities. If a city has a number of growth centers, the power-law relation of urban density distribution usually breaks. Third, the parameters cannot reveal all the components of a city's system. Even in a monocentric city, not all of the elements' distribution density obeys the power law. Smeed's model can be used to describe the spatial density of the traffic network within a metropolitan area (Batty and Longley 1994; Smeed 1963). In empirical studies, the inverse power law cannot be well fitted to urban population density and urban land use density data (Chen 2010).

Multifractal theory provides a powerful tool for spatial correlation of urban morphology. If the effect of the 1-dimension correlation function is not good because of polycentric growth of cities, or if we want to explore the spatial correlation based on the 2-dimension space, two approaches can be considered. One is to employ the 2-dimension-based correlation function (Chen and Jiang 2010), and the other is to utilize multifractal geometry (Grassberger 1985; Hentschel and Procaccia 1983; Mandelbrot 1999; Stanley and Meakin 1988). The second approach is well known for fractal scientists, but it is difficult to grasp the ideas from multifractals. In fact, the generalized correlation dimension of multifractals is based on correlation function. Based on the functional box-counting method, the generalized dimension is formulated as

$$D_q = \frac{1}{q-1} \lim_{\varepsilon \to 0} \frac{\ln \sum_{i=1}^{n} \sum_{j=1}^{n} P_i(\varepsilon)^q}{\ln \varepsilon}, \tag{34}$$

in which D_q denotes the qth order correlation dimension, $P(\varepsilon)$ is the growth probability of fractal elements in the ith box with linear size ε at the mth level ($m = 1, 2, 3, \ldots$), generally, $\varepsilon = 1, 1/2, 1/4, \ldots, 1/2^{m-1}$, thus $n = 1/\varepsilon$, and $N(\varepsilon) = n * n$ is the number of the boxes at the given level. The parameter q is termed "moment order" in statistics ($-\infty < q < +\infty$). If $q = 0$, $D_q = D_0$ denotes the *capacity dimension*; if $q = 1$, $D_q = D_1$ indicates the *information dimension*; if $q = 2$, $D_q = D_2$ represents the *correlation dimension*. Capacity dimension tells us whether or not there exists a fractal element in a box, information dimension tells how many fractal elements can be found in a nonempty box, and correlation dimension tells us how many fractal elements can be found around a given fractal element (within certain distance). In theory, $q(-\infty, +\infty)$ represents a continuous variable. But in empirical analysis, q is often a discrete sequence, and we can set an ordered set of numbers such as $q = \ldots, -100, -99, \ldots, -2, -1, 0, 1, 2, \ldots, 99, 100, \ldots$. Thus, we will have a pair of multifractal spectrums for urban analysis, including generalized dimension spectrum (D_q vs. q) and local dimension spectrum ($f(\alpha(q))$ vs. $\alpha(q)$) (Chen and Wang 2013). In practice, the conventional box-counting method can be substituted with the grid methods of fractal dimension estimation (Frankhauser 1998). One of the special grid methods is just the functional box-counting method, which is useful in urban studies on fractals (Chen and Wang 2013; Feng and Chen 2010).

4.2 *Model Generalization*

Spatial correlation modeling is related to spatial autocorrelation and spatial interaction models in human geography. There are two important classical theories of spatial analysis for geography: one is spatial autocorrelation analysis (Cliff and Ord 2009; Haggett et al. 1977), and the other, spatial interaction modeling (Haggett et al. 1977;

Wilson 2010). Spatial autocorrelation coefficient can be defined by correlation function, and spatial interaction models are associated with gravity models. The spatial correlation analysis can be generalized to the gravity models and spatial autocorrelation functions. As indicated above, the spatial correlation function can be linked to the gravity models defined at the micro-level. Based on the power-law density function of cities, Eq. (1), a micro-gravity model can be constructed as follows:

$$I(x, r) = K\frac{\rho(x)\rho(x+r)}{r^b} = \frac{\rho_1^2}{r^b}[x(x-r)]^{D_f-d}, \tag{35}$$

where K denotes the gravity coefficient. According to the above definition, if $r = 0$ as given, then $\rho(0) = \rho_0$, which can be observed but cannot be predicted by Eq. (1). For the interaction between the city center and the periphery area, Eq. (35) will change to

$$I_0(r) = K\frac{\rho(0)\rho(r)}{r^b} = K\frac{\rho_0\rho_1}{r^b}r^{D_f-d} = kr^{D_f-d-b}, \tag{36}$$

in which the proportionality coefficient is $k = K\rho_0\rho_1$. In fact, Eq. (36) is a distance-decay function based on the power-law gravity model. Obviously, the distance-decay function is not in the linear proportionality to the urban density function. This suggests that, where power law distribution is concerned, urban density does not merely result from spatial interaction, and the gravity decay is much faster than urban density decay. Integrating Eq. (35) yields a spatial correlation function based on the gravity model as below:

$$F(r) = K\int_{-\infty}^{\infty}\frac{\rho(x)\rho(x+r)}{r^b}\mathrm{d}x = 2K\rho_1^2\int_0^{\infty}\frac{x^{D_f-d}(x+r)^{D_f-d}}{r^b}\mathrm{d}x, \tag{37}$$

which follows the scaling law. A scaling analysis of Eq. (37) can be made as follows:

$$\begin{aligned} F(\xi r) &= 2K\rho_1^2\int_0^{\infty}\frac{x^{D_f-d}(x+\xi r)^{D_f-d}}{(\zeta r)^b}\mathrm{d}x \\ &= 2K\rho_1^2\int_0^{\infty}\frac{(\xi y)^{D_f-d}(\xi y+\xi r)^{D_f-d}}{(\xi r)^b}\mathrm{d}(\xi y) \\ &= \xi^{2(D_f-d)+1-b}2K\rho_1^2\int_0^{\infty}\frac{y^{D_f-d}(y+r)^{D_f-d}}{r^b}\mathrm{d}y \\ &= \xi^{2(D_f-d)+1-b}F(r). \end{aligned} \tag{38}$$

A solution to this functional equation is

$$F(r) = F_1 r^{2(D_f-d)+1-b}, \tag{39}$$

where F_1 refers to the proportionality constant associated with the parameters K and ρ_1. This suggests that the spatial correlation function based on the power law gravity model still follows the power law. However, its scaling exponent is significantly less than the power exponent of the density–density correlation function. The Fourier transform of the gravity-based correlation function, Eq. (39), also follows the scaling law, that is,

$$\begin{aligned} S(\xi k) &= \int_{-\infty}^{\infty} F(r)e^{-i2\pi\xi kr}\mathrm{d}r \\ &= C_1 \int_{-\infty}^{\infty} r^{1-2(d-D_f)-b}e^{-i2\pi\xi kr}\mathrm{d}r \\ &= \xi^{-2+2(d-D_f)+b}C_1 \int_{-\infty}^{\infty} (\xi r)^{1-2(d-D_f)-b}e^{-i2\pi k(\xi r)}\mathrm{d}(\xi r) \\ &= \xi^{-2(D_f-1)+b}S(k), \end{aligned} \tag{40}$$

which suggests a spectral exponent such as $\beta = 2(D_f - 1) - b$. Due to $\beta > 0$, we have $b < 2$.

In urban geography, there are two types of density distributions. One is the spatial distribution with characteristic scales, which can be described by Clark's model and its variants (Batty and Longley 1994; Clark 1951); the other is the distribution without a characteristic scale, which can be described with Smeed's model and the similar models (Batty and Longley 1994; Smeed 1963). The former can be termed scale distribution, and the latter, termed scale-free distribution or scaling distribution. One of the typical scale distributions is exponential distribution, and the mathematical model for this distribution is the well-known Clark's model (Clark 1951). The model can be expressed as

$$\rho(r) = \rho_0 e^{-r/r_0}, \tag{41}$$

where ρ_0 refers to the proportionality constant indicative of the urban density of central location, and r_0 to the characteristic radius of urban density distribution. Based on Eq. (41), the density–density correlation function can be constructed as

$$\begin{aligned} C(r) &= \int_{-\infty}^{\infty} \rho(x)\rho(x+r)\mathrm{d}x \\ &= 2\rho_0^2 \int_{0}^{\infty} e^{-x/r_0}e^{-(x+r)/r_0}\mathrm{d}x \\ &= -r_0\rho_0^2 e^{-r/r_0} \int_{0}^{\infty} e^{-2x/r_0}\mathrm{d}(-2x/r_0) \\ &= r_0\rho_0^2 e^{-r/r_0}, \end{aligned} \tag{42}$$

which implies that $C(r) = r_0\rho_0\rho(r)$. This suggests that, for the exponential distribution, the spatial correlation function is linearly proportional to the density function.

In other words, the exponential density function is in essence a spatial correlation function. This type of correlation function does not follow the scaling law. The micro-gravity model based on the exponential density distribution is as follows:

$$I(x,r) = K\frac{\rho(x)\rho(x+r)}{r^b} = K\rho_0^2\frac{e^{-(2x+r)/r_0}}{r^b}. \tag{43}$$

If x is fixed to the city center, Eq. (43) will be reduced to a gamma function

$$I_0(r) = K\frac{\rho(0)\rho(r)}{r^b} = K\rho_0^2 r^{-b} e^{-r/r_0}, \tag{44}$$

which is another distance-decay function. This suggests the gravity decay is significantly faster than urban density decay.

Spatial autocorrelation can be theoretically associated with spatial correlation analysis. Based on centralized variable, the spatial autocorrelation function of urban density can be transformed into a spatial correlation function. Suppose that urban density is modeled by Eq. (1) or Eq. (41). The formula of spatial autocorrelation function is as below:

$$C(x,r) = \frac{\sum_{x-r}(\rho(x)-\bar{\rho})(\rho(x+r)-\bar{\rho})}{\sum_{x}(\rho(x)-\bar{\rho})}. \tag{45}$$

where ρ bar denotes arithmetic mean of urban density. If the variable is centralized, the average value equals 0; if the variable is further standardized, the variance equals 1. Based on standardized density variable, Eq. (45) can be reduced to

$$C(x,r) = \sum_{x-r}(\rho(x)\rho(x+r)). \tag{46}$$

For the continuous variables x and r, Eq. (45) can be expressed as integral form, and we have a generalized spatial correlation function as follows:

$$C(x,r) = \lim_{R\to\infty}\frac{1}{R-r}\int_0^{R-r}\rho(x)\rho(x+r)\mathrm{d}x, \tag{47}$$

where R represents the length of sample path measured by urban maximum radius (Liu and Chen 2007). This implies that the spatial autocorrelation function is actually a spatial correlation function.

5 Conclusions

Fractals suggest the optimum structure of physical and human systems, and a fractal object can occupy its space in the most efficient way. Using the ideas from fractals to plan or design cities will possibly help human beings make the best of geographical space. The preconditions of applying fractal geometry to city design and planning are as below: the theoretical framework of fractal cities must be constructed, the mechanism of emergence of urban evolution must be revealed, and the self-organizing city planning methods must be developed. Many important mathematical models of cities can be associated with some type of spatial correlation functions. Spatial correlation is a ubiquitous process in urban evolution. This paper is a theoretical and methodological research of fractal cities by using ideas from scaling, symmetry, correlation, and fractal dimension. The scaling analysis, spectral analysis, and spatial correlation analysis are integrated into a 3S framework for urban studies. The theoretical standpoint of this work is monocentric cities, but possibly the conclusions can be generalized to polycentric cities and even systems of cities. The main points of this paper can be summarized as follows.

First, a new analytical process termed 3S analysis of urban morphology can be developed by means of fractal parameters. Geographical analysis is mainly spatial analysis. The traditional concept of geographical space is based on distance, and the spatial measurement and mathematical modeling are chiefly based on characteristic lengths. Now, new space concept of geographical systems based on fractal dimension has emerged, and the corresponding description and modeling methods are principally based on scaling notion. A series of fractal parameters of cities can be combined with each other by the radial dimension and the spectral exponent. The fractal parameters comprise the one-point correlation dimension (self-similar dimension), the point–point correlation dimension, the Hurst exponent, the self-affine record dimension, and the spatial autocorrelation coefficient. Thus, based on the inverse power law of urban density, scaling analysis, spectral analysis, and spatial correlation analysis can be integrated into a new analytical framework of urban morphology.

Second, the kernel method of the 3S analysis of urban morphology is the scaling analysis, despite the fact that the model foundation is correlation function. The 3S analytical framework comprises scaling analysis, spectral analysis, and spatial correlation analysis. The spatial correlation analysis forms the basis of this analytical process because many important models of cities are or associated with correlation function. The scaling analysis is the critical technology. A great number of nonlinear equations cannot be solved by the conventional approaches. However, if a nonlinear equation is based on scale-free relations, a parameter solution such as fractal dimension relation can be found for it through scaling analysis. The parameter solution can displace the common variable solution. A correlation function is always connected with Fourier transform, and thus the spectral analysis can be employed to open new ways of finding solutions to urban problems and simplify the analytical procedure of urban studies.

Third, the proper ranges of fractal parameter values can be found by using the 3S analysis. The radial dimension of urban morphology is the central correlation dimension (one-point spatial correlation dimension), while the scaling exponent based on the wave spectral density is the density–density correlation dimension (point–point correlation dimension). Using the 3S analysis, we can derive a set of equations of fractal parameters. Using these formulae, we can convert a fractal parameter such as spectral exponent and central correlation dimension into another parameter such as the radial dimension and point–point correlation dimension. The conversion relations reveal the reasonable numerical ranges of the fractal parameters. In particular, the reasonable scale of the radial dimension value ranges from 1.5 to 2. If the value of the radial dimension is greater than 2, the relation between the central correlation and the density–density correlation will fall into confusion. On the other hand, if the radial dimension value is less than 1, the density–density correlation dimension will become invalid. If the value of the radial dimension is less than 1.5, the relation between the macro-level and the micro-level of urban structure will be inharmonious. If and only if the radial dimension comes between 1.5 and 2, various fractal parameters of cities will be logical and thus valid meantime, and this suggests that varied relations of urban morphology become consistent with one another.

Acknowledgements This research was sponsored by the National Natural Science Foundation of China (Grant No. 41671167). The support are gratefully acknowledged. The author would like to thank my students, Xinye Tan and Yuqing Long for providing the essential data and figures on Beijing's urban land use.

References

Ariza-Villaverde AB, Jimenez-Hornero FJ, De Rave EG (2013). Multifractal analysis of axial maps applied to the study of urban morphology. *Computers, Environment and Urban Systems*, 38: 1–10

Arlinghaus SL (1985). Fractals take a central place. *Geografiska Annaler B*, 67(2): 83–88

Arlinghaus SL, Arlinghaus WC (1989). The fractal theory of central place geometry: a Diophantine analysis of fractal generators for arbitrary Löschian numbers. *Geographical Analysis*, 21(2): 103–121

Bak P (1996). *How Nature Works: the Science of Self-organized Criticality.* New York: Springer-Verlag

Batty M (1995). New ways of looking at cities. *Nature*, 377: 574

Batty M (2005). *Cities and Complexity: Understanding Cities with Cellular Automata, Agent-Based Models, and Fractals*. London, England: The MIT Press

Batty M (2008). The size, scale, and shape of cities. *Science*, 319: 769–771

Batty M, Longley PA (1994). *Fractal Cities: A Geometry of Form and Function*. London: Academic Press

Batty M, Xie Y (1999). Self-organized criticality and urban development. *Discrete Dynamics in Nature and Society*, 3(2–3): 109–124

Benguigui L, Czamanski D, Marinov M, Portugali J (2000). When and where is a city fractal? *Environment and Planning B: Planning and Design*, 27(4): 507–519

Benguigui L, Daoud M (1991). Is the suburban railway system a fractal? *Geographical Analysis*, 23(4): 362–368

Casti JL (1996). *Would-Be Worlds: How Simulation Is Changing the Frontiers of Science*. New York: John Wiley and Sons
Chen YG (2008). A wave-spectrum analysis of urban population density: entropy, fractal, and spatial localization. *Discrete Dynamics in Nature and Society*, vol. 2008, Article ID 728420, 22 pages
Chen YG (2009). Urban gravity model based on cross-correlation function and Fourier analyses of spatio-temporal process. *Chaos, Solitons & Fractals*, 41(2): 603–614
Chen YG (2010). Exploring fractal parameters of urban growth and form with wave-spectrum analysis. *Discrete Dynamics in Nature and Society*, vol. 2010, Article ID 974917, 20 pages
Chen YG (2011). Modeling fractal structure of city-size distributions using correlation functions. *PLoS ONE*, 6(9):e24791
Chen YG (2012). Fractal dimension evolution and spatial replacement dynamics of urban growth. *Chaos, Solitons & Fractals*, 45 (2): 115–124
Chen YG (2013). Fractal analytical approach of urban form based on spatial correlation function. *Chaos, Solitons & Fractals*, 49(1): 47–60
Chen YG, Jiang SG (2010). Modeling fractal structure of systems of cities using spatial correlation function. *International Journal of Artificial Life Research*, 1(1): 12–34
Chen YG, Wang JJ (2013). Multifractal characterization of urban form and growth: the case of Beijing. *Environment and Planning B: Planning and Design*, 40(5): 884–904
Chen YG, Zhou YX (2006). Reinterpreting central place networks using ideas from fractals and self-organized criticality. *Environment and Planning B: Planning and Design*, 33(3): 345–364
Chen YG, Zhou YX (2008). Scaling laws and indications of self-organized criticality in urban systems. *Chaos, Solitons & Fractals*, 35(1): 85–98
Clark C (1951). Urban population densities. *Journal of Royal Statistical Society*, 114(4): 490–496
Cliff AD, Ord JK (2009). What were we thinking? *Geographical Analysis*, 41(4): 351–363
De Keersmaecker M-L, Frankhauser P, Thomas I (2003). Using fractal dimensions for characterizing intra-urban diversity: the example of Brussels. *Geographical Analysis*, 35(4): 310–328
Feder J (1988). *Fractals*. New York: Plenum Press
Feng J, Chen YG (2010). Spatiotemporal evolution of urban form and land use structure in Hangzhou, China: evidence from fractals. *Environment and Planning B: Planning and Design*, 37(5): 838–856
Frankhauser P (1994). *La Fractalité des Structures Urbaines (The Fractal Aspects of Urban Structures)*. Paris: Economica
Frankhauser P (1998). The fractal approach: A new tool for the spatial analysis of urban agglomerations. *Population: An English Selection*, 10(1): 205–240
Frankhauser P, Sadler R (1991). Fractal analysis of agglomerations. In: *Natural Structures: Principles, Strategies, and Models in Architecture and Nature*. Ed. M. Hilliges. Stuttgart: University of. Stuttgart, pp 57–65
Frankhauser P (2014). From fractal urban pattern analysis to fractal urban planning concepts. In: *Computational Approaches for Urban Environments*. Eds. M. Helbich, J.J Arsanjani, M. Leitner. New York: Springer, pp 13-48
Grassberger P (1985). Generalizations of the Hausdorff dimension of fractal measures. *Physics Letters A*, 107(3): 101–105
Grassberger P, Procaccia I (1983). Measuring the strangeness of strange attractors. *Physica D*, 9(1–2): 189–208
Haggett P, Cliff AD, Frey A (1977). *Locational Analysis in Human Geography (2nd edition)*. London: Arnold
Haynes AH (1975). Dimensional analysis: some applications in human geography. *Geographical Analysis*, 7(1): 51–68
Hentschel HGE, Procaccia I (1983). The infinite number of generalized dimensions of fractals and strange attractors. *Physica D*, 8(3): 435–444
Hurst HE, Black RP, Simaika YM (1965). *Long-term Storage: An Experimental Study*. London: Constable

Liu JS, Chen YG (2007). Spatial autocorrelation and localization of urban development. *Chinese Geographical Science*, 17(1): 34–39

Longley PA, Batty M, Shepherd J (1991). The size, shape and dimension of urban settlements. *Transactions of the Institute of British Geographers (New Series)*, 16(1): 75–94

Makse H, Havlin S, Stanley H E (1995). Modelling urban growth patterns. *Nature*, 377: 608–612

Makse HA, Andrade Jr. JS, Batty M, Havlin S, Stanley HE (1998). Modeling urban growth patterns with correlated percolation. *Physical Review E*, 58(6): 7054–7062

Mandelbrot BB (1989). Fractal geometry: what is it, and what does it do? *Proceedings of the Royal Society of London A: Mathematical and Physical Sciences*, 423 (1864): 3–16

Mandelbrot BB (1999). *Multifractals and 1/f Noise: Wild Self-Affinity in Physics (1963–1976)*. New York: Springer

Mandelbrot BB. 1983. *The Fractal Geometry of Nature*. New York: W. H. Freeman and Company

Murcio R, Masucci AP, Arcaute E, Batty M (2015). Multifractal to monofractal evolution of the London street network. *Physical Review E*, 92, 062130

Peitgen H-O, Jürgens H, Saupe D (2004). *Chaos and Fractals: New Frontiers of Science (Second edition)*. New York: Springer

Portugali J (2000). *Self-Organization and the City*. Berlin: Springer

Shen G (2002). Fractal dimension and fractal growth of urbanized areas. *International Journal of Geographical Information Science*, 16(5): 419–437

Smeed RJ (1963). Road development in urban area. *Journal of the Institution of Highway Engineers*, 10(1): 5–30

Stanley HE, Meakin P (1988). Multifractal phenomena in physics and chemistry. *Nature*, 335: 405–409

Takayasu H (1990). *Fractals in the Physical Sciences*. Manchester: Manchester University Press

Thomas I, Frankhauser P, Biernacki C (2008). The morphology of built-up landscapes in Wallonia (Belgium): A classification using fractal indices. *Landscape and Urban Planning*, 84(2): 99–115

Thomas I, Frankhauser P, De Keersmaecker M-L (2007). Fractal dimension versus density of built-up surfaces in the periphery of Brussels. *Papers in Regional Science*, 86(2): 287–308

Thomas I, Frankhauser P, Frenay B, Verleysen M (2010). Clustering patterns of urban built-up areas with curves of fractal scaling behavior. *Environment and Planning B: Planning and Design*, 37(5): 942–954

Wang XS, Liu JY, Zhuang DF, Wang LM (2005). Spatial-temporal changes of urban spatial morphology in China. *Acta Geographica Sinica*, 60(3): 392–400 [In Chinese]

White R, Engelen G (1993). Cellular automata and fractal urban form: a cellular modeling approach to the evolution of urban land-use patterns. *Environment and Planning A*, 25(8): 1175–1199

White R, Engelen G (1994). Urban systems dynamics and cellular automata: fractal structures between order and chaos. *Chaos, Solitons & Fractals*, 4(4): 563–583

Wilson A (2010). Entropy in urban and regional modelling: retrospect and prospect. *Geographical Analysis*, 42 (4): 364–394

Central Place Theory and the Power Law for Cities

Wen-Tai Hsu and Xin Zou

Abstract This chapter provides a review of the link between central place theory and the power laws for cities. A theory of city size distribution is proposed via a central place hierarchy a la Christaller (1933) either as an equilibrium results or an optimal allocation. Under a central place hierarchy, it is shown that a power law for cities emerges if the underlying heterogeneity in economies of scale across good is regularly varying. Furthermore, we show that an optimal allocation of cities conforms with a central place hierarchy if the underlying heterogeneity in economies of scale across good is a power function.

Keywords Central place theory · Zipf's law · City sizes · Dynamic programming · Optimal city hierarchy

1 Introduction

City size distribution is known to be well approximated by a power law with a tail index around 1, i.e., a Zipf's law. To visualize it, we first rank cities by city size (population): #1 is New York, #2 Los Angeles etc., then we plot the city ranks against city sizes on a log-log scale using U.S. 2000 census data for all Metropolitan Statistical Areas (MSAs). The relationship for 362 MSAs, as shown in Fig. 1, is close to a straight line ($R^2 = 0.9857$), and the slope is close to -1 (-0.9491). Even when the smaller cities and towns are consider and the *overall* city size distribution does not follow the power law (Eeckhout 2004), the right tail of the distribution can still be well approximated by a power law. This empirical regularity has been widely

W.-T. Hsu (✉) · X. Zou
School of Economics, Singapore Management University, 90 Stamford Road, Singapore 178903, Singapore
e-mail: wentaihsu@smu.edu.sg

X. Zou
e-mail: xin.zou.2014@phdecons.smu.edu.sg

L. D'Acci (ed.), *The Mathematics of Urban Morphology*,
Modeling and Simulation in Science, Engineering and Technology,
https://doi.org/10.1007/978-3-030-12381-9_3

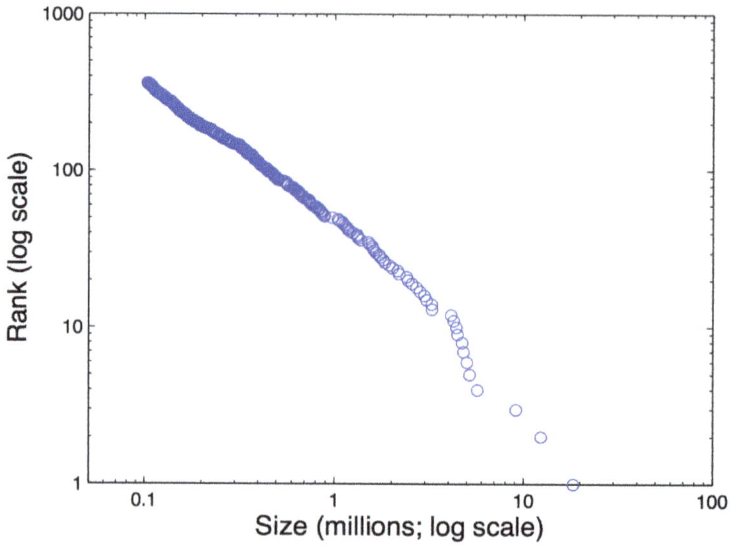

Fig. 1 Zipf plot for all 362 MSAs from 2000 U.S. Census

documented using data spanning across many countries and time periods (see Gabaix and Ioannides (2004) and Dittmar (2011)).

Formally, a power-law can be expressed as the tail probability of a city size distribution following a power function such that

$$P(S > s) = a/s^{\zeta}, \ s \geq \underline{s},$$

where a, ζ, and $\underline{s}$ are some positive constants.

One popular explanation for power laws is proportional random growth (see Champernowne (1953), Simon (1955)), based on which (Gabaix 1999) provides an explanation for why the exponent ζ should be close to 1. In this chapter, we introduce a very different theory for the power laws that is proposed by Hsu (2012) and Hsu et al. (2014) and built on the insight of *central place theory*. The key difference from the random growth approach is that central place theory explicitly accounts for the spatial pattern of different sized cities based on geography and the heterogeneity of industries/goods. In such a theory, cities of different sizes play different roles by hosting different industries.

Central place theory was first proposed by Christaller (1933), who described how a city hierarchy emerges from a uniformly populated plane of farmers as consumers. Cities and towns provide a wide variety of goods to these farmers, as these cities and towns have their market areas, they are referred to "central places" of these market areas. The key feature is that goods differ in terms of their degrees of scale economies, and the *hierarchy property* states that if a city provides a good with certain degree of scale economies, it provides all goods with lower scale economies.[1] Put it differently,

[1]This is often called "hierarchy principle" in the literature.

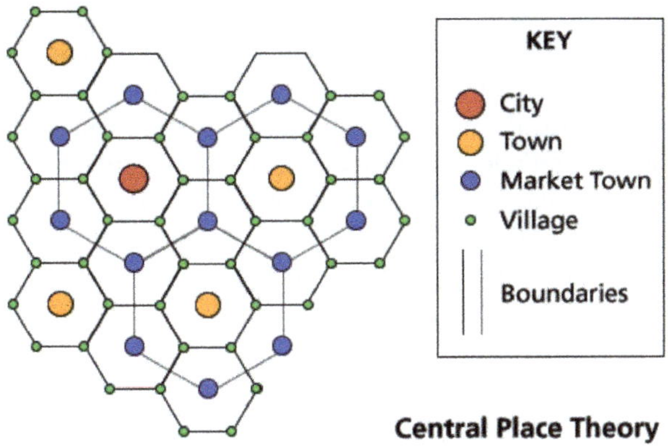

Fig. 2 Central place hierarchy on the plane

larger cities provide all goods that are provided by smaller ones. Some economists started to pay their attention to geography and spatial economics around early 80s. Although central place theory has developed into a main building block in economic geography, its theoretical foundation is weak in economists' point of view. The major problem is that it largely remains a collections of assumptions and statements without rigorous logical deduction, and how the *central place hierarchy* can emerge from an equilibrium with economic agents optimizing given their constraints is unclear. Hence, a microfoundation is needed.

What the picture below shows is a depiction of central place theory in Christaller's form. That is, on a uniformly populated plane of farmers, with uniform soil productivity, market areas of each central place is a hexagon. There are different layers of central places, with central places of the same layer having the same size of market area. A recently proven "honeycomb conjecture" in Hales (2001) provides a ground for why the shape of market areas should be hexagon, as regular hexagons are the most efficient way to partition the 2-D plane with least perimeter used. The hierarchy property is implicitly shown here. When the larger central places provides all goods and services provided by the smaller ones, we also see that the larger central places are selected from a subset of the locations of the smaller central places in a regular way. Of course reality wouldn't conform the central place pattern as in Fig. 2, but it provides a way to comprehend the complex locational patterns of cities and towns on the maps that we see. Put it differently, the seemingly pattern-less, complex map of cities and towns may indeed have some kind of sensible reasoning behind them, only to be obscured by many other factors that distorted the underlying pattern.

Unfortunately, central place theory in Christaller's form is too difficult the deal with, mainly due to the 2-dimensional geographic space. Economists have made several attempts to try to provide a "formal" central place theory in various ways, and these include Eaton and Lipsey (1982), Quinzii and Thisse (1990), Fujita et al. (1999), Tabuchi and Thisse (2006, 2011), Hsu (2012) and Hsu et al. (2014). The common task of the above-mentioned papers are again the hierarchy property and the locational patern. In all of these "modern" central place theories, the geographic space is either two-location space or a one-dimensional space. What distinguishes

Hsu (2012) and Hsu et al. (2014) from other economic models of central place theory is that these two explain the power laws for cities. However, note that these papers are not the first that link central place hierarchy with the power laws. Indeed, Beckmann (1958) was the first to point out the link between central place theory and the power law. Nevertheless, he provides no economic model for his hierarchical structure, and his hierarchy lacks the dimension of industries, which is crucial in the theory by Hsu (2012) and Hsu et al. (2014).

Similar to Beckmann (1958), there is a large literature in geography that examines the relationships between central place systems and power laws via the angle of fractal analysis. See, for examples, Arlinghaus (1985), Arlinghaus and Arlinghaus (1989), Batty and Longley (1994), Frankhauser (1998), Chen and Zhou (2006), Chen (2011, 2014). This strand of analysis is interesting as it examines different variants of fractals that could link the system of cities with the power laws. Compared with economic theories described above, we can thus see richer morphology of the system of cities without the hands being tied with the need for a microfoundation. In this sense, this strand of literature in geography and the attempts in economics to provide microfoundation for central place theory complement each other.

Before we formally introduce the theory, we first peek at the result—a central place hierarchy either as an equilibrium outcome or as an optimal allocation. Here, we formally define a central place hierarchy as an allocation of production locations such that both the *hierarchy property* and *central place property* hold. A central place hierarchy in a one-dimensional geographic space can be illustrated by Fig. 3. As the vertical axis is the commodity space, the hierarchy property is clearly seen in this figure. The central place property is that any city is located in the middle between two neighboring larger sized cities.

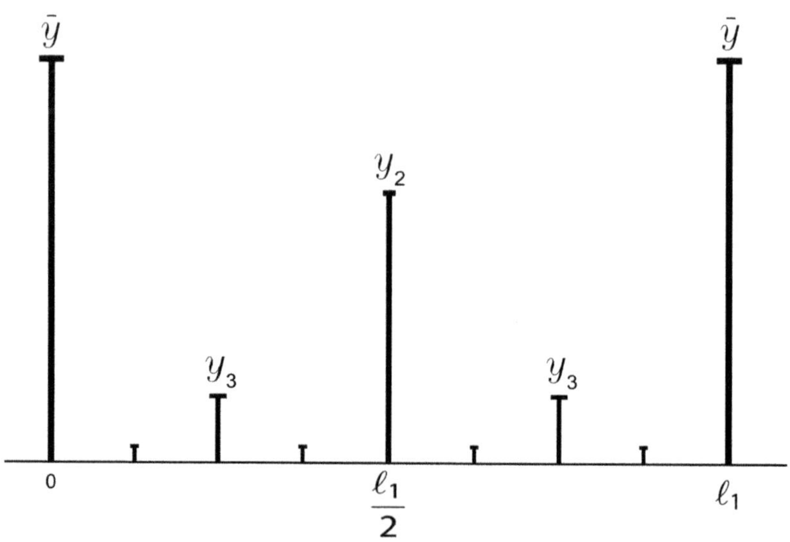

Fig. 3 Central place hierarchy on the line

This chapter is organized as follows: Sects. 2 and 3 explain the city size distribution by formalizing central place theory as an equilibrium outcome. With the hierarchy property imposed, Sect. 4 show that how a central place hierarchy can emerge as an optimal allocation. In particular, we apply dynamic programming to study the social planner's problem. We conclude this chapter in Sect. 5.

2 Central Place Theory

We first lay out a basic model and derive the central place hierarchy as an equilibrium outcome.

2.1 *Model and One-Good Equilibrium*

The geographic space is the real line, and the location is indexed as $x \in \mathbb{R}$. There is a continuum of consumption goods $z \in [0, z_1]$, where z_1 is exogenously given. There are two types of agents: farmers and firms. The farmers are immobile and are uniformly distributed on the real line with a density of 1. Each farmer demands one unit of each good z in $[0, z_1]$ inelastically.

For any production of good $z \leq z_1$, a fixed cost $y = \phi(z)$ is needed, and the (cumulative) distribution function of y is denoted as $F : [\underline{y}, \bar{y}] \subset \mathbb{R}_+ \rightarrow [0, 1]$. Besides the fixed cost, producing one unit of each good requires constant marginal cost c. Moreover, it occurs a transportation cost t per unit per mile traveled. For each good, there is an infinite pool of potential firms. The firms and farmers play the following two-stage game (Lederer and Hurter 1986).

1. Entry and location stage
   The potential firms simultaneously decide whether to enter. Upon entering, each entrant chooses a location and pays the setup cost for the good it produces. Assume the tie-breaking rule: if a potential firm sees a nonnegative opportunity, then it enters.
2. Price competition stage
   The firms deliver goods to the farmers. Given its own and other firms' locations, each firm sets a delivered price schedule over the real line. For each good, each location on the real line is a market in which the firms engage in Bertrand competition. Each farmer decides the specific firm from which to buy each good.

Now let's consider the subgame perfect equilibrium (SPNE) based on the two-stage game setting above. Denote the firm on the left-hand side as A located at $x = 0$ and that on the right-hand side as B located at $x = L$. The marginal costs of delivering the good to a consumer who is x distance from A are thus:

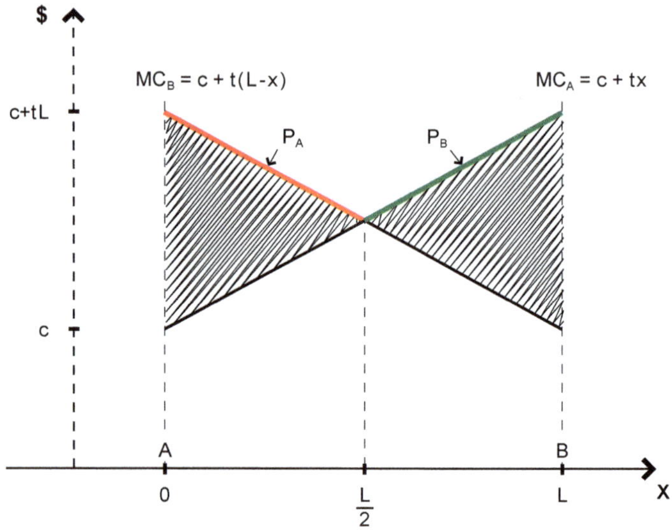

Fig. 4 Second-stage competition: prices and gross profits

$$MC_A = c + tx,$$
$$MC_B = c + t(L - x).$$

Bertrand competition results in firm with the lower marginal cost dominating the market and charging the price of its opponent's marginal cost. Thus, the equilibrium prices at each x on $[0, L]$ can be written as

$$p(x) = \begin{cases} c + t(L - x) & x \in \left[0, \frac{L}{2}\right], \\ c + tx & x \in \left[\frac{L}{2}, L\right]. \end{cases}$$

Figure 4 shows the marginal costs, the equilibrium prices, and the gross profits from the market area between A and B.

For a given good whose fixed cost is y, the firms producing the same given good should be apart equally. Furthermore, since the gross profit of any of these given firms with a market area of L is $\frac{tL^2}{2}$, and firms enter with nonnegative profits, we present the following lemma as the derivation of an SPNE for any arbitrary good.

Lemma 2.1 *Fix the level of fixed cost* y *and define* $\underline{L}(y)$ *as the solution to the zero-profit condition* $t\left[\underline{L}(y)\right]^2/2 = y$. *Thus,* $\underline{L}(y) = \sqrt{2y/t}$. *There is a continuum of equilibria in which one firm is located at every point in* $\{x + nL\}_{n=-\infty}^{\infty}$, *where* $L \in [\underline{L}(y), 2\underline{L}(y))$ *and* $x \in [0, \underline{L}(y))$.

There exists a continuum of equilibria because any L in the interval $\left[\underline{L}(\bar{y}), 2\underline{L}(y)\right)$ is an equilibrium distance; $L \geq \underline{L}(y)$ implies that all firms earn a nonnegative profit (no exit), whereas $L < 2\underline{L}(y)$ implies that any new entrant between any two existing firms must earn a negative profit (no entry).

2.2 Hierarchy Equilibrium

An equilibrium is a collection of locations of firms, delivered price schedules, and farmers' consumption choices such that the allocation for each good y is an SPNE. In this section, we consider a *hierarchy equilibrium* in which the *hierarchy property* holds.

Definition 2.1 A *hierarchy equilibrium* is an equilibrium in which, at any production location, the set of goods produced must take the form $[\underline{y}, y]$ for some level of fixed cost y.

In a hierarchy equilibrium, there exists a decreasing sequence $\bar{y} = y_1 > y_2 > \cdots > y_I \geq \underline{y}$, for some $I \in \mathbb{N} \cup \{\infty\}$, denoting the cutoffs. A hierarchy equilibrium is said to satisfy the *central place property* if the market area of the firms producing $(y_{i+1}, y_i]$ is half that of the firms producing $(y_i, y_{i-1}]$.

Definition 2.2 A hierarchy equilibrium that satisfies the central place property is called a *central place hierarchy*.

In fact, any hierarchy equilibrium is a central place hierarchy. Let $L_1 = m\underline{L}[\bar{y}]$, $m \in [1, 2)$, and $L_i = L_1/2^{i-1}$. Due to the hierarchy property, any production location produces goods in the range of $[\underline{y}, y_i]$ for some y_i so that it is called a layer-i city, and the cutoff y_i is given by the zero-profit condition:

$$y_i = \frac{tL_i^2}{2} \quad \forall\ 1 \leq i \leq I, \tag{2.1}$$

where the number of layers is:

$$I = \left\lfloor \frac{2\ln(m) + \ln(\bar{y}/\underline{y})}{2\ln(2)} + 1 \right\rfloor, \tag{2.2}$$

Figure 5 depicts four layers of such location configuration in which we can see that both the hierarchy and central place property are satisfied and the foregoing construction is an equilibrium.

Proposition 2.2 (Central place hierarchy) *For each* $L_1 = m\underline{L}(\bar{y})$, $m \in [1, 2)$, *let* $L_i = L_1/2^{i-1}$, $i \in [1, I]$, y_i *be given by the zero-profit condition (2.1), and the number of layers* I *be given by (2.2). Fix an* $x \in \mathbb{R}$, *and set the grid for* $(y_{i+1}, y_i]$ *as* $\{x + nL_i\}_{n=-\infty}^{\infty}$. *Then, for each* $m \in [1, 2)$, *the location configuration so constructed is the unique hierarchy equilibrium and satisfies the central place property.*

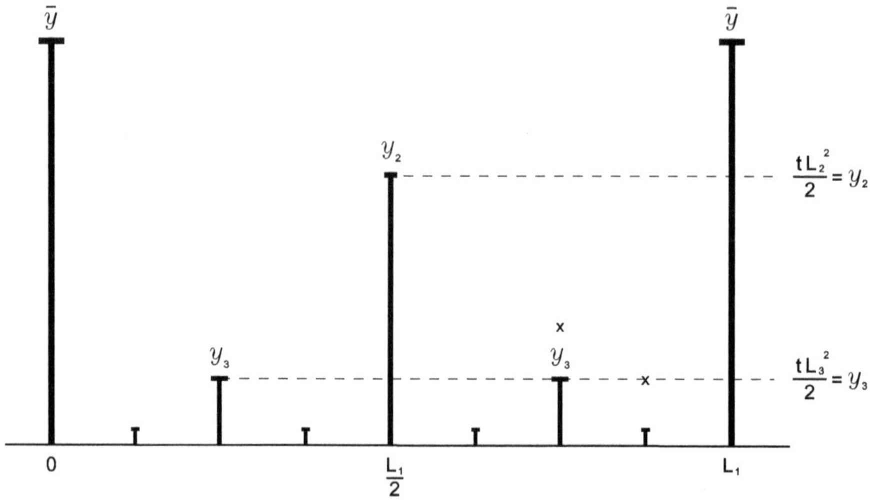

Fig. 5 A central place hierarchy. *Notes* The layer-i cities produce goods in $[\underline{y}, y_i]$. The cutoffs y_i are determined by the zero-profit conditions. The market areas for goods $(y_{i+1}, y_i]$ are half of that for $(y_i, y_{i-1}]$

3 Power Law for Cities

We now explain how a power law for cities emerges. In a central place hierarchy, the output of the firms in range $(y_{k+1}, y_k]$ is $L_k[F(y_k) - F(y_{k+1})]$. Define the size of a layer-i city by the total units produced in that city (as a measure of the level of economic activity):

$$Y_i = \sum_{k=i}^{I} L_k[F(y_k) - F(y_{k+1})]$$

Figure 6 illustrates the definitions of Y_i. The green (shaded with lines) and red (shaded with dots) areas represent the total quantity produced in a layer-1 and layer-2 city, respectively.

For every layer-1 city, there is one layer-2 city and 2^{i-2} layer-i cities. Thus, the total number of cities up to layer-i is

$$R_i = 1 + 1 + \sum_{k=3}^{i} 2^{k-2} = 2^{i-1}$$

Note that R_i represents the rank, by the rank-size rule, since the rank doubles from layer-i to the next layer-$i + 1$, Zipf's law can be approximated if city size shrinks by around half from layer-i to the next layer. Similarly, if city size shrinks by an approximately constant fraction from any layer-i to the next, then the power law is

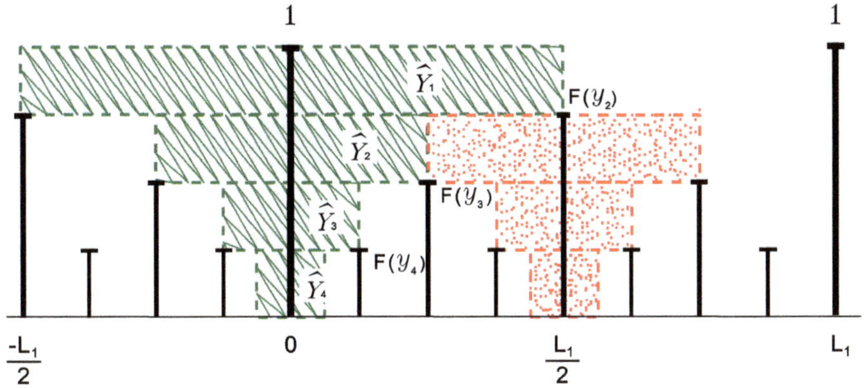

Fig. 6 City size. *Notes* The green (shaded with lines) and red (shaded with dots) areas denote the size of a layer-1 and layer-2 city, respectively. Both shaded areas are composed of rectangles, each of which represents the total production of the respective range of goods

approximated. It means that power laws can be generated by the fractal structures.[2] Therefore, a natural question arises: under what conditions can we guarantee such fractal structures?

There is, indeed, a simple but powerful condition that directly links central place hierarchies and the power law, regardless of the underlying economics behind that hierarchy. Given a central place hierarchy, the location patterns of cities of different layers are fixed, and different underlying economics matter only in relation to how the fractions of goods ($z_i = F(y_i)$) in the different layers are determined. The following proposition specifies the condition for the fractions of goods that renders the central place hierarchy a fractal structure.

Proposition 3.1 (Bounds on fraction ratios) *Suppose that there are infinitely many layers in a central place hierarchy. Let z_i denote the fraction of goods produced in a layer-i city, and let $\Delta_k = z_k - z_{k+1}$. Suppose that there is a $\delta > 0$ and a $\rho > 1$, such that for all $i \in \mathbb{N}$,*

$$\frac{\delta}{\rho} \leq \frac{\Delta_{i+1}}{\Delta_i} \leq \rho\delta.$$

Then,

$$\frac{1}{2}(\rho^{-1} - 1)\delta \leq \frac{Y_{i+1}}{Y_i} - \frac{\delta}{2} \leq \frac{1}{2}(\rho - 1)\delta. \tag{3.1}$$

Observe that δ is approximately the ratio of the increments (Δ_k) when ρ is approaching 1, hence, the slope is

---

[2]A fractal structure is a structure in which smaller parts of it resemble the entire structure.

$$\frac{\ln(R_{i+1}/R_i)}{\ln(Y_{i+1}/Y_i)} \approx \frac{\ln(2)}{\ln(\delta/2)} = -\frac{\ln(2)}{\ln(2) - \ln(\delta)}.$$

The Zipf's law requires only that the increments of the fraction of goods of two adjacent layers do not vary too much ($\delta \approx 1$), whereas the power law relaxes the ratio of increments between two layers from 1.

Besides the constrains on $z_i = F(y_i)$, the further question is how the behavior of $F(.)$ translates into a power law for cities. In order to answer this formally, we first need to introduce a few basic concepts of regular variation.

Definition 3.1 A measurable, positive function g is said to be regularly varying at zero (at infinity) if, for any $u > 0$, and for some $\alpha \in \mathbb{R}$,

$$\lim_{y \downarrow 0\ (\to \infty)} \frac{g(uy)}{g(y)} = u^{\alpha}.$$

If $\alpha = 0$, then g is said to be slowly varying. A function g is regularly varying with index α if and only if there exists a slowly varying function $\ell(y)$ such that

$$g(y) = y^{\alpha} \ell(y).$$

In what follows, $g \in RV_{\alpha}$ denotes that g is regularly varying at zero with index α. Suppose that $\underline{y} = 0$, and hence there are infinitely many layers. Recall from Proposition 2.2 that $y_{k+1} = y_k/4$ for all $k \geq 2$. Observe that the ratio between the increments between two layers can be written as

$$\delta_k \equiv \frac{\Delta_{k+1}}{\Delta_k} = \frac{F(y_{k+1}) - F(y_{k+2})}{F(y_k) - F(y_{k+1})} = \frac{1 - \frac{F(y_{k+1}/4)}{F(y_{k+1})}}{\frac{F(4y_{k+1})}{F(y_{k+1})} - 1}.$$

According to Definition 3.1, if $F \in RV_{\alpha}$, then in a small enough neighbourhood of 0, there are infinitely many k's such that

$$\delta_k \approx \frac{1 - \left(\frac{1}{4}\right)^{\alpha}}{4^{\alpha} - 1} = \left(\frac{1}{4}\right)^{\alpha}. \tag{3.2}$$

By Proposition 3.1, the power law is approximated with a tail index close to $1/(1 + 2\alpha)$.

A distribution function F on $(0, \bar{y}]$ can be regularly varying only with a nonnegative index α because an $F \in RV_{\alpha}$ on $(0, \bar{y}]$ with $\alpha < 0$ must be decreasing in a small neighbourhood of 0, which violates the requirement of a distribution function. However, a distribution function can be defined via a transformation of a non-increasing function $G \in RV_{\alpha}$ with $\alpha < 0$:

$$F(y) \equiv \frac{G(\underline{y}) - G(y)}{G(\underline{y}) - G(\bar{y})}, \tag{3.3}$$

where the domain of F is $[\underline{y}, \bar{y}]$ for some $\underline{y} > 0$. For a $\underline{y}$ that is close enough to zero, such an F behaves like a regularly varying function with a negative index α. This is because, for a $\underline{y}$ close enough to 0, there exists a sufficiently small neighbourhood of $\underline{y}$ such that, for all y_{k+1} in that neighbourhood:

$$\delta_k = \frac{F(y_{k+1}) - F(y_{k+2})}{F(y_k) - F(y_{k+1})} = \frac{G(y_{k+2}) - G(y_{k+1})}{G(y_{k+1}) - G(y_k)} = \frac{1 - \frac{G(y_{k+1}/4)}{G(y_{k+1})}}{\frac{G(4y_{k+1})}{G(y_{k+1})} - 1} \approx \frac{1 - \left(\frac{1}{4}\right)^{\alpha}}{4^{\alpha} - 1} = \left(\frac{1}{4}\right)^{\alpha}.$$

In any case, when the index α associated with the distribution function is positive (negative), then the slope of the Zipf plot is smaller (greater) than 1. The following proposition summarises the foregoing discussion and provides statements based on the density functions.

Proposition 3.2 (Regularly varying distributions) *Let $\delta = (1/4)^{\alpha}$ and fix any $\rho > 1$. Then, for a sufficiently small $\underline{y} \geq 0$, there exists an integer $K > 0$ such that condition (3.1) holds for all layers $I \geq i \geq K$ (with the possibility that $I = \infty$), if one of the following conditions is met:*

(a) the distribution function of fixed cost $F \in RV_{\alpha}$ with $\alpha \in [0, \infty)$;
(b) $G \in RV_{\alpha}$ with $\alpha \in (-1/2, 0)$ such that F is defined by (3.3);
(c) the density function of fixed cost $f \in RV_{\alpha-1}$, for $\alpha \in (-1/2, \infty)$.

In all cases, the approximate slope of the Zipf plot, i.e., the plot of log of rank on log of size, is $-1/(1+2\alpha)$.

Therefore, the power law arises when $F(.)$ has a regularly varying right tail, which is rather general, as it includes several well-known, commonly used distributions, such as the Pareto, Weibull, F, Beta (which subsumes the uniform), and Gamma, which subsumes the Chi-square, exponential, and Erlang.

Note that the sizes of cities of the same layer are equal and such city size distribution is understood as an approximation to the power laws. To fully conform with the relatively smooth city size distribution, some random factors to create differences among cities of the same layer is needed. Indeed, all central-place models for the city size distribution would need this random disturbance. See, for example, Beckmann (1958, p. 245).

4 A Dynamic Programming Approach to Central Place Theory

In this section, taking hierarchy property as given, we consider the social planner's problem about how to construct the optimal city hierarchy via a dynamic programming formulation (Hsu et al. 2014). There must be one and only one immediate smaller city between two neighboring larger-sized cities in any optimal solution. Furthermore, if the fixed cost of setting up a city is a power function, then the immediate smaller city will lie in the middle, which confirms the locational pattern suggested by Christaller, and further provides a rationale for central place theory of city hierarchy.

4.1 The Sequence Problem

The basic environment is similar to Sect. 2.1, we further assume the *hierarchy property*: at any location x, if a good $z \leq z_1$ is produced, then all $\tilde{z} \in [0, z]$ are also produced. We label a location that produces all goods up to z as a z-city. Denote the cost of setting up a z-city as $\Phi(z) \equiv \int_0^z \phi(u)\, du$. According to the hierarchy property, z also refers to a city's size. Assume that two *neighboring* z_1-cities are located at 0 and ℓ_1, respectively. Thus, the social planner needs to determine ℓ_1 first,[3] then taken ℓ_1 as given, to find the optimal *city hierarchy*, i.e., the locations and sizes of cities on the interval $(0, \ell_1)$.

Given ℓ_1, let the discrete set of cities on $(0, \ell_1)$ be denoted as

$$W \equiv \left\{ \begin{matrix} \left(z_i, L_{z_i}, I\right) | z_i \in (0, z_1],\ i = 1, 2, \ldots, I,\ I \in \mathbb{N} \cup \{\infty\},\ z_i > z_{i+1}, \\ L_{z_i} \text{ is the set of locations of } z_i\text{-cities} \end{matrix} \right\}.$$

That is, z_i is the i-th largest among all cities on $(0, \ell_1)$. For now, there may be multiple z_i cities, and L_{z_i} and $|L_{z_i}|$ denote the set of locations and the number of z_i -cities on $(0, \ell_1)$, respectively. The number I is the number of *layers* of cities, and I can be (countably) infinite. The optimization problem, given ℓ_1, is to search for a city hierarchy W that minimizes the per capita cost of production to serve every consumer a unit of each good in $[0, z_1]$:

$$C^*(\ell_1, z_1) \equiv \inf_W \frac{1}{\ell_1} \left[\sum_{z_i} |L_{z_i}| \Phi(z_i) + \text{total transport cost} \right], \qquad (4.1)$$

4.2 The Dynamic Programming Problem

The following two lemmas provide key characteristics of an optimal hierarchy that enables us to set up the planner's problem as a dynamic programming problem.

Lemma 4.1 *It is never optimal to have an interval without any city in it.*

[3]We do not discuss how to decide the optimal distance ℓ_1 here, since our focus is the city hierarchy between any two largest cities. Nevertheless, this is very crucial question for presenting a complete and meaningful model, please see Sect. 3.5 in Hsu et al. (2014) for the solution.

Proof Consider adding a z'-city in the middle in between with $z' \leq z$. Then, the savings in transport cost per good is

$$2\int_0^{\ell/2} txdx - 4\int_0^{\ell/4} txdx = t\ell^2/8$$

so the he net saving from having a z'-city is given by

$$S\left(z';\ell\right) \equiv \int_0^{z'} \left[\frac{t\ell^2}{8} - \phi(y)\right] dy$$

Because ϕ is continuous and strictly increasing, and $\phi(0) = 0$, $S\left(z';\ell\right) > 0$ for sufficiently small $z' > 0$, given ℓ. The result follows from the fact that there always exists sufficiently small z' such that adding a z'-city improves the allocation. □

Lemma 4.2 *It is never optimal to have two cities of the same size* $z' < z_1$ *without a larger city in between.*

The intuition behind this lemma is that, without a larger city in between, the two neighboring cities cannot be the same size in the optimal solution. We can produce another good more with a infinitesimally higher set-up cost at one of the two cities in order to save the transport cost, in which case there always exists a better allocation whose net cost is less. A full proof can be found in Sect. 2.3 in Hsu et al. (2014).

Lemmas 4.1 and 4.2 indicate that in between two z_1-cities it is optimal to place one and only one immediate sub-city, which is denoted as a z_2-city. Notice that z_2-city is not necessarily located in the middle.[4] Let $\ell_{2,1}$ and $\ell_{2,2}$ represent the distances from the z_2-city to the z_1-city on the left and right side, respectively. When the values of z_2, $\ell_{2,1}$ and $\ell_{2,2}$ are chosen, the recursive nature of the problem becomes apparent: given z_2, $\ell_{2,1}$ and $\ell_{2,2}$, we search for the optimal solutions for endless iterated bifurcations as the one given z_1 and ℓ_1. Figure 7 illustrates the city building process for the first three rounds of bifurcations. In general, the i-th round of bifurcation involves setting up cities of sizes $z_{i+1,1}, z_{i+1,2}, \ldots, z_{i+1,K_i}$, where $K_i = 2^{i-1}$, which divides intervals of length $\ell_{i,1}, \ell_{i,2}, \ldots, \ell_{i,K_i}$, respectively. Formally, let $\mathbf{z}_1 \equiv \{z_1\}$, and for all $i \in \mathbb{N}$, let $\boldsymbol{\ell}_i \equiv \left\{\ell_{i,k}\right\}_{k=1}^{K_i}$ and $\mathbf{z}_{i+1} \equiv \left\{z_{i+1,k}\right\}_{k=1}^{K_i}$, where $\ell_{1,1} \equiv \ell_1$ and $z_{2,1} \equiv z_2$. We define

$$\Gamma_1\left(\ell_1, z_1\right) \equiv \Gamma\left(\ell_1, z_1\right) \equiv \left\{\left(\boldsymbol{\ell}_2, \mathbf{z}_2\right) | z_2 \in \left[0, z_1\right], \ell_{2,1}, \ell_{2,2} \in \left(0, \ell_1\right) \text{ and } \ell_{2,1} + \ell_{2,2} = \ell_1\right\}. \tag{4.2}$$

and for $i \geq 2$,

$$\Gamma_i\left(\boldsymbol{\ell}_i, \mathbf{z}_i\right) \equiv \left\{ \begin{array}{c} \left(\boldsymbol{\ell}_{i+1}, \mathbf{z}_{i+1}\right) | z_{i+1,2k-1}, z_{i+1,2k} \in \left[0, z_{i,k}\right] \text{ for all } k = 1, 2, \ldots, K_{i-1}, \\ \ell_{i+1,2k-1}, \ell_{i+1,2k} \in \left(0, \ell_{i,k}\right) \text{ and } \ell_{i+1,2k-1} + \ell_{i+1,2k} = \ell_{i,k}, \text{ for all } k = 1, 2, \ldots, K_i. \end{array} \right\}$$

[4] For example, let $t = z_1 = \ell_1 = 1$. Consider a discontinuous setup cost requirement function: for an arbitrarily small $e \in (0, 1)$, $\phi(y) = 1/13$ for $y \in [0, e]$ and $\phi(y) = 1$ for $y \in (e, 1]$. It is readily verified that, in between two z_1-cities, the per capita cost is minimized by evenly placing two immediate sub-cities with $z' = e$.

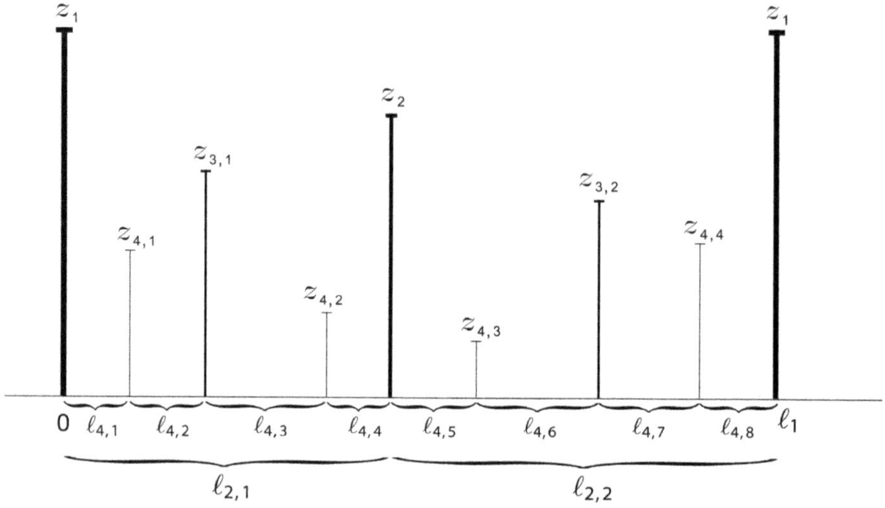

Fig. 7 Illustration of sequence problem

Then, define

$$\Pi(\ell_1, z_1) \equiv \left\{(\boldsymbol{\ell}_i, z_i)_{i=1}^{\infty} \mid (\boldsymbol{\ell}_{i+1}, z_{i+1}) \in \Gamma_i(\boldsymbol{\ell}_i, z_i), \text{ for all } i = 1, 2, \ldots\right\}.$$

Any $(\boldsymbol{\ell}, z) \equiv (\boldsymbol{\ell}_i, z_i)_{i=1}^{\infty} \in \Pi(\ell_1, z_1)$ is called a *feasible* sequence, given (ℓ_1, z_1).

Let ℓ be the distance between the two neighboring larger-sized z-cities, hence the distances of a immediate sub-city z' to the two cities are $\alpha\ell$ and $(1-\alpha)\ell$, where $\alpha \in (0, 1)$. The savings in transport costs for each good in $[0, z']$ is

$$s^1(\ell, \alpha) \equiv 2\int_0^{\frac{\ell}{2}} txdx - \left(2\int_0^{\frac{\alpha\ell}{2}} txdx + 2\int_0^{\frac{(1-\alpha)\ell}{2}} txdx\right) = \frac{t\ell^2}{2}\alpha(1-\alpha), \tag{4.3}$$

Then, the optimal magnitude of z' is determined by

$$s^1(\ell, \alpha) = \frac{t\ell^2}{2}\alpha(1-\alpha) = \phi(z'). \tag{4.4}$$

The left-hand side of (4.4) is the savings in transport costs when increasing z' marginally, whereas the right-hand side is the corresponding setup cost. If z' is low such that $\phi(z') < \frac{t\ell^2}{2}\alpha(1-\alpha)$, it incurs positive net savings (savings in transport costs net of setup costs) by increasing z'. Similarly, when $\phi(z') > \frac{t\ell^2}{2}\alpha(1-\alpha)$, one can improve the allocation by decreasing z'. In sum, Lemmas 4.1 and 4.2 and (4.4) imply that in any optimal city hierarchy the following constraint holds:

$$\left\{\begin{array}{l} z_{i+1,2k-1}, z_{i+1,2k} \in \left(0, z_{i,k}\right) \\ \ell_{i+1,2k-1}, \ell_{i+1,2k} \in \left(0, \ell_{i,k}\right), \ell_{i+1,2k-1} + \ell_{i+1,2k} = \ell_{i,k} \\ z_{i+1,k} = \phi^{-1}\left(\frac{t}{2}\ell_{i+1,2k-1}\ell_{i+1,2k}\right) \end{array}\right\}. \tag{4.5}$$

Equivalently, any optimal city hierarchy is associated with a sequence $\boldsymbol{\alpha} = \left\{\alpha_{i,k}\right\}$ such that $\ell_{i+1,2k-1} = \alpha_{i,k}\ell_{i,k}$ (hence $\ell_{i+1,2k} = \left(1 - \alpha_{i,k}\right)\ell_{i,k}$) and (4.5) holds.

Note that in defining the choice set of $(\boldsymbol{\ell}, \boldsymbol{z}) \equiv (\boldsymbol{\ell}_i, \boldsymbol{z}_i)_{i=1}^{\infty}$ by Γ_i and Π above, we leave (4.4) implicit and take the closure of $\left(0, z_{i,k}\right)$. According to Lemmas 4.1 and 4.2, we know that situations in which $z_{i+1,2k-1}$ or $z_{i+1,2k}$ equals 0 or $z_{i,k}$ are never optimal (except possibly for $i = 1$), but we do not lose any generality by including this possibility. When the choice of $z_{i+1,2k-1}$, according to (4.4) and given $\ell_{i,2k-1}$, is such that $z_{i+1,2k-1} > z_{i,k}$, one can always relabel i, k to ensure that the constraint $z_{i+1,2k-1}, z_{i+1,2k} \in \left[0, z_{i,k}\right]$ is obeyed. Thus, the choice set defined by Π encompasses all possible candidates for an optimal city hierarchy. In other words, any sequence $(\boldsymbol{\ell}, \boldsymbol{z})$ that satisfies all constraints in (4.5) is included in $\Pi(\ell_1, z_1)$. If one would like to make the constraint (4.4) explicit, one could redefine Γ_i by replacing $z_{i+1,2k-1}, z_{i+1,2k} \in \left[0, z_{i,k}\right]$ with

$$z_{i+1,2k-1} = \min\left\{\phi^{-1}\left(\frac{t}{2}\ell_{i+1,4k-1}\ell_{i+1,4k-2}\right), z_{i,k}\right\}$$

and

$$z_{i+1,2k} = \min\left\{\phi^{-1}\left(\frac{t}{2}\ell_{i+1,4k-3}\ell_{i+1,4k-4}\right), z_{i,k}\right\}.$$

Suppose the social planner has two z-cities with distance ℓ and nothing in between them, then the total cost in this interval of ℓ is

$$A(\ell, z) \equiv \Phi(z) + \frac{zt\ell^2}{4}.$$

Note that only one setup cost of a z-city is counted in this definition. When a z'-city divides an interval of ℓ bounded by two cities producing at least up to z, the total cost for the range of goods $(z', z]$ is given by $A(\ell, z) - A\left(\ell, z'\right) = \Phi(z) - \Phi\left(z'\right) + \left(z - z'\right)t\ell^2/4$. We can view the per capita cost for the goods $[0, z_1]$ on ℓ_1 as the sum of the per capita cost of different ranges of goods on different market areas within ℓ_1. Namely, the sequence problem is

$$C^*(\ell_1, z_1) \equiv \inf_{\substack{(\boldsymbol{\ell}, \boldsymbol{z}) \in \Pi(\ell_1, z_1), \\ z_1 > 0 \text{ given.}}} \frac{1}{\ell_1}\left[\begin{array}{l} A(\ell_1, z_1) - A(\ell_1, z_2) \\ +\sum_{i=2}^{\infty}\sum_{k=1}^{K_{i-1}}\left[\begin{array}{l} A\left(\ell_{i,2k-1}, z_{i,k}\right) - A\left(\ell_{i,2k-1}, z_{i+1,2k-1}\right) \\ +A\left(\ell_{i,2k}, z_{i,k}\right) - A\left(\ell_{i,2k}, z_{i+1,2k}\right) \end{array}\right] \end{array}\right]. \tag{SP}$$

Besides sequence problem, it can also be represented by a dynamic programming problem. Given state variables ℓ and z, the social planner needs to decide the size and location of the immediate sub-city, z'-city. Denote the length of the intervals to the left/right of z'-city as ℓ_l/ℓ_r. Then, $\ell_l + \ell_r = \ell$. Alternatively, let $\ell_l = \alpha\ell$ and $\ell_r = (1-\alpha)\ell$ for $\alpha \in (0, 1)$. We present the following dynamic programming problem.

$$\begin{aligned} C(\ell, z) &= \inf_{\ell_l, \ell_r \in (0,\ell), \ell_l + \ell_r = \ell, z' \in [0,z]} \frac{1}{\ell}\left[A(\ell, z) - A\left(\ell, z'\right) + \ell_l C(\ell_l, z') + \ell_r C(\ell_r, z')\right] \\ &= \inf_{\alpha \in (0,1), z' \in [0,z]} \frac{1}{\ell}[A(\ell, z) - A\left(\ell, z'\right)] + \alpha C(\alpha\ell, z') + (1-\alpha)C((1-\alpha)\ell, z'). \end{aligned} \tag{DP}$$

Equivalently, we can consider infimum of the total cost of all the goods $[0, z_1]$ on the interval of length ℓ_1. By defining $D^*(\ell_1, z_1) = \ell_1 C^*(\ell_1, z_1)$, and $D(\ell, z) = \ell C(\ell, z)$, it transforms (SP) and (DP) to (SP^D) and (DP^D) as following:

$$D^*(\ell_1, z_1) = \inf_{\substack{(\boldsymbol{\ell}, \mathbf{z}) \in \Pi(\ell_1, z_1), \\ z_1 > 0 \text{ given.}}} \begin{matrix} A(\ell_1, z_1) - A(\ell_1, z_2) \\ +\sum_{i=2}^{\infty} \sum_{k=1}^{K_{i-1}} \begin{bmatrix} A\left(\ell_{i,2k-1}, z_{i,k}\right) - A\left(\ell_{i,2k-1}, z_{i+1,2k-1}\right) \\ +A\left(\ell_{i,2k}, z_{i,k}\right) - A\left(\ell_{i,2k}, z_{i+1,2k}\right) \end{bmatrix}, \end{matrix} \tag{SP^D}$$

and

$$D(\ell, z) = \inf_{\alpha \in (0,1), z' \in [0,z]} A(\ell, z) - A\left(\ell, z'\right) + D\left(\alpha\ell, z'\right) + D\left((1-\alpha)\ell, z'\right). \tag{DP^D}$$

For showing the equivalence between the sequence problem(SP^D) and the respective dynamic programming problem (DP^D), i.e., the *principle of optimality*, we present the following corollary.

Corollary 1 *For any two positive real numbers ℓ_1 and z_1, let $X = [0, \ell_1] \times [0, z_1]$, and let $\mathcal{D}(X)$ denote the set of all real-valued continuous functions $d : X \to \mathbb{R}_+$ such that*

$$0 \le d(\ell, z) \le A(\ell, z). \tag{4.6}$$

Then a feasible sequence $(\boldsymbol{\ell}^*, \mathbf{z}^*) \in \Pi(\ell_1, z_1)$ attains the infimum in (SP^D) if and only if it satisfies (DP^D) recursively, i.e.,

$$D^*(\ell^*_{i,k}, z^*_{i,k}) = A\left(\ell^*_{i,k}, z^*_{i,k}\right) - A\left(\ell^*_{i,k}, z^*_{i+1,k}\right) + D^*(\ell^*_{i+1,2k-1}, z^*_{i+1,k}) + D^*(\ell^*_{i+1,2k}, z^*_{i+1,k}).$$

Now, the question left is: under what conditions can we be sure that there is a unique solution to (SP^D) as well as (DP^D)?

Corollary 2 *Let $\mathcal{D}(X)$ be given by Corollary 1. Let $T : \mathcal{D}(X) \to \mathcal{D}(X)$ be given by, for each $d \in \mathcal{D}(X)$,*

$$Td(\ell, z) \equiv \inf_{\alpha \in (0,1), z' \in [0,z]} A(\ell, z) - A\left(\ell, z'\right) + d(\alpha\ell, z') + d((1-\alpha)\ell, z'). \tag{4.7}$$

Then,

(i) Td is continuous. Hence, T is a self-mapping on $\mathcal{D}(X)$.
(ii) The minimum is attained; so inf *in the definition of T in (4.7) can be replaced with* min. *Moreover, the set of minimizers is an upper hemi-continuous correspondence on X.*
(iii) D^* is the unique solution to (DP^D) in $\mathcal{D}(X)$ and hence the unique fixed point of the mapping T on $\mathcal{D}(X)$.
(iv) For any $d \in \mathcal{D}(X)$, the sequence $\{T^n d\}$ converges to D^*.

Corollary 2 allows the numerical solution for any arbitrary ϕ/or any initial guess. We have implemented this iterative method via Matlab in which the approximation works well. The proofs of corollary 1 and 2 can be seen from Sect. 3.2 and 3.3 in Hsu et al. (2014), which starts from the routine pursue of principle of optimality as in almost all the dynamic programming problems. For a very intriguing and thorough introduction of such recursive problems in economics, please refer to Ljungqvist and Sargent (2012).

4.3 The Central Place Property

It turns out that the central place property holds when the setup cost is a power function: $\phi(z) = az^b$, for $a > 0$ and $b > 0$. Under this functional form, the total setup cost for a z-city is $\Phi(z) = \frac{a}{b+1} z^{b+1}$. The power function assumption of ϕ, in fact, means that the distribution of setup costs across goods is also a power function. Let Y denote the random variable of setup cost for a good. Then, for $y \in [0, \phi(z_1)]$,

$$\Pr[Y \leq y] = \frac{\phi^{-1}(y)}{z_1} = \frac{1}{z_1}\left(\frac{y}{a}\right)^{\frac{1}{b}}.$$

As shown in the hierarchy equilibrium, Sect. 2.2, this distribution of setup cost is a prototype of a class of distributions that leads to a power law distribution of city size.

Recall that it is possible that the optimal $z_2 = z_1$ if ℓ_1 is too large. Note from (4.4) that savings $s^1(\ell_1, \alpha)$ is bounded by $s^1(\ell_1, 1/2) = t\ell_1^2/8$. Define $\bar{\ell}(z)$ by the solution of ℓ in the following equation.

$$\frac{t\ell^2}{8} = \phi(z). \tag{4.8}$$

Then, for any $\ell_1 < \bar{\ell}(z_1)$, optimal $z_2 < z_1$, and thus the two z_1-cities with distance ℓ_1 are neighboring. For the rest of the analyses in this paper, we impose the condition that $\ell_1 < \bar{\ell}(z_1)$.

Proposition 4.3 *Suppose that $\ell_1 < \bar{\ell}(z_1)$, where $\bar{\ell}(z)$ is defined as the solution to (4.8). Suppose that the setup cost function $\phi(y) = az^b$, for positive constants a and b. Then, the central place property holds.*

Proof For ease of presentation, let $a = 1$. A general $a > 0$ does not change the result. From (4.4),

$$z' = \left(\frac{t\ell^2}{2}\alpha(1-\alpha)\right)^{\frac{1}{b}} = \left(\frac{t}{2}\alpha(1-\alpha)\right)^{\frac{1}{b}} \ell^{\frac{2}{b}} \equiv \kappa(\alpha)\ell^{\frac{2}{b}}. \tag{4.9}$$

Equation (4.9) implicitly assumes that $z' < z$. Recall that Lemma 4.2 rules out $z_{i+1,2k-1} = z_{i,k}$ or $z_{i+1,2k-1} = z_{i,k}$ as an optimal solution, and hence (4.9) is necessarily true for all optimal choices of $z_{i,k}$, except possibly for $i = 2$. However, the constraint $\ell_1 < \bar{\ell}(z_1)$ ensures that optimal $z_2 < z_1$.

Recall from (4.5) that there is a sequence $\boldsymbol{\alpha} = \{\alpha_{i,k}\}$ associated with any sequence $(\boldsymbol{\ell}, \mathbf{z}) \in \Pi(\ell_1, z_1)$. The fact that the optimal solution of z' is separable in ℓ and α implies that, except for z_1, we can write $z_{i,k} = \ell_1^{2/b} h_{i,k}(\boldsymbol{\alpha})$ and $\ell_{i,k} = \ell_1 g_{i,k}(\boldsymbol{\alpha})$, for some functions $h_{i,k}$ and $g_{i,k}$. Thus, both $A(\ell_{i,2k-1}, z_{i,k}) - A(\ell_{i,2k-1}, z_{i+1,2k-1})$ and $A(\ell_{i,2k}, z_{i,k}) - A(\ell_{i,2k}, z_{i+1,2k})$ are multiplicatively separable in $\ell_1^{2(b+1)/b}$ and some functions of $\boldsymbol{\alpha}$. Thus, for some function H, (SP) can be rewritten as

$$C^*(\ell_1, z_1) = \frac{z_1^{b+1}}{(b+1)\ell_1} + \frac{z_1 t\ell_1}{4} + \ell_1^{\frac{b+2}{b}} H(\boldsymbol{\alpha}^*).$$

By Corollarys 1 and 2, an optimal $\boldsymbol{\alpha}^*$ exists, and as such, $H(\boldsymbol{\alpha}^*)$ is well defined. Note that $H(\boldsymbol{\alpha}^*) < 0$, and $\ell_1^{(b+2)/b} |H(\boldsymbol{\alpha}^*)|$ is the per capita savings from building the optimal city hierarchy. Given the equivalence between (SP) and (DP), observe that the negative of per capita savings from having an optimal city hierarchy in an interval of ℓ is given by

$$\tilde{S}(\ell, z) \equiv C(\ell, z) - \frac{A(\ell, z)}{\ell} = C(\ell, z) - \frac{z^{b+1}}{(b+1)\ell} - \frac{zt\ell}{4} = \ell^{\frac{b+2}{b}} H(\boldsymbol{\alpha}^*). \tag{4.10}$$

This says that the $\tilde{S}$ function is homogenous of degree $(b+2)/b$ in ℓ and independent of z. With a little abuse of notation, we write $\tilde{S}(\ell) = \tilde{S}(\ell, z)$. Given (4.9) and (4.10), (DP) can be rewritten as

$$\tilde{S}(\ell) = \min_{z' \in (0,z), \alpha \in (0,1)} \frac{A(\alpha\ell, z') + A((1-\alpha)\ell, z') - A(\ell, z')}{\ell} + \left[\alpha^{\frac{2(b+1)}{b}} + (1-\alpha)^{\frac{2(b+1)}{b}}\right] \tilde{S}(\ell) \tag{4.11}$$

Thus,

$$\tilde{S}(\ell) = \frac{-b}{b+1}\left(\frac{t}{2}\right)^{\frac{b+1}{b}} \ell^{\frac{b+2}{b}} \max_{\alpha \in (0,1)} \frac{[\alpha(1-\alpha)]^{\frac{b+1}{b}}}{1 - \alpha^{\frac{2(b+1)}{b}} - (1-\alpha)^{\frac{2(b+1)}{b}}}. \tag{4.12}$$

We show in the separate appendix in Hsu et al. (2014) that the unique solution to the maximization problem in (4.12) is $\alpha = 1/2$. □

Observe that the optimal sequence $\boldsymbol{\alpha}^*$ does not depend on ℓ_1. The recursive nature implies that for all i, k, the optimal sequence in the interval of $\ell_{i,k}$, i.e., $\{\alpha_{i',k}\}_{i' \geq i}$, does not depend on the magnitude of $\ell_{i,k}$. Thus, under this power function distribution of setup costs, the optimal city hierarchy in any interval of $\ell_{i,k}$ resembles that of the entire one in ℓ_1. As Sect. 3 shows that this *scale-free* property gives the city hierarchy a *fractal structure*; specifically, the structure of the smaller part of the hierarchy resembles that of the larger.

5 Concluding Remarks

This chapter presents a parsimonious model in which central place hierarchies arise from both equilibria and optimal allocations. In the equilibrium model, we show that hierarchy property can arise as an equilibrium outcome. Even though there are possible deviations from the hierarchy property that could still constitute an equilibrium, Hsu (2012) shows that such deviations can be ruled out by adding in the home market effect of the central places. In the social planner's problem, the problem of location choice is complex, but we show that if the distribution of the fixed cost follows a power function, the central place property holds. As such distribution is also regularly varying, a power law also emerges from an optimal allocation.

One potential criticism is that the primary industry (agriculture, fishing, forestry, mining, etc.) and its employment have become a small fraction of the economy in developed countries, and thus central place theory is not quite relevant for these countries. Such a criticism is, however, not well grounded because the reason why a central place hierarchy emerges has nothing to do with the number/fraction of immobile consumers in the economy. It is the fact that they are immobile and dispersed over the entire geographic space that prompts the central places that serve them. A smaller relative size of the primary industry may make the cities and towns more sparse in spacing and smaller in their sizes, but it does not qualitatively alter the fractal structure of the central place hierarchy. Hence, it does not affect the power law for cities either.

Recall that the hierarchy property posits that larger cities are more diverse not only because they have more industries than smaller cities but also because they specialise in industries with more scale economies. This is arguably a reasonable view, especially when we look at industries in broader classifications. Mori et al. (2008) and Landman et al. (2011) find that the hierarchy property holds well for 3-digit JSIC but dissipates for 4-digit JSIC (the finest Japanese industrial classification). These findings hint that, at the finest levels, specialisation matters to the extent that heavy industries can be located in small towns due to numerous factors outside central place theory. The development of a comprehensive theory that produces more realistic patterns of diversity and specialisation is a desirable direction for future research.

References

Arlinghaus, S. L. (1985). Fractals take a central place. *Geografiska Annaler: Series B, Human Geography*, 67(2):83–88.

Arlinghaus, S. L. and Arlinghaus, W. C. (1989). The fractal theory of central place geometry: a diophantine analysis of fractal generators for arbitrary löschian numbers. *Geographical Analysis*, 21(2):103–121.

Batty, M. and Longley, P. A. (1994). *Fractal cities: a geometry of form and function*. Academic press.

Beckmann, M. J. (1958). City hierarchies and the distribution of city size. *Economic Development and Cultural Change*, 6(3):243–248.

Champernowne, D. G. (1953). A model of income distribution. *The Economic Journal*, 63(250):318–351.

Chen, Y. (2011). Fractal systems of central places based on intermittency of space-filling. *Chaos, solitons & fractals*, 44(8):619–632.

Chen, Y. (2014). Multifractals of central place systems: Models, dimension spectrums, and empirical analysis. *Physica A: Statistical Mechanics and its Applications*, 402:266–282.

Chen, Y. and Zhou, Y. (2006). Reinterpreting central place networks using ideas from fractals and self-organized criticality. *Environment and Planning B: Planning and Design*, 33(3):345–364.

Christaller, W. (1933). *Central Places in Southern Germany*, translated by C. W. Baskin (1966), Englewood Cliffs, NJ: Prentice-Hall.

Dittmar, J. (2011). Cities, markets, and growth: the emergence of Zipf's law. Working paper.

Eaton, B. C. and Lipsey, R. G. (1982). An economic theory of central places. *The Economic Journal*, 92(365):56–72.

Eeckhout, J. (2004). Gibrat's law for (all) cities. *American Economic Review*, 94(5):1429–1451.

Frankhauser, P. (1998). The fractal approach. A new tool for the spatial analysis of urban agglomerations. *Population: An English Selection*, pages 205–240.

Fujita, M., Krugman, P. R., Venables, A. J. and Fujita, M. (1999). *The Spatial Economy: Cities, Regions and International Trade*, volume 213. Wiley Online Library.

Gabaix, X. (1999). Zipf's law for cities: an explanation. *The Quarterly Journal of Economics*, 114(3):739–767.

Gabaix, X. and Ioannides, Y. M. (2004). The evolution of city size distributions. In *Handbook of Regional and Urban Economics*, volume 4, pages 2341–2378. Elsevier.

Hales, T. C. (2001). The honeycomb conjecture. *Discrete & Computational Geometry*, 25(1):1–22.

Hsu, W.-T. (2012). Central place theory and city size distribution. *The Economic Journal*, 122(563):903–932.

Hsu, W.-T., Holmes, T. J. and Morgan, F. (2014). Optimal city hierarchy: A dynamic programming approach to central place theory. *Journal of Economic Theory*, 154:245–273.

Landman, B. A., Huang, A. J., Gifford, A., Vikram, D. S., Lim, I. A. L., Farrell, J. A., Bogovic, J. A., Hua, J., Chen, M., Jarso, S., et al. (2011). Multi-parametric neuroimaging reproducibility: a 3-t resource study. *Neuroimage*, 54(4):2854–2866.

Lederer, P. J. and Hurter, A. P. Jr. (1986). Competition of firms: Discriminatory pricing and location?, *Econometrica*, 54(3):623–640

Ljungqvist, L. and Sargent, T. J. (2012). *Recursive Macroeconomic Theory*. The MIT Press.

Mori, T., Nishikimi, K. and Smith, T. E. (2008). The number-average size rule: a new empirical relationship between industrial location and city size. *Journal of Regional Science*, 48(1):165–211.

Quinzii, M. and Thisse, J.-F. (1990). On the optimality of central places. *Econometrica: Journal of the Econometric Society*, pages 1101–1119.

Simon, H. A. (1955). On a class of skew distribution functions. *Biometrika*, 42(3/4):425–440.

Tabuchi, T. and Thisse, J.-F. (2006). Regional specialization, urban hierarchy, and commuting costs. *International Economic Review*, 47(4):1295–1317.
Tabuchi, T. and Thisse, J.-F. (2011). A new economic geography model of central places. *Journal of Urban Economics*, 69(2):240–252.

Distribution of City Size: Gibrat, Pareto, Zipf

Marco Bee, Massimo Riccaboni and Stefano Schiavo

Abstract The exact shape of the distribution of city size is subject to considerable scholarly debate, as competing theoretical models yield different implications. The alternative distributions being tested are typically the Pareto and the log-normal, whose finite sample upper tail behavior is very difficult to tell apart. Using data at different levels of aggregation (census blocks and cities) we show that the tail behavior of the distribution changes upon aggregation, and the final result depends crucially on the shape of the distribution of the number of elementary units associated with each aggregate element.

Keywords Zipf distribution · Log-normal distribution · Maximum entropy · Cities · Size distribution

JEL Classification: C14 · C51 · C52

M. Bee
Department of Economics and Management, University of Trento,
via Inama 5, 38122 Trento, Italy
e-mail: marco.bee@unitn.it

M. Riccaboni (✉)
IMT School for Advanced Studies, Piazza San Francesco 19, 55100 Lucca, Italy
e-mail: massimo.riccaboni@imtlucca.it

M. Riccaboni
Department of Managerial Economics, Strategy and Innovation (MSI), K.U. Leuven,
Leuven, Belgium

S. Schiavo
School of International Studies and Department of Economics and Management,
University of Trento, via Inama 5, 38122 Trento, Italy
e-mail: stefano.schiavo@unitn.it

S. Schiavo
OFCE-DRIC, Paris, France

L. D'Acci (ed.), *The Mathematics of Urban Morphology*,
Modeling and Simulation in Science, Engineering and Technology,
https://doi.org/10.1007/978-3-030-12381-9_4

1 Introduction

Among the different phenomena in economics that follow a Pareto distribution, such as wealth (Pareto 1896; Champernowne 1953; Benhabib et al. 2011), firm size (Gibrat 1931; Ijiri and Simon 1977; Axtell 2001; Cabral and Mata 2003; Luttmer 2007), and city size (Zipf 1949; Gabaix 1999), the last one has been at the center of a lively debate on whether it is better approximated by a Pareto or by a log-normal distribution (Eeckhout 2004; Levy 2009; Eeckhout 2009; Malevergne et al. 2009; Rozenfeld et al. 2011; Berry and Okulicz-Kozaryn 2012; Hsu 2012; Ioannides and Skouras 2013; González-Val et al. 2015; Fazio and Modica 2015).

On the empirical side, investigations for various countries have been carried out by Ioannides and Skouras (2013), Puente-Ajovín and Ramos (2015), Cieślik and Teresiński (2016), Ramos (2017), Li and Nam (2017), Luckstead et al. (2017) and Calderín-Ojeda (2016). Scholars are engaged in an ongoing debate on whether such a distribution displays a Pareto tail and whether this conforms to a Zipf's law (i.e., a power-law distribution with shape parameter equal to one) or not. Beside the specific intellectual curiosity the issue may raise, there are broader theoretical reasons for investigating the matter, as competing models yield different implications (Masahisa et al. 2001). Indeed, while the seminal paper by Gabaix (1999) predicts a Zipf's law, and a recent contribution by Hsu (2012) shows that a central place hierarchy gives rise to a power-law, Eeckhout (2004) proposes an equilibrium theory from which he derives a log-normal distribution of cities.

The contention is fueled partly by the difficulty in distinguishing log-normal from Pareto tails (Embrechts et al. 1997; Bee et al. 2011), and partly by the difficulty of properly defining what a city is and, empirically, what is the correct measure to use.[1] Indeed, while early studies focus on the largest US Metropolitan Statistical Areas (MSA) only (Gabaix 1999), Eeckhout (2004) uses data for all the US populated places identified by the US Census in years 2000 and 2001. By so doing the author shows that the size distribution of cities is log-normal, not power-law as previously thought (at least since Zipf 1949). A few years later, Levy (2009) acknowledges that the body of the city-size distribution is well approximated by a log-normal, but claims that there are significant departures in the upper tail. Specifically, the top 0.6% of the distribution, i.e., the MSA, appear to fit better a power-law. Eeckhout (2009) replies to these new findings by highlighting potential problems associated with the procedures used by Levy (2009) to identify the power-law tail. Specifically, Eeckhout (2009) suggests that the graphical procedure based on visual inspection of a log-log plot introduces significant biases in the right tail of the distribution. Moreover, he warns against the practice of estimating a truncated subsample of

[1]This point is particularly stressed in Rozenfeld et al. (2011), who propose a new methodology to define cities based on microdata and a clustering algorithm that identifies a city as the maximal connected cluster of populated sites. By applying this methodology to both US and UK data, the authors find that a Zipf's law approximates well the distribution of 1,947 US cities with more than 12,000 inhabitants (1,007 cities with more than 5,000 inhabitants for the UK). An alternative approach entails the use of high-resolution night-light satellite image data (Bagan and Yamagata 2015; Zhou et al. 2015).

the distribution only, while testing its significance against a complete distribution with specific parameters. Malevergne et al. (2011) have suggested that the debate rests on the small power of the tests employed by both Eeckhout (2004) and Levy (2009). They claim the issue can be definitely settled by adopting a better testing procedure, namely the Uniformly Most Powerful Unbiased test of the exponential versus truncated normal distribution in log-scale developed by Del Castillo and Puig (1999). Hence, methodological issues related to the proper way of identifying a power-law tail seem at least as relevant as sample coverage.

In this chapter, we summarize some empirical evidence that points to the fact that the tail behavior of the distribution changes upon aggregation, giving rise to a more pronounced (longer) power-law tail. Furthermore, we investigate three different mechanisms that can generate this phenomenon, namely sample size, correlation between the number of elementary units and their average size, and the aggregation function (i.e., the shape of the distribution of the number of elementary units associated to each aggregate element). We find the last factor exerts the largest influence and needs to be carefully scrutinized when proposing models aiming at explaining the shape of the distribution of particular phenomena (Rozenfeld et al. 2011). However, sample size appears to crucially determine the length of the power-law tail found in the data.

More in general, our work casts new light on the fact that upon aggregation one observes the emergence of a thicker tail, so that extreme events become more likely (Perline 2005; Growiec et al. 2008). Such a behavior is counterintuitive since the common wisdom based on the Central Limit Theorem assumes idiosyncratic shocks to cancel out upon aggregation, and reinforces the claim recently made by Gabaix (2011).

2 Methodology and Data

In what follows, we analyze the distribution of US city sizes. We decompose aggregate entities (i.e., cities) into constituent parts, that is census blocks. We study the size distribution of aggregate entities, $P(S_i)$, and the size of constituent parts $P(s_j)$, with $S_i = \sum_{j=1}^{K_i} s_j$ where K_i is the number of parts of aggregate i and s_j is the size of its building blocks.

A traditional argument states that if an integrated entity of larger size had higher unit costs, then it should be possible to split it into independent and separately managed units so that any such disadvantage is eliminated (Sutton 1997). Thus, the presence of non-diminishing returns at the level of aggregated entities and constituent units implies that at the aggregate level economic organizations are made by an uneven number of units of different sizes.[2]

[2] Aggregate economies can thus be represented and as (random) partitions (Sutton 2002; Aoki and Yoshikawa 2011) or as sums of a random number of random variables.

2.1 Testing for a Power-Law Tail

Discriminating between power-law (Pareto) and log-normal tail behavior is a difficult task. The methodological reference is Extreme Value Theory (EVT), which studies the statistical properties of the distributions of upper order statistics. It is well-known that they belong to the domain of attraction of one of three distributions, namely Fréchet, Gumbel or Weibull (Embrechts et al., 1997). Whereas the distributions in the Fréchet domain of attraction are definitely heavy-tailed and the distributions in the Weibull domain are light-tailed, the Gumbel domain includes both distributions with a relatively light tail (exponentially decreasing, such as the normal) and with a relatively heavy tail (such as the log-normal). Things are further complicated by the fact that there exist several definitions of "heavy-tailed distributions", corresponding to different degrees of tail heaviness (see Embrechts et al. 1997, pp. 49–50).

For the purposes of testing between Pareto and log-normal, the main result is that the upper order statistics of the log-normal converge to the Gumbel distribution, whereas the upper order statistics of the Pareto converge to the Fréchet. This implies that the asymptotic tail behavior of the two distributions are mathematically different. However, the convergence of the log-normal to the asymptotic distribution is extremely slow (Perline 2005), so that the difference may be very small, at the extent that they are often practically indistinguishable for any finite sample size.

A similar conclusion is reached by recalling that a continuous random variable (r.v.) is in the domain of attraction of the Fréchet if and only if its density is a regularly varying function (Embrechts et al. 1997, pp. 131–132). Although the Pareto density is regularly varying and the log-normal is not, Malevergne et al. (2009) point out that, when the variance is large, the log-normal probability density function (pdf) can be rewritten in a form similar to the Pareto pdf, the only difference being that the exponent of the log-normal, unlike the Pareto one, varies with x. However, the log-normal exponent is almost constant with respect to x, so that in practice, unless the sample size is huge and/or the variance is very small, discriminating between a constant and an "almost constant" exponent is problematic.

Given these difficulties, several tests have been proposed, in an attempt to find the one that guarantees the best performance. We mention here, and employ in the following, the Uniformly Most Powerful Unbiased (UMPU) test based on the clipped sample coefficient of variation developed by Del Castillo and Puig (1999) and used by Malevergne et al. (2009), the Maximum Entropy (ME) test by Bee et al. (2011) and a test proposed by Gabaix and Ibragimov (Gabaix 2009; Gabaix and Ibragimov 2011; Rozenfeld et al. 2011) (GI henceforth).

The UMPU test is uniformly most powerful, but only in the class of unbiased tests. A more serious drawback is that it is a test of the null of power-law against the alternative of log-normal, and rejects the null hypothesis for small values of the coefficient of variation c. Implicitly, this implies that it works well (i.e., its power is high) in cases such as the log-normal-Pareto mixture, namely when the data generating process is such that $c \geq 1$ above the threshold that separates the log-normal and the Pareto and $c < 1$ below the threshold (Bee et al. 2011). However, if the distribution below the threshold is not power-law but nonetheless has $c \geq 1$, as happens, for example, for the Weibull with shape parameter equal to 1, UMPU is completely unreliable (see Bee et al. 2013).

The ME approach entails maximizing the Shannon's information entropy under k moment constraints $\mu^i = \hat{\mu}^i$ $(i = 1, \ldots, k)$, where $\mu^i = \mathrm{E}[T(x)^i]$ and $\hat{\mu}^i = \frac{1}{n}\sum_j T(x_j)^i$ are the i-th theoretical and sample moments and n is the number of observations. This can be solved by introducing $k+1$ Lagrange multipliers λ_i $(i = 0, \ldots, k)$, so that the solution (that is, the ME density) takes the form $f(x) = e^{-\sum_{i=0}^{k} \lambda_i T(x)^i}$. The Pareto distribution is an ME density with $k = 1$, whereas the log-normal is ME with $k = 2$.

A log-likelihood ratio (llr) test of the null hypothesis $k = k^*$ against $k = k^* + 1$ is given by

$$\mathrm{llr} = -2n \left(\sum_{i=0}^{k^*+1} \hat{\lambda}_i \hat{\mu}^i - \sum_{i=0}^{k^*} \hat{\lambda}_i \hat{\mu}^i \right),$$

where n is the population size. From standard limiting theory the llr test is asymptotically χ_1^2 and is optimal (Cox and Hinkley 1974; Wu 2003).

When the whole distribution is of interest, the method can be used for fitting the best approximating density, with the optimal k found by the log-likelihood ratio (llr) criterion. The procedure is based on the following steps: (1) estimate sequentially the ME density with $k = 1, 2, \ldots$; (2) perform the test for each value of k; (3) stop at the first value of k (k_0, say) such that the hypothesis $k = k_0$ cannot be rejected and conclude that $k^* = k_0$. If the aim consists in testing a power-law against a log-normal tail, we just test $k = 1$ against $k = 2$.

When ME tests for the optimal value of k, it is computed iteratively starting from $k = 1$ and stopping only when the p-value is sufficiently small. When the true distribution might be neither Pareto nor log-normal, the test should be carried out for some values of k larger than 2, even though the p-value for $k = 2$ against $k = 1$ may be relatively small. Typically, a very small p-value will be obtained for the optimal value of k, which is expected to be larger than 2. In other words, in such a case it may be worth to use a rather high level of α, such as 10%, in order to avoid accepting the null hypothesis when $k = 2$. It is also recommended to look at the graphs of the ME densities for various values of $k \geq 2$, superimposed on the histogram of the data, in order to ascertain whether the rejection of $k = 2$ was the correct decision.

Finally, the GI test is based on the following procedure. Estimate by OLS the regression

$$\log\left(r - \frac{1}{2}\right) = \text{constant} - \xi \log(x_r) + q[\log(x_r) - \gamma]^2,$$

where ξ is the Pareto shape parameter, q is the quadratic deviation from a Pareto, r is the rank, and x_r is the r-th order statistic. Asymptotically, for the Pareto distribution, $q = 0$, so that a large value of $|q|$ points towards rejection of the null hypothesis of power-law. Gabaix and Ibragimov (2011) show that, under the null of a Pareto, the statistic $\sqrt{2nq_n}/\xi^2$ converges to a standard normal distribution, which can therefore be used to find the critical points of the test.

2.2 Data Description

Information on the population of US cities is derived from the 2010 census data collected by the US Census Bureau. The elementary unit of analysis, corresponding to disaggregate data, is the population of each city block: we have data for 6 127 259 blocks. These figures are then aggregated into administrative units that represent populated places. As in Eeckhout (2004) we take populated places as the unit of analysis at the aggregate level.[3]

The focus of this chapter is on the statistical properties of the city size distribution. As such, we do not discuss the proper definition of a city. In fact, we are well aware that administrative definitions often do not properly capture the social or economic interactions that determine the tendency of people to cluster and ultimately define city boundaries.[4]

The use of census blocks and populated places is therefore rather unsatisfactory, at least from the point of view of urban studies, and should not be read as an endorsement of administrative definitions of cities. On the other hand, census data have two main advantages that make them a natural starting point for our work. First of all, administrative data provide us with an immediate mapping between elementary and aggregate units, a feature that is necessary if one aims at assessing the role of aggregation in shaping the distribution of city size. Second, using the same data employed in Eeckhout (2004) enhances the comparability of our results, especially with respect to a study that finds no power-law tail in the distribution and, thus, works "against" our hypothesis.[5]

Table 1 Test results on disaggregate data ($n = 1\,547\,203$): rank (percentile) after which the power-law hypothesis is rejected

	5%	1%
ME	3600 (0.06)	3800 (0.06)
UMPU	3600 (0.06)	3800 (0.06)
GI	1870 (0.03)	2953 (0.05)

[3]In the rest of the chapter the terms city and populated place are used interchangeably.

[4]We thank an anonymous referee for highlighting this important point.

[5]In Bee et al. (2013) we perform our analysis on the clusters (cities) identified by Rozenfeld et al. (2011) and find results that are in line with those reported in the original paper.

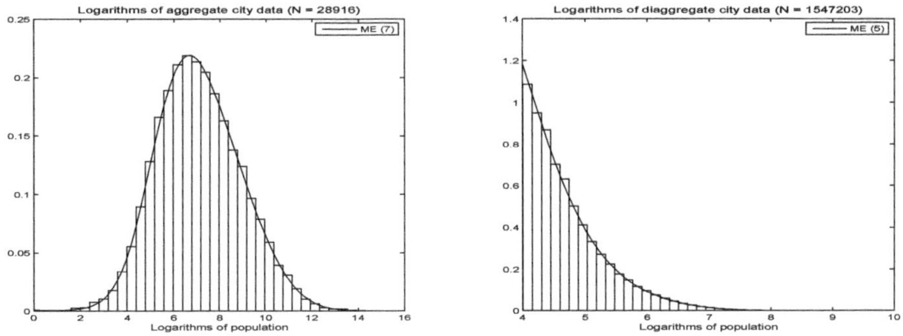

Fig. 1 Maximum entropy estimates of empirical distributions for aggregate and disaggregate data. The sample sizes reported above the panels stands for the number of observations used to fit the ME distribution and generate the plots (k value into parentheses)

3 Empirical Results

We start the empirical analysis by fitting the maximum entropy density to the empirical distributions of both aggregate $P(S)$ and disaggregate $P(s)$ data. Results are displayed in Fig. 1. Small observations are characterized by jumps and discontinuities that make estimation problematic, thus we have to truncate the distributions below a certain threshold. This is particularly true in the case of city-size data at the disaggregate level, where we exclude all blocks with a population smaller than 54 (4 in natural logs). The fit with the ME distribution reveals that $k > 2$ for all the distributions, thus suggesting that the best fit for the whole distributions is significantly different both from Pareto ($k = 1$) and Log-normal ($k = 2$).

3.1 Test Results

We first look at disaggregated data, i.e., block sizes in the US. Table 1 reports results for the three tests (UMPU, ME, GI) at the 5 and 1% level. The table reports the highest numerical values of the rank associated with rejection of the null hypothesis of a power-law tail, as well as the associated percentile in parentheses. This means that the figures in the table represent the length of the Pareto tail in terms of number of observations (and in percentage of the sample size). We report the rank at which the tests start staying in the critical region and never go back to the acceptance region, so that we disregard instances where a test goes in the critical region but then is unable to reject the null hypothesis once we increase the sample size. In so doing we are giving more chances to the null hypothesis, which implies a possible overestimation of the length of the power-law tail.

Irrespective of the chosen level of significance, the power-law tail appears to be limited to the very top of the distribution, and all tests show good agreement on this.

Table 2 Test results on aggregate data ($n = 28\,916$): rank (percentile) after which the power-law hypothesis is rejected

	5%	1%
ME	1030 (3.56)	1250 (4.32)
UMPU	990 (3.42)	1050 (3.63)
GI	1759 (6.08)	2159 (7.47)

Moving to aggregate data at the level of cities, we see that the length of the Pareto tail significantly increases in terms of percentile. Again, there is good agreement among the ME and UMPU tests, whereas the GI test identifies a longer tail. While focusing on aggregate rather than disaggregate data does imply a marked difference, the power-law tail is still limited to the top 1000 (ME and UMPU) or 2200 (GI) cities. These correspond to a population of about 37 600 and 19 200 inhabitants and represent between 3.6 and 7.5% of the whole sample. Moreover, they are in line with previous findings by Malevergne et al. (2009) and Rozenfeld et al. (2011).

Taken together, Tables 1 and 2 provide evidence that the level of aggregation at which phenomena are studied plays an important role in determining the results and, in particular, the findings about the shape of the upper tail of the empirical distributions. In what follows we explore three different mechanisms through which such a behavior may emerge in the data.

3.2 Estimates of the Shape Parameter

Before moving to discuss the mechanisms that drive the change in the tail behavior upon aggregation, we take a look at the estimates of the shape parameter of the power-law portions of the distributions. Its magnitude has played an important part in the debate, since Gabaix's model implies a shape parameter equal to 1 (Zipf's law). This prediction finds empirical support both in Gabaix (1999) and more recently in Rozenfeld et al. (2011); on the other hand, Eeckhout (2004) finds that the value of shape parameter changes significantly at different cutoffs, inferring from this that the distribution cannot be truly power-law. Finally, Malevergne et al. (2009) report a coefficient significantly larger than 1.

Table 3 reports the estimates of the shape parameter obtained using the methodologies associated with the three tests. The estimation is performed at the cutoff identified by each of the tests. In particular, the estimate of the shape parameter is a byproduct of both the ME and GI testing procedures, whereas in the case of UMPU we rely on the Hill estimator, as done by Malevergne et al. (2009). We can see that the estimates are in line with results presented by Malevergne et al. (2009): the shape parameter takes values around 1.3. Different tests yield different estimates, with ME

Table 3 Estimates of the shape parameter

	5%	1%
ME	1.30	1.27
UMPU	1.32	1.30
GI	0.92	0.94

and UMPU being rather close to each other, whereas the GI result is more in line with the findings in Rozenfeld et al. (2011).

3.3 *Emergence of a Power-Law Upon Aggregation*

We investigate three candidate mechanisms that could explain the length of the power-law tail upon aggregation: the sample size, the shape of the aggregation function (i.e., the number of elementary units of which aggregate entities are made), and the correlation between the number of elementary units comprised in each aggregate element and their average size.

3.3.1 Sampling

To verify the impact of sample size on the results, we run the tests on a sample of the same size of the aggregate datasets, obtained by simple random sampling from disaggregate data. Since detecting the difference between a log-normal and a Pareto tail is difficult and the tests have low power, in particular when n is small, the tests might well suffer the smaller sample size associated with more aggregate datasets and therefore have more troubles rejecting the null Pareto hypothesis. Our sampling exercise aims precisely at investigating the impact of such an effect.

Sampled data display—at least according to the UMPU and ME tests—a much longer power-law tail than the one found for actual aggregate observations. Indeed, as reported in Table 4, these two tests identify a power-law tail spanning roughly 3000–3200 observations, i.e. more than 10% of the sample. This seems to imply that the reduction in the power of the tests associated with smaller sample size plays a major role in determining the power-law tail observed in the data.[6]

[6]Such a conclusion is partially tempered by results of the GI test, which finds a power-law tail limited to the top 655 observations in the sampled dataset.

Table 4 Test results on synthetic datasets obtained by random sampling from disaggregate data with the size of aggregate ones ($n = 28\,916$): rank (percentile) after which the power-law hypothesis is rejected

	5%	1%
ME	3000 (10.37)	3200 (11.07)
UMPU	3000 (10.37)	3200 (11.07)
GI	655 (2.27)	885 (3.06)

3.3.2 Aggregation Rule

By noting that aggregates are obtained by summing the size of the elementary units associated with each aggregate element, it is fairly easy to conclude that a very simple mechanism giving rise to a power-law tail is the shape of the aggregating function. Indeed, calling K_i the number of disaggregate elements comprised in aggregate object i, then a power-law distribution for K gives rise to a power-law distribution of aggregate sizes, if disaggregate units are (approximately) of the same size. Denoting aggregate size with S_i we have that $S_i = K_i \times \bar{s}_i$ where $\bar{s}_i$ is the average size of the elementary units of aggregate i. If K_i is Pareto, and $\bar{s}$ is independent of K_i for sufficiently large K, S_i will also be Pareto, and this holds true even if the sizes of elementary units are not themselves Pareto distributed (see Growiec et al. 2008, for a detailed explanation).

Figure 2 shows the complementary cumulative distribution (CCDF) of the number of blocks associated with each city (in double logarithmic scale). We can see that the distribution is approximately Pareto in the upper tail. Table 5 reports the results of the tests applied to the distribution of K. The intuition coming from the visual inspection of the CCDFs is confirmed: a Pareto tail is present.

However, $\bar{s}$ is neither constant nor independent of K, so that the presence of a power-law aggregation rule is not sufficient to generate a power-law in aggregate data. When looking at the relationship between the number of elementary units (K) and their average size ($\bar{s}$), we find they are positively correlated.

A positive correlation between the number of elementary units assigned to each aggregate element and their average size (which makes the 'identity' of units relevant) may result from a very skewed distribution for disaggregate data: this makes convergence to the central limit theorem rather slow upon aggregation. This correlation may end up stretching the upper tail of the aggregate distribution as it is more probable that aggregate elements made up of a larger number of units (i.e., featuring a large K), contain units of larger (average) size.

3.3.3 Correlation

To clean this effect from the data, we run the usual tests on synthetic datasets obtained by aggregating elementary units according to a random rule. Table 6 shows the

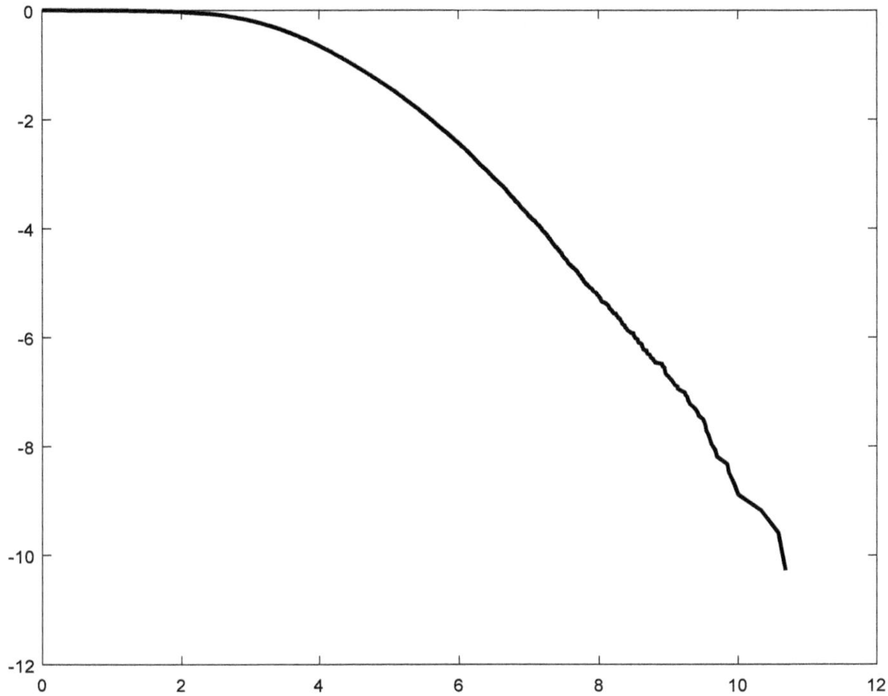

Fig. 2 Complementary cumulative distribution of the number of blocks in each city. Double logarithmic scale

Table 5 Test results on the aggregation function, $P(K)$ ($n = 28\,916$): rank (percentile) after which the power-law hypothesis is rejected

	5%	1%
ME	2350 (8.12)	2600 (8.99)
UMPU	1300 (4.49)	1650 (5.70)
GI	1890 (6.53)	2900 (10.02)

average length of the power-law tail detected in 100 synthetic samples obtained by reshuffling the disaggregate data before assigning them to the aggregate elements. Comparing these results with those reported in Table 2 we can see that the power-law tends to be longer after reshuffling than in the original aggregate datasets. More in details, ME and UMPU report longer power-law tails in the synthetic data, whereas GI reports a shorter one.

The overall message we derive from this exercise is that very large aggregate elements are rarer than one would expect following a purely random assignment of elementary units (given the number of units assigned to each aggregate). So, for instance, if blocks were randomly assigned to cities, we should find more large cities

Table 6 Test results on synthetic datasets obtained by reshuffling data ($n = 28\,916$): rank (percentile) after which the power-law hypothesis is rejected

	5%	1%
ME	2231 (7.72)	2227 (7.70)
UMPU	1361 (4.71)	1354 (4.68)
GI	1278 (4.42)	1278 (4.42)

than we actually see. One possible way to rationalize these results is by means of agglomeration diseconomies or congestion effects. More plainly, not all the blocks of a city are made up of densely populated skyscrapers.

4 Discussion and Conclusion

The exact shape of the distribution of city size has been the object of intense scrutiny in recent years. However, despite it is well-known that it is difficult to discriminate between log-normal and Pareto extreme values, most of the literature has applied a single testing strategy, typically assumed to be the best one. In this chapter we take a broader perspective considering alternative tests, and different levels of aggregation. We find that the tail behavior changes upon aggregation, with the emergence of a longer power-law upper tail.

A number of considerations are in order. First, it is worth noticing that when testing log-normal versus Pareto we are deliberately excluding possible alternative distributions both for the tail and for the whole distribution.

Second, the validity of the results is severely constrained by sample size. This is particularly important whenever the variance of the distribution is large. Based on this consideration, we cast new doubts as for the existence of a genuine Pareto tail in the case of the US city-size distribution and conclude that further analysis should be performed on a larger sample of world cities.

Third, since a Pareto tail emerges upon aggregation, we argue that the shape of the aggregation function is critically important. This has been stressed by Rozenfeld et al. (2011), who use a clustering algorithm to define cities, instead of the standard administrative definition. In particular, when the aggregation function is Pareto and the size of the elementary units is independent from their number in each aggregate entity, the shape of the aggregate size distribution is essentially the same as the shape of the distribution of the number of units. City size is composed by a number of units that is Pareto distributed, at least in the upper tail. In case of no relationship between the number of elementary units and their average size, this would be sufficient to generate a power-law tail in aggregate data. However, we find evidence of a sizable relationship between the number and size of constituent parts of aggregate entities. In such a case, the functional relationship between the two matters.

Even if it is difficult to derive analytical results for the distribution of the sum of dependent heavy-tailed distributed random variables,[7] numerical experiments suggest that when the size of the units is a power-law function of the number of elements of which an aggregate entity is composed and the number of units is Pareto distributed, the size of aggregate entities is also Pareto distributed.

The dependence of the size of elementary units on their number deserves further scrutiny, as it may signal the presence of non-constant returns to scale which, only under specific conditions, give rise to a power-law distribution.[8]

The level at which the various phenomena are investigated has a great influence on results, that is on the length of the power-law tail found in the data. Theoretical models that aim at explaining the shape of the distribution should devote more attention to this aspect. Our results suggest that to adequately explain the emergence of a power-law tail one should focus on the factors that determine the shape of $P(K)$, i.e., the number of blocks of a city or, more in general, the number of elementary units in each aggregate. When the main cause of a Pareto tail at the aggregate level is the skewed shape of the aggregation rule, the usual argument that assumes idiosyncratic shocks to cancel out upon aggregation breaks down. Hence, the aggregate distribution is still heavy-tailed.

References

Aoki, M. and Yoshikawa, H. (2011). *Reconstructing macroeconomics: a perspective from statistical physics and combinatorial stochastic processes*. Cambridge University Press.

Asmussen, S. and Rojas-Nandayapa, L. (2008). Sums of dependent lognormal random variables: Asymptotics and simulation. *Statistics & Probability Letters*, 78:2709–2714.

Axtell, R. L. (2001). Zipf distribution of U.S. firm sizes. *Science*, 293(5536):1818–1820.

Bagan, H. and Yamagata, Y. (2015). Analysis of urban growth and estimating population density using satellite images of nighttime lights and land-use and population data. *GIScience & Remote Sensing*, 52(6):765–780.

Bee, M., Riccaboni, M., and Schiavo, S. (2011). Pareto versus lognormal: A maximum entropy test. *Physical Review E*, 84:026104.

Bee, M., Riccaboni, M., and Schiavo, S. (2013). The size distribution of US cities: Not Pareto, even in the tail. *Economics Letters*, 120(2):232–237.

Benhabib, J., Bisin, A., and Zhu, S. (2011). The distribution of wealth and fiscal policy in economies with finitely lived agents. *Econometrica*, 79(1):123–157.

Berry, B. J. and Okulicz-Kozaryn, A. (2012). The city size distribution debate: Resolution for US urban regions and megalopolitan areas. *Cities*, 29(Supplement 1):S17–S23.

Cabral, L. M. B. and Mata, J. (2003). On the evolution of the firm size distribution: Facts and theory. *American Economic Review*, 93(4):1075–1090.

Calderín-Ojeda, E. (2016). The distribution of all French communes: A composite parametric approach. *Physica A: Statistical Mechanics and its Applications*, 450:385–394.

---

[7] See Asmussen and Rojas-Nandayapa (2008) for some asymptotic results in the special case of log-normal random variables with dependence structure given by the Gaussian copula.

[8] For a discussion of the role of increasing and decreasing returns to scale in determining the size of cities, see Masahisa et al. (2001, p. 225).

Champernowne, D. G. (1953). A model of income distribution. *The Economic Journal*, 63(250):318–351.

Cieślik, A. and Teresiński, J. (2016). Does Zipf's law hold for Polish cities? *Miscellanea Geographica*, 20(4):5–10. Cited By 0.

Cox, D. R. and Hinkley, D. V. (1974). *Theoretical Statistics*. Chapman and Hall.

Del Castillo, J. and Puig, P. (1999). The best test of exponentiality against singly truncated normal alternatives. *Journal of the American Statistical Association*, 94:529–532.

Eeckhout, J. (2004). Gibrat's law for (all) cities. *American Economic Review*, 94(5):1429–51.

Eeckhout, J. (2009). Gibrat's law for (all) cities: Reply. *American Economic Review*, 99(4):1676–83.

Embrechts, P., Klüppelberg, C., and Mikosch, T. (1997). *Modelling Extremal Events for Insurance and Finance*. Springer.

Fazio, G. and Modica, M. (2015). Pareto or log-normal? best fit and truncation in the distribution of all cities. *Journal of Regional Science*, 55(5):736–756.

Gabaix, X. (1999). Zipf's law for cities: An explanation. *Quarterly Journal of Economics*, 114(3):739–67.

Gabaix, X. (2009). Power laws in economics and finance. *Annual Review of Economics*, 1:255–93.

Gabaix, X. (2011). The granular origins of aggregate fluctuations. *Econometrica*, 79:733–772.

Gabaix, X. and Ibragimov, R. (2011). Rank-1/2: A simple way to improve the OLS estimation of tail exponents. *Journal of Business and Economic Statistics*, 29(1):24–39.

Gibrat, R. (1931). *Les Inegalites Economiques*. Sirey, Paris.

González-Val, R., Ramos, A., Sanz-Gracia, F., and Vera-Cabello, M. (2015). Size distributions for all cities: Which one is best? *Papers in Regional Science*, 94(1):177–196.

Growiec, J., Pammolli, F., Riccaboni, M., and Stanley, H. E. (2008). On the size distribution of business firms. *Economics Letters*, 98(2):207–212.

Hsu, W.-T. (2012). Central place theory and city size distribution. *The Economic Journal*, 122(563):903–932.

Ijiri, Y. and Simon, H. A. (1977). *Skew Distributions and the Sizes of Business Firms*. North Holland, Amsterdam.

Ioannides, Y. and Skouras, S. (2013). US city size distribution: Robustly Pareto, but only in the tail. *Journal of Urban Economics*, 73(1):18–29.

Levy, M. (2009). Gibrat's law for (all) cities: Comment. *American Economic Review*, 99(4):1672–75.

Li, X. and Nam, K.-M. (2017). One country, two "urban" systems: focusing on bimodality in China's city-size distribution. *Annals of Regional Science*, 59(2):427–452.

Luckstead, J., Devadoss, S., and Danforth, D. (2017). The size distributions of all Indian cities. *Physica A: Statistical Mechanics and its Applications*, 474:237–249. Cited By 1.

Luttmer, E. G. (2007). Selection, growth, and the size distribution of firms. *Quarterly Journal of Economics*, 122(3):1103–1144.

Malevergne, Y., Pisarenko, V., and Sornette, D. (2009). Gibrat's law for cities: uniformly most powerful unbiased test of the Pareto against the lognormal. *Swiss Finance Institute Research Paper Series*, pages 09–40.

Malevergne, Y., Pisarenko, V., and Sornette, D. (2011). Testing the Pareto against the lognormal distributions with the uniformly most powerful unbiased test applied to the distribution of cities. *Physical Review E*, 83(3):036111.

Masahisa, F., Krugman, P., and Venables, A. J. (2001). *The Spatial Economy: Cities, Regions, and International Trade*. The MIT Press.

Pareto, V. (1896). *Cours d'économie politique professé à l'Université de Lausanne*. F. Rouge, Lausanne.

Perline, R. (2005). Weak and false inverse power laws. *Statistical Science*, 20:68–88.

Puente-Ajovín, M. and Ramos, A. (2015). On the parametric description of the French, German, Italian and Spanish city size distributions. *The Annals of Regional Science*, 54(2):489–509.

Ramos, A. (2017). Are the log-growth rates of city sizes distributed normally? empirical evidence for the USA. *Empirical Economics*, 53(3):1109–1123.

Rozenfeld, H., Rybski, D., Gabaix, X., and Makse, H. (2011). The area and population of cities: New insights from a different perspective on cities. *American Economic Review*, 101(5):2205–25.
Sutton, J. (1997). Gibrat's legacy. *Journal of Economic Literature*, 35(1):40–59.
Sutton, J. (2002). The variance of firm growth rates: the scaling puzzle. *Physica A: Statistical Mechanics and its Applications*, 312(3):577–590.
Wu, X. (2003). Calculation of maximum entropy densities with application to income distribution. *Journal of Econometrics*, 115:347–354.
Zhou, Y., Smith, S., Zhao, K., Imhoff, M., Thomson, A., Bond-Lamberty, B., Asrar, G., Zhang, X., He, C., and Elvidge, C. (2015). A global map of urban extent from nightlights. *Environmental Research Letters*, 10(5):054011. https://doi.org/10.1088/1748-9326/10/5/054011.
Zipf, G. K. (1949). *Human Behavior and the Principle of Last Effort*. Addison-Wesley, Cambridge: MA.

The Signature of Organic Urban Growth

Degree Distribution Patterns of the City's Street Network Structure

Leonard Nilsson and Jorge Gil

Abstract Cities are complex, perhaps one of the most complex kinds of structure created by humans. Some cities have been planned to a large extent while others have grown organically. The outcome of planning and growth processes is the diverse morphological built form and street patterns observed in cities today. So far, in urban planning and history research, cities are classified as planned or organic, largely based on the visual assessment of their urban morphology. To understand cities and their characteristics, it is necessary to develop methods that quantitatively describe these characteristics and enable objective comparisons between cities. Using graph representations of the street network of cities, it is possible to calculate measures that seem to exhibit patterns with scale-free properties (Jiang in Phys A Stat Mech Appl 384(2):647–655, 2007). And thanks to progress in the field of complexity studies, it is also possible to test if and to what extent the distribution of a measure in a collection of elements fits a power law distribution (Clauset et al. in SIAM Rev 51:43, 2009). In this chapter, we show that the degree distribution of a city's street network seems to fit a power law distribution for cities that are considered to have mostly grown organically. On the other hand, cities that are planned to a large extent do not exhibit such a good fit. This result is relevant since a power law distribution is a signature of multiplicative growth processes. Furthermore, the result of the quantitative classification method correlates well with the results of earlier qualitative morphological classifications of cities.

Keywords Street networks · Spatial morphology · Growth patterns · Power law · Degree distribution

---

L. Nilsson (✉) · J. Gil
Department of Architecture and Civil Engineering, Chalmers University of Technology, Gothenburg, Sweden
e-mail: leonard.nilsson@chalmers.se

J. Gil
e-mail: jorge.gil@chalmers.se

L. D'Acci (ed.), *The Mathematics of Urban Morphology*,
Modeling and Simulation in Science, Engineering and Technology,
https://doi.org/10.1007/978-3-030-12381-9_5

1 Introduction

Cities have been a very important part of human civilization since agriculture-based society emerged. They have been built independently from each other in different parts of the world and by different cultures. In order to better understand and study cities, researchers in sociology, economy, physics and geography have for long resorted to quantitative methods based on functional characteristics (Harris 1943; Nelson 1955; Beaverstock et al. 1999) and population size (Clark 1951; Bettencourt et al. 2010), to describe, rank and produce classifications of cities. From an urban morphology studies perspective, a basic way of classifying cities has been by their growth process and the resulting morphological pattern, using a simple dichotomy: organic cities that have grown gradually, versus planned cities that have emerged from larger planning interventions. Another way of approaching this dichotomy has been using the concepts of bottom-up growth (organic) versus top-down development (planned) (Batty 2013).

Some cities are entirely planned in one sweep, while others have mainly grown organically in a piecemeal process governed by local decisions. But often these processes have alternated back and forth over time (Kostof and Tobias 1991). Hence, growth does not manifest itself in an absolutely distinct dichotomy, but it is rather a question of different degrees of planning resulting in different morphological patterns that combine these two main forces. Nevertheless, it is still a dominant morphological classification of cities and urban areas.

A peculiar aspect is that these organic patterns have been for the past decades regarded as desirable in cities, and therefore efforts to mimic them have often been made in planning (Kostof and Tobias 1991). Despite research on the topic of their growth, design and generation, there are still few consistent ways of classifying and describing these morphologic patterns. The classification approach used is largely based on the visual judgement of the perceived patterns. It is, therefore, necessary to develop methods that quantitatively measure their characteristics and enable the classification of cities and parts thereof, towards their objective comparison and the understanding of city growth.

The street network of a city is one of its most durable structures, with a crucial influence on its land use and building development over time. It is therefore important to understand this complex network. Using graph representations of the street network it is possible to calculate topological measures that describe different aspects of its configuration (Hillier 1996), and efforts to find properties of entire cities based on such measures show that these seem to exhibit patterns with scale-free properties. Nevertheless, the results are diverging and sometimes contradictory. In some studies, the various network centrality measures like degree, betweenness and others, reveal a scale-free behaviour in many cities (Porta et al. 2006a, b; Jiang 2007; Volchenkov and Blanchard 2008; Shpuza 2014). Other studies refute this and have shown that there are differences and exceptions from these patterns (Jiang and Claramunt 2004; Crucitti et al. 2006a, b; Porta et al. 2006a, b).

By studying the statistical distribution characteristics of some individual centrality measure, the emerging distribution can tell us something about the processes shaping this measure. One of the most common distributions is the normal distribution (hence the name 'normal'). This distribution emerges as a result of numerous random factors added together. Exponential distributions, on the other hand, are the result of time-dependent processes. A distribution that has caught the interest of many scientists in the last 20 years is the power law distribution, which has scale-free properties and is the result of a multiplicative growth process. This is often called a complex or self-organizing process.

In this study, we measure the degree centrality distribution of the street network's topology in a number of cities with different morphological patterns and test if they fit the power law distribution, to uncover whether there are universal mathematical patterns of cities as found in earlier studies, or if these vary between cities as other studies suggest. The difference in morphological structure and growth could explain the contradictory results found in those latter studies. The hypothesis is that these mathematical patterns are indeed not universal, but rather depend on the extent to which the cities are organic or planned. If that holds true, then these measures provide a signature of organic urban growth for the objective categorization of cities, which can extend our understanding of cities in general.

In the next section, we provide an overview of relevant background theories and related work, followed by a description of the two case studies—a longitudinal study of the growth of Paris and a cross-sectional study of several world cities—and the methodology used, leading to the presentation of the studies' results, and finally wrapping up with a summary of findings and pointers to further work.

2 Of Street Networks and Degree Centrality Distributions

Urban morphology has been studied as an academic discipline for a long time, but is has often been done in qualitative studies based on perceived patterns of urban morphology (Kostof and Tobias 1991). Those patterns are usually derived from the geometric properties of urban elements, such as buildings, parcels and streets, and their spatial arrangements. Since the 1990s, with increasing computing power and availability of spatial analysis software such as Geographic Information Systems (GIS), we have witnessed the exploration of new approaches to identify and quantify urban morphological patterns. Webster (1995) proposes the analysis of urban satellite images to extract a set of urban texture measures of streets and blocks, thus identifying the fingerprint and profile of different neighbourhoods. Gil et al. (2012) pursue a similar goal by looking at individual elements of the city, i.e. urban streets and blocks, and classifying them according to a set of geometric and topological measures to create summary profiles of urban areas. Louf and Barthelemy (2014) also use geometric attributes of urban elements, but instead develop indices based on the ratios of the distribution of said attributes in order to classify different cities across the world.

Parallel to this work, other research groups started in the 1990s to analyse cities by means of growth simulations, often using cellular automata (Batty and Longley 1994), measuring characteristics of the boundary and general shape of the city. In this field, cities are often said to be fractal, which means that certain measures have scale-free properties. This property is claimed to be the result of a bottom-up growth process, and by analogy, organic cities might also be called fractal and possess scale-free properties.

2.1 The Topology of Street Networks

Around the 1980s a new method for representation and quantitative analysis of spatial structures was developed called space syntax (Hillier and Hanson 1984). As the name indicates, it addresses the configuration of spaces: architectural, urban streets or open spaces. Spatial configuration. is best defined as how spaces, and in the case of cities usually streets, are connected and related to each other (Hillier 1996; Marshall 2015). A basic concept of the original space syntax theory is that it analyses the topology of the street network, and leaves out metric information (Hillier 1999). The geometry of space only plays a role in defining its representation as discrete elements and in constraining their connections.

Since the late 1990s, parallel to the growth of space syntax as a research field, another important field of research that has been rapidly evolving is that of network science (Watts 2004). Topological analysis of graphs is also the fundamental model for this scientific field, where many kinds of networks are studied, including city street networks (Barthélemy 2011). As computers have become more powerful, methods appeared for calculating properties of large networks based on existing theories, such as small-world network properties and scale-free properties (Watts and Strogatz 1998). This offered the possibility to study cities from an entire new perspective (Batty 2013).

Based on this progress in complex network analysis in general, new methods for analysing road and street network patterns have been developed, evolving the research field pioneered by space syntax (Jiang et al. 1999; Porta et al. 2006a; Jiang 2007). These methods use the results from topological network analysis and thereafter analyse various properties, both individual measures and statistical distributions (Jiang 2007; Volchenkov and Blanchard 2008; Shpuza 2017). These elaborations of the space syntax research field, combined with the general network science field, might enable the quantification of cities in a richer way.

2.2 Street Network Representation

When the space syntax research field started in the 1980s, the main element of analysis of street configurations was the axial line (Hillier and Hanson 1984). An axial line is an unobstructed line of sight, and in order to analyse an area or a city, an axial map is created made of the fewest and longest unobstructed lines of sight. The axial map is converted into an unweighted undirected graph, and analysed based on its topology (Hillier 1996). This model has been dominating the research field, but in the early 2000s efforts were made to integrate space syntax in ordinary GIS research and make it possible to use the more common road centreline maps for larger scale studies (Jiang et al. 1999; Dalton et al. 2003).

One approach widely used today is to convert the road centreline maps to a model of street segments and analyse its network centrality. This can be achieved using a so-called ‘primal’ approach (Porta et al. 2006b) with geographic weight on the links between street junctions that are nodes in the graph; or using a so-called ‘dual’ approach (Porta et al. 2006a), where the nodes are the street segments and the links have angular (fractional) weight based on the degree of direction change between street segments, instead of topological (discrete) steps (Dalton 2001; Turner 2007). The latter model is similar to the segment model derived from an axial map and equivalent in terms of analysis results (Krenz 2017; Kolovou et al. 2017). However, both street network models based on street segments lose topological structural properties of the network as a whole.

Another approach for using road centreline maps and replace axial lines has been to aggregate street segments based on the concept of ‘Natural roads’ (Jiang et al. 2008), which is similar to that of ‘Strokes’ (Thomson 2003) and of ‘Continuity lines’ (Figueiredo and Amorim 2005). These concepts are based on the Gestalt principle of good continuation (Jiang 2007), and applying improved street segment aggregation algorithms results in a network model that reveals the structure of the city (Jiang and Liu 2009). Its correlation with pedestrian and vehicular movement is on par with the other methods from the space syntax research field, such as angular segment analysis and axial integration (Jiang 2009; Jiang and Liu 2009), which supports its validity for representing urban structure.

The different approaches of representing streets as entities continuing across intersections open up a particular view of urban street networks. Since streets are often planned and built in their entire length and seldom subject to changes over time, a method that reconstructs the street as an entity makes it possible to represent and understand the city as a collection of streets added together as the city grows. In this way, space syntax and similar topological approaches to street network representation can be used to understand urban growth and development.

2.3 *Degree Distribution and Power Law Fitting*

The most basic measure of centrality obtained from network analysis is degree centrality, also called connectivity in space syntax studies, i.e. the number, per node, of connections with other nodes. When using the street network models based on street segments described in the previous section, degree centrality only classifies street junctions by type, which is considered uninteresting for studying urban configuration (Porta et al. 2006a). However, when using a street network model based on 'Natural roads' or axial lines degree centrality provides an interesting measure of street hierarchy.

When analysing the distribution of degree centrality values of street networks, a certain pattern often occurs, namely that there are far more low connected elements than highly connected ones. This pattern seems to follow a power law distribution, which has scale-free properties (Porta et al. 2006a; Jiang 2007) and is a signature of a natural process governing the creation of the distribution. In other fields of network science, such as social networks, flight connections or power grid structures, the same pattern occurs, and many studies rely on degree centrality as the single measure of distribution to test for power law fitting (Clauset et al. 2009; Barabasi and Albert 1999; Volchenkov and Blanchard 2008). In urban studies, this aspect has not been extensively researched, but seems to offer the possibility of new and interesting discoveries.

2.4 Related Work

Some efforts to find properties of cities as a whole using street network models have been made (Crucitti et al. 2006a, b; Porta et al. 2006a, b; Jiang 2007; Volchenkov and Blanchard 2008; Shpuza 2014), although many of these studies were carried out on a small sample area of a city or on a small number of cities, which makes it harder to extract general conclusions from the results. The results are also diverging and sometimes incompatible. It has been shown in some studies that various network measures like degree, betweenness and other centrality measures, reveal a scale-free behaviour in many cities (Porta et al. 2006a; Jiang 2007; Volchenkov and Blanchard 2008). While other studies refute this observation, and have shown that there are differences and exceptions from these patterns (Jiang and Claramunt 2004; Crucitti et al. 2006a, b; Porta et al. 2006b). To further support or refute these claims, a more extensive study is needed. The results of these earlier works can be discussed in light of the results from the present study, even though the preconditions and methods are not exactly identical.

3 Datasets and Methods

In this work we carry out two experiments, one longitudinal and one cross-sectional, in which we analyse the degree centrality distribution pattern of cities and test to what extent they fit a power law distribution, to test the hypothesis that these patterns are not universal, but rather depend on the extent to which the cities are organic or planned. In the following sections we explain in detail the 'what' and the 'how' of this work.

3.1 Objects of Study

The first experiment is a longitudinal study of Paris, France, testing how the signature of degree centrality, and thus the scale-free nature of cities, changes over time as the city grows. Growing from a medieval walled core like many European cities, the city of Paris has undergone periods of profound transformation to its urban structure and character, namely: through the building of new bridges and public spaces in the seventeenth century; the famous Haussman plan in the nineteenth century that introduced a network of avenues and boulevards; and the urban expansion of its periphery in the twentieth century with projects like La Défense.

The second experiment is a cross-sectional study of a selection of world cities, testing how the signature of degree centrality relates to the planning process, and changes across geographic and cultural contexts, to try to confirm or refute the existence of universal patterns. There are no standardized or consistent ways of classifying cities, therefore the selection of cities is mainly based on earlier classifications focused on urban patterns resulting from different development processes (Kostof and Tobias 1991; Batty and Longley 1994; Hillier 1996). In addition, other criteria for selection have been chosen in order to have variation in the sample of cities.

The first and foremost criterion used has been the degree of planning or self-organization, aiming to have a somewhat balanced collection of organic and planned cities. The second criterion has been the dominant culture that has shaped the city morphology, since different cultures may play a large role in the cities' street network structure (Hillier 1996). This is probably because of the different social structure, organization and interaction found in different cultures (Kostof and Tobias 1991). The third criterion has been geographic location, since the preconditions for urban development differ considerably in different parts of the world, namely the geology, climate, building materials supply and cost. Therefore, the cities selected for this study are chosen from different continents or continental areas. Finally, the age of the city can also play a role, in that organic urban transformation is a slow process that takes long time to manifest itself, whereas building a planned city can be done quite fast in comparison.

Based on these criteria, ten cities have been chosen for this study (Table 1). Including a larger number of cities in the study could bring benefits, but since the aim is not to reach a conclusive answer, but rather to explore the path for new theories, the number of cities chosen (10) can be considered sufficient. This selection should be seen as a starting point for further and deeper studies.

3.2 Sources of Street Network Data

The street network of the selected cities is the main object of study, therefore it is essential to use a reliable data source. For the first experiment, a data set of the historical evolution of the Paris street network was obtained from the GeoHistorical-

Table 1 Summary of properties for the cities in the cross-sectional analysis

City	Degree of planning	Shaped by culture and religion	Location	Age	Geographical characteristics
Rio de Janeiro	Many self-organized favelas	Catholic South American	Brazil	Modern (60s)	Coastal city
Venice	Self-organized villages grown together	Catholic European	Europe	Middle age	Several islands
Damascus	Old roman city organically, reshaped when under Arab government	Originally roman, later Arabic/Islamic	Middle east	Roman age	Desert/plain
London	Low, self-organizing	Anglo-Saxon, Lutheran	Europe	Middle age	River city
Tokyo	Low, self-organizing	Japanese all the time, although became capital later	Eastern Asia	Middle age	Coastal city
Beijing	High, since it became capital	Chinese all the time, although it became capital later	Eastern Asia	Antiquity	Topographical constraints
Moscow	High, as an expression of power	Russian Orthodox, capital all the time	Russia	Middle ages	River city
New York	High, the original grid is virtually unchanged	American all the time	North America	About 1600	Several islands
Brasilia	Very high, entire city planned at once	Very new, built as capital	Brazil	60s	Exploited jungle
Paris	High, since both the Baroque planning and Haussmann's boulevards	Catholic European	Europe	Middle age	River city

Data project (http://www.geohistoricaldata.org), with 10 maps spanning from 1300 to the present day. Figure 1 shows a temporal sequence of maps of the street network of Paris. Historical maps are relatively scarce and are produced at irregular time intervals. Therefore, it is troublesome to use some kind of strict selection criterion, because the resulting selection would be too sparse. That is the reason for using many maps although some of them are quite close in time.

For the second experiment, in order to make a worldwide comparison, it is essential to obtain a common data source with relatively consistent data classification and collection. Publication restrictions and copyright issues make the usage of official land surveys or proprietary navigational street data (e.g. TomTom, Navteq, Google) hard. Instead, Open Street Map (OSM) was used as the source for the street network data. OSM is one of the most successful Volunteered Geographic Information (VGI) initiatives that aim to create global public domain map data. It is open for anyone to view, use, edit and create new features; and it has a worldwide coverage, although the level of detail and classification criteria might differ between different areas.

There has been some debate around the data quality of OSM, since anyone can modify it. However, the process is moderated, and previous research has suggested that the errors in this map data are on par with those in other data sets from authoritative sources (Haklay 2010; Dhanani et al. 2012; Krenz 2017). Therefore, the benefits of being able to use a common map source for all the cities in this study, without copyright or other legal restrictions, outweigh the possible data quality drawbacks of using OSM. The OSM data in shape file format was obtained from the download section of Geofabrik (www.geofabrik.de). Excerpts of the street network of the ten selected cities (listed in Table 1) are shown in Fig. 2.

3.3 Street Network Model Considerations

In this study, the street network data set is converted to a street network model, which is then translated to a graph to analyse its degree centrality. Other centrality measures can be calculated, and they offer possible directions of investigation on their distribution patterns, but to produce results that are easily comparable with other studies, also from other network research fields, only degree centrality will be used.

To obtain discrete and meaningful degree centrality results for each street, it is necessary to use an appropriate street network model. This rules out the street segment based models widely used today, presented in Sect. 2.2. As for aggregate street network model options, the axial line model would require the manual digitizing of the network, which is impractical for this extensive study. Since we have access to road centre line data of the selected cities, the 'Natural roads' street network model and analysis method, based on aggregating road centre line segments, is the most suitable.

A parameter that has to be taken into consideration is the extent of the studied areas in each city. In this work, the main goal is to capture the structure of the city

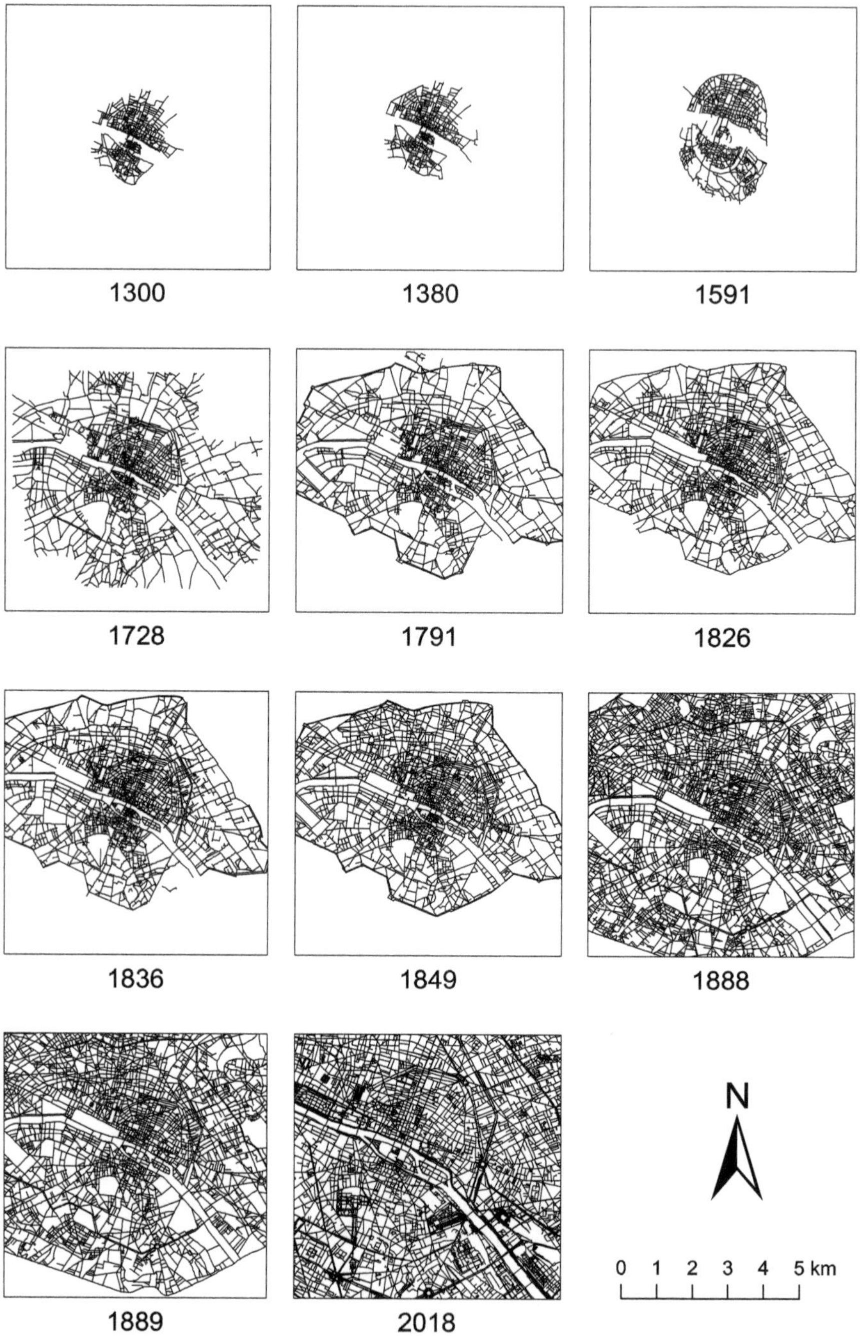

Fig. 1 Sequence of street network growth and change in Paris, from 1300 to 2018

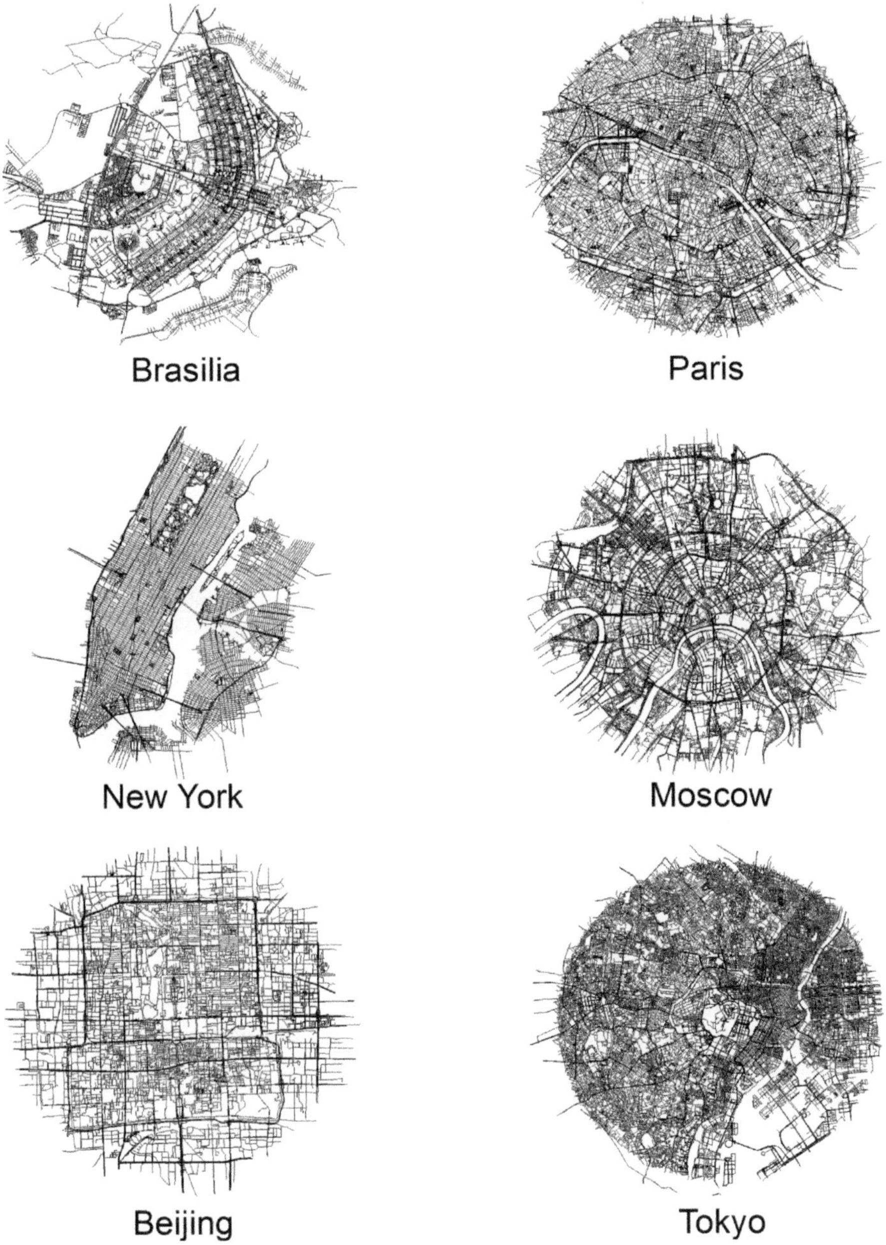

Fig. 2 The street networks of the ten cities in the cross-sectional analysis

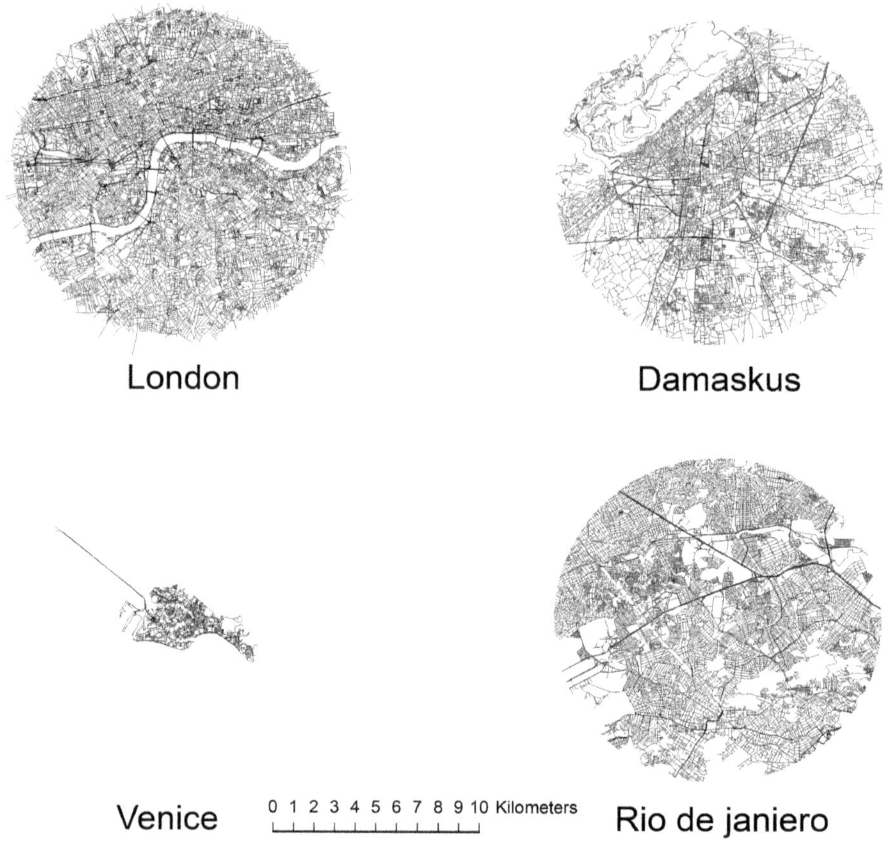

Fig. 2 (continued)

resulting from its historical growth to get an indication of whether the hypothesis holds. It can be argued that the larger the study area are better, in order to have a larger amount of data to analyse, and a certain number of cases is required to fit power law to the degree distributions. On the other hand, cities with organic patterns are often small, or have started small and possess small areas with organic pattern. To capture the organic parts of cities, and to make cities with different sizes more comparable, we use a circular sample area with a radius of 8 km. These areas are centred on the historical origin of the city (where applicable), like the Île de la Cité in Paris.

Finally, one aspect that must be considered is the boundary of the network model. In some cases, one can argue that this delimitation is arbitrary, while in other cases, like Venice, the city limits are very clear. Nevertheless, in every case we extract a significant section of the city. In addition, the degree centrality measure used is a local measure, meaning that edge effects of this boundary are much less pronounced (Gil 2017).

3.4 Methodological Process

Having defined the case studies, obtained the data and defined the street network model, the analytic methodology needs to be carried out.

The first step involves a series of data set extraction, cleaning and validation operations to enable the creation of the 'Natural roads' street network model. This includes clipping the data sets to extract the street segments in the 8 km study area; transforming the data to the WGS84 World Mercator projected coordinate system; breaking the road centre lines at every intersection, because by mistake some street segments continue across intersections, and this is a requirement for the natural streets interconnection algorithm to work; finally, identifying and deleting any isolated street segments or small 'islands' of streets, in order to obtain a network with a single connected component, again a requirement of the centrality algorithm.

The next step is to generate 'Natural roads', by joining the street segments at the junctions using the 'every best fit' algorithm with a threshold angle of 45° (Jiang et al. 2008). Once the 'Natural roads' have been created, we calculate degree centrality for each of the cities' street networks.

The final step is to statistically analyse the degree centrality results of the cities to check if their distribution seems to fit a power law distribution. First, we do a power law fit, where the best-fit alpha and x-min values are calculated. Next, we test whether the data can fit a power law doing a Kolmogorov–Smirnoff (K-S) test (Clauset et al. 2009). This test generates many samples based on an ideal power law distribution, using the x-min and alpha values derived from the best fit of the degree distribution being tested. Then, the samples are compared against the distribution by doing a regression analysis to derive a significance test. If the resulting *p*-value is above 0.1, the distribution has a probability of following a power law, but it is not certain. Even though a lot of empirical data looks like it follows a power law distribution, it is not always the case. The results of these tests (the *p*-value, x-min and alpha) are then used in the final assessment and judgement of the fit.

A parameter that deserves some attention is alpha, not only for the statistical tests but also as a distribution's characteristic. It tells us the scaling factor in the distribution, which is the relation between roads with low degree (minor roads) and roads with a high degree (main roads). A high alpha represents a steep slope of the best-fit line, showing that there are relatively fewer high degree roads, compared to the low degree roads. A low alpha represents the opposite. Having this information about the proportion of different road categories can be used in the interpretation and understanding of the results.

After a distribution passes the K-S test (p-value > 0.1) it has to pass a final set of criteria in order to claim that the distribution is a power law. The distribution under test is compared to other possible distributions such as exponential and log-normal using a log-likelihood test. To meet the criteria the power law has to obtain a higher likelihood than these other distributions. This test is done because heavy-tailed distributions are often quite similar and can give (false) positive results on the

K-S test, especially in distinguishing between power law and log-normal. Based on these tests' results we are able to put all cities into various categories.

When all these measures are extracted, it is time to do a categorical comparison between the cities to see how the quantitative properties are distributed among the studied cities. A categorization is made based on the measures in order to investigate whether the measures might have some relation to their qualitative morphological characteristics.

3.5 Software Used in the Process

To carry out a largely automated and reproducible study, several pieces of software are required for data preparation, validation and analysis. First, the FME software is used for cleaning and validating the data sets. Then, the software Axwoman 6.3 (Jiang 2015) for ArcGIS is used to create 'Natural roads' and calculate degree centrality for each city. The degree distribution of the streets is imported, processed and analysed using the poweRlaw package (Gillespie 2015) in R (R Core Team 2016), which is also based on the work by Clauset (2007). Finally, Microsoft Excel is used to collect and visualize the statistical results. The use of several pieces of software was necessary to reach the desired results, although it would have been simpler if a single application with all these capabilities combined existed.

4 Results

Following the geographical and topological analysis steps described in Sect. 3, applied to the objects of study, the result is a street network model of 'Natural roads' classified according to the measure of degree centrality. The result on the set of cities of the longitudinal study can be seen in Fig. 3 and of the cross-sectional study can be seen in Fig. 4. The results of the statistical analyses are presented in next sections, starting with a description of the different statistical parameters used. The number of street segment refers to the number before the natural roads' aggregation across intersections while the degree numbers refers to the natural roads (Tables 2 and 3).

4.1 Derived Parameters of Degree Distribution

The results of the statistical analysis of degree centrality distributions is summarized in Table 4 for the longitudinal study of Paris and Table 5 for the cross-sectional study Table 4 is sorted chronologically to understand the evolution and its expansion and restructuring. Table 5 is sorted on the p-value in ascending order, since the purpose is to understand the power law properties. The next parameter in these tables is Alpha,

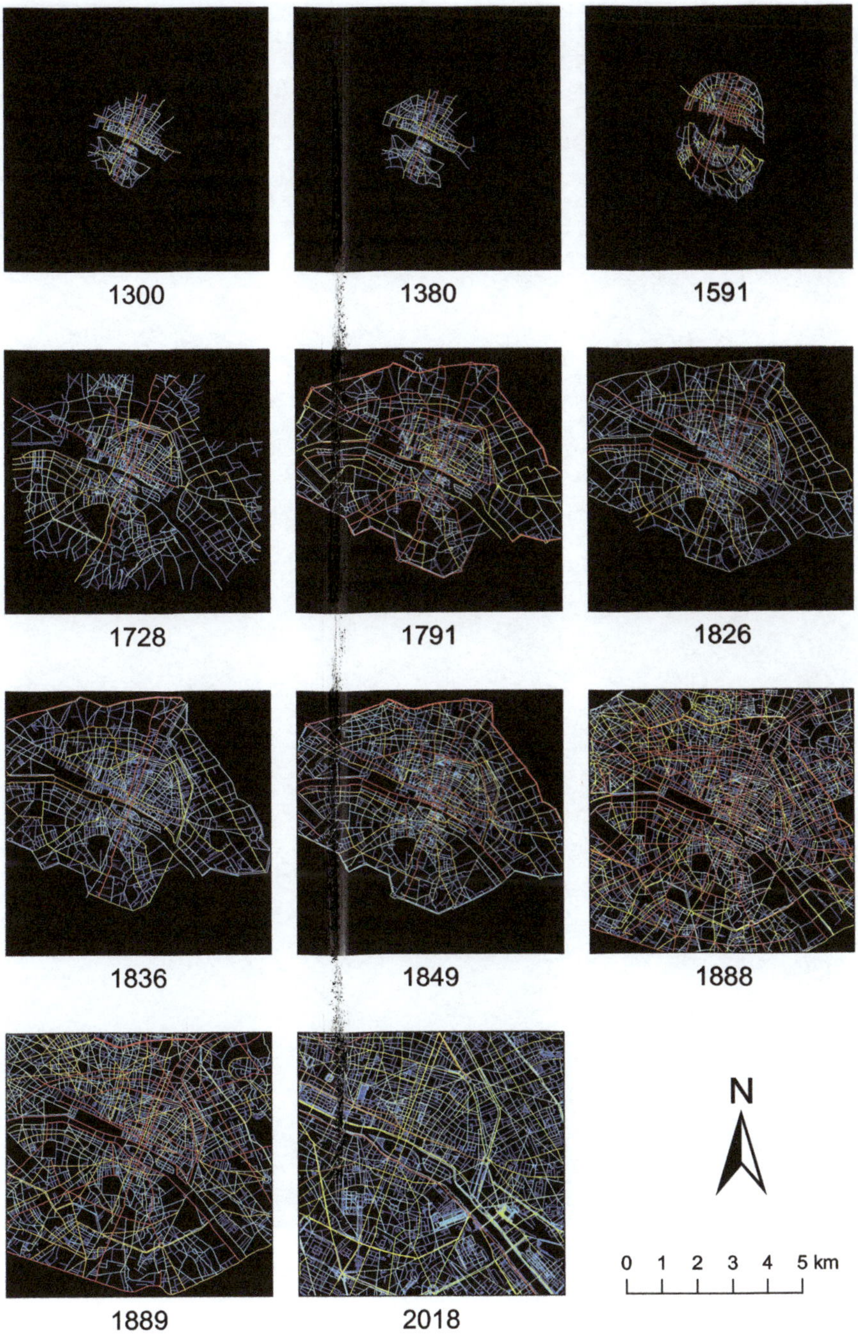

Fig. 3 The longitudinal analysis cities' dual graph network coloured according to degree centrality (red represents a high degree value, blue low)

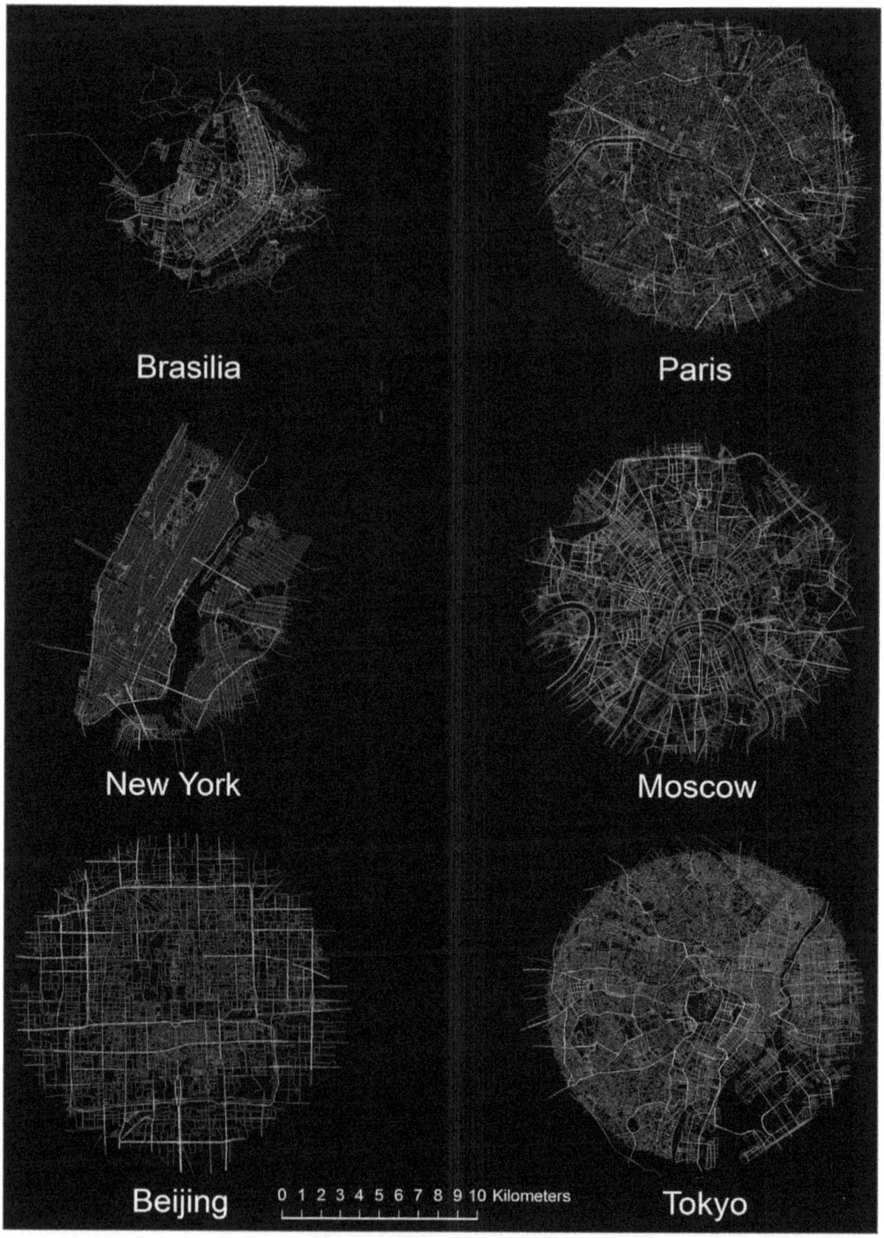

Fig. 4 The cross-sectional analysis cities' dual graph network coloured according to degree centrality (red represents a high degree value, blue low)

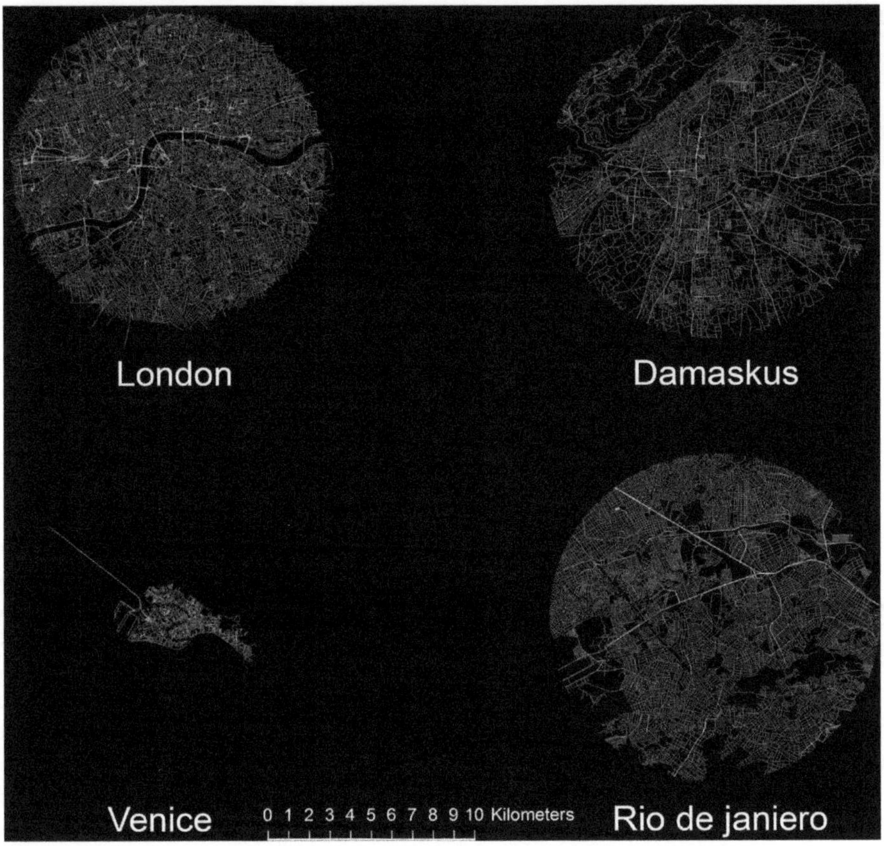

Fig. 4 (continued)

Table 2 Descriptive statistics of the natural roads models of the longitudinal study

Year	Number of street segments	Number of natural roads	Mean degree	Max degree	Median degree	Length in kilometres
1300	1125	353	3.84	57	3	73.8
1380	1065	350	3.94	56	3	69.2
1591	1251	386	3.98	28	3	105.2
1728	3255	961	4.02	73	3	350.3
1791	3463	994	4.24	82	3	393.2
1826	3820	1111	4.35	61	3	407.6
1836	4368	1330	4.00	122	2	434.8
1849	4836	1276	4.79	93	3	497.5
1888	15,749	3511	5.03	88	3	1230.6
1889	9726	2413	5.43	109	3	1587.7

Table 3 Descriptive statistics of the natural roads models of the cross-sectional study

Name	Number of street segments	Number of natural roads	Mean degree	Max degree	Median degree	Length in kilometres
Beijing	4862	4861	4.72	154	3	2575.8
Brasilia	7646	7645	3.57	189	2	2242.2
Damaskus	6337	6336	3.77	92	3	2189
London	16,786	16,785	3.83	217	2	3762.4
Moscow	14,471	14,470	4.35	210	3	4177.3
Newyork	3449	3448	6.59	173	3	2049
Paris	13,043	13,042	4.82	180	3	3985.8
Rio	8240	8239	4.02	118	3	2841.9
Tokyo	19,007	19,006	4.93	327	3	5212
Venice	2145	2144	3.42	50	2	319.3

which is the scaling factor of the hypothetical power law. Following that is x-min, which is the starting value from where the distribution can fit a power law. Finally, the number of vertices and their mean degree value are in the next two columns for reference.

The comparison between possible distributions was made for two purposes. The first purpose is, as mentioned above, for those that pass the K-S test to exclude that other distributions could provide a better fit. The second purpose was to obtain an indication of which kind of distribution the non-power law objects could possibly fit. Those results can provide some clues for further research regarding the cities where power laws are ruled out. The results of these log-likelihood tests are commented in the last column. In some cases, the results of the test were not significant, which means that some other distributions cannot be ruled out.

One thing that needs to be mentioned about these tests is the finite-size effect in the street network. Due to the geographical constraints, city limits or sample area limit, the maximum possible degree for a road is limited. This maximum size limit leads to that all possible power laws will in reality be truncated power laws (power laws with an exponential tail). Almost all geographical networks with power law properties are in fact truncated power laws due to finite-size effects. Since this issue is something common to all the cities studied and the study is comparative, it is left out of the analysis. Another reason to leave it out is that the study is aimed at understanding the main part of the street network, not the small (exponential) tail part.

Table 4 Results from the longitudinal study of Paris

Year	*p*-value	Alpha	x-min	Number of vertices	Mean degree	Comment	Plausible other distribution
1300	0.58	2.66	3	353	3.84	Medieval core	
1380	0.95	2.62	3	350	3.94	Medieval core	
1591	0.49	2.83	4	386	3.98	Medieval core	Log-normal cannot be ruled out
1728	0.77	2.83	5	961	4.02		
1791	0.95	2.69	4	994	4.24		
1826	0.31	2.70	4	1111	4.35		
1836	0.42	2.58	4	1330	4.00		
1849	0.72	2.61	5	1276	4.79		
1888	0.00	2.27	3	3511	5.03	Post-Hausmann	
1889	0.01	2.42	4	2413	5.43	Post-Hausmann	Possibly exponential
2018	0.00	2.44	4	13,043	4.82	Today	

4.2 *The Evolution of Paris*

The results from the evolution of Paris seem to indicate, as expected, that there are two distinct categories of city structures: Pre-Hausmann (1300–1849) and post-Hausmann (1888 until today). The main difference is observed in the *p*-value for power law probability. From medieval times until the mid-nineteenth century the results indicate that it is possible that the road network structure adheres to a power law, while three post-Hausmann maps fail the K-S test. Maybe the first map from 1889 could be a candidate since 0.1 is the test threshold, but when compared to all of the other maps' *p*-value the result seems to be distinctly non-power law. Looking at the alpha values reveals that the four post-Hausmann maps have the lowest values while the other maps have a consistently higher alpha value. Altogether, these results show a clear distinction between the organically grown medieval city and the post-Hausmann Paris, subject to a very comprehensive restructuring.

Table 5 Results of the calculations on the cities in the cross-sectional study

City	*p*-value	Alpha	x-min	Number of vertices	Mean degree	Growth process	Comparisons to other distributions
Brasilia	0.00	2.60	3	7646	3.566	Entirely planned from start	
Paris	0.00	2.44	4	13,043	4.822	Planned to a large extent	
New York	0.01	2.32	6	3449	6.590	Entirely planned grid	Log-normal seems to be a good fit
Moscow	0.03	2.52	3	14,471	4.354	Planned to express power	
Beijing	0.10	2.49	4	4862	4.736	Planned	
Tokyo	0.29	2.55	7	19,007	4.934	Organically grown	
London	0.39	2.73	6	16,786	3.832	Organically grown	
Damascus	0.40	3.09	8	6337	3.765	Organically grown on Roman grid	Log-normal cannot be ruled out
Venice	0.46	3.50	8	2145	3.425	Several cities organically grown together	Log-normal cannot be ruled out
Rio de Janeiro	0.88	3.13	9	8240	4.021	Many organically grown favelas	

4.3 Comparison of Cities

The results in the cross-sectional study can be seemingly grouped into two rather distinct categories, those with p-value < 0.1 and those with p-value > 0.1. This means that approximately half of the selected cities' street network degree distribution does not fit a power law, while the other half has a decent probability of doing that. This finding is interesting since this statistical distribution can tell something about the growth process that has created it. If we identify what differs in the two categories according to Sect. 3.1, there is one main factor that differs: top-down planning or governing during city growth, versus organic growth.

The cities with a p-value < 0.1 have all been subject to rigorous planning and structuring, even though the purpose, time and type of planning vary (Kostof and Tobias 1991). The other cities that have a p-value > 0.1 are all to a varying extent what is called organic cities (Kostof and Tobias 1991; Batty and Longley 1994; Hillier 1996), or cities with large areas of organic patterns. These so-called organic cities are those where the city to a large extent has slowly grown over a long period of time and have not been not subject to any comprehensive planning effort. There is a city on the edge of the p-value limit, Beijing. The value means that it cannot be ruled out that it fits a power law, although it is just 0.01 above the limit. It is unclear why this is the case, but it would be surprising if all data could perfectly fit the dichotomies of way of growth.

Even though the differences in alpha values for the cross-sectional study are not as striking as with the p-values, the two groups have a significantly ($p<0.05$) different mean value. Alpha values are on average higher in the group of organic cities than on the group of planned ones, which is in line with the results from the longitudinal study.

Finally, an observation that also deserves some attention is the part of the distribution below x-min. This part is not included in the distributional analysis, hence the name x-min for the lower threshold. What is striking is that all cities seem to share the same kind of pattern in this part of the distribution. On average, streets with degree centrality of two (about 30% of total) are around twice as many as the streets with degree centrality of one (roughly 15–20% of total). Then streets with degree three (about 15–20% of total) are approximately half as many as the ones with degree two. Thereafter, the pattern of half as many streets as with the preceding degree number is continuing up until approximately x-min. This finding (illustrated in Fig. 5) indicates that although cities seem to differ regarding the kind of distribution and scaling factor (alpha), they still share a universal pattern for the finest grained street network, the streets that are connecting directly to the buildings. Whatever the reasons behind this phenomenon, these are so far unexplored.

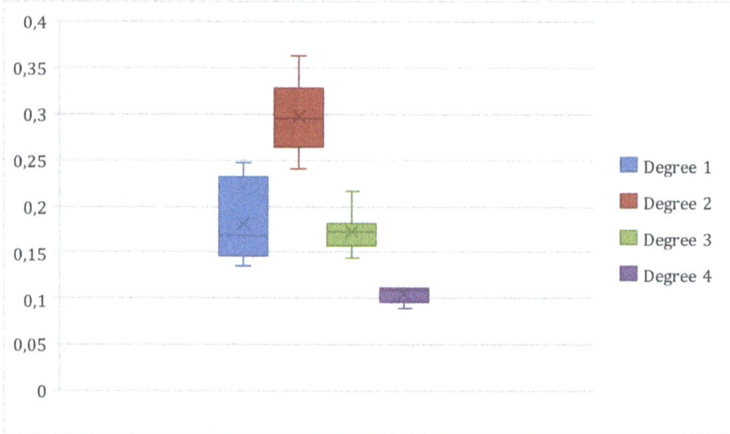

Fig. 5 Proportion of total number for the streets with degree 1–4

5 Discussion

5.1 Organic and Planned Growth Processes

It seems that, based on these results, whether the degree distribution fits a power law says something about the level of planning in a city. This result can be explained by considering the city as a self-organized scale-free network (Barabasi and Albert 1999). If a city is not subject to large-scale planning, it tends to organize itself according to some growth principle, e.g. preferential attachment or similar processes (Volchenkov and Blanchard 2008). Preferential attachment is an example of a kind of multiplicative growth (Batty and Longley 1994) that results in a scale-free network. This explanation is in line with previous research (Jiang 2007; Volchenkov and Blanchard 2008).

A hypothesis for organic growth could be that new roads attach first and foremost to the existing roads, and these existing roads gain higher degree up to a certain limit (when there is no more space). This means that the organic growth process results in the evolution of the fine-grained minor road (background) network connecting houses to the closest street, while the high degree main roads (foreground) network is more stationary or undergoes a very slow development. The higher alpha values (steeper slope) found in the organic street networks supports this explanation. Since there are mainly local bottom-up forces that shape these (organic) cities, relatively few primary highly connected streets are built because these require a more central initiative and a larger amount of resources (Wang 2015). This also makes sense since it is a more effective use of land (less proportion of roads relative other land use), according to the space filling principle. (Kuffner 2009), which is logical from the perspective that the forces driving the bottom-up growth (often residential and firms) are acting economically.

But, what the result in this study also indicates is that not all cities' development can be explained in this way; there seems to be a dividing line between planning and self-organizing processes. In planned cities, a common trait is the creation of many high degree streets such as highways, boulevards and main streets. This can have various reasons: political, e.g. it is said that Haussmann's boulevards were created to facilitate the movement of police in case of unrest; traffic management, e.g. new high degree roads are often arterial roads that can handle increased traffic; or to make the road network more redundant. It is probably the case that many of the planned cities started with an organic growth process or have grown on an organic core, but then had a modern road system imposed on them, like in the case of Paris.

5.2 *The Classification of Cities*

Planned cities can consist of many different kinds and structures. The selection of cities in this study includes those that are planned in one go (e.g. Brasilia, Fig. 5), as well as cities that have grown on a planned substrate (e.g. New York grown on a regular grid), or cities that have got a new structured superimposed (e.g. Hausmann's Paris). The city of Moscow (Fig. 6) was also subject to repeated and extensive restructuring by centralized regimes. The results from this study are inconclusive regarding planned cities; some seem to be exponential while others might be log-normal, and some do not even seem to have a good fit for any heavy-tailed distribution.

This is not surprising if we consider the kinds of distributions that some ideal city forms have. In a perfect grid, all streets in one direction will have the same degree value while the perpendicular streets will all have another degree value, which gives a distribution with only two values. The same occurs in the case of radial cities with concentric rings. A deeper and more systematic investigation into the many outcomes of city planning actions and their consequences for the street network structure could probably present very fruitful work (Fig. 7).

Whether a city is organic or planned is not a black and white classification either. Cities can be founded and grow an organic network and then suddenly planning measures are put in place affecting the organic pattern, or further growth outside the core is subject to new planning measures. This is quite common for European cities, that a medieval city core is surrounded with more planned patterns. Especially areas built during the twentieth century are to a very large extent planned. The opposite, organic growth on a planned pattern, can also happen. A good example is Herat in Afghanistan (Kostof and Tobias 1991). Where a more fine-grained organic network has grown on the original roman grid. In this way, cities alternate back and forth between these two types of structure. This temporal alternation is often slow for the organic pattern while planning interventions and new planned areas often represent fast change.

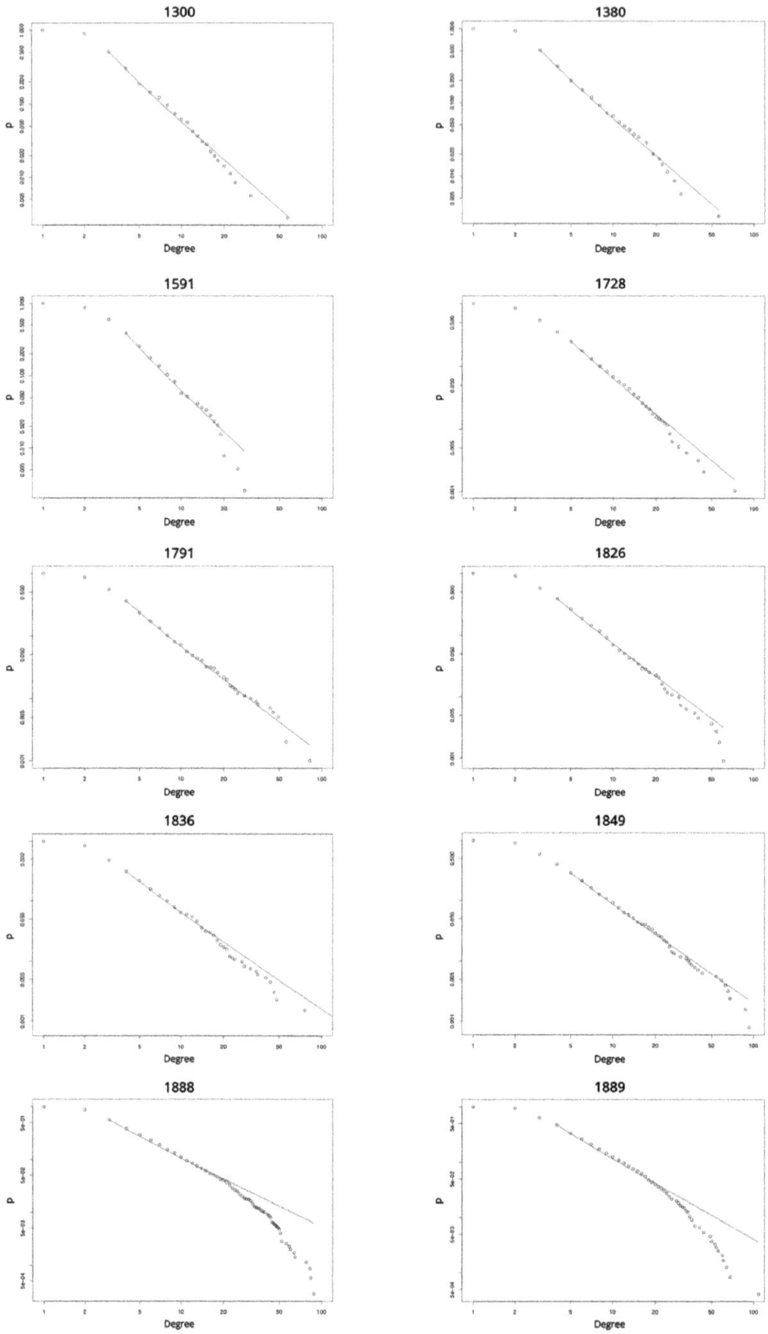

Fig. 6 CCDF of degree distribution in the longitudinal study (best-fit line drawn)

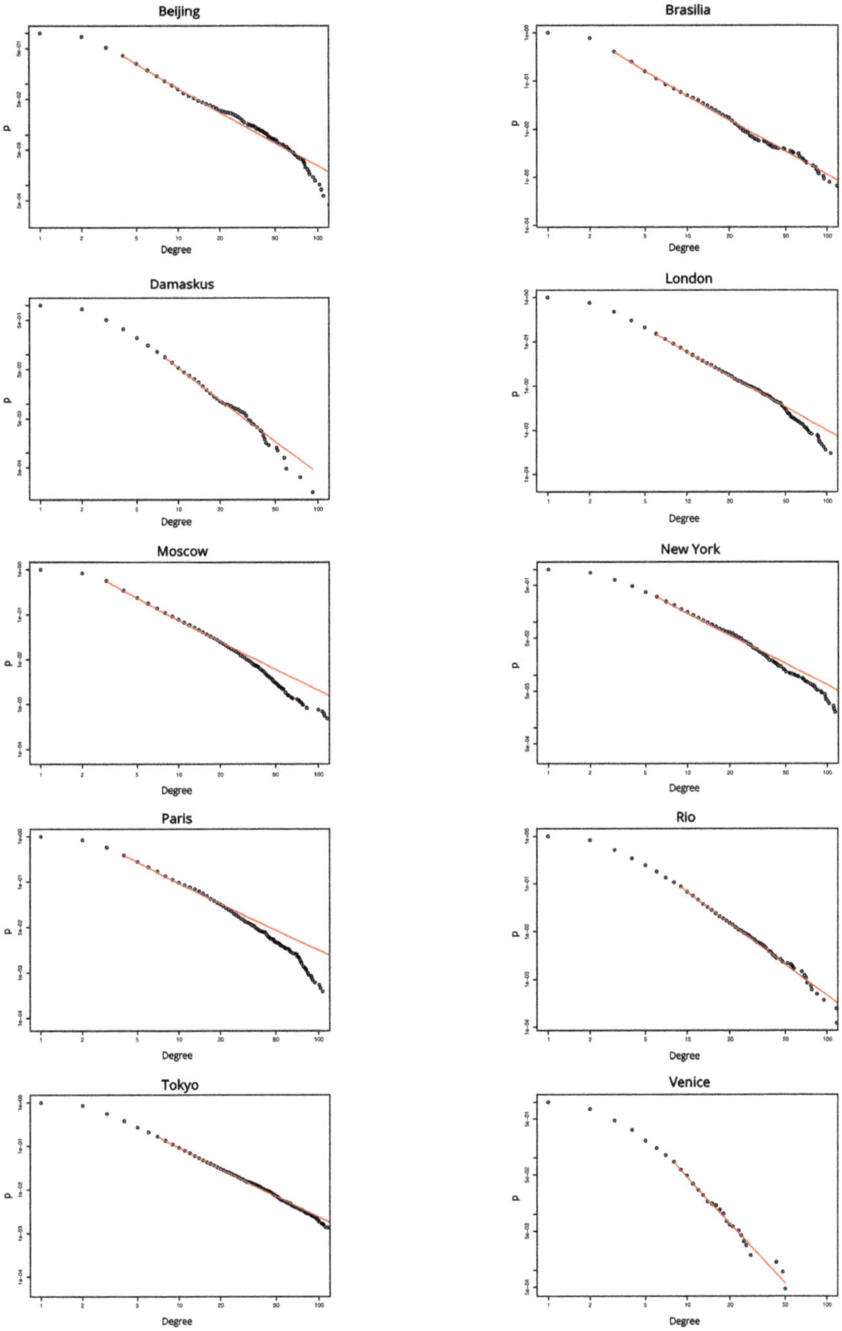

Fig. 7 CCDF of degree distribution in the cross-sectional study (best-fit line drawn)

The temporal alternation also goes in hand with spatial alternation of areas grown organically and other planned. As a consequence, the degree distribution of cities suffers disruptions in the middle range from an increased number of planned streets representing smaller areal interventions, as opposed to the few whole city interventions. While these disruptions deviate from a power law distribution representative of organic growth they do not completely rule out this fit, as occurs with entirely planned cities.

Overall, these results are interesting because they can provide an explanation to the earlier, sometimes contradictory, results (Crucitti et al. 2006a, b; Porta et al. 2006a, b; Jiang 2007; Volchenkov and Blanchard 2008), where some conclusions stated that cities adhered to a power law while others did not. It can be that the studied cities simply were of different kinds (i.e. organic/planned). Another possible explanation is in the methods used; as described in the methods section, there are several criteria that can be used to judge whether a distribution is a power law or not. All together these results show that more studies are required in order to make firmer inferences.

6 Conclusions

This study has revealed similarities and differences in the degree distribution of the street network among cities with varying cultural, topographical, geographical or societal organizational circumstances. It has also revealed a striking pattern in the longitudinal study of Paris, which is to a large extent a different city, before and after Hausmann's rebuilding of the city. The factor that seems to explain the striking differences found in the degree distributions is the growth process, organic bottom-up or planned top-down. It seems that cities with a resulting organic street pattern are likely to follow a power law in their degree distribution. One the other hand, the extensively planned cities do not follow a power law distribution. These results can be useful in several ways.

First, they offer an explanation for the disparate results from other studies on the properties of the street networks, which have tried to find universal patterns in cities, often power law distributions. The answer to the question whether a universal pattern exists is: no, because it depends on the growth process of the city. Rather, it seems that there are at least two kinds of patterns.

Second, they elegantly verify the categorization and judgment of cities that urban planning historians and theorists such as Kostof and Tobias have done on a qualitative basis, through visual analysis of the morphology. This opens up the possibility for urban historians, theorists and morphologists to support their work with quantitative analysis on a larger scale than the usual qualitative judgements allow.

Third, for a long time there have been attempts to plan cities or neighbourhoods with an organic morphology. These attempts have often been unsuccessful or failed to produce the kind of urban fabric that was intended. The insights in this study might help such attempts, opening up the possibility to test various designs. The results can be used to see if the designs have the same properties as the organically grown cities that serve as models.

Although this study is too small to draw conclusive lessons, it provides clues and starting points for further research. It also raises the question of whether there is some differentiated universal pattern in cities. It has shown that one measure of the street network can reveal such patterns and relations that seem to depend on the city's history of growth and planning.

References

Barabasi, A. L. & Albert, R. (1999), Emergence of scaling in random networks. *Science*, 286(5439), 509–512.

Barthélemy, M. (2011), Spatial networks. *Physics Reports*, 499(1–3), 1–101.

Batty, M. (2013), *A new science of cities*.

Batty, M. & Longley, P. (1994), *Fractal Cities: A Geometry of Form and Function* (Vol. 1). San Francisco and London: Academic Press.

Beaverstock JV, Smith RG, Taylor PJ (1999) A roster of world cities. *Cities* 16:445–458. https://doi.org/10.1016/s0264-2751(99)00042-6.

Bettencourt LMA, Lobo J, Strumsky D, West GB (2010) Urban Scaling and Its Deviations: Revealing the Structure of Wealth, Innovation and Crime across Cities. *PLoS ONE* 5:e13541. https://doi.org/10.1371/journal.pone.0013541.

Clark C (1951) Urban Population Densities. *Journal of the Royal Statistical Society Series A (General)* 114:490–496. https://doi.org/10.2307/2981088.

Clauset, A. (2007), Fitting a power-law distribution. http://tuvalu.santafe.edu/~aaronc/powerlaws/

Clauset, A., Rohilla Shalizi, C. & Newman, M. E. J. (2009), Power-law distributions in empirical data. 51, 43.

Crucitti, P., Latora, V. & Porta, S. (2006a), Centrality in networks of urban streets. *Chaos*, 16(1), 015113.

Crucitti, P., Latora, V. & Porta, S. (2006b), Centrality Measures in Spatial Networks of Urban Streets.

Dalton N (2001) Fractional configurational analysis and a solution to the Manhattan problem. In: Peponis J, Wineman J, Bafna S (eds) *Proceedings of the 3rd International Space Syntax Symposium*. Georgia Institute of Technology, Atlanta, Georgia, USA, pp 26.1–26.13.

Dalton N, Peponis J, Conroy Dalton R (2003) To tame a TIGER one has to know its nature: extending weighted angular integration analysis to the description of GIS road-centerline data for large scale urban analysis. In: Hanson J (ed) *Proceedings of the 4th International Space Syntax Symposium*. London, UK, pp 65.1–65.10.

Dhanani, A., et al. (2012), FROM THE AXIAL LINE TO THE WALKED LINE: evaluating the utility of commercial and user-generated street network datasets in space syntax analysis. *Paper presented at the Eighth International Space Syntax Symposium*, Santiago de Chile.

Figueiredo, L. & Amorim, L. (2005), Continuity lines in the axial system.

Gil J (2017) Street network analysis "edge effects": Examining the sensitivity of centrality measures to boundary conditions. *Environment and Planning B: Urban Analytics and City Science* 44:819–836. https://doi.org/10.1177/0265813516650678.

Gil J, Beirão JN, Montenegro N, Duarte JP (2012) On the discovery of urban typologies: data mining the many dimensions of urban form. *Urban Morphology* 16:27–40.

Gillespie, C. S. (2015), Fitting Heavy Tailed Distributions: The poweRlaw Package. *Journal of Statistical Software*, 64(2).

Harris CD (1943) A Functional Classification of Cities in the United States. *Geographical Review* 33:86–99. https://doi.org/10.2307/210620.

Haklay, M. (2010), How good is volunteered geographical information? A comparative study of OpenStreetMap and Ordnance Survey datasets. Environment and Planning B-Planning & Design, 37(4), 682–703.

Hillier, B. (1996), *Space is the Machine: a configurational theory of architecture*. Cambridge: Cambridge University Press.

Hillier, B. (1999), The hidden geometry of deformed grids: or, why space syntax works, when it looks as though it shouldn't. *Environment and Planning B-Planning & Design*, 26(2), 169–191.

Hillier, B. & Hanson, J. (1984), *The social logic of space*. Cambridge: Cambridge University Press.

Jiang, B. (2007), A topological pattern of urban street networks: Universality and peculiarity. *Physica A: Statistical Mechanics and its Applications*, 384(2), 647–655.

Jiang, B. (2009), Ranking spaces for predicting human movement in an urban environment. *International Journal of Geographical Information Science*, 23(7), 823–837.

Jiang, B. (2015). Axwoman 6.3: An ArcGIS extension for urban morphological analysis. University of Gävle, Sweden. Retrieved from http://fromto.hig.se/~bjg/Axwoman/.

Jiang, B. & Claramunt, C. (2004), A structural approach to the model generalization of an urban street network. *Geoinformatica*, 8(2), 157–171.

Jiang, B., Claramunt, C. & Batty, M. (1999), Geometric accessibility and geographic information: extending desktop GIS to space syntax. *Computers, Environment and Urban Systems*, 23, 127–146.

Jiang, B. & Liu, C. (2009), Street-based topological representations and analyses for predicting traffic flow in GIS. *International Journal of Geographical Information Science*, 23(9), 1119–1137.

Jiang, B., Zhao, S. & Yin, J. (2008), Self-organized natural roads for predicting traffic flow: a sensitivity study. *Journal of Statistical Mechanics: Theory and Experiment*, 2008(07), P07008.

Kolovou I, Gil J, Karimi K, Law S, Versluis L (2017) Road Centre Line Simplification Principles for Angular Segment Analysis. In: *Proceedings of the 11th Space Syntax Symposium*. Universidade de Lisboa, IST, Lisbon, Portugal, pp 163.1–163.16.

Kostof, S. K. & Tobias, R. (1991), The city shaped: urban patterns and meanings through history. London: Thames & Hudson.

Krenz K (2017) Employing Volunteered Geographic Information in Space Syntax Analysis. In: *Proceedings of the 11th Space Syntax Symposium*. Universidade de Lisboa, IST, Lisbon, Portugal, pp 150.1–150.26.

Kuffner, J. J. L., S.M. (2009). Space-filling trees, Carnegie Mellon University.

Louf R, Barthelemy M (2014) A typology of street patterns. *Journal of The Royal Society Interface* 11:20140924. https://doi.org/10.1098/rsif.2014.0924.

Nelson HJ (1955) A Service Classification of American Cities. *Economic Geography* 31:189–210. https://doi.org/10.2307/142045.

Marshall, S. (2015), Line structure representation for road network analysis. *Journal of Transport and Land Use*, 8(2).

Porta, S., Crucitti, P. & Latora, V. (2006a), The network analysis of urban streets: A dual approach. *Physica a-Statistical Mechanics and Its Applications*, 369(2), 853–866.

Porta, S., Crucitti, P. & Latora, V. (2006b), The Network Analysis of Urban Streets: A Primal Approach. *Environment and Planning B: Planning and Design*, 33(5), 705–725.

R Core Team (2016), R: A language and environment for statistical computing. R Foundation for Statistical Computing, Vienna, Austria. https://www.R-project.org/.

Shpuza, E. (2014), Allometry in the Syntax of Street Networks: Evolution of Adriatic and Ionian Coastal Cities 1800–2010. *Environment and Planning B: Planning and Design*, 41(3), 450–471.

Shpuza, E. (2017), Relative size measures of urban form based on allometric subtraction. *Environment and Planning B: Urban Analytics and City Science*, 44(1), 141–159.

Thomson, R. C. (2003), Bending the axial line: Smoothly continuous road centre-line segments as a basis for road network analysis.

Turner, A. (2007), From axial to road-centre lines: a new representation for space syntax and a new model of route choice for transport network analysis. *Environment and Planning B-Planning & Design*, 34(3), 539–555.
Volchenkov, D. & Blanchard, P. (2008), Scaling and universality in city space syntax: Between Zipf and Matthew. *Physica a-Statistical Mechanics and Its Applications*, 387(10), 2353–2364.
Wang, J. (2015), Resilience of Self-Organised and Top-Down Planned Cities–A Case Study on London and Beijing Street Networks. *PLoS ONE*, 10(12), e0141736.
Watts, D. J. (2004), The "new" science of networks. *Annual Review of Sociology*, 30, 243–270.
Watts, D. J. & Strogatz, S. H. (1998), Collective dynamics of 'small-world' networks. *Nature*, 393(6684), 440–442.
Webster, C. J. (1995), Urban Morphological Fingerprints. *Environ Plann B Plann Des* 22:279–297. https://doi.org/10.1068/b220279.

A Fractal Approach to Explore Australian Urban Form and Its Impacting Factors at Neighbourhood Scale

Mengyuan Jia, Yan Liu, Scott N. Lieske and Tian Chen

Abstract This chapter demonstrates an application of the *correlation fractal* method for exploring place diversity at neighbourhood scale. An empirical case study was performed on 130 neighbourhoods in the state capital city of Brisbane in Queensland, Australia. We used Google Maps as the data source to capture the neighbourhood building footprints for fractal analysis and compared the fractal results (fractal dimension and α) of Brisbane with those from similar studies of European cities. Spatial correlation analysis was employed to explore key factors impacting place diversity including different characterisation of density, metrics of land-use mix and accessibility. Results show that the urban form in Brisbane lacks place diversity and is homogenized, with no remarkable difference between inner city and suburban neighbourhoods, and its fractal dimension is lower than European cities. The fractal results of neighbourhoods are influenced by modernist planning principles including low-density housing, functional zoning and hierarchical street networks. Our application confirms that the *correlation fractal* method is suitable for describing urban form at the neighbourhood scale in the Australia context.

1 Introduction

Planning is changing the way cities grow, such that '(naturally) growing cities develop much more slowly than those which are planned' (Batty and Longley 1994: 8). Mod-

M. Jia · Y. Liu (✉) · S. N. Lieske
School of Earth and Environmental Sciences, The University of Queensland, Brisbane, Australia
e-mail: yan.liu@uq.edu.au

M. Jia
e-mail: jiamengyuan@tju.edu.cn

S. N. Lieske
e-mail: scott.lieske@uq.edu.au

M. Jia · T. Chen
School of Architecture, Tianjin University, Tianjin, China
e-mail: teec@tju.edu.cn

L. D'Acci (ed.), *The Mathematics of Urban Morphology*,
Modeling and Simulation in Science, Engineering and Technology,
https://doi.org/10.1007/978-3-030-12381-9_6

ernist planning follows the motto of Modernist Architecture which says ‘form ever follows function’ (Sullivan 1896: 408). This functional design principal satisfies the general requirements of industrialization and rapid urbanization, which has been influential to planners for generations. Yet, an increasing number of planners, sociologists and ecologists have realized the social and environmental problems caused by modernist planning in cities (Forman and Wu 2016). For instance, under the modernist planning model, single-use areas are developed to separate residential uses from industrial or commercial uses, and large number of single-family houses are built at suburban areas with the goal of making a safer and healthier living environment. The result, however, is urban sprawl, work-residence separation and high energy consumption (Levy 1999). Streets are designed for high-efficiency motorized transportation, while pedestrian-friendly streets with shops and restaurants are disappearing (Jacobs 1993). The planning and design of urban areas are being repeated, following similar design methods. As a result, cities lose diversity, becoming simpler, similar and modularized. However, diversity is an essential feature of cities, as cities are viewed as complex systems where people and communities interact with one another and with the objects in cities (such as roads, buildings and parks) (Batty and Torrens 2005; Batty 2013).

Diversity in cities include social (such as different races, ethnicities, genders, ages, households), economic (such as mixed in income, mixed in function), as well as spatial aspects. Spatial diversity in cities is also called *place diversity* by New Urbanists, smart growth advocates, creative-class adherents, and sustainability theorists (Talen 2006). These groups point out that a lack of place diversity has resulted in a range of issues including social segregation (Pendall 2001; Hall 2002), loss of vitality (Jacobs 1961) and reduced level of sustainability (Steiner 2002; Pickett et al. 2004). The concept of *place diversity* includes two types of spatial characteristics—variety and connectivity. Variety includes buildings, plots, blocks and streets in various sizes and shapes that ensure the mix of functions, households as well as social and economic requirements (Talen 2006). Connectivity suggests that spatial elements should be well connected from the building scale to the city scale. While modernist planning separates land use and develops ‘strict hierarchical subdivision’ (Batty 2013: 175) to organize the functions and street networks of cities, the result is the homogenization of urban space and the loss rather than connection of urban spatial *tails* (that is links between buildings, streets and open spaces) (Alexander 1967).

The study of urban form focuses on describing and interpreting the main physical elements that structure and shape a city—buildings, and their related open spaces, blocks, plots and streets (Moudon 1997). Position, outline and internal arrangement are typical characteristics used to describe urban form (Kropf 1996). Thus, urban form needs to be measured and described not only using the visible shape and pattern of the buildings, streets and other external space, but also the organizations and relationships between these elements. A number of quantitative analyses have been developed to evaluate the urban form of modern planned cities, such as the use of Space Syntax to describe the internal structure of urban form (Hillier et al. 1976;

Hillier 1999), and spatial metrics to analyse the shape of urban form (Frank et al. 2004; Frenkel and Ashkenazi 2008; Lieske et al. 2012; Lowry and Lowry 2014).

Fractal methods emerged as a practical approach to describe urban form in the late 1980s as they are able to describe both the shape and structure of urban built environment (Arlinghaus 1985; Batty and Longley 1986; Goodchild and Mark 1987; Batty and Longley 1994; Batty and Torrens 2005; Batty 2013). Fractal methods have been applied to quantify both linear features (such as streets) or areal features (such as built-up areas) at scales ranging from regional to cities, towns and neighbourhoods (Lucien et al. 2000; Shen 2002; Keersmaecker et al. 2003; Thomas et al. 2012; Thomas and Frankhauser 2013). For instance, Keersmaecker et al. (2003) use fractal-based parameters to describe the intra-urban diversity in Brussels. Their findings show associations between fractal dimensions and population density, housing rent, distance to the central business district (CBD), household income and the age of buildings. A series of studies conducted by Thomas and colleagues (2007, 2008, 2012), Thomas and Frankhauser (2013) applied fractal methods at the neighbourhood scale and tested the usefulness of fractal dimension when describing the urban pattern as uniform or contrasting. They found that the fractal dimensions are affected by population density and historical process of urban development. Through a comparative study of neighbourhoods in 18 European cities, they reported the fashion of the modernist planning theories that result in highly similar urban form across different countries (Thomas et al. 2012). While most of the current studies of urban form using fractal methods consider European cities which have a long histories of self-organized growth, few studies have been done in other regions using this method. Therefore, there is a need to evaluate the urban forms in other cities, especially those developed under the modernist planning theories.

In terms of data source used for fractal analyses of urban form, digitized topographical maps (Thomas et al. 2007, 2008, 2012) or remotely sensed images (Sun et al. 2006) are commonly used to acquire building footprints. In some countries, the availability of this type of data source is limited, and may have a coarse spatial resolution. In addition, this type of maps often requires sophisticated processing techniques for fractal analyses. Therefore, it would be useful to explore other source of data at higher spatial resolution such as the building scale covering large geographic areas.

This chapter uses the City of Brisbane, Australia as a case study to quantify its place diversity at neighbourhood scale by *correlation fractal* method, using high-resolution spatial data extracted from Google Maps. Spatial correlation analysis between the fractal results and 13 spatial metrics representing different characterizations of density, metrics of land-use mix and accessibility is conducted to explore factors impacting on place diversity. We aim at addressing two inter-related research questions; (1) how does the fractal dimension of the urban form in Brisbane compare to its European counterparts? (2) how do modernist planning principles, as reflected by different characterizations of density, land-use mix and accessibility, impact on the fractal dimension of urban form in the Brisbane context? The outcome from this study can provide additional insights to understand the place diversity of Australian cities. It also evaluates the application of the fractal methods in quantifying urban form through the use of new data source at the neighbourhood scale citywide.

The rest of this chapter is organized as follow: Sect. 2 introduces the *correlation fractal* method as well as the spatial correlation analysis between the fractal dimension measures and 13 spatial metrics representing different characterizations of density, land-use mix and accessibility. A description of the study context and data is also presented in this section. Section 3 presents the results from the fractal and correlation analyses and Sect. 4 discusses the impacts of modern planning principles on place diversity and insights for further planning practice. The conclusions are drawn in the final section.

2 Methodology, Study Context and Data

2.1 The Fractal Methods

Fractal geometry has been employed since the 1960s to quantify shapes that are irregular but self-similar such as coastlines, mountains and a variety of other physical phenomena (Mandelbrot 1967). As Mandelbrot (1983) described, such shapes, called fractals, appear similar at different scales such that 'apart looks like the whole' (Batty and Longley 1986: 1148). Fractals exist both in the natural world as well as in artificial phenomena, such as in cities. Various aspects of cities are confirmed to be fractals, such as urban growth patterns and rank-size distribution at regional scale (Batty and Longley 1986; Frankhauser 1998), the built environment at the building scale (Keersmaecker et al. 2003; Thomas et al. 2007, 2008; Thomas and Frankhauser 2013), the street network (Ariza-Villaverde et al. 2013; Thomas and Frankhauser 2013), and land use (Feng and Chen 2010).

The property of fractals and methods for computing fractal dimension can be described by a classical idealized model termed the Sieprinski's tree (Fig. 1) (Batty and Longley 1994). The black square is an initiator of base length L at the initial stage ($n = 0$ and is consistently replaced by five small black squares of base length $\frac{1}{3}L$ at each subsequent stage (i.e. $n = 1, 2 \ldots n$). Sieprinski's tree constitutes smaller and smaller similar Sieprinski's trees. The total surface area of the black squares can be computed following a procedure described below.

At the initial stage ($n = 0$), the original surface of black squares is

$$M_0 = N_0 \times l_0^2 \tag{1}$$

where M_0 is the surface area of black squares at $n = 0$; N_0 is the number of black squares ($N_0 = 1$), l_0 is the base length of black squares ($l_0 = L$).

Using the same convention with $M_1, \ldots, M_n$ representing the surface areas of black squares, $N_1 \ldots, N_n$ the number of black blocks and $l_1 \ldots l_n$ the length of the block at each subsequent step (*i.e.* $n = 1, 2 \ldots n$), the total surface area of the black squares at each step can be calculated by applying the generator to the initiator as

$$\begin{aligned} M_1 &= N_1 \times (l_1)^2 \\ M_2 &= N_2 \times (l_2)^2 \\ &\vdots \\ M_n &= N_n \times (l_n)^2 \end{aligned} \tag{2}$$

As each subsequent step follows the same generator, the number of black squares N_n can be generated from the length of black squares (l_n). When l_n decreases at each step, N_n increases. Therefore, N_n can be calculated by the following (Batty and Longley 1994)

$$N_n = (l_n)^{-D} \tag{3}$$

where D represents a scaling factor, or the fractal dimension of the shape. By transforming Eq. 3, D can be calculated as

$$\mathrm{D} = -\frac{\log N_n}{\log l_n} = \frac{\log N_n}{\log(1/l_n)} \tag{4}$$

Assuming that the original base length of a shape $L = 1$ in Fig. 1, the fractal dimension value D of Sieprinski's tree can be calculated to be 1.465.

In a typical two-dimensional space, the fractal dimension ranges from 1 to 2. This can be demonstrated by combining Eqs. 2 and 3 to get the total surface area of black squares of

$$M_n = N_n \times (l_n)^2 = (l_n)^{(2-D)} \tag{5}$$

If $D < 1$, M_n would be larger than l_n. Given that $l_n = \left(\frac{1}{3}\right)^n L$ in Sieprinski's tree, it is impossible that the surface value would be smaller than its base length. When $D = 1$, the surface of the black squares in Sieprinski's tree equals to the length of the black squares, representing a line. When $D = 2$, the black surface in Sieprinski's tree equals the original square (n = 0).

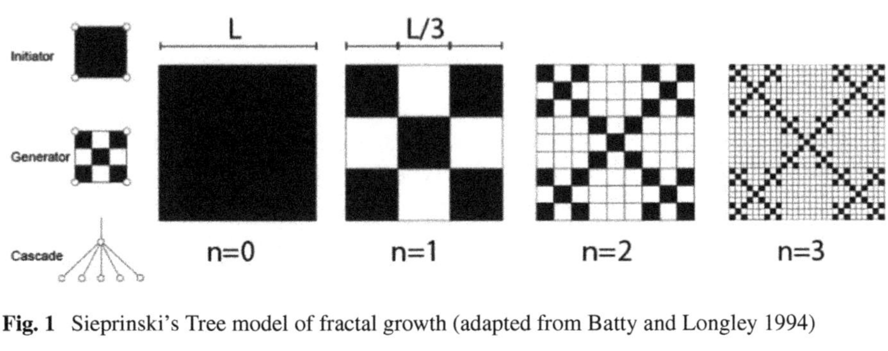

Fig. 1 Sieprinski's Tree model of fractal growth (adapted from Batty and Longley 1994)

This fractal structure also appears in cities. At the regional scale, one black square can represent one city. At the city scale, a black square represents a building cluster and at neighbourhood scale, a black square can represent a building. Even though cities are not well-defined fractals like the Sieprinski's tree, we can nevertheless explore the spatial distribution of urban built-up environment using the fractal methods. However, in the fractal analyses of cities, D cannot be derived mathematically, but computed empirically (Thomas et al. 2012). By fitting the empirical curve to a theoretical fractal law curve, the fractal dimension of a real-world shape will be computed (Keersmaecker et al. 2003). There are several methods for building an empirical curve that are suitable for measuring the fractal dimension of the urban built environment, including the *Dilation*, *Grid* and *Correlation* methods (Shen 2002; Keersmaecker et al. 2003). These methods share the same logic; they enable us to establish an empirical curve of the calculating areas with base length ε and the number of measured objects $N_{(\varepsilon)}$. The definition of the initial calculation of areas and how it changes at each subsequent step are different in each fractal method. For instance, in the *Dilation* method, the initial calculation of areas are windows with base length ε covering all the black pixels, and the number of windows is considered as $N_{(\varepsilon)}$; at each subsequent step, the window base length ε increases incrementally and the number of windows $N_{(\varepsilon)}$ covering all the black pixels would change correspondingly. In the *Gird* method, we cover the map with a grid (the size of the squares in the grid is ε), and then count the number of squares of the grid $N_{(\varepsilon)}$ that are covering part of the building footprints; gradually enlarging the squares size of the grid at each subsequent step, we will get a series of ε and $N_{(\varepsilon)}$ to establish the empirical curve.

We choose the *correlation fractal* method (also called the *Correlation Dimension*) in this study due to its advantages of being straightforward, and the fractal measures can be calculated efficiently and is less noisy when only a small number of points are available (Decoster and Mitchell 1991). We first tessellated the map of building footprint (in raster format, tiff or bmp) into same size windows of base length ε. Then, we counted the average number of black pixels within each window $N_{(\varepsilon)}$. This procedure was repeated by increasing the window base length ε, resulting in a series of $N_{(\varepsilon)}$ values. Next, an empirical curve was generated where the X-axis represents the window base length $\varepsilon = (2i + 1)$ (i being the iteration step) and the Y-axis represents $N_{(\varepsilon)}$. This curve was fitted to a theoretical fractal law curve as (Batty and Longley 1994; Keersmaecker et al. 2003; Thomas et al. 2008, 2012):

$$N_{(\varepsilon)} = \alpha\,\varepsilon^{D} + c \tag{6}$$

where α is a 'form factor' or 'prefactor of shape' (Thomas et al. 2012: 192) that corresponds to the size of the buildings. When the window size ε equals to or is smaller than the size of a building or courtyard, the empirical $N_{(\varepsilon)}$ value would be un-stable. Thus, a constant c is introduced to the fractal law curve in order to avoid local effects and hence incorrect estimations (Frankhauser 2008).

The process of generating an empirical curve and fitting to a theoretical fractal law curve can be done by using the *correlation* function in the software program

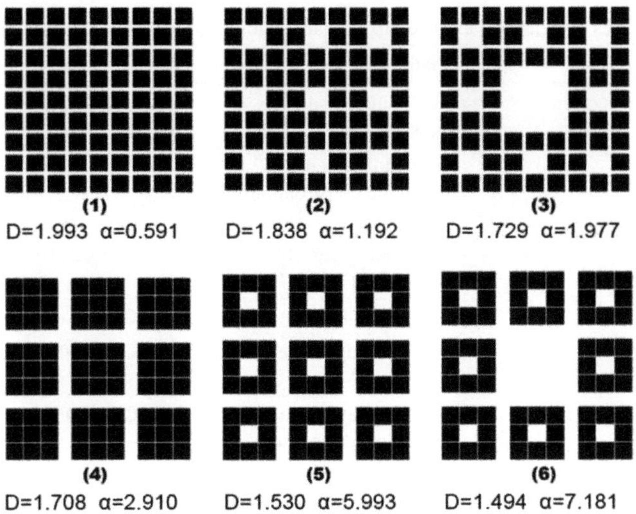

Fig. 2 The fractal dimension (D) and α of six artificial black distribution patterns

Fractalyse.[1] This software program is dedicated to conduct fractal analyses of raster files (tiff or bmp) and has been applied in built environment studies (Keersmaecker et al. 2003; Thomas et al. 2007, 2008, 2012; Thomas and Frankhauser 2013). In *Fractalyse*, the three parameters (D, α, and *c*) are estimated along with the determination coefficient (R^2) which shows the fitness of the estimated theoretical curve to the empirically derived curve. We consider $R^2 > 0.99$ to indicate that the generated fractal dimension value is reliable; otherwise, the pattern is not fractal or is multifractal (Thomas et al. 2012).

To illustrate how *correlation fractal* method can be applied to measure the shape and hierarchical structure of urban form, we hypothesize six different patterns as illustrated in Fig. 2. When comparing pattern (1) with (2) and (3), we find *D* is higher when the black squares are distributed uniformly (1). This is also the case in patterns (4–6). When we compare pattern (1) with (4) (or 2–5, 3–6) where the black squares are occupied at the same area in each pattern, the more hierarchically structured pattern in (4) results in a lower *D* value. The illustration also shows that α is higher when the size of black square mass is larger. Therefore, a higher D value indicates more uniformity and a less hierarchical distribution of buildings while higher α relates to larger building size.

In this study, we first apply *correlation fractal* method to explore the fractal dimension and α of the urban form in our case study site at the neighbourhood scale. The results are compared with European counterparts as evidenced in existing studies (Thomas et al. 2007, 2008; Thomas and Frankhauser 2013) to explore the differences of place diversity between self-organized cities and modernist planned cities.

[1] This is a free software program downloadable from the website at http://www.fractalyse.org/.

Table 1 Measures of density, land-use mix and accessibility to correlate with fractal dimensions

Category	Spatial metric	Data source
Density	A_1 Population density (persons/km^2)	Census 2016[a]
	A_2 Building density (occupied cells/total land cells)	Google maps
	A_3 Housing density (units/km^2)	Census 2016
	A_4 Median land lot size (m^2)	DCDB 2017[b]
Land-use mix	B_1 Percentage of residential land-use (%)	ACLUMP 2012[c]
	B_2 percentage of public institutions land-use (%)	ACLUMP 2012
	B_3 Percentage of environmental land-use (%)	ACLUMP 2012
	B_4 Percentage of industry land-use (%)	ACLUMP 2012
	B_5 SD of land lot size	DCDB 2017
	B_6 SD of perimeter of blocks	Brisbane City Plan 2014[d]
Accessibility	C_1 Street connectivity (ratio of streets to intersections)	Brisbane City Plan 2014
	C_2 Dendritic street pattern (ratio of cul-de-sacs to streets)	Brisbane City Plan 2014
	C_3 Percentage of transportation land-use (%)	DCDB 2017

[a]The Census of Population and Housing (Census) is Australia's largest statistical collection undertaken by the Australian Bureau of Statistics (ABS). More information is available at http://www.abs.gov.au/websitedbs/censushome.nsf/home/2016
[b]The Digital Cadastral Database (DCDB) contains the property boundaries and related property description of all land parcels in Queensland. Data is available at https://data.qld.gov.au/dataset/cadastral-data-queensland-series
[c]The Australian Collaborative Land Use and Management Program (ACLUMP) coordinates the development of nationally consistent land use and land management practices information for Australia. ACLUMP land use data is available at http://www.agriculture.gov.au/abares/aclump/land-use/data-download
[d]Brisbane City Plan 2014 is Brisbane City Council's plan for the future development of Brisbane. The dataset is available at https://www.brisbane.qld.gov.au/planning-building/planning-guidelines-tools/brisbane-city-plan-2014

2.2 Spatial Correlation Analysis

To explore the impact of modernist planning principles on urban form, we applied the Pearson correlation coefficient to quantify the relationships between the fractal measures and 13 classic metrics of urban form (Table 1) selected based on existing literature (Keersmaecker et al. 2003; Frank et al. 2004; Thomas et al. 2007, 2012; Frenkel and Ashkenazi 2008; Lowry and Lowry 2014). These spatial metrics reflect three aspects of modernist planning principles (Jacobs 1961): *density, land-use mix and accessibility*. *Density* is the most intuitive measurement of the occupation of space and is often applied in measuring urban sprawl (Tsai 2005; Ewing 2008). Population density (A_1) is the most commonly used index expressed by the number of inhabitants per square kilometre. Building density (A_2) is also used to reflect land use and development intensity, often expressed as a percentage of urban built-up area in the total areas. Housing density (A_3) is also considered as a good measure of the physical condition of land use and urban form, which is quantified by the amount of housing per square kilometre. Median land lot size (A_4) can affect the housing type and quantity, and influence density. Measures of *land-use mix* quantify the mixture of land uses and functions in cities, which is regarded as an important feature of vitality according to Jacobs (1961), Lynch (1995) and planning theories such as New Urbanism (Talen 2005) and Smart Growth (Daniels 2001). Four types of land uses ($B_{1,}$ $B_{2,}$ B_3 and B_4) related to the functions of living, social service, leisure and working are measured to describe the land-use mixture. The standard deviation (SD) of land lot sizes (B_5) and SD of the perimeter of blocks (B_6) show the variance in land lots and blocks, which would support different functions. *Accessibility* refers to the ability to access one destination from a location. It considers access by pedestrians more than by motor vehicles. The street connectivity (C_1) and dendritic street pattern (C_2) indices are used to quantify whether the street network is pedestrian-friendly (Lowry and Lowry 2014). They also relate to the hierarchical structure of the street network. More intersections and fewer cul-de-sacs usually mean that the streets are well connected, while the highly hierarchical street network has fewer intersections and more cul-de-sacs, which are less accessible. Furthermore, the percentage of transportation land-use (C_3) measures the quantity and width of streets, which also quantifies accessibility.

2.3 Study Context: Brisbane, Australia

Brisbane is the state capital city of Queensland and the third most populous city in Australia. Its metropolitan area has a population of 2.42 million and an area of 15, 826 km^2 (Australia Bureau of Statistics 2016a, b). The city was founded in 1824 as a penal colony, boomed in the 1900s due to its abundant mining resource and pleasant environment. From the late twentieth century onward immigration and international tourism have resulted in remarkable growth in population and peripheral developments (Hamnett 1984; Lieske 2017). As Stimson and Taylor (1999) state, Brisbane experienced a low-density, car-focused suburbanization era. In the 1990s, there were considerable debates about suburbanization in Australia, and some city plans as well as urban renewal projects were implemented to contain urban sprawl. However, the Australian dream for detached single family houses led to strong trend for urban expansion (Gleeson 2006). Siksna (2006: 89) states that the urban form of Australian cities 'share similarities only with the United States' but different from cities in European and Asian countries. However, the place diversity in Brisbane and how it may differ from the European self-organized cities are yet to be explored.

We used Statistical Area Level 2 (SA2) defined by the Australian Bureau of Statistics in the Australian Statistical Geography Standard (ASGS) as our base unit of analyses. SA2s are medium-sized general purpose areas, typically with a population size ranging from 3000 to 25,000 habitants, and representing a community that interacts together socially and economically (ABS 2016b). There are 133 SA2 units in Brisbane with three SA2s—Lake Manchester-England Creek, Mount Coot-tha and Enoggera Reservoir—having few resident population. As such, only 130 SA2 were analysed in this study. These SA2 are termed neighbourhoods thereafter. Figure 3 illustrates the study area with SA2 boundaries and one insert map showing the building footprint of one SA2 unit in its CBD region.

2.4 Data

To conduct fractal analysis in *Fractalyse*, the base data are required to be raster images with the urban built-up areas shown as black colour pixels and non-built-up areas as white colour pixels (Keersmaecker et al. 2003; Thomas et al. 2008, 2012). Using the styling map service of Google Maps (https://mapstyle.withgoogle.com/), we were able to edit the display of maps to illustrate buildings in black while removing other factors (roads, parks, rivers, POIs). After editing the display of the maps, the styling map service generated a URL based on Google Static Maps API service from which we downloaded the map tiles (in zoom level 17[2]) covering the study area. Subsequently, we merged all map tiles to produce one raster data layer, and used the SA2 boundary file to split the raster to produce neighbourhood building footprint maps. One pixel on the map of a neighbourhood building footprint represents a 2 × 2 m area on the ground. Using *Fractalyse*, we define a rectangular area covering the whole building footprint map of each neighbourhood, and then calculate the fractal dimensions using the *correlation fractal* method described earlier. Although the size of the rectangular areas for each neighbourhood varies, which can lead to different number of pixels on the map boundary with no-data value, the impact of these edge

[2]Zoom level is an integer on Google Maps that defines the resolution of the current view. Zoom levels between 0 and 21, and building outlines appear on the map around zoom level 17. Detail information of Google Maps Statistic APIs see https://developers.google.com/maps/documentation/static-maps/.

effect on fractal results is limited (Keersmaecker et al. 2003; Thomas et al. 2007, 2008, 2012).

Other data used in this study include the 2016 Australian Census of Population and Housing (ABS 2016a), the Digital Cadastral Database (DERM 2017), the land-use map in 2012 from the Australian Collaborative Land Use and Management Program (Department of Agriculture and Water Resources 2012) and the Brisbane City Plan 2014 (Brisbane City Council 2014). These data sources are employed to develop spatial metrics listed in Table 1. All spatial metrics are calculated according to the same boundaries with fractal analysis as shown in Fig. 3.

Fig. 3 The 130 SA2 neighbourhoods in Brisbane, Australia, with one insert map illustrating its building footprints. The SA2 boundary data were extracted from the ABS Census 2016 and the building footprints data were extracted from Google Maps

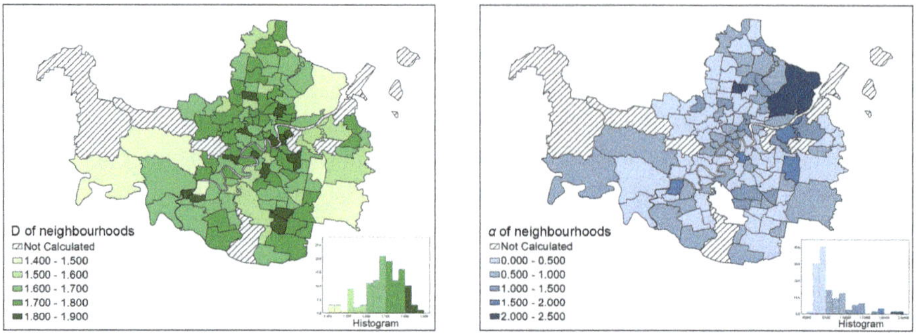

Fig. 4 Spatial distribution of the fractal dimension (D) and α across 126 neighbourhoods in Brisbane

3 Results

3.1 D *and* α *Values of Brisbane Neighbourhoods*

Using the *correlation fractal* method, only four neighbourhoods—Pallara-Willawong, Brisbane Port-Lytton, Tingalpa and Morningside-Seven Hills—resulted in a R^2 less than 0.99. These neighbourhoods correspond to port and industrial areas, where buildings are mostly large factories, plants, or workshops. These neighbourhoods were measured with insignificant D values because they are not fractals or they are multifractal (Thomas et al. 2012) and are therefore excluded from further analyses. Therefore, we presented results for the rest 126 Brisbane neighbourhoods. Univariate choropleth maps were drawn to show the spatial distribution of fractal results (Fig. 4). We classified D and α values into five categories using equal interval at 0.1 and 0.5, respectively.

Batty and Longley (1994), Frankhauser and Sadler (1991), and Frankhauser (2008) computed the fractal dimensions of several cities around the world, and for most cities they measured, the D value ranges between 1.6 and 1.8. Assuming that $D = 1.6 \sim 1.8$ as the best value range, we note that 71.4% of Brisbane neighbourhoods are within this range. 11.1% of neighbourhoods have a high fractal dimension value ($D > 1.8$) which are scattered throughout city, including the one with the highest value ($D = 1.875$) in the city centre. On the other hand, 17.5% of neighbourhoods have a D value lower than 1.6, with the lowest $D = 1.401$ being located at the southeast suburban area. There are 106 neighbourhoods (84.1%) whose α value is less than 1.0, with the neighbourhood having the highest α being located in the suburb areas. Figure 5 illustrates the relationship between D and α for Brisbane neighbourhoods, indicating a negative correlation between α and D.

A comparison with fractal dimension measures in cities of some European countries by Thomas et al. (2012) is listed in Table 2. The mean D of Brisbane neighbourhoods is 1.684, less than the mean D of European neighbourhoods. The standard

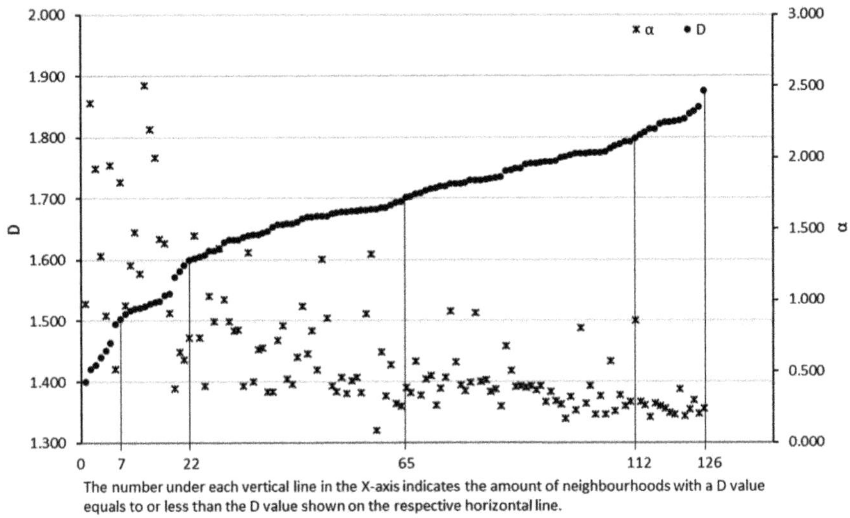

Fig. 5 Scatterplot of D and α. By ranking the neighbourhood according to D value in ascending order, the corresponding α value for the neighbourhood is shown in descending order, indicating a negative correlation between the two measures

Table 2 Number of observed neighbourhoods (n), mean of fractal dimension (D) and standard deviation (σ)

City/Country	*n*	*D*	*σ*
Brisbane	126	1.684	0.102
Italy	5	1.768	0.153
Finland	5	1.787	0.124
France	45	1.818	0.133
Belgium	21	1.835	0.103
Germany	16	1.859	0.075
Switzerland	5	1.866	0.042

deviation (σ) of *D* in Brisbane is 0.102, less than cities in Italy, Finland, France and Belgium, which shows the variation of urban form among neighbourhoods is by comparison, small. Yet, the number of observed neighbourhoods in Thomas et al. (2012) is less than the number of neighbourhoods analysed here. Fewer neighbourhoods may bias the observed *D* values. Thomas et al. (2012) selected neighbourhoods in various cities by considering their location, morphology, function, history and other factors for analyses of fractal dimensions, rather than all neighbourhoods across each city as is done here.

We compare the distribution of *D* in Brisbane with the Wallonia region of Belgium (Thomas et al. 2008) and the City of Antwerp, Belgium (Thomas and Frankhauser 2013). Figure 6 shows that the spatial distribution of neighbourhoods with high (or low) *D* value in Brisbane is not remarkably different between inner city and suburban areas, except in industrial areas. In contrast, the Wallonia and Antwerp examples

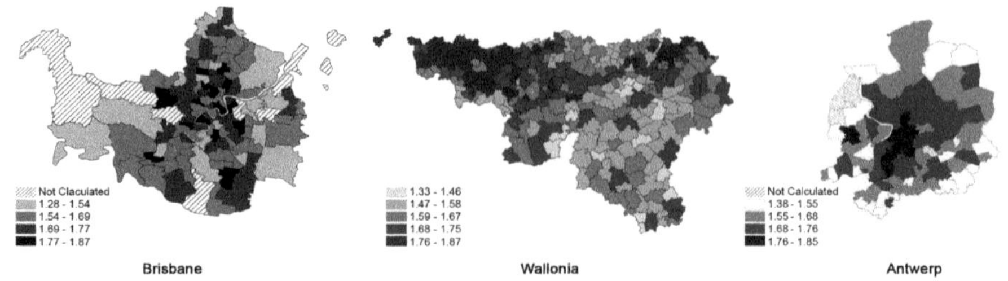

Fig. 6 Comparison of the fractal dimension of neighbourhoods in Brisbane, Wallonia and Antwerp (Data classified using natural break method). The map of Wallonia is from Thomas et al. (2008) and Antwerp from Thomas and Frankhauser (2013) (with permission)

reveal apparent transitions from high D in city centre (or urban areas) to low D in the outskirts (or rural areas). For example, the central business districts, such as Brisbane City ($D = 1.671$), South Brisbane ($D = 1.729$), and residential areas in inner city or suburban area, like Toowong ($D = 1.707$), Eight Mile Plains ($D = 1.729$), show little differences in fractal dimension. The urban form between inner city and suburban areas in Brisbane is highly similar compared to those in European cities or regions.

3.2 *Correlation Between Fractal Measures and Classical Spatial Form Metrics*

The correlation coefficients between the fractal measures and classic spatial form metrics are listed in Table 3. Except for public institutions land-use percentage (B_2) and street connectivity (C_1), all other spatial metrics quantifying density, land-use mix and accessibility have significant correlations with D. The three types of density (A_1, A_2, and A3) and residential land-use percentage (B_1) have correlation coefficients larger than 0.45, and are positively correlated with D. For α, its correlations with four spatial metrics (A_1, B_1, B_2, and B_5) are significant ($p < 0.01$). Among them, residential land-use (B_1) has the highest negative correlation with α, whereas accessibility measures show little correlation with α. The relatively small values of correlation coefficients, although significant at various p scale, indicate the weak linear relationships between the spatial form metrics and D or α (Table 3).

We selected six neighbourhoods with similar fractal dimensions (1.680–1.682) to further analyse the relationships between fractal dimension and density. The comparison shows all selected neighbourhoods have similar morphological characteristics that there are large scale buildings and some vacant spaces among the similar size residential houses. Some neighbourhoods have similar urban densities like (b), (d) and (e) in Fig. 7, but others (a, c and f) have large differences in densities even though they may have the same or similar fractal dimensions. This is evident in the Pinjarra Hills Pullenvale area (Fig. 7a) with low urban density compared with an inner city

Table 3 Correlation between fractal measures (D and α) and classical spatial form metrics

Category	Spatial form metric	D	α
Density	A_1 Population density	0.50***	−0.22***
	A_2 Building density	0.49***	−0.03
	A_3 Housing density	0.46***	−0.18**
	A_4 Median land lot size	−0.36***	0.03
Land use mix	B_1 Percentage of residential land-use	0.49***	−0.59***
	B_2 percentage of public institutional land-use	−0.03	0.30***
	B_3 Percentage of environmental land-use	−0.43***	0.16**
	B_4 Percentage of industry land-use	−0.18**	0.20**
	B_5 SD of land lot size	−0.24***	0.35***
	B_6 SD of perimeter of blocks	−0.44***	0.20**
Accessibility	C_1 Street connectivity	−0.12	0.04
	C_2 Dendritic street pattern	−0.22***	−0.08
	C_3 Percentage of transportation land-use	0.43***	−0.17**

***$p < 0.01$ (2-tailed); **$p < 0.05$ (2-tailed)

neighbourhood in Albion (Fig. 7b). It appears that density does not directly affect the level of scale disparity between buildings which is measured by fractal dimension. As Thomas et al. (2007: 303) report in their study of Brussels, 'For a given value of density, the fractal dimension can be very different and this variation can be explained by the internal organization, the morphology of the built-up area'.

Among the spatial metrics measuring land-use mix, residential land-use (B_1) is relatively highly correlated with D and α. The neighbourhoods of larger residential areas tend to have higher D and lower α. Environmental land-use (B_3) has a negative correlation with D. More open spaces such as parks can enlarge the contrasts between built-up and non-built-up areas and reduce the highly homogeneous urban spatial environment. For instance, a park between the homogenized single-family houses can change the landscape scenery and provide open space for the residents. Neighbourhoods with more public institutional land-use (B_2) have higher α. This confirms that α can describe the building size because the buildings for administrative, business, commercial or cultural functions are usually in large scale. Comparing the land use of different neighbourhoods (Fig. 8), neighbourhoods with high D value are predominately residential areas with little environmental or industrial land-use. Lower D corresponds to more diverse land-use patterns and lower percentage of residential land-use. It may be due to the nature of public institutional and environmental land-uses that are more flexible and diversified, which also provide the variety of building functions.

Our results also show some correlations between the accessibility measures and D. A dendritic street pattern (C_2) correlates negatively with D, and more dead-end streets in a neighbourhood tends to make D lower. Figure 9 illustrates two types

Fig. 7 Examples of selected neighbourhoods with similar fractal dimensions. The building footprints data were extracted from Google Maps

of street networks, one is dendritic in the suburban neighbourhood of the Eight Mile Plains which is highly hierarchical, lacking interactions and having more cul-de-sacs; the other is of a rectilinear street pattern in an inner neighbourhood of New Farm, with each street well connected with other streets and less dead-end streets. Neighbourhoods in Brisbane developed in the rapidly suburbanizing period are usually featured with hierarchical street networks while streets in the inner city neighbourhoods are more networked in grid. The different street network patterns reflect different level of accessibility for pedestrians which have somehow impacts on the fractal dimension of its urban form.

4 Discussion

This study applies the *correlation fractal* method to quantify urban form in Brisbane at the neighbourhood scale. The average value of the fractal dimension across all neighbourhoods is lower than those observed in European cities, indicating a lower

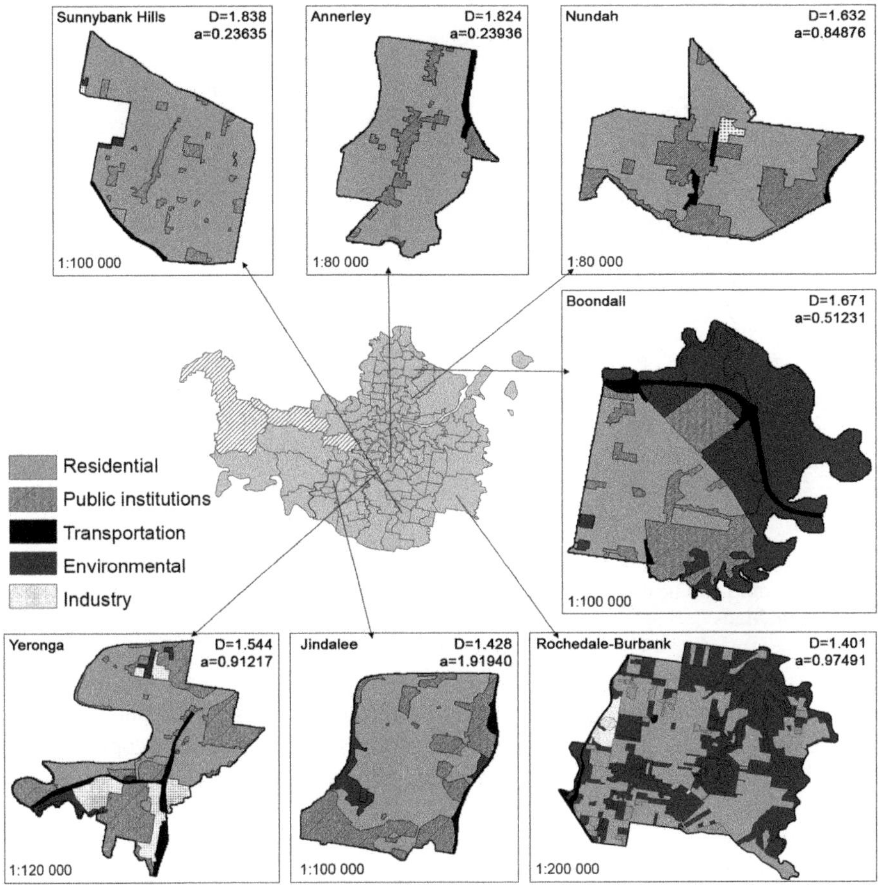

Fig. 8 Comparison of neighbourhoods with different land-uses and their D values. Land use data used to produce this figure were extracted from ACLUMP

level of place diversity in Brisbane compared with its European counterparts. This can be possibly explained by three reasons: *low-density housing, functional zoning*, and *the hierarchical layout of street networks*. First, our results show that there is a stronger relationship between *D* value and residential land development, including housing density (A_3), population density (A_1), and percentage of residential land-use in all land-use types (B_1). Limited variations exist in the urban form due to the single detached family houses being the dominant residential building type in Australian cities. This results in the drabness and monotonous space (Adams et al. 2013). Second, land lots in Brisbane are subdivided based on lot functions, where *functional zoning* limits the use of land, type of buildings to be constructed and diversity of occupants. Variation in lot size plays a critical role in maintaining certain level of diversity in a city. Different lot dimensions could contain various functions,

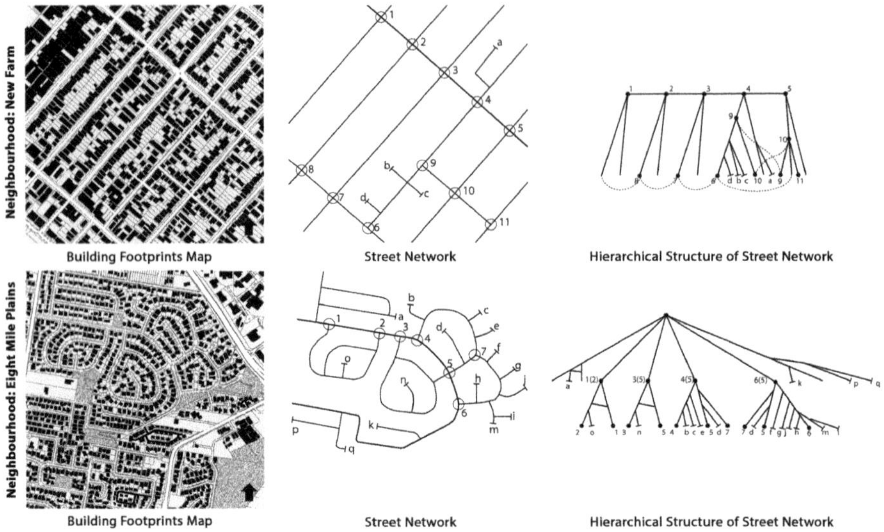

Fig. 9 Two exemplars of street network in Brisbane neighbourhoods: dendritic (Eight Mile Plains, D = 1.729) and rectilinear (New Farm, D = 1.827). The building footprints data were extracted from Google Maps

such as large size land lots for apartments or commercial complexes, and small land lots for single-family houses. Differently sized lots would offer more opportunities to developers with different finance abilities (Shelton et al. 2013), which can increase social-economic prosperity and vitality.

The third reason for the low place diversity is the *hierarchical layout of street networks*. Modern cities in hierarchical structures are more like a tree structure while cities developed in the nineteenth century are more similar to a fishnet with more complex interactions. European colonists planned Brisbane in a rectilinear fashion (Siksna 2006), similar to American cities developed in the late eighteenth century to early nineteenth century. Gridded streets usually contain two arrays of lots in a block, and with homogenous block sizes within the neighbourhood. However, under the influence of the modernist planning theories (Brody 2013), suburban neighbourhoods are designed to be functional, self-contained and hierarchical. Local streets are designed to be substantially different from the arterial streets in order to enhance neighbourhood safety and speed-up traffic flow, with more dead-end streets and the disappearance of urban street life.

Although the modernist planning methods have been debated in Australia since the 1990s, their impacts are long-lasting due to people's preference for the Australian lifestyle of living in detached house with gardens rather than in high-rise apartments. Nevertheless, planners and local governments have been working on increasing urban diversity in Brisbane in recent years. As stated in the latest Brisbane City Plan 2014, 'Brisbane's major new housing opportunities will be provided within the existing urban area'(Brisbane City Council 2014: 3.4.1). Multiple redevelopment and infill development projects are planned, which can lead to the increase in social, economic and spatial diversity of the neighbourhoods. To this end, some implications can be drawn from our fractal analyses to contribute to increasing place diversity of the city.

First, there are opportunities to boost place diversity by increasing population density in the inner city and controlling more strictly urban expansion to the suburban region. As Jacobs (1961) claimed, sufficient population density is essential for urban diversity. More apartment buildings can be developed in inner-city areas, along public transportation lines, or in existing or planned sub-centres. The long-last trend of urban sprawl needs to be controlled so that the city can grow more vertically rather than horizontally.

Second, it would be beneficial to encourage more land-use mix and flexible subdivision of land. The lacunae areas such as parks and urban squares are important to ensure diversity of the urban form (Thomas et al. 2012; Frankhauser 2015). More urban green space and public space can reduce the highly homogeneous urban form of residential areas, especially in neighbourhoods that are planned and rapidly constructed. The subdivision of land parcels can be more flexible by taken into account different functional use, which can offer more opportunities to developers of differing financial means.

Finally, we can also design more pedestrian-friendly streets and improve connectivity of our neighbourhoods. Given the highly hierarchical urban form in Brisbane with limited interactions between urban spatial tails, it is beneficial to create smaller blocks and grid-type of street networks in our neighbourhood in order to increase the connectivity and its place diversity. Well-connected canyon style streets can be more attractive to pedestrians and can also be safe because there are more 'eyes on the street' that watch over the neighbourhoods (Jacobs 1961: 35).

5 Conclusion

Drawing on an empirical application of fractal methods to the capital city of Brisbane in Queensland, Australia, this paper confirms that the *correlation fractal* method is capable of quantifying urban form at the neighbourhood scale. Higher fractal dimensions correspond to more uniformly distributed buildings and less hierarchical street networks. The results show that α is related to the size of buildings and is negatively correlated with the fractal dimension. Two urban areas—a residential area and an industrial area—can have similar fractal dimension values, but the α value of the industrial area would be higher, as the size of buildings is generally larger than residential buildings.

Our analysis shows that the fractal dimension of Brisbane neighbourhoods is lower than the European cities, and there is less variation between inner city areas and suburban neighbourhoods. This is due to the low-density dwellings or townhouses built on similar size land lots as the dominant form of residential housing in most neighbourhoods, with the exception of a few older neighbourhoods with historical settings. The highly hierarchical structure of street network with many dead-end roads also contributes to lower fractal dimensions. Our results concur with the criticism of modernist planning leading to the loss of place diversity.

Further research is needed to explore the optimal range of fractal dimension for an urban neighbourhood to be considered as of good place diversity. Although existing literatures tend to suggest that $D = 1.6 \sim 1.8$ is ideal, this value range is generated at citywide or regional scale, which may not be applicable at a much smaller neighbourhood scale. Furthermore, we also need to bear in mind that fractal dimension is a measure of locational characteristics, therefore, an optimal value may be different for a neighbourhood in urban areas compared to those in a more rural or suburban setting. Further research can help to identify optimal urban diversity measure which can be used to guide neighbourhood redevelopment and planning.

Our work also demonstrates the value of using Google Maps as a new data source to study urban form. Google Maps are freely available and provide worldwide coverage. By extracting the building footprints data of different cities from Google Maps, comparative studies across cities, regions or countries can be conducted which can yield further insights on the driving forces of urban form including urban governance, traditions and living habit of people from different geographical and cultural background. Such comparison would be more useful when data from the same source are used. We conclude by encouraging scholars to continue advancing the fractal methods as well as its wide applications in studies of urban form.

References

Adams D, Tiesdell S and White JT (2013) Smart Parcelization and Place Diversity: Reconciling Real Estate and Urban Design Priorities. *Journal of Urban Design* 18: 459–477.

Alexander C (1967) A City is not a Tree. *Ekistics* 23: 344–348.

Ariza VAB, Jiménez HFJ and Ravé EGD (2013) Multifractal analysis of axial maps applied to the study of urban morphology. *Computers, Environment and Urban Systems* 38: 1–10.

Arlinghaus SL (1985) Fractals Take a Central Place. *Geografiska Annaler. Series B, Human Geography* 67: 83–88.

Australian Bureau of Statistics (2016a) Australian Census of Population and Housing, Canberra. http://www.abs.gov.au/websitedbs/censushome.nsf/home/2016. accessed 15 Jan 2018

Australian Bureau of Statistics (2016b) Australian Statistical Geography Standard (ASGS): Volume 1 – Main structure and greater capital city statistical areas (2016b) Australian Bureau of Statistics, Canberra. http://www.abs.gov.au/ausstats/abs@.nsf/mf/1270.0.55.001. accessed 15 Jan 2018

Batty M (2013) *The new science of cities,* Cambridge: Mit Press.

Batty M and Longley PA (1986) The Fractal Simulation of Urban Structure. *Environment and Planning A: Economy and Space* 18: 1143–1179.

Batty M and Longley PA (1994) *Fractal cities: a geometry of form and function,* Cambridge: Academic press.

Batty M and Torrens PM (2005) Modelling and prediction in a complex world. *Futures* 37: 745–766.

Brisbane City Council (2014) Brisbane City Plan 2014, Brisbane. http://eplan.brisbane.qld.gov.au/. accessed 15 Jan 2018

Brody J (2013) The Neighbourhood Unit Concept and the Shaping of Land Planning in the United States 1912–1968. *Journal of Urban Design* 18: 340–362.

Daniels T (2001) Smart growth: A new American approach to regional planning. *Planning practice and research* 16: 271–279.

Decoster GP and Mitchell DW (1991) The efficacy of the correlation dimension technique in detecting determinism in small samples. *Journal of Statistical Computation and Simulation* 39: 221–229.

Department of Agriculture and Water Resources (2012) Australian Collaborative Land Use and Management Program, Canberra. http://www.agriculture.gov.au/abares/aclump/land-use/data-download. accessed 15 Jan 2018

Ewing RH (2008) Characteristics, causes, and effects of sprawl: A literature review. In: Marzluff J, Shulenberger E, Endlicher W, et al. (eds) *Urban ecology.* Berlin: Springer, 519–535.

Feng J and Chen Y (2010) Spatiotemporal evolution of urban form and land-use structure in Hangzhou, China: evidence from fractals. *Environment and Planning B: Planning and Design* 37: 838–856.

Forman RT and Wu J (2016) Where to put the next billion people. *Nature* 537: 608–611.

Frank LD, Schmid TL, Sallis JF, et al. (2004) Linking objectively measured physical activity with objectively measured urban form. *American Journal of Preventive Medicine* 28: 117–125.

Frankhauser P (1998) The fractal approach. A new tool for the spatial analysis of urban agglomerations. *Population: An English Selection*: 205–240.

Frankhauser P (2008) Fractal geometry for measuring and modelling urban patterns. In: Albeverio S, Andrey D, Giordano P, et al. (eds) *The dynamics of complex urban systems.* Berlin: Springer, 213–243.

Frankhauser P (2015) From Fractal Urban Pattern Analysis to Fractal Urban Planning Concepts. In: Helbich M, Jokar Arsanjani J and Leitner M (eds) *Computational Approaches for Urban Environments.* Berlin: Springer, 13–48.

Frankhauser P and Sadler R (1991) Fractal analysis of agglomerations. *Natural Structures: Principles, Strategies, and Models in Architecture and Nature*: 57–65.

Frenkel A and Ashkenazi M (2008) Measuring Urban Sprawl: How Can We Deal with It? *Environment and Planning B: Planning and Design* 35: 56–79.

Gleeson B (2006) Towards a new Australian syburbanism. *Australian Planner* 43: 10–13.

Goodchild MF and Mark DM (1987) The Fractal Nature of Geographic Phenomena. *Annals of the Association of American Geographers* 77: 265–278.

Hall P (2002) *Cities of tomorrow: an intellectual history of urban planning and design in the twentieth century,* Oxford: Blackwell.

Hamnett S (1984) Brisbane. *Cities* 1: 442–448.

Hillier B (1999) The hidden geometry of deformed grids: or, why space syntax works, when it looks as though it shouldn't. *Environment and Planning B: Planning and Design* 26: 169–191.

Hillier B, Leaman A, Stansall P, et al. (1976) Space syntax. *Environment and Planning B: Planning and Design* 3: 147–185.

Jacobs J (1961) *The death and life of great American cities,* New York: Vintage Books.

Jacobs AB (1993) *Great Streets,* Cambridge: MIT Press.

Keersmaecker ML, Frankhauser P and Thomas I (2003) Using Fractal Dimensions for Characterizing Intra-urban Diversity: The Example of Brussels. *Geographical analysis* 35: 310–328.

Kropf K (1996) Urban tissue and the character of towns. *Urban Design International* 1: 247–263.

Levy A (1999) Urban morphology and the problem of the modern urban fabric: some questions for research. *Urban Morphology* 3: 79–85.

Lieske SN (2017) Urban, suburban and rural household density trends. *Australian Population Studies* 1: 69–72.

Lieske SN, McLeod DM, Coupal RH, et al. (2012) Determining the relationship between urban form and the costs of public services. *Environment and Planning B: Planning and Design* 39: 155–173.

Lowry JH and Lowry MB (2014) Comparing spatial metrics that quantify urban form. *Computers Environment and Urban Systems* 44: 59–67.

Lucien B, Daniel C, Maria M, et al. (2000) When and Where is a City Fractal? *Environment and Planning B: Planning and Design* 27: 507–519.

Lynch K (1995) *City sense and city design: writings and projects of Kevin Lynch*: MIT press.

Mandelbrot BB (1967) How long is the coast of Britain? Statistical self-similarity and fractional dimension. *Science* 156: 636–638.
Mandelbrot BB (1983) *The fractal geometry of nature,* New York: W.H. Freeman.
Moudon AV (1997) Urban morphology as an emerging interdisciplinary field. *Urban Morphology* 1: 3–10.
Pendall R (2001) Exploring connections between density, sprawl, and segregation by race and income in the US metropolitan areas, 1980–1990. *International seminar on segregation in the city.* Cambridge.
Pickett STA, Cadenasso ML and Grove JM (2004) Resilient cities: meaning, models, and metaphor for integrating the ecological, socio-economic, and planning realms. *Landscape and Urban Planning* 69: 369–384.
Queensland Government Department of Natural Resources, Mines and Energy (2017) Digital Cadastral Database of Queensland, Brisbane. https://data.qld.gov.au/dataset/cadastral-data-queensland-series. accessed 15 Jan 2018
Shelton B, Karakiewicz J and Kvan T (2013) *The making of Hong Kong: from vertical to volumetric,* London: Routledge.
Shen G (2002) Fractal dimension and fractal growth of urbanized areas. *International Journal of Geographical Information Science* 16: 419–437.
Siksna A (2006) The study of urban form in Australia. *Urban Morphology* 10: 89.
Steiner FR (2002) *Human ecology: Following nature's lead,* Washington D.C.: Island Press.
Stimson RJ and Taylor SP (1999) City profile: Brisbane. *Cities* 16: 285–295.
Sullivan LH (1896) The tall office building artistically considered. *Lippincott's Magazine* 57: 403–409.
Sun W, Xu G, Gong P, et al. (2006) Fractal analysis of remotely sensed images: A review of methods and applications. *International Journal of Remote Sensing* 27: 4963–4990.
Talen E (2005) *New urbanism and American planning: the conflict of cultures,* London: Psychology Press.
Talen E (2006) Design That Enables Diversity: The Complications of a Planning Ideal. *Journal of Planning Literature* 20: 233–249.
Thomas I and Frankhauser P (2013) Fractal Dimensions of the Built-up Footprint: Buildings versus Roads. Fractal Evidence from Antwerp (Belgium). *Environment and Planning B: Planning and Design* 40: 310–329.
Thomas I, Frankhauser P and De Keersmaecker ML (2007) Fractal dimension versus density of built-up surfaces in the periphery of Brussels. *Papers in Regional Science* 86: 287–308.
Thomas I, Frankhauser P and Biernacki C (2008) The morphology of built-up landscapes in Wallonia (Belgium): A classification using fractal indices. *Landscape and Urban Planning* 84: 99–115.
Thomas I, Frankhauser P and Badariotti D (2012) Comparing the fractality of European urban neighbourhoods: do national contexts matter? *Journal of Geographical Systems* 14: 189–208.
Tsai YH (2005) Quantifying Urban Form: Compactness versus 'Sprawl'. *Urban Studies* 42: 141–161.

Part II
Cellular Automata

Geographic Cellular Automata for Realistic Urban form Simulations: How Far Should the Constraint be Contained?

Jean-Philippe Antoni, Gilles Vuidel, Hichem Omrani and Olivier Klein

Abstract Cellular automata (CA) are discrete models that are being ever more widely used to study urban forms and, more broadly, to understand, simulate, and forecast land use changes (LUC). But LUC models are not based on CA dynamics alone and so they are not fully consistent with mathematical definitions of CA. Accordingly, to study urbanization, authors often use "constraint CA" or "geographic CA" (GCA), i.e., CA which are coupled with other models in order to integrate geographical assumptions related to urban form and to provide more realistic results. These complementary models are usually calibrated according to expert knowledge and do not lead to reproducible deterministic results. Consequently, there is often a sizeable gap between the theory of CA as defined in mathematics and their practical use for LUC. In this chapter, cellular automata are constrained by a Markovian process helping to determine the number of cells that can change from one land use category to another. Second, a potential model is used to create a suitability map and define the probability of a cell changing from one category to another. Finally, all these additional constraints lead to a suite of models which is clearly more complex than classical CA as it can be considered mathematically. Nevertheless, as far as possible, it presents GCA as a mathematical adaptation of CA integrating the geographical assumptions necessary for studying urban forms in a realistic way.

Keywords Cellular automata · Markov chain · Potential model · Urban form · Spatial modeling

---

J.-P. Antoni (✉) · G. Vuidel
Laboratoire ThéMA, UMR 6049—CNRS et Université Bourgogne Franche-Comté, 6 Boulevard Gabriel, 21000 Dijon, France
e-mail: jean-philippe.antoni@u-bourgogne.fr

H. Omrani · O. Klein
Luxembourg Institute of Socio-Economic Research, Maison des Sciences Humaines, 11, Porte des Sciences, 4366 Esch-sur-Alzette, Belval, Luxembourg

L. D'Acci (ed.), *The Mathematics of Urban Morphology*, Modeling and Simulation in Science, Engineering and Technology, https://doi.org/10.1007/978-3-030-12381-9_7

1 Introduction

Urban form may be defined as the relationship between buildings and open spaces within agglomerations or different types of urban aggregates according to the specificities of local context. It refers more specifically to the outer envelope or contours of the city (Antoni 2008). This form, which is characterized by developments at different scales (from the entire agglomeration to a single building), is the result of human occupation of the territory. As a result of individual behaviors, it reflects urban lifestyles generated through several factors such as the urban fabric, the built environment, density/compactness, and the spatial distribution of activities and facilities.

Geographers tackling urban forms seek primarily to understand the mechanisms that lead to the current form of a given urban context, to provide procedures for designing optimal forms, and to simulate future developments. Such approaches rely on adapted modeling tools that integrate features based on explanatory and predictive models, which can be used as tools to support reflection and decision-making. Among the mathematical models, many computer-based solutions attempt to simulate the evolution of cities and specially to understand how urban forms change over time, past and/or future. Among them, cellular automata (CA) stand out as a form of mathematical computation models based on a discrete dynamic modeling system. They are structured into procedures based on the nesting of simple rules that reflect the complexity of real systems. This approach is attractive because it relies on a generic development principle that fits very well with the way systems in general, and urban systems in particular, evolve. In this framework, LucSim[1] was designed and developed by ThéMA Laboratory (Antoni et al. 2017) as a cellular automata model specially designed for geographical analysis and spatial simulation for both researchers and advanced planning institutes. This user-friendly software is well adapted for analyzing and simulating land use changes and spatial dynamics at different scales for decision-making in urban and land planning.

After having recalled some definitions of CA as used in geography, the next section shows why they remain difficult to apply directly to concrete urban form planning or studies. Section 3 then presents two major constraints (temporal and spatial) that can improve CA simulation results by refining basic assumptions related to land use change. Section 4 presents the results in a new theoretical CA formalization, leading to more realistic urban form simulations illustrated by the example of Wroclaw in Poland. These results are then discussed in Sect. 5 questioning the extent to which these constraints need to be contained.

[1]See https://sourcesup.renater.fr/lucsim/ for more details.

2 CA-Based Discrete Modeling

Starting from a formal definition of CA-based discrete modeling, this section focuses on the strengths and limitations of this type of approach when considering urban form simulations.

2.1 CA Formal Definition

CA are discrete computer models composed of a grid of regular cells assigned to one particular state (among a finite number of states) which may change into another state over time. They were invented in the 1940s through the works of S. Ulam and J. von Neumann (1963) and popularized in the 1970s by John Conway's *Game of Life* (Conway 1970). Initially, they were of interest only to a few theorists of mathematics or computer science, who used them to solve puzzles or to build mathematical games in scientific journals. In the 1980s, a number of papers, especially those of Wolfram (1983, 2002) made CA fashionable, or rather showed them in a new light with a multitude of possible applications for very different fields. Having been focused initially on problems in physics and chemistry, several innovative experiments opened up biology, medicine, and ecology to CA before they were introduced into spatial studies, particularly geography and urban planning. In geography, the use of CA indeed echoes the cellular conception of geographical space defended by Tobler (1979) and Couclelis (1985) and reveals the deeply geographical character of this kind of tool (Couclelis 1988). For these authors, this cellular conception is more advantageous than considering space through the irregular spatial polygons defined by political and administrative jurisdictions. It provides a notational simplification allowing a cell of an array to be indexed in the same way as in matrix algebra. In such a notation, g_{ij}^{t} is a cell characterized by a land use category (urban, forest, industry, etc.) at the location *i, j* at time *t*, and $g_{ij}^{t+\Delta t}$ corresponds to the change in the land use category at the same location at time $t + t$.

From this basis, Tobler (1979) was probably the first geographer to envisage and describe all the formal possibilities of cell transitions according to different processes involving their neighborhood (Fig. 1):

1. An independent model where $g_{ij}^{t+\Delta t}$ is not related to g_{ij}^{t} in any way.
2. A dependent model where the land use at location *i, j* at time $t + t$ depends on the previous land use at that location, such that $g_{ij}^{t+\Delta t} = f(g_{ij}^{t})$.
3. An historical model where the land use at position *i, j* in the future depends on the initial land uses at that location, such that $g_{ij}^{t+\Delta t} = f\left(g_{ij}^{t}, g_{ij}^{t-\Delta t}, g_{ij}^{t-2\Delta t}, \ldots, g_{ij}^{t-k\Delta t}\right)$.
4. A multivariate model where the land use at location *i, j* is dependent on several other variables at that location, such that $g_{ij}^{t+\Delta t} = f\left(u_{ij}^{t}, v_{ij}^{t}, w_{ij}^{t}, \ldots, z_{ij}^{t}\right)$.

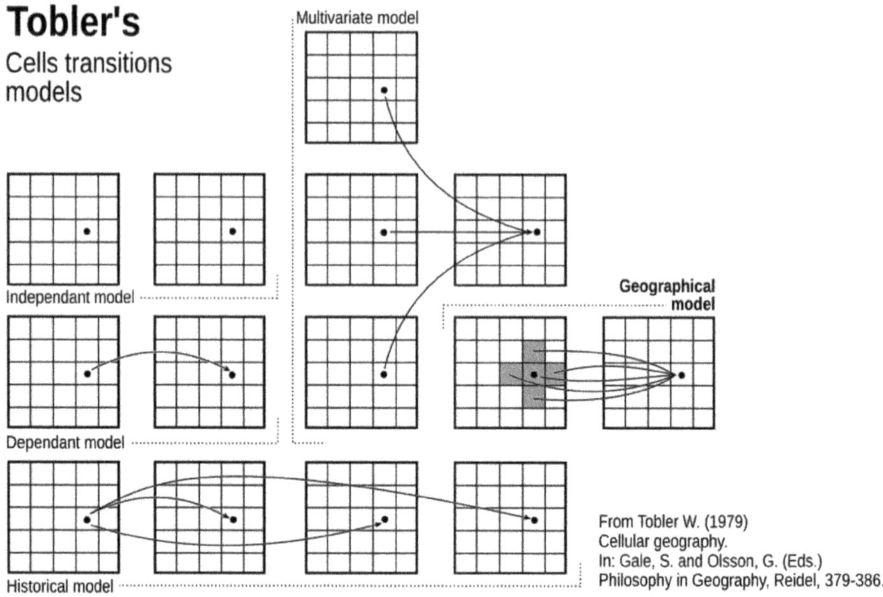

Fig. 1 Tobler's cells transitions models

5. A geographical model where the land use at location *i, j* is dependent on the land use at other locations, such that $g_{ij}^{t+\Delta t} = f\left(g_{i+p,\,j+q}^{t}\right)$.

This fifth model clearly corresponds to the process implemented in most CA models. Nevertheless, in the field of spatial studies and geographical sciences, formal definitions remain rare except for Tobler's former theoretical formalization. Researchers using CA seldom take the time to describe the mathematical form of the model they are using and refer only to other fundamental papers (White and Engelen 1993; Benenson and Torrens 2004), or describe CA as *if-then-else* algorithms (Batty 1997). Torrens (2000) is one of the rare geographers to use a mathematical notation to define the principles of CA transition according to the geographical process (fifth model) defined by Tobler.

2.2 CA Limits

A Tobler-like geographical notation is clearly pleasant mathematically and correctly describes how a transition can operate from one cell state to another according to theoretical neighboring configurations. But despite this advantage, it does not model land use change in an operative way, nor does it reproduce or create realistic simulations. An illustrative example based on a case study of Wroclaw (Poland) helps to explain why.

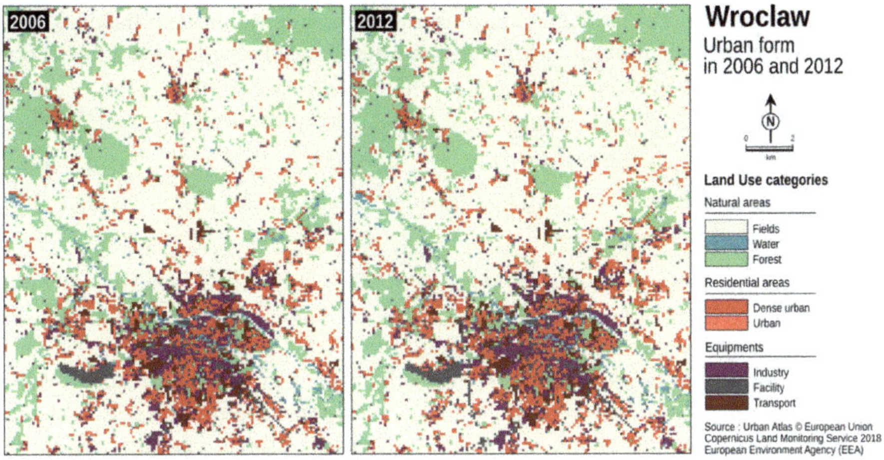

Fig. 2 The urban form of Wroclaw in 2006 and 2012

This example is built using data from the Urban Atlas (Copernicus Programme)[2] which describes land use in 2006 and 2012. As shown in Fig. 2, land use is classified into eight categories: water, fields, forest, dense urban, urban, industry, facilities, and transport. To simulate the evolution of this territory in the future, we use four transition rules implemented in the LucSim CA software (Antoni et al. 2017). These rules are constructed from expert knowledge and are supposed to reproduce urban expansion based on simple principles that have largely determined the evolution of urban form in the past. Rules are here expressed in two ways: verbal language and their conversion into a computer-based language specific to LucSim software (dashes):

1. Fields will become urban if, within a neighborhood of two cells, the current cell is surrounded by at least 10% of urban and dense urban cells, if, within a neighborhood of fifteen cells, there is at least one cell of facilities, if there is no direct connection to forest (within a neighborhood of one cell), and if within a neighborhood of three cells less than 50% of the cells are urban and dense urban:

   – *Fields -> Urban: pCellCir(Urban,2) + pCellCir(Urban_dense,2) + pCellCir(New,2) >= 10% and nbCellCir(Facilities,15) >= 1 and nbCellCir(Forest,1) <= 1 and pCellCir(Urban,3) + pCellCir(Urban,3) <= 50%;*

2. Urban cells will be densified if they are completely surrounded by urbanized areas within a neighborhood of one cell:

   – *Urban -> Urban_dense: pCellCir(Urban,1) = 100%;*

[2]See https://land.copernicus.eu/about for more details.

3. Urban parks (forests) will be created from urban or dense urban if the density of urbanized areas is more than 90% within a small radius (within a neighborhood of 2 cells for an urban category and within a neighborhood of 1 cell for a dense urban category):

   – *Urban -> Forest: pCellCir(Urban,2) + pCellCir(Urban_dense,2) >= 90%;*
   – *Urban_dense -> Forest: pCellCir(Urban,1) + pCellCir(Urban_dense,1) >= 90%;*

However, the strict application of these rules within LucSim produces results that have nothing to do with the current land use, nor with any logical development from a town planning or land use planning perspective.

Figure 3 indeed shows resulting spatial configurations, which are supposed to reproduce an urban sprawl process. It clearly shows that the urban sprawl process simulated by the model leads to a credible expansion of the urban form during the first iterations of the CA run, with an expansion of the city taking the form of an oil slick. But very quickly, the number of newly urbanized cells produced by the

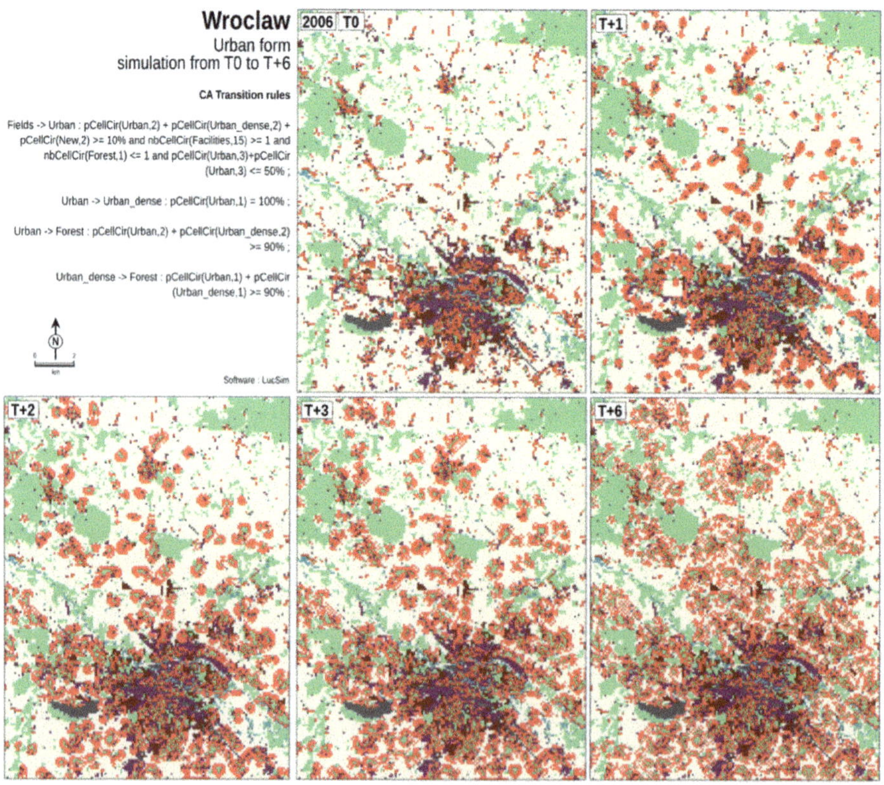

Fig. 3 Simulation of the urban form of Wroclaw from $T0$ to $T + 6$, (To make the images easier to read, newly urbanized areas appear in black before returning to their original color (red))

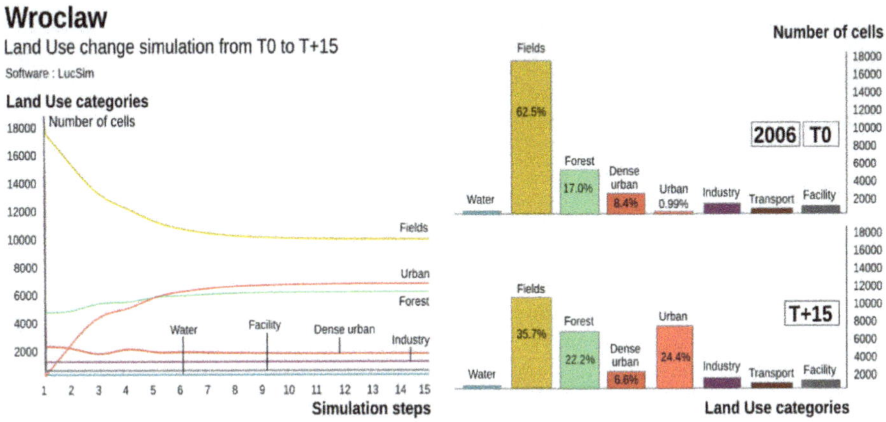

Fig. 4 Land use changes in Wroclaw from $T0$ to $T + 15$

model generates a snowball effect that far exceeds any realistic forecast of future urbanization for a city like Wroclaw.

From this example, we can conclude that the simple use of CA with transition rules is not sufficient for forecasting realistic and operational simulations of urban forms. On the one hand, the recognition of spatial patterns leading to the application of a transition rule corresponds only very partially to the reality of the urbanization process. Indeed, the number of possible transitions appears to be much greater than the real needs in terms of new housing and population growth. On the other hand, the results give no information about the timing of this urbanization. Therefore, although it seems clear that a time step does not correspond to a regular and stable duration, we are unable to say whether the images produced from $t + 1$ to $t + 16$ lead us to 2020, 2050, 2200, or 3500 (Fig. 4).

It appears clearly, then, that a CA cannot be directly applied to simulate city growth and more generally land use change. It must necessarily be constrained to answer more precise questions about space and time, so that the results produced can be integrated more easily, and realistically within a range of analysis and decision-making for planning. Only these kinds of constraints, based on assumptions about urban form, enable CA to be used in geography and allow for the fundamental difference between classical mathematical cellular automata and geographic cellular automata (GCA).

3 Suitable Constraints for Urban Modeling

There are many methods by which to constrain CA and the literature abounds with examples using several methods and models. In this section, we shall focus on just two kinds of fundamental constraints. The first is a time constraint to situate the results produced by the automaton over time. As in many publications (Arsanjani et al. 2013), it relies on a Markov chain process. The second constraint is a space constraint, which aims to reduce the number of neighboring configurations for possible transitions and focuses on the most realistic of them. It is based on a potential model.

3.1 Temporal Constraint

The first step uses a Markov chain to constrain the process of land use change in quantitative terms. Comparison of two static land use images (2006, 2012) can be used to determine what has happened between each image and so formulate a transition process. By comparing the land use categories date by date and cell by cell, it is possible to determine cellular changes between t and $t + 1$ and to identify the land use dynamics. Theoretically, each cell can either change from one land use category to another or remain in its initial category. The dynamics of the model can therefore be presented as a series of possible transitions from one land use category k at time t to another land use category l at $t + 1$. For a given cell N_i, a transition Δ can be written as:

$$\Delta \mathrm{N}_{\mathrm{i,kl}} = 1 \quad \text{if} \quad N_{\mathrm{i,k}}(t) = 1 \quad \text{and} \quad N_{\mathrm{i,l}}(\mathrm{t} + 1) = 1$$

To simplify the complexity resulting from the large number of cells and possible transitions, changes can be aggregated by land use categories. The aggregate transition for the complete system is then

$$\Delta \mathrm{N}_{\mathrm{kl}} = \sum_{\mathrm{i}=1}^{n} \Delta \mathrm{N}_{\mathrm{i,kl}}$$

This formulation allows us to build a contingency matrix indicating the number of cell transitions from a category k to a category l between t and $t + 1$ (i.e., between 2006 and 2012). This matrix can be easily converted into a transition matrix indicating the probability of change between all land use categories (Table 1). When associated with the previous vectors, this matrix provides all the elements needed for the construction of a Markov chain (MC). In the literature, an MC is defined as a mathematical process where transition probabilities are conditional on the past, and express the state of a variable at a time t as a function of observations of this variable at t − 1 (Feller 1968, Berchtold 1998). It relies on the connection of three items: (i) the description of the relative values associated with an initial state (land uses visualized as a vector for example); (ii) a transition matrix expressing the transition probabilities of different groups of observations from one category to another; and (iii) a diachronic transformation by an operator in the form of a matrix multiplication iteration.

If we follow this procedure, land use at time $t + 1$ can be simulated by multiplying the corresponding vector at time t by the corresponding contingency matrix, after the

Table 1 The transition matrix for Wroclaw between 2006 and 2012

	Water	Fields	Forest	Dense U	Urban	Industry	Facilities	Transp.
Water	**98.784**	0.608	0	0	0.304	0	0	0.304
Fields	0.08	**96.849**	0.017	0.028	2.178	0.364	0.034	0.049
Forest	0	0.063	**99.561**	0	0.104	0.125	0	0.146
Dense U	0	0	0	**99.831**	0.042	0.085	0.042	0
Urban	0	5.755	0.719	11.151	**69.784**	6.475	0.719	5.396
Industry	0	0.558	0	0	1.275	**98.088**	0	0.08
Facilities	0	0	0.159	0.638	1.115	0.159	**97.448**	0.478
Transp	0	0	0	0	0.662	0	0	**99.338**

transformation of the latter into transition probabilities from one land use category *k* to another *l*. To transform observed contingencies into transition probabilities, we use the following:

$$p_{\mathrm{kl}}(\mathrm{t}) = \frac{\Delta \mathrm{N_{kl}}}{N_k(t)} \quad \text{and} \quad \sum_{\mathrm{k}=1}^{m} p_{\mathrm{kl}}(t) = 1$$

We then consider the MC as follows:

$$N_{\mathrm{i}}(\mathrm{t}+1) = \sum_{\mathrm{k}=1}^{m} p_{\mathrm{kl}} \cdot N_k(t)$$

$$\text{where} \quad p_{\mathrm{kl}} = \frac{\Delta \mathrm{N_{kl}}}{N_k(t)} = \frac{\Delta \mathrm{N_{kl}}}{\sum_l \Delta \mathrm{N_{kl}}} \quad \text{and} \quad \sum_l p_{\mathrm{kl}} = 1$$

According to this formulation, the MC process gives us the chance to prospectively calculate future states from known past states, based on observation of past trends and probabilities. According to the method, this calculation is based on the assumption that future changes will follow the trend of past changes, but as it is based on a matrix calculation, this trend is not necessarily linear. Moreover, the values of the transition matrix can also be modified by users of the model to integrate different parameters for the quantification of future land use changes. In our case, LucSim uses the original transition matrix to calculate the number of cells in each land use category in 2018, 2024, 2030, etc., on the basis of 2006 and 2012 land uses (same interval of 6 years between each date). This system gives us a more plausible picture of urban dynamics by calculating land use vectors for each future date, as presented in Table 2.

This table also indicates that the total number $n_{l,t}$ of cells that should be urbanized in 2030 must not exceed 984 "urban" cells and 2571 "dense urban" cells.

Table 2 Expected future land use vectors

	Water	Fields	Forest	Dense U	Urban	Industry	Facilities	Transp.
2018	339	17058	4771	2401	613	1322	620	1006
2024	348	16568	4758	2474	836	1407	615	1121
2030	357	16107	4747	2571	984	1504	612	1245

3.2 Spatial Constraint

Among other methods, MCs are a way to quantify future land use changes when space is considered through cells. However, they say nothing about the location of those changes. The places where changes occur are strictly determined by the transition rules of the CA. To integrate information known elsewhere about the spaces most likely to be urbanized quickly (or on the contrary not to be if they are protected), it is therefore mandatory to add a second constraint capable of determining the most suitable locations. This second constraint is relatively conventional using GCA and is usually based on expert knowledge. It consists of constructing a *suitability map* based on *driving factors*, namely geographical features that are supposed to influence urbanization (Clarke 2008).

It seems obvious that working on locations requires a theoretical framework for geographical space. It is not surprising, then, that geographers have developed many models for this, often dedicated to residential location (Putman 1983). Diffusion models, for example, are spatial models that make it possible to locate certain elements on the assumption that they are generated through the diffusion of other elements. Fractal models are other models that can also simulate urban growth (Batty 2007). The city is then considered as a system that maximizes interactions between the elements it contains. Spatial interaction models are also another family of models derived from Newton's law of universal gravitation. They are based on identical principles and make it possible to locate changes where they are complementary to those around them by minimizing the distances between them. They have been used for calculating areas of traffic or influence (Helvig 1964) and for estimating residential or industrial locations (Abler et al. 1972).

Among spatial interaction models, potential models indicate that the probability of there being a relationship between places decreases with distance. Basically, they are used to measure "accessibility" aiming to evaluate the variation of the relative amount of relationship opportunity depending on the position of all places. Generally, the potential of a place is calculated from the analysis of the importance of all the other points of the system, an importance that is termed "mass" in reference to the Newtonian gravity model. The potential of a cell is usually the sum of all the potentials created at that location by the set of individual masses that make up the system (i.e., all the other cells). The calculation of the potential P of each point i therefore consists in applying to them a formula simultaneously taking into account the mass value m of all the points j located in a geographical area as a function of the

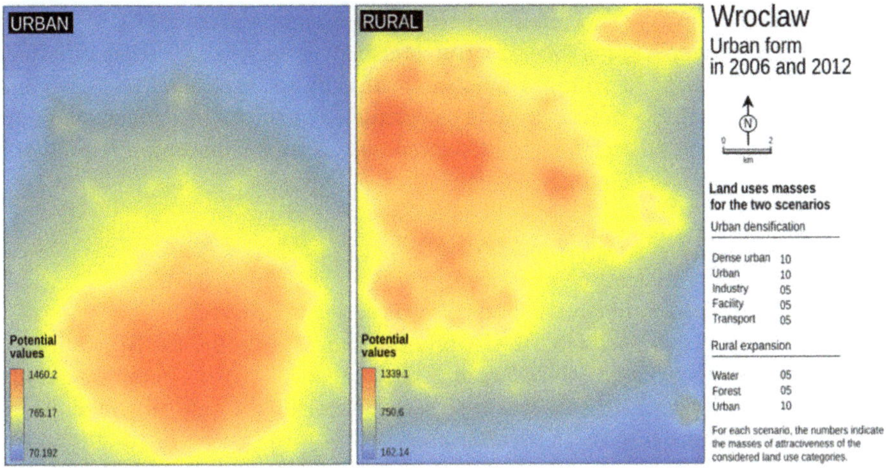

Fig. 5 Two contrasting realistic potential models

distance d_{ij}^{α} separating those points from the one for which the calculation is made. The operative formula is defined as follows:

$$P_i = \sum_{j=1}^{n} \frac{m_j}{d_{ij}^{\alpha}}$$

Spatial constraints based on potential results were applied to the case study of Wroclaw by distinguishing two contrasting scenarios. The first scenario (S1) focuses on "Urban densification" and assigns significant masses to *Dense urban* and *Urban* (m = 10) categories and medium masses to *Industry*, *Facility*, and *Transport* (m = 5) categories. As a result, the potential map (Fig. 5, left) shows high potential for land use change around the heart of the city of Wroclaw. Areas with high potential for change—urbanization—are limited to very closed urban areas. The second scenario (S2) deals with "Rural expansion" and assigns significant masses to *Urban* (m = 10) and medium masses to *Forest* and *Water* (m = 5) categories. Therefore, in this case, the potential map (Fig. 5, right) highlights more sprawling areas with high potential for change, mainly in the northwestern part of the study area, relatively far from Wroclaw city center.

4 Constraint Geographical CA

Based on the spatiotemporal constraints presented in Sect. 3, a new more integrated formalization of CA can be proposed. In this new formula, the state of a cell i at step $t + 1$ still depends on the state of the cell at step t ($c_{j,\,t}$) and the state of the cells in the neighborhood $\left(V_{i,t}^{r}\right)$. This relation is clearly based on the classical CA definition defined par Tobler or Torrens (Sect. 2). But, it also integrates the MC results limiting land use development to within a number n_t of cells for each land use category. Moreover, the suitability of the simulations calculated by the CA is dependent on the cell's potential P_i based on land use attractiveness masses and distances. A synthetic expression of the model could be written

$$c_{i,t+1} = f\left(c_{i,t}, V_{i,t}^{r}, n_{l,t}, P_i\right)$$

where $n_t =$ was defined in Sect. 3.1 and P_i was defined in Sect. 3.2.

This formula corresponds to the CA process integrated in the LucSim software and was applied to produce two contrasting and realistic scenarios in the case study of Wroclaw in 2006 and 2012. Scenario S1 seeks to densify the more urbanized areas in a pronounced manner. This urban development is thus concentrated around the previously densely built-up areas and seeks to fill the open space corresponding to fields or less dense urban categories. S2 is a peri-urban development scenario which takes place around villages relatively far from the most urbanized areas, where the fields have high potential for change into the less dense urban category.

Moreover, for each scenario, MCs make it possible to quantify future urbanization by 2030 by estimating the number of cells that change from a nonurban to an urban state with a distinction between two categories: urban and dense urban categories for the years 2018, 2024, and 2030 (Table 2). In a second step, two contrasting suitability maps (Fig. 5) constrain the spatial development according to the weighting of each land use category. The resulting suitability maps based on the potential model (Fig. 5) show two contrasting potentials for development: one that is more concentrated around the city core for S1, and the other that is more dispersed in the center and northern part of the case study for S2. Then in a third step, based on the results of steps 1 and 2, the AC could be run according to the three rules set out in Sect. 2.2.

As expected, S1 concentrates on urban development in the southern part of the study area around the core of Wroclaw. This concentration around the core is accompanied by a few outgrowths mainly in the northwestern and northeastern parts of the city. By contrast, S2 reveals a marked expansion in the urban category in the more rural northwestern part of the study area. Urbanization there is less dense and takes on more the form of urban sprawl. As can be seen from the example, LucSim makes it easy to simulate urban development scenarios and their consequences for urban forms, on the one hand reinforcing the compactness of the city and on the other fostering its expansion in rural areas. This kind of modeling process helps in inter-

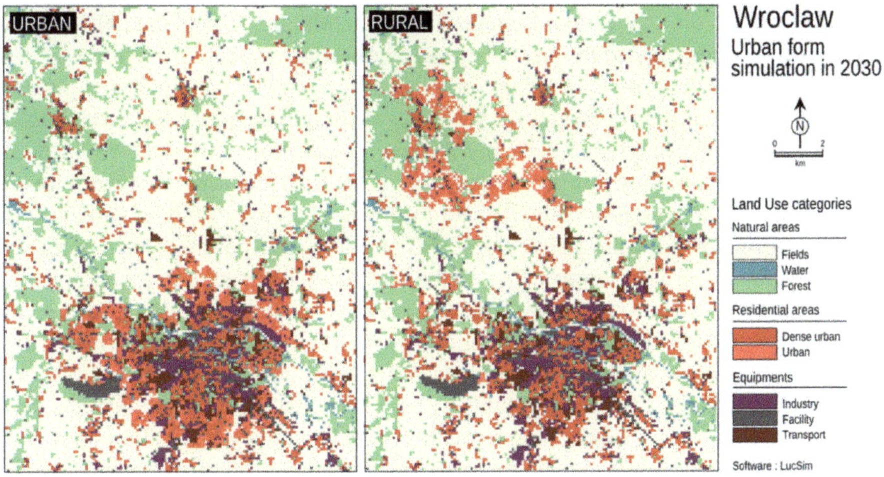

Fig. 6 Simulation of two contrasting scenarios by 2030

actively analyzing the direct consequences of spatial planning policies. Finally, it is also important to emphasize that by integrating space and time constraints, the simulation results are made much more realistic than in the total absence of constraints (Fig. 6).

5 Discussion

The results presented in the previous section are certainly consistent, but they raise a number of questions that call for discussion. First, the CA is driven by a dual constraint system applied to the initial transition rules. However, although there is no prior theoretical, conceptual, and formal incompatibility between the Markovian model and the potential model, some inconsistencies may appear in its actual use. For example, constraints derived from MCs, such as the transition rules themselves, apply at the level of the land use categories. They are dependent on each transition involving two-to-one states. On the other hand, the potential model produces a result applicable for all of these categories. It is therefore independent of transitions and less precise. One solution to overcome this problem would be not to calculate one single potential model, but as many potential models as there are transitions. All these potential models should then be calibrated on the masses of attractiveness corresponding to each transition. Such a solution might be attractive in theory, but in practice, it makes modeling increasingly more complex by multiplying the problems of calibration, which were already questionable in the example discussed here.

The calibration of models also raises a second set of questions. Among the models used here, only MCs can be considered "autonomous" since they automatically produce results from two original images. The potential model and the definition of the transition rules require the use of expert knowledge. In the current state of our knowledge about the growth of urban forms, such recourse is a qualitative input that does not guarantee the reproducibility of the results produced by the modeling in any way. Moreover, insofar as these parameters are not directly derived from mathematical calculation, they appear questionable. This fundamental point is obviously a limitation for the modeling exercise and for the application of mathematical tools for forecasting urban forms. To overcome this problem, many authors have proposed to use "machine learning" approaches so that transition rules are automatically generated based on known past states. Indeed, recent work simulates transitions using decision trees (Samardžić-Petrović et al. 2015) or artificial neural networks (Li and Yeh 2002; Almeida et al. 2008; Tayyebi et al. 2011). Although still very exploratory, the results obtained so far seem promising and, through artificial intelligence processes, they offer an additional step to mathematical and geographical modeling.

But finally, these recent developments also raise the question of the importance of expert knowledge in forecasting urban growth. The results produced in this field are often considered as "images of the future" leading to a collective reflection on the future of territories, rather than final results. Given our difficulty in predicting the future, and the fact that it is very unlikely that this future will be a mere reproduction of the past, it is clear that these results will probably be wrong in the long run (Antoni 2016). Consequently, using expert knowledge to involve local actors in defining a common future does not seem completely absurd. In essence, these reflections on CA calibration ask how far the constraint should be contained. Depending on their objectives, anyone can define the level of constraint they wish to apply to simulate land use change, from a calibration entirely defined by mathematical models or totally derived from expert knowledge. The best way might be a mixed approach combining both machine learning and expert knowledge.

6 Conclusion

After having shown the necessity for constraining CA in the study of urban forms, this paper has proposed a mathematical formalization for specific geographic cellular automata (GCA) implemented in LucSim software. At this level of detail, such a formalization adapted to the social sciences is rare in the literature and therefore appears as one of the fundamental originalities of this chapter. In particular, it aims to link the cellular design of geographical space, the Markovian approach to transition processes, the distance weighting included in gravity models (potential), and the phenomenon of emergence that defines artificial intelligence models in a single formal notation. That's not so bad! In addition, the results produced using this set of methods and models appear quite realistic and are able to correctly reproduce a credible process of urban growth. But at the same time, this reproduction remains open to the intricacy of planning scenarios and allows us to consider a wider use of CA in the framework of a more operational territorial forecast. And that's even better!

Finally, in a more general way, this chapter also shows that it is worth transferring methods developed in mathematics, physics, computer sciences, or mathematics to the social sciences. This transfer obviously requires a substantial effort of abstraction and what may be considerable investment for researchers or developers who are not immediately comfortable with mathematical tools. But since the quantitative revolution started by geographers in the 1960s (Burton 1963), this approach is currently enabling us to work the latest advances in artificial intelligence for decision support into urban and land use planning.

References

Abler, R., Adams, J.S., Gould, P. (1972) *Spatial Organization. The Gepgrapher's Viaw of the World*, Prentice/Hall International, 587 p.

Almeida, C., Gleriani, J., Castejon, E., Soares-Filho, B. (2008) Using neural networks and cellular automata for modelling intra-urban land-use dynamics. *International Journal of Geographical Information Science*, 22(9), 943–963.

Antoni, J.P. (2008) *Lexique de la ville*, Ellipses, 184 p.

Antoni, J.P. (2016) *Concepts, méthodes et modèles pour l'aménagement et les mobilités: l'aide à la décision face à la transition eco-energetique*. Economica, 250 p.

Antoni, J.P., Judge, V., Vuidel, G., Klein, O. (2017) "Constraint Cellular Automata for Urban Development Simulation: An Application to the Strasbourg-Kehl Cross-Border Area." In: Camacho Olmedo M.T., Paegelow M., Mas J.F., Escobar F. (Eds.) *Geomatic Approaches for Modeling Land Change Scenarios*, Springer, pp. 293–306.

Arsanjani, J.J., Helbich, M., Kainz, W., Ali Darvishi, B. (2013) Integration of logistic regression, Markov chain and cellular automata models to simulate urban expansion. *International Journal of Applied Earth Observation and Geoinformation*, 21, 265–275.

Batty, M. (1997) Cellular automata and urban form: A primer. *Journal of the American Planning Association*, 63:2, 266–274.

Batty, M. (2007) Cities and Complexity: Understanding Cities with Cellular Automata, Agent-Based Models, and Fractals, MIT Press, 592 p.

Benenson, I., Torrens, P.M. (2004) *Geosimulation: Automata-Based Modelling of Urban Phenomena*. John Wiley, London, UK.

Berchtold, A. (1998) *Chaînes de Markov et modèles de transition. Applications aux sciences sociales*. Hermes Science Publications, Paris.

Burton, I. (1963) The quantitative revolution and theoretical geography. *The Canadian Geographer*, 7, 151–162.

Clarke, K.C. (2008) *A Decade of Cellular Urban Modeling with SLEUTH: Unresolved Issues and Problems*, Ch. 3 in Planning Support Systems for Cities and Regions (Ed. Brail R. K., Lincoln Institute of Land Policy), Cambridge, MA, pp 47–60.

Conway, J. (1970) Mathematical games, *Scientific American*, October, 120–127.

Couclecis, H. (1985) Cellular world: A framework for modelling micro-macro dynamics. *Environment and Planning A*, 17, 585–596.

Couclecis, H. (1988) Of mice and men: What rodent populations can teach us about complex spatial dynamics. *Environment and Planning A*, 20, 99–109.

Feller, W. (1968) *An Introduction to Probability Theory and Its Applications*. 3rd edition. Wiley.

Helvig, M. (1964) *Chicago's external Truck Movements: Spatial Interactions between the Chicago Area and its Hinterland*, University of Chicago, Department of Geography, Research paper, 90, 132 p.

Li, X., Yeh, A.G.O. (2002) Neural-network-based cellular automata for simulating multiple land use changes using GIS. *International Journal of Geographical Information Science*, 16(4), 323–343.

Neumann (von), J. (1963) The General and Logical Theory of Automata. *Collected Work*, no. 5, Taub, A.H.

Putman, S.H. (1983) *Integrated Urban Models. Policy Analysis of Transportation and Land-Use*. Pion Limited.

Samardžić-Petrović, M., Dragićević, S., Bajat, B., Kovačević, M. (2015) Exploring the decision tree method for modelling urban land use change. *Geomatica*, 69(3), 313–325.

Tayyebi, A., Pijanowski, B.C., Tayyebi, A.H. (2011) An urban growth boundary model using neural networks, GIS and radial parameterization: An application to Tehran, Iran. *Landscape and Urban Planning*, 100(1), 35–44.

Tobler, W.R. (1979) "Cellular Geography." In: Gale S., Olsson G., *Philosophy in Geography*. Reidel Pub., Dordrecht, Holland, pp. 379–386.

Torrens, P.M. (2000) How cellular models of urban systems work (1. Theory), *CASA Working Paper Series*.

White, R., Engelen, G. (1993) Cellular automata and fractal urban form: A cellular modelling approach to the evolution of urban land use patterns. *Environment and Planning A*, 25, 1175–1199.

Wolfram, S. (1983) Statistical mechanics of cellular automata. *Review of Modern Physics*, 55, 601–644.

Wolfram, S. (2002) *A New Kind of Science*. Champaign, 549 p.

Mathematical Foundations of Cellular Automata and Complexity Theory

Keith C. Clarke

Abstract This chapter covers the mathematical foundations of cellular automata (CA) and complexity theory as the basis for models of urban forms and processes. CA have been used for both urban growth models and for more general simulation of land use changes, especially those driven by urban expansion. The chapter covers the origins of CA in the work on von Neumann and others, of complexity and emergence, some theory of CA behavior, and of the general form of CA models of urban systems. Particularly, research has paid attention to the methods for calibration of urban models using real data and of using such data to extract the rules governing state changes in CA. CA models have been shown to be a good combination of being light in data demands, yet accurate and robust in modeling and forecasting.

Cellular automata (CA) are discrete models of practical value in computer science, mathematics, physics, complexity science, theoretical biology and geography. Their applied mathematical nature largely hides a significant amount of theory and the mathematical foundation upon which CA are based, whose origins lie firmly in the work of John von Neumann as formal models of self-reproducing organisms (Sarkar 2000). Von Neumann was a masterful architect of the theory behind applied mathematics, with major contributions in quantum theory, game theory, nuclear weapons, economics, and weather prediction, to name but a few (Ulam 1958). His work on CA and his theory of self-reproducing automata date from the 1940s, and represent five of his publications covering the period from 1949 to 1957. Important among these was a series of lectures delivered at the University of Illinois, which unfortunately remained unpublished at the time of his death at the age of 53 in 1957. Two of the unpublished works were edited and completed by Arthur W. Burks and published as the "Theory of Self-Reproducing Automata" (von Neumann 1963). Part 1 of this book introduced modern complexity theory under the title "The Role of High and Extremely High Complication," while Part 2 (the original lecture 5) advanced a for-

K. C. Clarke (✉)
University of California, Santa Barbara, USA
e-mail: kclarke@geog.ucsb.edu

L. D'Acci (ed.), *The Mathematics of Urban Morphology*,
Modeling and Simulation in Science, Engineering and Technology,
https://doi.org/10.1007/978-3-030-12381-9_8

mal model of self-reproducing automata. These writings remain as the true origin point of CA.

Of the author, Wolfram (2002, p. 876) notes "Von Neumann appears to have believed—presumably in part from seeing the complexity of actual biological organisms and electronic computers—that something like this level of complexity would inevitably be necessary for a system to exhibit sophisticated capabilities such as self-reproduction." There is no doubt that Von Neumann was interested in how complex biological systems were capable of reproduction, and how a formalization of such a system would be the mathematical equivalent of Watson and Crick's discovery of the structure of deoxyribonucleic acid (DNA) (Watson 1968) in 1953. However, most accounts offer Alan Turing's work on self-reproducing machines in 1936 and finite difference approximation methods dating from the early 1900s as pre-cursors to Von Neumann's work. Turing machines were based on arbitrary operations on sequences of discrete elements, an abstract machine which manipulates symbols on a strip of tape according to rules, with the eventual goal of creating its own assembly instructions. As Wolfram (2002, p. 876) notes, however, "von Neumann was interested in something more profound: construction universality and evolution" with the eventual goal being a demonstration of the logical requirements for self-replication. Von Neumann described a universal constructor as an abstract mathematical model of a physical universal assembler, anticipating general-purpose computer programming languages. Thus, CA's origins are also embedded within the invention and improvement of a general-purpose computer, and hence computer science.

Important in Von Neumann's motivation was his dissatisfaction with the prevalent mathematical methods that used partial differential calculus to solve problems such as the aerodynamics of air flow. In his foreword, Burk (Von Neumann 1966, p. 2) notes that Von Neumann "soon found that existing analytical methods were inadequate for obtaining even qualitative information about the solutions of non-linear partial differential equations in fluid dynamics." While his own experience was in hydrodynamics, the use of the Navier–Stokes equations governing flight shared the common problem and invoked a common solution—the use of physical (and eventually digital) simulation. Von Neumann considered flight and water flow simulators to be computers, all be it analog approximations. Burk (Von Neumann 1966, p. 3) further notes that this approach replaces or augments experimentation in science. Subsequent history has proven the power of this assessment.

Von Neumann's work on the universal constructor led him to models based on 3D component "factories" with processes governed by partial differential equations. Von Neumann's self-replicating machine had three parts: a blueprint for itself, a process to read any blueprint and construct the machine described by the blueprint, and a copy machine to make copies of any blueprint. After the blueprint reader is used to construct the machine specified by the blueprint, the copy machine creates a copy of the blueprint. The copy is then placed into the new machine, yielding a perfect replication of the original. In 1951, Von Neumann followed a recommendation by Stanislaw Ulam to simplify his abstraction into a 2D cellular grid with 29 possible classes for each cell, and rules that simulated the components of an electronic computer and other machines (Wolfram 2002). Von Neumann's lecture had suggested a 200,000 cell configuration that could be proven to self-replicate, though the definitive proof remained to be demonstrated. Such proof remained elusive in the productive phase of CA research that followed in the 1970s and 80s. Nobili and Pesavento (1996) finally published the first completely self-reproducing CA in 1995, using a 32-state CA instead of von Neumann's original 29-state specification. Eight years later, Mange et al. (2004) reported an implementation of a self-replicator that is fully consistent with von Neumann's original intent.

There are many formal specifications of CA models. While 2D CAs have dominated applications, Wolfram (2002) has argued that only a 1D CA is necessary to replicate almost any mathematical formalization, process, or rule. The dominance of GIS and its raster data structures have allowed a large amount of gridded data to be compiled that is highly suitable for 2D CA, and the simple elegance of the array data structure in common programming languages is a natural format for coding CA. Using the formulation of Şalap-Ayça et al. (2018) and assuming a 2D model, within a set of regular cells, the cellular automaton, A, is defined by the set of states, S, that changes in discrete time steps according to transition rules, T, applied within the invariant neighborhood, N. The CA then begins with an initial configuration of the states S within the set of cells, and applies the change rules for each additional time step.

The basic formulation of a cellular automaton can be written as:

$$\mathrm{A} \sim (\mathrm{S}, \mathrm{T}, \mathrm{N})$$

To simulate growth, the cell's state is guided by the transition rules, which quantify the local environment and link the spatial patterns to the underlying CA process. The transition rules are applied within the cell's neighborhood with uniform temporal increments. The rules need not necessarily be applied uniformly to all cells since the evolution does not encapsulate all states (for example water bodies in land use change modeling). It is the rules that capture the intrinsic variability of the model's nature, and CA model design is highly sensitive to the transition rules (White and Engelen 1993; Torrens and O'Sullivan 2001; Ménard and Marceau 2005; Kocabas and Dragićević 2006; Samat 2006; Yeh and Li 2006; Pan et al. 2010; Pontius and Neeti 2010). Numerous studies have used techniques from data mining and statistics to extract the rules from observed changes, which remains the topic of intense research (Clarke 2014).

Another consequence of applications and mathematical developments in CA, like Conway's game of life (Gardner 1972), was to provide a simple platform for experiments in the new field of Complex Systems (Waldrop 1993). Foremost among scholars working on the computational aspects of CA was John Holland at the University of Michigan (Holland 1998). Holland described bottom-up self-organized systems called complex adaptive systems. A complex adaptive system is a system in which an understanding of its individual parts does not allow a perfect understanding of the whole system's behavior. Such systems have been successful in modeling both natural and human systems because they allow for heterogeneous agents, phase transitions, and emergent behavior. Emergence is the observation of system behavior that was not a logical outcome of the inputs and structures of the model itself. In one CA model, for example, urban development is attracted to roads and allowed to move along them. Emergence is revealed when road intersections become attractors of growth from multiple directions, and then grow as new cities around the intersections.

Holland's thesis was "that the study of emergence is closely tied to [the] ability to specify a large, complicated domain via a small set of laws." Holland shows that the language of constrained generating procedures (CGPs), and in particular those that allow for variable inter-connections (CGP-Vs) can formally describe these diverse models. Using this approach, models such as CA can simulate extremely complex behavior from the simplest of models. Agents translate directly to cells in CA. For example, in the game of life, aggregate behavior can be random then become stable, as structures such as oscillators are created, indicating a system-wide phase transition. Similarly, extremely complicated and subtle behavior can emerge, such as the CA game of life form Gemini which creates a copy of itself while destroying its parent (Aron 2010). This pattern replication takes 34 million iterations, and uses an instruction tape made of gliders which oscillate between two stable configurations.

Stephen Wolfram has also worked extensively to further the mathematical theory behind CA, almost exclusively with the simplest 1D CAs, where each cell has only two neighbors. He proposed a set of possible CA forms with their behavior types. In Class 1, most patterns rapidly evolve into a stable set, eliminating randomness. In Class 2, most patterns rapidly evolve into an oscillating structure, with some remaining randomness. In Class 3, most patterns evolve into pseudorandom or chaotic structures. Regular structures in Class 3 are quickly broken up by randomness, which quickly propagates through the system. In Class 4, most initial distributions evolve into structures that interact in complex and interesting ways. Wolfram has proposed that many Class 4 cellular automata are capable of universal computation. More recently, in a literature survey of CA models of urban growth, Sante et al. (2010) classified geographical CA transition rules. Type I rules are those of classical CA, where transitions can only occur based on the states of neighboring cells. Type II rules are based on the probabilities of change being altered by the status of a cell. Type III rules change the CA states based on shape or the existence of a network, such as roads. Type IV rules use machine learning methods to determine the rules from prior system behavior, such as Case-Based Reasoning, neural networks, data mining, and kernel-based methods. Type V rules use fuzzy logic and reasoning about uncertainty, while Type VI rules are those not compatible with Types I–V.

The spread of CA theory into geography was primarily in modeling and simulating patterns of urban growth and later, land use changes. Tobler (1979) introduced "cellular geography", and important contributions from geographers followed (Tobler 1979; Couclelis 1997; Batty et al 1997). The 1990s saw a rapid increase in formal models based on CA for simulating urban growth (Batty et al. 1999; Clarke et al. 1997; Clarke and Gaydos 1998) and later, land use change, for example those models reviewed by Agarwal et al. (2001) and Verburg et al. (2004). Application of CA models to various cities worldwide was paralleled by developments in the CA models themselves, variations in their assumptions and input data, and in particular, derivation of the rules governing transitions, and the move toward more objective calibration of the models. Beyond these areas, CA have been applied in remote sensing, transportation modeling, economics, evacuation and pedestrian modeling, quantum dot particle theory, metallurgy, encryption, electrical engineering, ecology, and many more disciplines. There are several journals and conferences that specialize in CA, including the *Journal of Cellular Automata*, and the conference series ACRI (International Conference on Cellular Automata for Research and Industry).

Considerable current research has focused on the means of automatically deriving CA rules from single or multiple images reflecting states at a given time. CA model calibration primarily consists of determining what rules govern change, and how, when and where the changes happen. For example, land use change has well-known drivers, and these are used to modify the rule behavior. Some methods use a single land use map, and calibrate the likelihood of different spatial combinations of cells to change states, based on geographic variables such as proximity and adjacency. Other methods use a land use pattern to simulate change probabilities based on one observed set of changes (computed from two maps at different dates) and relate what changes occur to the factors enumerated at that cell or in its vicinity. Methods that determine and calibrate rules include multi-criterion evaluation (Wu and Webster 1998; Wu 1998), logistic regression (Wu 2002), boosted regression trees (Altartouri and Jolma 2013; Altartouri et al. 2015), decision trees (Li and Yeh 2004), and neural networks (Yang and Li 2007). Some models use neural networks not just for calibration but as the determinants for the land use change model itself (e.g., Li and Yeh 2002). Logistic regression has been used to evaluate and choose CA transition rules during model design (Long et al. 2009; Hu and Lo 2007; Liu and Phinn 2003). Artificial neural networks have also been used (Guan et al. 2005; Basse et al. 2014), and support vector machines for supervised learning to forecast cell states (Yang et al. 2008). Others have used neural networks to optimize CA control parameters (Li and Yeh 2004). In addition, particle swarm optimization (Feng et al. 2011), naïve Bayes (Altartouri and Jolma 2013) and ensemble learning (Gong et al. 2012) have also been applied to both model calibration and rule selection.

CA modeling, especially of land use and land cover change in geography and environmental science, has undergone a surge of interest and application over the past decades (Batty 2005; Torrens and O'Sullivan 2001). The CA models have proven versatile, robust, and accurate. While major advances have been made in model design, and in calibration and validation, the central mathematical core of CA remains very much in the tradition of its simple mathematical foundations and origins in complexity theory. Some themes remain active in research: synchronous versus nonsynchronous update, varying the shape of the tessellation to include both regular (triangles, hexagons, and other shapes) and irregular cells, such as census units and counties. Applications abound in many fields of inquiry, and in urban and land use change modeling only agent-based models have seen as many applications. Few fields of applied mathematics have generated such an extraordinarily varied suite of applications, or such links across disciplines as this relatively recent branch of applied mathematics.

References

Aaron, Jacob (2010). First replicating creature spawned in life simulator. *New Scientist*. June 12th. Retrieved 12 October 2013.

Agarwal C., Green G.M., Grove J.M., Evans T.P., Schweik C.M. (2001). *A Review and Assessment of Land-Use Change Models. Dynamics of Space, Time, and Human Choice.* Bloomington and South Burlington, Center for the Study of Institutions, Population, and Environmental Change, Indiana University and USDA Forest Service. CIPEC Collaborative Report Series 1.

Altartouri, A., Jolma, A., (2013) A naive Bayes classifier for modeling distributions of the common reed in Southern Finland. Modelling and Simulation Society of Australia and New Zealand. In: Piantadosi, J., Anderssen, R.S., Boland, J. (Eds.), *Proceedings of MODSIM2013, the 20th International Congress on Modelling and Simulation*, December 2013, Adelaide, Australia, 1645–1651. ISBN: 978-0-9872143-3-1.

Altartouri, A., Nurminen, L., and Jolma, A. (2015) Spatial neighborhood effect and scale issues in the calibration and validation of a dynamic model of Phragmites australis distribution—A cellular automata and machine learning approach. *Environmental modeling and software.* 71, 15–29. https://doi.org/10.1016/j.envsoft.2015.04.010.

Basse, R. M, Omrania, H, Charifa, O, Gerber, P., and Bódisc, K. (2014) Land use changes modelling using advanced methods: Cellular automata and artificial neural networks. The spatial and explicit representation of land cover dynamics at the cross-border region scale. *Applied Geography*, 53, 160–171. https://doi.org/10.1016/j.apgeog.2014.06.016.

Batty, M., Xie, Y., and Sun, Z. (1999) Modeling urban dynamics through GIS-based cellular automata, *Computers, Environment and Urban Systems*. 23, 205–233.

Batty, M. (2005) *Cities and Complexity: Understanding Cities with Cellular Automata, Agent-Based Models, and Fractals*. Cambridge, MA: MIT Press.

Batty, M., Couclelis, H, and Eichen, M. (1997) Editorial: urban systems as cellular automata'. *Environment and Planning B: Planning and Design* 24 159–164.

Clarke, K. C. (2014) Cellular Automata and Agent-Based Models. Chapter 62 in Fischer, M. M., and Nijkamp, P. (eds) *Handbook of Regional Science*. Springer-Verlag, Berlin Heidelberg.

Clarke, K. C., Hoppen, S., and Gaydos, L. J. (1997). A self-modifying cellular automaton model of historical urbanization in the San Francisco Bay area. *Environment and Planning B*, 24, 247–261.

Clarke, K. C., & Gaydos, L. J. (1998). Loose-coupling of a cellular automaton model and GIS: long-term growth prediction for the San Francisco and Washington/Baltimore. *International Journal of Geographical Information Science*, 12, 699–714.

Couclelis, H. (1997) From cellular automata to urban models: new principles for model development and implementation. *Environment and Planning B*. 24 (2), 165–174.

Feng, Y., Liu, Y., Tong, X., Liu, M., and Deng, S. (2011) Modeling dynamic urban growth using cellular automata and particle swarm optimization rules. *Landscape and Urban Planning* 102(3): 188–196. http://dx.doi.org/10.1016/j.landurbplan.2011.04.004.

Gardner, M. (1972) Mathematical Games. *Scientific American*, 226, 114–118.

Gong, Z., Tang, W., Thill, J.C. (2012) Parallelization of ensemble neural networks for spatial land-use modeling. In *Proceedings of the 5th ACM SIGSPATIAL international workshop on location-based social networks*. ACM, 48–54.

Guan, Q., Wang, L., and Clarke, K. C. (2005) An Artificial-Neural-Network-based, Constrained CA Model for Simulating Urban Growth. *Cartography and Geographic Information Science*. 32(4):369–380.

Holland J. H. (1998) *Emergence: From Chaos to Order*. Addison-Wesley, Redwood City, CA.

Hu, Z., and Lo, C. P. (2007) Modeling urban growth in Atlanta using logistic regression. *Computers, Environment and Urban Systems*. 31, 6, 667–688.

Kocabas, V., and Dragićević, S., 2006. Assessing cellular automata model behaviour using a sensitivity analysis approach. *Computers, Environment and Urban Systems*, 30, 6, 921–953. https://doi.org/10.1016/j.compenvurbsys.2006.01.001.

Li, X., and Yeh, A.G.O. (2002) Neural-network-based cellular automata for simulating multiple land use changes using GIS. *International Journal of Geographic Information Science* 16, 4, 323–343.

Li, X, Yeh, A.G.O. (2004) Data mining of cellular automata's transition rules. *International Journal of Geographical Information Science* 18, 723–744.

Liu, Y., and Phinn, S. R. (2003) Modelling urban development with cellular automata incorporating fuzzy-set approaches. *Computers, Environment, and Urban* Systems 27:637–658.

Long, Y., Mao, Q., and Dang, A. (2009) Beijing urban development model: Urban growth analysis and simulation. Tsinghua Science and Technology 14(6):782–794.

Mange, D., Stauffer, A., Peparaolo, L., and Tempesti, G. (2004), A Macroscopic View of Self-replication, *Proceedings of the IEEE*, 92 (12): 1929–1945. https://doi.org/10.1109/jproc.2004.837631.

Ménard, A., and Marceau, D.J. (2005) Exploration of spatial scale sensitivity in geographic cellular automata. *Environment and Planning B*, 32, 5, 693–714. https://doi.org/10.1068/b31163.

Nobili, R., and Pesavento, U. (1996), Generalised von Neumann's Automata, in Besussi, E.; Cecchini, A., *Proceedings, Artificial Worlds and Urban Studies*, Conference 1 (PDF), Venice: DAEST.

Pan, Y., Roth, A., Yua, Z., and Doluschitz, R. (2010) The impact of variation in scale on the behavior of a cellular automata used for land use change modeling. *Computers, Environment and Urban Systems*, 34 (5), 400–408. https://doi.org/10.1016/j.compenvurbsys.2010.03.003.

Pontius, R.G., and Neeti, N., 2010. Uncertainty in the difference between maps of future land change scenarios. Sustainability Science, 5 (1), 39–50. https://doi.org/10.1007/s11625-009-0095-z.

Şalap-Ayça, S., Jankowski, P., Clarke, K. C., Kyriakidis, P. C., and Nara, A. (2018): A meta-modeling approach for spatio-temporal uncertainty and sensitivity analysis: an application for a cellular automata-based urban growth and land-use change model, *International Journal of Geographical Information Science*, https://doi.org/10.1080/13658816.2017.1406944.

Samat, N. (2006) Characterizing the scale sensitivity of the cellular automata simulated urban growth: a case study of the Seberang Perai Region, Penang State, Malaysia. *Computers, Environment and Urban Systems*, 30, 6, 905–920. https://doi.org/10.1016/j.compenvurbsys.2005.11.002.

Sante, I., Garcia, A. M., Miranda, D., and Crecente, R. (2010) Cellular automata models for the simulation of real-world urban processes: A review and analysis. *Landscape and Urban Planning*. 96, 108–122.

Sarkar, P. (2000) A Brief History of Cellular Automata. *ACM Computing Surveys*, 32, 1, 80–107.

Tobler, W. R. (1979), Cellular Geography, pp 379–386 of Olsson & Gale, eds., *Philosophy in Geography*, Riedel, Dortrecht.

Torrens, P. T., and O'Sullivan, D. (2001) Cellular automata and urban simulation: where do we go from here? *Environment and Planning B*, 28, 2, 163–68.

Ulam, S. (1958) John Von Neumann 1903–1957. *Bulletin of the American Mathematical Society.* 64, 1. https://doi.org/10.1090/s0002-9904-1958-10189-5.

Verburg, P. H., Schot, P. P., Dijst, M. J. Paul P., and Veldkamp, A. (2004) Land use change modelling: current practice and research priorities. *Geojournal*. 61: 309. https://doi.org/10.1007/s10708-004-4946-y.

von Neumann, John; with Burks, Arthur W. (1966), *Theory of Self-Reproducing Automata*. (online), University of Illinois Press, retrieved 2017-02-28.

von Neumann, John (1963) *Collected Works of John von Neumann*, Taub, A. H., ed., Pergamon Press. ISBN 0-08-009566-6.

Wagner D. F., (1997) Cellular automata and geographic information systems. *Environment and Planning B*, 24, 219–234. 1997.

Waldrop. M. M. (1993) *Complexity: The Emerging Science at the Edge of Order and Chaos*. Simon and Schuster, New York.

Watson, J. D. (1968) *The Double Helix: A Personal Account of the Discovery of the Structure of DNA* Atheneum, 1980, ISBN 0-689-70602-2, OCLC 6197022.

White, R., Engelen, G. (1993) Cellular automata and fractal urban form: a cellular modelling approach to the evolution of urban land-use patterns" *Environment and Planning A* 25 1175–1199.

Wolfram, S. (2002). *A New Kind of Science.* (online). Champaign, IL: Wolfram Media, Inc. ISBN 1-57955-008-8. OCLC 47831356.

Wu, F., and Webster, C. J. (1998), Simulation of land development through the integration of cellular automata and multicriteria evaluation, *Environment and Planning B*, 25, 1, 103–126.

Wu, F. (1998) Simulating urban encroachment on rural land with fuzzy-logic-controlled cellular automata in a geographical information system. *Journal of Environmental Management.* 53, 4, 293–308.

Wu, F. (2002) Calibration of stochastic cellular automata: the application to rural-urban land conversions. *International Journal of Geographical Information Science* 16, 8, 795–818. https://doi.org/10.1080/13658810210157769.

Yang Q.S., Li, X. (2007) Calibrating urban cellular automata using genetic algorithms. *Geographical Research* 26, 2, 229–237.

Yang Q, Li X, Shi X (2008) Cellular automata for simulating land use changes based on support vector machines. *Computers and Geosciences* 34(6):592–60.

Yeh, A. G.-O., and Li, X., (2006) Error propagation and model uncertainties of cellular automata in urban simulation with GIS. *Computers, Environment and Urban Systems*, 30 (1), 10–28. https://doi.org/10.1016/j.compenvurbsys.2004.05.007.

Part III
Spatial Networks and Space Syntax

Assessing Complexity of Urban Spatial Networks

Dimitri Volchenkov

Abstract In the present chapter, we discuss the graph representations of urban spatial patterns (maps) and suggest a computationally feasible technique for understanding urban forms based on scale-dependent random walks that can be used in order to spot the relatively isolated locations and neighborhoods, to detect urban sprawl, and to illuminate the hidden community structures in complex urban textures. The approach may be implemented for the detailed expertise of any urban pattern and the associated transport networks that may include many transportation modes.

1 "We Shape Our Buildings; Thereafter They Shape Us"

A belief in the influence of the built environment on humans was common in architectural and urban thinking for centuries.[1] Cities generate more interactions with more people than rural areas because they are central places of trade that benefit those who live there. Spatial organization of a place has an extremely important effect on the way people move through spaces and meet other people by chance (Hillier 1984). Compact neighborhoods can foster casual social interactions among neighbors, while creating barriers to interaction with people outside a neighborhood. Spatial

[1] "*We Shape our Buildings; thereafter They Shape Us*". In his speech to the meeting in the House of Lords, October 28, 1943, W. Churchill had requested that the House of Commons bombed out in May 1941 be rebuilt exactly as before, since the configuration of space and even its scarcity in the House of Commons played a greater role in effectual parliament activity. In his view, "giving each member a desk to sit at and a lid to bang" would be unreasonable, since "the House would be mostly empty most of the time; whereas, at critical votes and moments, it would fill beyond capacity, with members spilling out into the aisles, giving a suitable sense of crowd and urgency" (Churchill 2019). The old House of Commons was rebuilt in 1950 in its original form, remaining insufficient to seat all its members.

D. Volchenkov (✉)
Department of Mathematics and Statistics, Texas Tech University, Broadway and Boston, Lubbock, TX 79409-1042, USA
e-mail: dimitri.volchenkov@ttu.edu

Sichuan University of Science and Engineering, Sichuan, Zigong, China

L. D'Acci (ed.), *The Mathematics of Urban Morphology*,
Modeling and Simulation in Science, Engineering and Technology,
https://doi.org/10.1007/978-3-030-12381-9_9

configuration promotes peoples encounters as well as making it possible for them to avoid each other, shaping social patterns (Ortega-Andeane et al. 2005). The phenomenon of clustering of minorities, especially that of newly arrived immigrants, is well documented since the work of (Wirth 1928) (the reference appears in Vaughan (2005)). Clustering is considered to be beneficial for mutual support and for the sustenance of cultural and religious activities.

The study of London's change over 100 years performed by Vaughan et al. (2005) has indicated that the creation of poverty areas is a spatial process: by looking at the distribution of poverty at the street, it is possible to find a relationship between spatial segregation and poverty. The patterns of mortality in London studied over the past century by Orford et al. (2002) also show that the areas of persistence of poverty cannot be explained other than by an underlying spatial effect.

In the following sections of this chapter, we propose the quantitative technique for describing urban forms that allows us to study the urban structure in detail, and to measure structural isolation of city neighborhoods, in particular. Our approach (Blanchard and Volchenkov 2009; Volchenkov 2018) is based on the analysis of *random walks*, *trajectories*, or *paths* available within the urban structure represented by a connected undirected graph. Humans live and act in Euclidean space which they percept visually as affine space, and which is present in them as a mental form. In another circumstance we spoke of fishes: they know nothing either of what the sea, or a lake, or a river might really be and only know fluid as if it were air around them. While in a complex environment, humans have no sensation of it, but need time to construct its affine representation so they can understand and store it in their spatial memory. Random walks help us to find such an affine representation of the environment, giving us a leap outside our Euclidean aquatic surface and opening up and granting us the sensation of new space (Blanchard and Volchenkov 2009).

2 Spatial Graphs of Urban Environments

In traditional urban researches, the dynamics of an urban pattern come from the landmasses, the physical aggregates of buildings delivering a place for people and their activities. The representation of urban spatial networks by connected graphs can be based on a number of different principles. In the present chapter, we take a "named-streets"-oriented point of view on the decomposition of urban spatial networks into spatial graphs following our previous works (Volchenkov and Blanchard 2007, 2008a). Being interested in the statistics of random walks defined on spatial networks of urban patterns, we assign an individual street ID code to each continuous segment of a street. The spatial graph of an urban environment is then constructed by mapping all continuous segments of streets (*spaces of motion*) into nodes and all intersections between *continuous segments of streets* into edges.

For each graph, there exists a unique *adjacency matrix* (up to permuting rows and columns). If we assume that the spatial graph $G(V, E)$ consisting of $i = 1, \dots, N$ spaces V, some of them are connected by edges $E \subseteq V \times V$, is *simple* (contains neither loops, nor multiple edges), the adjacency matrix is a $\{0, 1\}$-matrix with zeros on its diagonal:

$$A_{ij} = \begin{cases} 1, \; i \sim j, \quad i \neq j, \\ 0, \; \text{otherwise}, \end{cases} \tag{1}$$

where $i \sim j$ means that the space i is directly connected to the space j.

Definition 1 The *degree* (of k-th order) of the vertex $i \in V$ is the number of walks of length n available from i,

$$\deg_n(i) = \#\{W_n(i)\} = \sum_{j \in V} A^n_{ij}, \quad k \geq 1. \tag{2}$$

The 0-order degree of a vertex $i \in V$ accounts for the walk of length 0 at the vertex $i \in V$, which we consider to be identical to the vertex itself, so that $\deg_0(i) = 1$. The 1-order degree of a vertex $i \in V$ is the number of vertices adjacent to i in the graph $G(V, E)$, $\deg_1(i) \equiv \sum_{j \in V} A_{ij}$. Summing over all vertices in the graph, we obtain $\sum_{i=1}^{N} \deg_1(i) = 2E$, where E is the number of edges in the graph.

Let us suppose that the graph G is undirected, so that its adjacency matrix is symmetric $\mathbf{A} = \mathbf{A}^\top$. Let $\mathbf{U}$ is the orthogonal matrix of eigenvectors of the adjacency matrix $\mathbf{A}$. Then the eigen decomposition of the adjacency matrix is $A^k{}_{ij} = \sum_{l=1}^{n} \alpha_l^k \psi_{il} \psi_{jl}$.

Theorem 1 *The number of all walks of length n in the graph G equals*

$$\#W_n \equiv \sum_{i,j \in V} A^n{}_{ij} = \sum_{l=1}^{N} \left(\sum_{i=1}^{N} \psi_{il} \right)^2 \alpha_l^n = \sum_{l=1}^{N} \gamma_l^2 \alpha_l^n, \quad \gamma_l^2 \equiv \left(\sum_{i=1}^{N} \psi_{il} \right)^2 . \tag{3}$$

Let $\mathbf{deg}_n \in \mathbb{Z}^N$ be the vector, whose elements $\deg_n(i)$, $i = 1, \ldots, N$, are the numbers of walks of length n available from $i \in V$. The adjacency operator defines the following (time-forward) dynamical system

$$\mathbf{deg}_{n+1} = \mathbf{A} \, \mathbf{deg}_n, \tag{4}$$

mapping the vector $\mathbf{deg}_n$ into the vector $\mathbf{deg}_{n+1}$, whose elements are the numbers of walks of length $n + 1$ available at the node $i \in V$ (Volchenkov 2018).

2.1 Locally Anisotropic Random Walks on Graphs

The stochastic matrices, viz.,

$$T_{ij}(t+1) = \frac{A_{ij} \, \deg_t(j)}{\deg_{t+1}(i)} = \frac{A_{ij} \sum_{s=1}^{N} A^t_{js}}{\sum_{s=1}^{N} A_{is} \sum_{r=1}^{N} A^t_{sr}}, \quad t \geq 0, \tag{5}$$

define the discrete time locally anisotropic (i.e., direction dependent) nearest neighbor random walks on the graph such that, at each node, the walker picks the available edges that are linked to the node with some (presumably unequal) probability.

Theorem 2 *All possible walks of length $t \geq 1$ starting at every node $i \in V$ of the graph G are chosen with the equal probability in the random walk $T_{ij}(t)$ (Volchenkov 2018).*

Proof The first-order random walk $T(1)_{ij} = A_{ij}/\deg_1(i)$ is *locally isotropic*, since all edges available for such a random walker from every vertex of the graph are chosen with equal probability.

For the random walks of order $t = 2$, the transition probability (5) reads as follows:

$$T_{ij}(2) = \frac{A_{ij}\sum_{s=1}^{N} A_{js}}{\sum_{s=1}^{N} A_{is}\sum_{r=1}^{N} A_{sr}}, \tag{6}$$

so that each walk of length 2 starting at the node i is chosen with equal probability. Although all walks of length 2 starting at the node i are equiprobable under the transition operator (6), the probabilities of transition to the nearest neighbors from the node i might be different. The nearest neighbors having more neighbors than others are preferable. The resulting random walk (6) is a direction-dependent random walk (locally anisotropic). The further conclusion is inductive and self-evident. □

Theorem 3 *The series of locally anisotropic random walks* (5) *converges as $n \to \infty$ to a random walk, in which all infinitely long walks starting at every node of the graph are equally probable, viz.,*

$$\lim_{n\to\infty} T_{ij}(n) = T_{ij}(\infty) = \frac{A_{ij}\psi_{1j}}{\alpha_1\psi_{1i}}, \tag{7}$$

where ψ_1 is the completely positive eigenvector of the graph adjacency matrix $\mathbf{A}$ *belonging to its maximal eigenvalue α_1.*

Proof The N eigenvalues of the graph adjacency matrix $\mathbf{A}$ are assumed to be ordered, such as $\alpha_1 > \alpha_2 \geq \ldots$. The nth-order degree of the node $i \in V$ is

$$\begin{aligned}\deg_n(i) \equiv \sum_j (A^n)_{ij} &= \sum_k \alpha_k^n \psi_{ik} \underbrace{\sum_j \psi_{kj}}_{\gamma_k} \equiv \sum_k \alpha_k^n \gamma_k \psi_{ik} \\ &= \alpha_1^n \gamma_1 \psi_{i1}\left(1 + \sum_{k>1}\left(\frac{\alpha_k}{\alpha_1}\right)^n \frac{\gamma_k}{\gamma_1}\frac{\psi_{kj}}{\psi_{1j}}\right).\end{aligned} \tag{8}$$

In the limit $n \gg 1$, the last sum in (8) is dominated by the largest eigenvalue α_1 of the adjacency matrix, so that $\lim_{n\to\infty} = \alpha_1^n\gamma_1\psi_{i1}$. and therefore,

$$\lim_{n\to\infty} T_{ij}(n) = \frac{A_{ij}\psi_{j1}\gamma_1\alpha_1^{n-1}}{\psi_{i1}\gamma_1\alpha_1^n} = \frac{A_{ij}\psi_{j1}}{\alpha_1\psi_{i1}}. \tag{9}$$

Finally, it is easy to check that the matrix (7) is a stochastic matrix, since $\sum_j A_{ij}\psi_{j1} = \alpha_1\psi_{i1}$. □

Locally anisotropic random walks allow to study the graph structure with respect to the walks of different lengths, $n = 1, \ldots, \infty$, i.e., on a variety of scales.

2.2 Stationary Distributions of Locally Anisotropic Random Walks

For a random walk defined on a connected undirected graph, the Perron–Frobenius theorem asserts the unique strictly positive probability vector $\boldsymbol{\pi} = (\pi_1, \ldots, \pi_N)$, which is the left eigenvector of the transition matrix $\mathbf{T}$ belonging to the maximal eigenvalue $\mu = 1$, viz., $\boldsymbol{\pi}\,\mathbf{T} = 1 \cdot \boldsymbol{\pi}$, is the *stationary distribution* of the random walk on the graph. The element π_i of the stationary distribution is called the *density of the node* in the graph G. The condition of *detailed balance*, $\pi_i T_{ij} = \pi_j T_{ji}$, asserts that a random walk defined on an undirected graph is *time reversible*: it is also a random walk if considered backward. And it is impossible to determine, which state came first and which state arrived later, after running the walk.

Theorem 4 *For the random walk $T_{ij}(n)$, $n \geq 1$, the stationary distribution is*

$$\pi_i(n) = \frac{\deg_n(i)\deg_{n-1}(i)}{2E_n}, \quad 2E_n \equiv \sum_i \deg_n(i)\deg_{n-1}(i), \tag{10}$$

with the assumption of $\deg_0(i) = 1$.

Proof

$$\begin{aligned} \textstyle\sum_i \pi_i(n)\frac{A_{ij}\deg_j(n-1)}{\deg_i(n)} &= \textstyle\sum_i \frac{\deg_n(i)\deg_{n-1}(i)}{2E_n}\frac{A_{ij}\deg_{n-1}(j)}{\deg_n(i)} \\ &= \frac{\deg_n(j)\deg_{n-1}(j)}{2E_n} = \pi_j(n). \end{aligned} \tag{11}$$

The detailed balance condition is satisfied by (10): $\pi_i(n)T_{ij}(n) = \pi_j(n)T_{ji}(n) = \frac{A_{ij}\deg_{n-1}(i)\deg_{n-1}(j)}{2E_n}$. □

For $n = 1$, the normalization factor $2E_1 = 2E$, since $\deg_0(i) = 1$.

Theorem 5 *For the limiting anisotropic random walk $T_{ij}(\infty)$, the stationary distribution is $\pi_i(\infty) = \psi_{1i}^2$ where ψ_1 is the major eigenvector of the graph adjacency matrix $\mathbf{A}$ belonging to the maximal eigenvalue α_1.*

Proof

$$\sum_i \pi_i(\infty)\frac{A_{ij}\psi_{1j}}{\lambda_1\psi_{1i}} = \psi_{1j}^2 = \pi_j(\infty). \tag{12}$$

The detailed balance condition is satisfied by (5)
$\pi_i(\infty)T_{ij}(\infty) = \pi_j(\infty)T_{ji}(\infty) = \frac{A_{ij}\psi_{1i}\psi_{1j}}{\alpha_1}$. □

2.3 Entropy of Anisotropic Random Walks

The number of possible walks of length n on a regular d-dimensional lattice grows up exponentially with the path length n, viz., $W_n = 2^{nd}$. If every walk of length n is chosen with an equal probability by a random walker, the probability to observe a particular walk w of length n decreases exponentially as $\Pr[w \in W_n] = 2^{-nd}$. The probability to observe a random walk passing through the node $i \in V$ in a finite- connected undirected graph is $\Pr[\to i \to] = 2^{\log_2 \pi_i}$, where π_i is the density of the node i with respect to the random walk. Therefore, the probability to find a particular walk $w = \{i_1, i_2, \ldots, i_n\}$ on the graph G is given by $\Pr[i_1, i_2, \ldots i_n] = 2^{\sum_{s=1}^{n} \log_2 \pi_{i_s}}$, and the probability to observe a long enough *typical* random walk of length n decreases asymptotically exponentially with $n \gg 1$, $2^{-n(H+\varepsilon)} \le \Pr[\{i_1, i_2, \ldots i_n\}] \le 2^{-n(H-\varepsilon)}$ where the entropy parameter

$$H = -\sum_{i=1}^{N} \pi_i \log_2 \pi_i, \tag{13}$$

assessing the uncertainty of walks by estimating the spread of random trajectories, and, therefore plays the role of the global physical dimension of the graph G generalizing the lattice dimension. In (13), we assume that $0 \cdot \log(0) = 0$. According to (10), the entropy (13) is also calculated for locally asymmetric random walks of the order $t \ge 1$ as

$$\begin{aligned} H(t) &= -\sum_{i=1}^{N} \pi_i(t) \log_2 \pi_i(t) \\ &= -\sum_{i=1}^{N} \frac{\deg_t(i)\deg_{t-1}(i)}{2E_t} \log_2 \frac{\deg_t(i)\deg_{t-1}(i)}{2E_t}. \end{aligned} \tag{14}$$

In the limit $t \to \infty$, the entropy (14) takes the following form: $H(\infty) = -\sum_{i=1}^{N} \psi_{i1}^2 \log_2 \psi_{i1}^2$, where ψ_1 is the completely positive major eigenvector of the graph adjacency matrix.

The entropy $H(t)$ asymptotically grows at a fixed rate called the *entropy rate*,

$$h(n) \equiv -\sum_{i \in V} \pi_i(n) \sum_{j \in V} T_{ij}(n) \log_2 T_{ij}(n). \tag{15}$$

2.4 The Relative Entropy Rate for Locally Anisotropic Random Walks

The Markov chain $T_{ij}(1)$ defines the locally isotropic random walk, in which the next node of the walk is chosen by a random walker uniformly, with the probability $\Pr = \deg_1^{-1}(i)$ among all nearest neighbors of the node $i \in V$. When performing the locally isotropic random walk, the random walker, metaphorically speaking, needs to know the number of immediate neighbors. In contrast to it, the probability of transition to a neighbor in the higher order random walks $T_{ij}(t)$, $t > 1$, is determined regarding the numbers of lengthy t-walks available from the neighbors. Therefore, performing the locally anisotropic random walk $T_{ij}(t)$, $t > 1$, in general, requires more information than involved in $T_{ij}(1)$.

We introduce the relative entropy rate parameter in order to compare the locally isotropic and anisotropic random walks defined on a finite connected undirected graph. The entropy rate determines the portion of information produced at each step of the locally isotropic random walk, viz.,

$$\begin{aligned} h(1) &= -\sum_i \frac{\deg_1(i)}{2E} \sum_{j\smile i} \frac{1}{\deg_1(i)} \log_2\left(\frac{1}{\deg_1(i)}\right) \\ &= \frac{1}{2E}\sum_i \deg_1(i)\log_2 \deg_1(i) \;=\; \langle \delta_i \rangle \end{aligned} \tag{16}$$

where $\langle \delta_i \rangle$ is the mean local physical dimension at the node $i \in V$, with respect to the locally isotropic random walks $T_{ij}(1)$. The entropy rates of the higher order random walks differ from (16).

Definition 2 The *relative entropy rate* between $T_{ij}(1)$ and $T_{ij}(t)$, $t > 1$, is defined as

$$\mathfrak{h}(t) \;\equiv\; \sum_i \pi_i(1) \sum_j T_{ij}(1)\, \log_2 \frac{T_{ij}(1)}{T_{ij}(t)}, \quad t \geq 1. \tag{17}$$

For $t = 1$, the relative entropy rate $\mathfrak{h}(1) = 0$, and $\mathfrak{h}(t) > 0$ for $t > 1$. The relative entropy rate (17) quantifies the *additional* amount of information generated at each step of the random walk $T_{ij}(t)$, $t > 1$ due to increased knowledge about the numbers of t-paths available at the nearest neighbors of the walker. The following result is self-evident from the definitions (17) and (5):

Theorem 6

$$\begin{aligned} \mathfrak{h}(t) &= \frac{1}{2E}\sum_{i,j\smile i} \log_2 \frac{\deg_t(i)}{\deg_{t-1}(j)\deg_1(i)} \\ &= \frac{1}{2E}\sum_{i,j\smile i}\left(\log_2\left(\frac{\deg_t(i)}{\deg_{t-1}(j)}\right) - \log_2 \deg_1(i)\right) \\ &\equiv \frac{1}{2E}\sum_{i,j\smile i}\left(\Delta_{ij}(t) - \delta_i\right), \quad t \geq 1. \end{aligned} \tag{18}$$

Here, we assume that $\deg_0(i) = 1$ for every $i \in V$, $\delta_i \equiv \log_2 \deg_1(i)$ is the local analog of the physical dimension of space, and

$$\Delta_{ij}(t) \equiv \log_2\left(\frac{\deg_t(i)}{\deg_{t-1}(j)}\right), \quad t \geq 1 \tag{19}$$

is the local *directional-dependent space dimension tensor* which coincides with δ_i for $t = 1$. The cumulative difference, viz., $\zeta_i(t) \equiv \sum_{j \smile i} \left(\Delta_{ij}(t) - \delta_i\right)$, $t \geq 1$, is a natural *measure of the scale-dependent anisotropy* of the higher order random walks (5). The relative entropy rate (18) is the *specific scale-dependent anisotropy* of the random walk per one step of the random walk $T_{ij}(t)$, $t > 1$, viz., $\mathfrak{h}(t) = \frac{1}{2E}\sum_{i=1}^{N} \zeta_i(t)$.

3 Information Decomposition for Markov Chains

Let us work out the example of a Markov chain with two states ('heads' or 'tails') that represents tossing of a unfair coin, in which each state repeats itself with the probability $0 \leq p \leq 1$, as shown in the diagram in Fig. 1 (Volchenkov 2018).

The transition probabilities in the Markov chain shown in Fig. 1 are determined by the matrix

$$\mathbf{T} = \begin{pmatrix} p & 1-p \\ 1-p & p \end{pmatrix}. \tag{20}$$

For $p = 1$ and $p = 0$, this Markov chain generates the constant sequences of symbols, viz., (i) the alternating sequence, $\ldots 0, 1, 0, 1, 0, \ldots$ when $p = 0$; (ii) the sequences of repeating elements, $\ldots 1, 1, 1, 1, 1, \ldots$, or $\ldots 0, 0, 0, 0, 0, \ldots$ when $p = 1$. The chain represents tossing of a fair coin when $p = 1/2$. The information measures for Markov chains depend on the time steps from the past and into the future only.

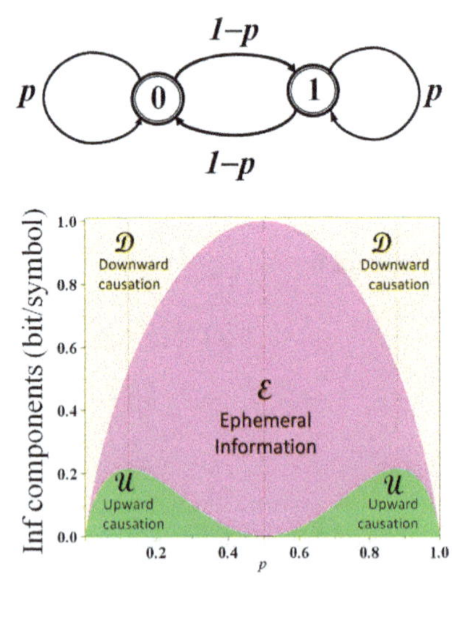

Fig. 1 Information decomposition for tossing a unfair coin

They are simply related to each other by the transition probabilities. Since all information shared between the past and future states in Markov chains goes only through the present, the mutual information between the past states of the chain and its future states conditioned on the present moment is always trivial: $I(X_{t-1}; X_{t+1}|X_t) = 0$. Throwing a unfair coin to choose between two alternative states reveals a single bit of information, for any value of p. The density of states in the Markov chain is determined by the major left eigenvector of the transition matrix (20) belonging to its largest eigenvalue 1, viz., $\pi = [1/2, 1/2]$, so that Pr['head'] = Pr['tail'] = 1/2, independently of the probability $0 \leq p \leq 1$. The Shannon entropy characterizing the uncertainty of a state in the Markov chain is defined by

$$H = -\sum_{i=1}^{2} \pi_i \log_2 \pi_i = -\log_2 \frac{1}{2} = 1 \text{ bit}. \tag{21}$$

Although the value of entropy (21) is independent of p. this single bit of information manifests itself in different forms according to the value of p.

3.1 *Conditional Information Measure for the Downward Causation Process*

When $p = 0$ or $p = 1$, the Markov chain (20) generates the constant sequences of symbols, so that the forthcoming state ("heads" or "tails") is determined by the past states of the chain, i.e., by the *downward causation* process. A single bit of information in the process manifests itself in the long-range structural correlations in the chain (when $p = 0$ or $p = 1$). The long-range structural correlation decays (as $p > 0$ or $p < 1$), so that the symbols may appear occasionally at the "improper" places. The amount of information generated *at each step* of the chain is defined by the conditional entropy per symbol, the *entropy rate*, viz.,

$$h \equiv H(X_{t+1}|X_t) = -\sum_{i=1}^{2} \pi_i \sum_{j=1}^{2} T_{ij} \log_2 T_{ij}. \tag{22}$$

For the Markov chain (20), we obtain $h(p) = -p \log_2 p - (1 - p) \log_2(1 - p)$, which is trivial for $p = 0$ and $p = 1$ when the forthcoming symbol is fully determined by the chain structure.

The strength of long-range structural correlations between blocks of symbols in the Markov chain for $0 < p < 1$ is quantified by the mutual information between the past and future segments of the chain called the *excess entropy*,

$$I(X_{t+1}; X_t) \equiv \mathcal{D} = H(X_t) - H(X_{t+1}|X_t) = H - h. \tag{23}$$

In (23) the letter $\mathcal{D}$ stands for "downward causation", since this measure is associated to the information component related to the *downward causation* process. The excess entropy quantifies the component of information, which is predictable from the long-range structural correlations in the chain for $0 < p < 1$, viz., $\mathcal{D}(p) = 1 + p\log_2 p + (1-p)\log_2(1-p)$. When tossing a fair coin ($p = 1/2$), the structural correlations in the Markov chain (20) vanish ($\mathcal{D}(1/2) = 0$) and the forecast of the future symbols based on observation of the long sequences of past symbols loses any predictive power (see Fig. 1).

3.2 Conditional Information Measure for the Upward Causation Process

Our capability to predict the forthcoming state while tossing a unfair coin weakens as $p > 0$ or $p < 1$ although every cloud has a silver lining. Indeed, we can guess the forthcoming symbol in the chain simply by alternating (or repeating) the current symbol with the probability $p > 0$ (or $p < 1$). The information component quantifying the goodness of such a guess is the *mutual information* between the present state of the chain and its future state conditioned on the past, viz.,

$$I(X_t;\, X_{t+1}|X_{t-1}) \equiv \mathcal{U} = H\left(X_{t+1}|X_{t-1}\right) - H\left(X_{t+1}|X_{t-1}\right). \tag{24}$$

In (24) the letter $\mathcal{U}$ stands for "upward causation" determining the influence of the present symbol on the forthcoming state of the chain: $\mathcal{U}(p) = p\log_2 p + (1-p)\log_2(1-p) - 2p(1-p)\log_2 2p(1-p) - \left(p^2 + (1-p)^2\right)$ $\log_2\left(p^2 + (1-p)^2\right)$. The mutual information $\mathcal{U}(p)$ grows as $p > 0$ ($p < 1$) until $p \approx 0.121$ ($p \approx 0.879$) when the effect of *destructive interference* between the obviously incompatible guesses on alternating the current symbol at the next step (for $p > 0$) and on repeating the current symbol (for $p < 1$) causes the attenuation and complete cancelation of this information component in the case of fair coin tossing (at $p = 1/2, \mathcal{U}(1/2) = 0$) (see Fig. 1).

3.3 Ephemeral Information in Markov Chains

The remaining conditional entropy $H\left(X_t|X_{t+1}, X_{t-1}\right)$ quantifies the portion of uncertainty of a state in the Markov chain (20) that can neither be predicted from the large-scale structure of the chain (i.e., from downward causation), nor be guessed by alternating (or repeating) the current state symbol (i.e., from upward causation). This conditional entropy belongs to the present moment only, viz.,

$$H\left(X_t|X_{t+1},X_{t-1}\right) \equiv \mathcal{E} = h - \mathcal{U}. \tag{25}$$

This information component is *ephemeral*, as existing only in the present moment and *released* at each transition of the chain, being neither a consequence of the past, nor of consequence for the future: $\mathcal{E}(p) = -2p\log_2 p - 2(1-p)\log_2(1-p) + 2p(1-p)\log_2 2p(1-p) + \left(p^2+(1-p)^2\right)\log_2\left(p^2+(1-p)^2\right)$.

We have decomposed one bit of information characterizing uncertainty of a state in the Markov chain of unfair coin tossing into three independent information components, viz.,

$$H \;=\; \mathcal{D}(p)+\mathcal{U}(p)+\mathcal{E}(p) \;=\; 1 \text{ bit}. \tag{26}$$

1. The first component, $\mathcal{D}(p)$, characterizes our capability to predict the forthcoming state of the chain from the past states.
2. The second component, $\mathcal{U}(p)$, characterizes our capability to guess it from the current symbol.
3. The third component, $\mathcal{E}(p)$, characterizes our incapability to predict it.

4 Exploring Graph Structures by Random Walks

The stationary distribution of random walks (5) defined with respect to the equiprobable walks $\{\gamma_t\}$ of length $t \geq 1$ in a connected undirected graph $G(V, E)$ determines a unique measure on V, $\mu_t = \sum_{j\in V}\pi_i(t)\delta_j$, with respect to which the transition operator becomes self-adjoint and is represented by a symmetric transition matrix,

$$\widehat{T_{ij}}(t) \;=\; \left(\pi(t)^{1/2}\,\mathbf{T}(t)\,\pi(t)^{-1/2}\right)_{ij} \tag{27}$$

where $\pi(t)$ is the diagonal matrix of the densities of nodes with respect to the random walk $T_{ij}(t)$. The matrix (27) corresponds to the *normalized Laplace operator*, $\widehat{\mathbf{L}}(t) = \mathbf{1} - \widehat{\mathbf{T}}(t)$ where $\mathbf{1}$ is the unit matrix. From now on, we shall omit the length of walks t from our notations, as the methods we discuss are equally suitable for any random walk operator defined on the graph. The use of self-adjoint operators (27) becomes now standard in spectral graph theory (Chung 1997), and in studies devoted to random walks on graphs (Lovász 1993).

Diagonalizing the symmetric matrix (27), we obtain $\widehat{\mathbf{T}} = \boldsymbol{\Psi}\,\mathbf{M}\,\boldsymbol{\Psi}^\top$, where $\boldsymbol{\Psi}$ is an orthonormal matrix, $\boldsymbol{\Psi}^\top = \boldsymbol{\Psi}^{-1}$, and $\mathbf{M}$ is a diagonal matrix with entries $1 = \mu_1 > \mu_2 \geq \ldots \geq \mu_N > -1$ (here, we do not consider bipartite graphs, for which $\mu_N = -1$). The rows $\boldsymbol{\psi}_k = \left\{\psi_{k,1},\ldots,\psi_{k,N}\right\}$ of the orthonormal matrix $\boldsymbol{\Psi} = \{\boldsymbol{\psi}_1,\boldsymbol{\psi}_2,\ldots\boldsymbol{\psi}_N\}^\top$ are the real eigenvectors of $\widehat{\mathbf{T}}$ that forms an orthonormal basis in Hilbert space $\mathcal{H}(V)$, $\boldsymbol{\psi}_k : V \to S_1^{N-1}$, $k = 1,\ldots N$, where S_1^{N-1} is the $N-1$-dimensional unit sphere. The first eigenvector $\boldsymbol{\psi}_1$ belonging to the largest eigenvalue $\mu_1 = 1$ (which is simple) is the Perron–Frobenius eigenvector that determines the stationary distribution of random walks over the graph nodes,

$\boldsymbol{\psi}_1 \widehat{\mathbf{T}} = \boldsymbol{\psi}_1$, $\psi_{1,i}^2 = \pi_i$, $i = 1, \ldots, N$. The squared Euclidean norm of the vector in the orthogonal complement of $\boldsymbol{\psi}_1$, $\sum_{s=2}^{N} \psi_{s,i}^2 = 1 - \pi_i > 0$, expresses the probability that a random walker is not in i.

4.1 Affine Probabilistic Geometry of Graphs

Discovering of important nodes and quantifying differences between them in a graph is not easy, since the graph does not possess a priori the structure of Euclidean space. However, we can use the algebraic properties of the self-adjoint operators in order to define an Euclidean metric on any finite connected undirected graph.

Geometric objects, such as points, lines, or planes, can be given a representation as elements in projective space based on *homogeneous coordinates*. Given an orthonormal basis $\left\{\boldsymbol{\psi}_k : V \to S_1^{N-1}\right\}_{k=1}^{N}$ in $\mathbb{R}^N$, any vector in Euclidean space can be expanded into $\mathbf{v} = \sum_{k=1}^{N} \langle \mathbf{v}|\psi_k\rangle \langle \psi_k|$. Provided $\{\boldsymbol{\psi}_k\}_{k=1}^{N}$ are the eigenvectors of the symmetric matrix of the operator $\widehat{\mathbf{T}}$, we can define the new basis vectors, $\boldsymbol{\Psi}' \equiv \left\{1, \frac{\psi_{2,2}}{\psi_{1,2}}, \ldots, \frac{\psi_{N,N}}{\psi_{1,N}}\right\}$, since we have always $\psi_{1,i} \equiv \sqrt{\pi_i} > 0$ for any $i \in V$. The basis vectors $\boldsymbol{\psi}'$ span the projective space $P\mathbb{R}_{\pi}^{(N-1)}$, so that the vector $\mathbf{v}$ can be expanded into $\mathbf{v}\boldsymbol{\pi}^{-1/2} = \sum_{k=2}^{N} \langle v|\psi_k'\rangle \langle \psi_k'|$. This transformation defines a stereographic projection on $P\mathbb{R}_{\pi}^{(N-1)}$ such that all vectors in $\mathbb{R}^N(V)$ collinear to the vector $|\,\psi_1\rangle$ corresponding to the stationary distribution of random walks are projected onto a common image point. If the graph $G(V, E)$ has some isolated nodes $\iota \in V$, for which $\pi_\iota = 0$, they play the role of the plane at infinity, away from which we can use the basis $\boldsymbol{\Psi}'$ as an ordinary Cartesian system. The transition to the homogeneous coordinates transforms vectors of $\mathbb{R}^N$ into vectors on the $(N-1)$-dimensional hyper-surface $\left\{\psi_{1,x} = \sqrt{\pi_x}\right\}$, the orthogonal complement to the vector of stationary distribution $\boldsymbol{\pi}$. The kernel of the generalized inverse operator in the homogeneous coordinates $\boldsymbol{\Psi}'$ is given by $\widehat{\mathbf{L}}^{\natural} = \sum_{k=2}^{N} \frac{|\psi_k'\rangle\langle\psi_k'|}{\lambda_k}$.

4.2 Probabilistic Interpretation of Euclidean Geometry

The Euclidean structure introduced in the previous section can be related to a length structure $V \times V \to \mathbb{R}_+$ defined on the class of all admissible paths $\mathcal{P}$ between pairs of nodes in G. It is clear that every path $\mathfrak{p}(i, j) \in \mathcal{P}$ is characterized by some probability to be followed by a random walker depending on the weights $w_{ij} > 0$ of all edges necessary to connect i to j. Therefore, the path length statistics is a natural candidate for the length structure on G.

Let us consider the vector $\mathbf{e}_i = \{0, \ldots 1_i, \ldots 0\}$ that represents the node $i \in V$ in the canonical basis as a density function. The basis vector $\mathbf{e}_i$ can be viewed as a pointwise distribution on the graph $G(V, E)$. The vector $\mathbf{e}_i$ has the squared norm of $\mathbf{e}_i$ associated to a random walk defined on the graph F, viz., $\|\mathbf{e}_i\|_T^2 = (\mathbf{e}_i, \mathbf{e}_i)_T = \frac{1}{\pi_i}\sum_{s=2}^{N}\frac{1}{\lambda_s}\frac{\psi_{s,i}^2}{\psi_{1,i}^2}$. It is remarkable that in the theory of random walks (Lovász 1993) this squared norm is the spectral representation of the *first-passage time* to the node $i \in V$, the expected number of steps required to reach the node $i \in V$ for the first time starting from a node randomly chosen among all nodes of the graph according to the stationary distribution π. The first-passage time, $\|\mathbf{e}_i\|_T^2$, can be directly used in order to characterize the level of accessibility of the node i.

The Euclidean distance between any two nodes of the graph G calculated in the $(N-1)$–dimensional Euclidean space associated to random walks, viz.,

$$K_{ij} = \|\mathbf{e}_i - \mathbf{e}_j\|_T^2 = \sum_{s=2}^{N}\frac{1}{\lambda_s}\left(\frac{\psi_{s,i}}{\psi_{1,i}} - \frac{\psi_{s,j}}{\psi_{1,j}}\right)^2, \tag{28}$$

also gets a clear probabilistic interpretation as the spectral representation of the *commute time*, the expected number of steps required for a random walker starting at $i \in V$ to visit $j \in V$ and then to return back to i (Lovász 1993). The commute time can be represented as a sum, $K_{ij} = H_{ij} + H_{ji}$, in which $H_{ij} = \|\mathbf{e}_j\|_T^2 - (\mathbf{e}_i, \mathbf{e}_j)_T$ is the *first-hitting time* which quantifies the expected number of steps a random walker starting from the node i needs to reach j for the first time (Lovász 1993).

The first-hitting time satisfies the equation $H_{ij} = 1 + \sum_{i\sim v} H_{vj}T_{vi}$ reflecting the fact that the first step takes a random walker to a neighbor $v \in V$ of the starting node $i \in V$, and then it has to reach the node j from there (Lovász 1993). The latter equation can be directly used for computing of the first-hitting times, however, H_{ij} are not the unique solutions of this equation; the correct definition requires an appropriate diagonal boundary condition, $H_{ii} = 0$, for all $i \in V$ (Lovász 1993). The spectral representation of H_{ij} given by

$$H_{ij} = \sum_{s=2}^{N}\frac{1}{\lambda_s}\left(\frac{\psi_{s,j}^2}{\psi_{1,j}^2} - \frac{\psi_{s,i}\psi_{s,j}}{\psi_{1,i}\,\psi_{1,j}}\right), \tag{29}$$

seems much easier to calculate. From the obvious inequality $\lambda_2 \leq \lambda_r$, it follows that the first-passage times are asymptotically bounded by the spectral gap, namely $\lambda_2 = 1 - \mu_2$.

From the spectral representation (29), we may conclude that the average of the first-hitting times with respect to its first index is nothing else, but the first-passage time to the node (Lovász 1993), $\|\mathbf{e}_i\|_T^2 = \sum_{j\in V}\pi_j H_{ji}$. The average of the first-hitting times with respect to its second index is called the *random target access time* (Lovász 1993). It quantifies the expected number of steps required for a random walker to reach a randomly chosen node in the graph (a target). The random target access time $\mathfrak{T}_G$ is independent of the starting node $i \in V$ being a global spectral characteristic of the graph, $\mathfrak{T}_G = \sum_{j\in V}\pi_j H_{ij} = \sum_{k=2}^{N}\lambda_k^{-1}$. The latter equation expresses the so-called *random target identity* (Lovász 1993). Several models were developed to study the mean first-passage time taken by a walker to move from an arbitrary source to a target in complex media. For instance, such situations were usually encountered in predatory animals and biological cells (Shlesinger 2007).

5 How a City Should Look?

On the one hand, the joint use of scarce space creates life in cities which is driven largely by microeconomic factors, which tend to give cities similar structures. The emergent street configuration creates differential patterns of occupancy, whereby some streets become, over time, more highly used than others (Iida and Hillier 2005). On the other hand, a background residential space process driven primarily by cultural factors tends to make cities different from each other, so that the emergent urban grid pattern forms a network of interconnected open spaces, being a historical record of a city creating process driven by human activity and containing traces of society and history Hanson 1989).

In the present section, we study and compare the spatial structures of a labyrinth and the different urban patterns by random walks.

5.1 Labyrinths

It is interesting to discuss the general structural properties of labyrinths that make them so difficult to navigate in and, at the same time, so mysteriously attractive to our minds. In Fig. 2a, we have presented the maze consisting of 45 interconnected rectangles providing the space for motion and displayed its spatial graph representing the connectivity pattern. Although it is clear that movements of a real human traveler are rather self-determined than random, the interpretation of a place with respect to the overall structure of the environment can be based on some random walks defined in that (Blanchard and Volchenkov 2009; Volchenkov and Blanchard 2007, 2008a, b). Humans are used to live in Euclidean space, and they form a *Euclidean mental model* of any urban environment by learning it. The Euclidean metric on urban spaces can be discovered by analyzing the first- passage times to these spaces in the urban spatial graphs.

It is known that, for a stationary discrete-valued stochastic process, the expected recurrence time to return back to a state is the reciprocal of the density of the state (Kac 1979). The expected *recurrence time* to a node REC_i which indicates how long a random walker must wait to revisit the site is inverse proportional to the stationary distribution of random walks over the graph. It follows from the spectral representation of first-passage time FPT_i that it is proportional to the expected recurrence time, viz., $\mathrm{FPT}_i = \|\mathbf{e}_i\|_T^2 = \frac{1}{\psi_{1,i}^2}\sum_{k=2}^{N}\frac{\psi_{k,i}^2}{\lambda_k} = \mathrm{REC}_i \cdot \sum_{k=2}^{N}\frac{\psi_{k,i}^2}{\lambda_k}$, where the proportionality coefficient is determined by the entire graph structure.

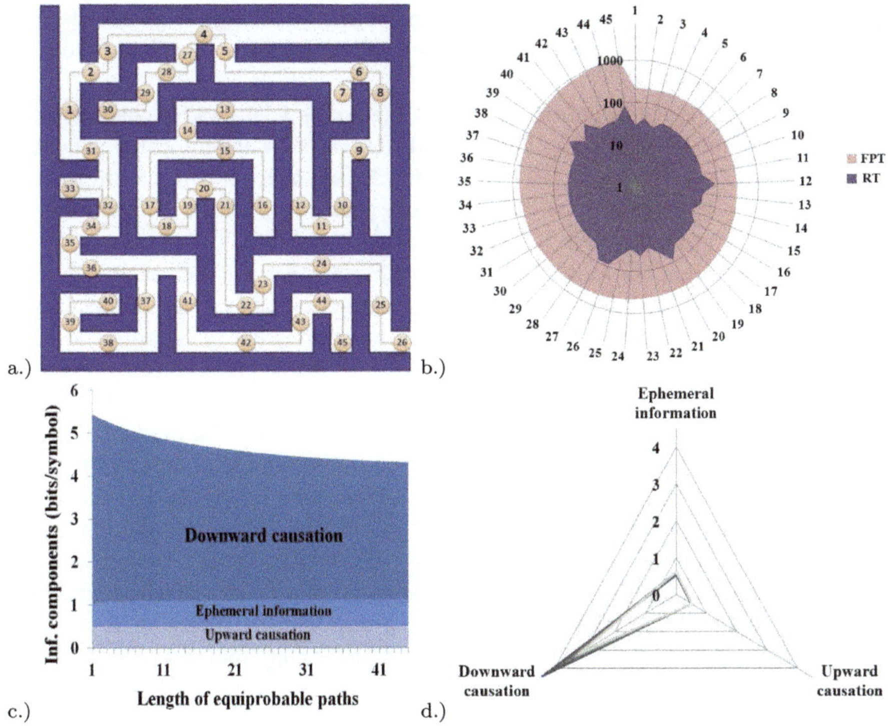

Fig. 2 **a** The maze consisting of 45 interconnected rectangles providing the space for motion, along its spatial graph representing the connectivity pattern. **b** The relation between the recurrence times (RT) and first-passage times (FPT) to the rectangular places of movement by the locally isotropic random walks **T**(1) defined in the labyrinth. **c** The information components of Shannon entropy for the random walks **T**(n) at the different lengths of equiprobable paths $n > 1$. **d** According to the information classification of complex systems and processes, random walks in the labyrinth is a fragile process, similar to survival in the model of mass extinction and subsistence and to tossing a unfair coin

For example, in Fig. 2b we have shown the relation between the recurrence times and first-passage times to the rectangular places of movement by the locally isotropic random walks **T**(1) defined on the labyrinth. Interestingly, first-passage times to places in this maze are up to one order of magnitude *longer* than the recurrence times to them.

Although it may take quite a long time for a random walker to reach a place for the first time (up to 1000 random steps) according to the data of Fig. 2b, the walker is doomed to revisit this place again and again, as the corresponding recurrence time determined by the *local connectivity* of the place might be ten times shorter than the first -passage time determined by the *global connectedness* of the entire graph, as well as the role of the given place in that. Metaphorically speaking, the random walker appears to be trapped within the labyrinth environment and might find it confusing.

In Fig. 2b, we can see that all spaces of movement (rectangles) in the labyrinth constitute such traps (as the first-passage times to all of them substantially exceed the recurrence times).

The use of the higher order random walks $\mathbf{T}(n)$, $n > 1$, allows for characterizing the structure of the labyrinth (Fig. 2a) on the different scales and information decomposition of uncertainty to visit a place. In Fig. 2c, we have shown the information components of Shannon entropy of the random walks $\mathbf{T}(t)$ for the different lengths $t > 1$ of equally probable paths.

Regarding to a random walk defined in the labyrinth, a space (i.e., each numbered rectangle in the labyrinth shown in Fig. 2a) is characterized by some uncertainty of being visited by a walker. The total amount of uncertainty decays steadily (Fig. 2c), as we employ the walks of increasingly long equiprobable paths since some places become more preferable for such a walker than others. The portion of information associated to downward causation obviously dominates the random walks in Fig. 2a at all scales, revealing the presence of strong structural correlation between the sequences of aligned places visited by a walker in the labyrinth. The remaining information (approximately 1 bit) is almost evenly shared between the ephemeral and upward information components (Fig. 2c). The walker in the labyrinth can turn back at any time, so that the information component of upward causation quantifies the amount of uncertainty related to the choice of the next step direction. Eventually, ephemeral information quantifies the portion of uncertainty that can neither be predicted from observing the structure of the labyrinth, nor be guessed by alternating the current direction of the walk at the present place.

The information decomposition of random walks in the labyrinth is dominated by the component associated to the downward causation process (Fig. 2d), so that according to the information classification of complex systems and processes, random walks in the labyrinth is a fragile process, similar to tossing a unfair coin. Learning the structure of labyrinths does not improve the quality of movements in them.

The relative entropy rate (bit/step) between the locally isotropic random walks $T_{ij}(1)$ and the locally anisotropic random walk $T_{ij}(\infty)$ on the spatial graphs of urban patterns quantifies the additional amount of information generated at each step of the anisotropic random walk due to increased knowledge about the numbers of lengthy paths available from the nearest neighbors of the present location of walker.

In Fig. 3, we have shown the radar diagram comparing the relative entropy rates (bit/step) between the locally isotropic random walks $T_{ij}(1)$ and the locally anisotropic random walk $T_{ij}(\infty)$ on the spatial graphs of urban patterns. Learning the real urban structures (such as the street grid in Manhattan, the city canal networks in Amsterdam and Venice, the organic German medieval cities of Bielefeld and Rothenburg) changes the quality of walks that is reflected by the certain amount of additional information generated at each step of the random walk $T_{ij}(\infty)$ due to increased knowledge about the numbers of infinitely long paths available from every node.

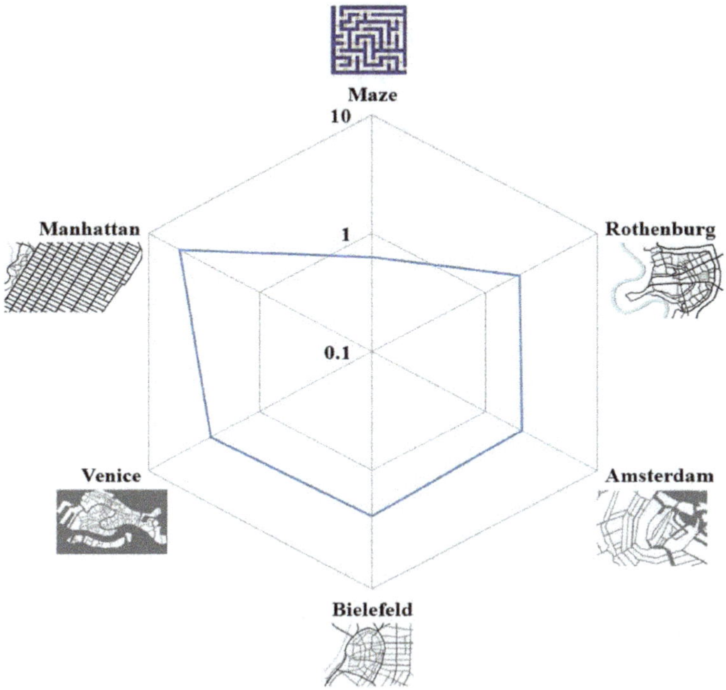

Fig. 3 The relative entropy rates (bit/step) between the locally isotropic random walks $T_{ij}(1)$ and the locally anisotropic random walk $T_{ij}(\infty)$ on the spatial graphs of urban patterns

However, learning of the numbers of infinite paths available from the nearest neighbors of the present location of walker makes almost no change to the character of random walks in labyrinths.

5.2 Manhattan's Grid

The most common pattern favored by the Ancient Greeks and the Romans, used for thousands of years in China, established in the south of France by various rulers, and being almost a rule in the British colonies of North America is the grid. In most cities of the world that did not develop and expand over a long period of time, streets are traditionally laid out on a grid plan. Manhattan, a borough of New York City, is a paradigmatic example (see Fig. 4a), with the standard city block, the smallest area that is surrounded by streets, of about 80 m by 271 m.

The diagram showing the relations between the recurrence and first-passage times to the spaces of movement in the spatial graph of Manhattan is presented on Fig. 4b. First, in contrast to the trapping structure of labyrinth (where first-passage times greatly exceed recurrence times, see Fig. 2b), the values of first-passage times by

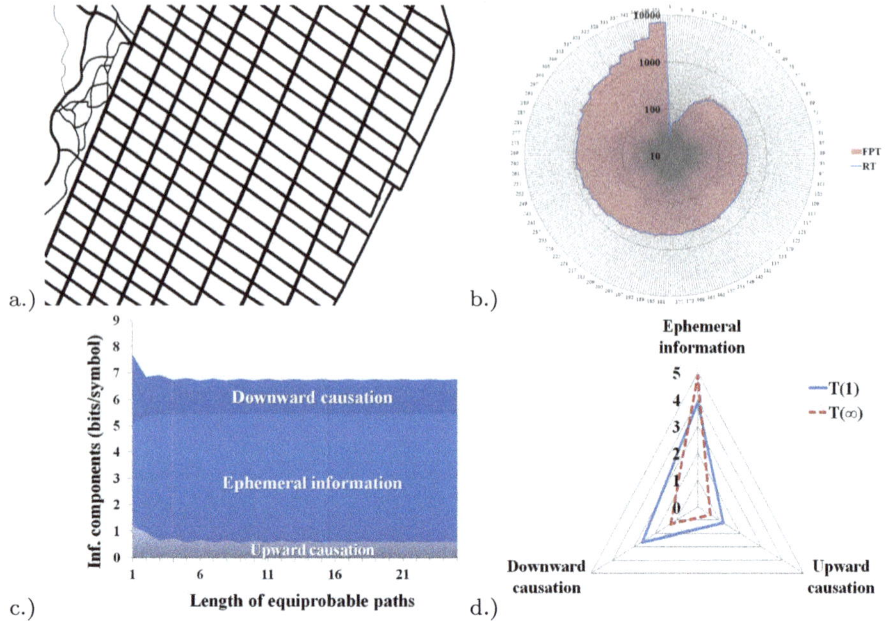

Fig. 4 **a** The Upper East Side (Manhattan), a 1.8 square miles (4.7 km^2) neighborhood in the borough of Manhattan in New York City, USA, between Central Park and the East River. **b** The relation between the recurrence times and first-passage times to the spaces of movement in Manhattan by the locally isotropic random walks $\mathbf{T}(1)$. **c** The information components of Shannon entropy for the random walks $\mathbf{T}(t)$ at the different lengths of equiprobable paths $t > 1$. **d** According to the information classification of complex systems and processes, random walks on the spatial grid of Manhattan is predominantly an ephemeral process, as random paths can hardly be predicted from the almost regular structure of urban grid, and each navigation step has almost no consequence for the future trajectory of the walk

random walks in the spatial graph of Manhattan are very close to recurrence times to the same places, so that random walkers are not trapped, but instead the following natural "*urban rule of thumb*" is applicable: *The better connected the place to its urban environment, the easier one can get there for the first time and the higher the chance it is visited again.*

Second, the striking feature of the spatial graph of Manhattan is the extreme diversity (up to four orders of magnitude) of the first-passage and recurrence times. We shall discuss the consequences of such a structural disparity of places in the forthcoming section in detail. In Fig. 4c, we have shown the information components of Shannon entropy of the random walks $\mathbf{T}(t)$ for the different lengths $t > 1$ of equally probable paths. The total amount of uncertainty related to a random walker visiting a place in the spatial graph of Manhattan oscillates with the scale of walks and tends to a fixed value for the very long lengths $t \gg 1$. Information decomposition of random walks on the spatial graph of Manhattan is dominated by the ephemeral information component. According to the information classification of complex sys-

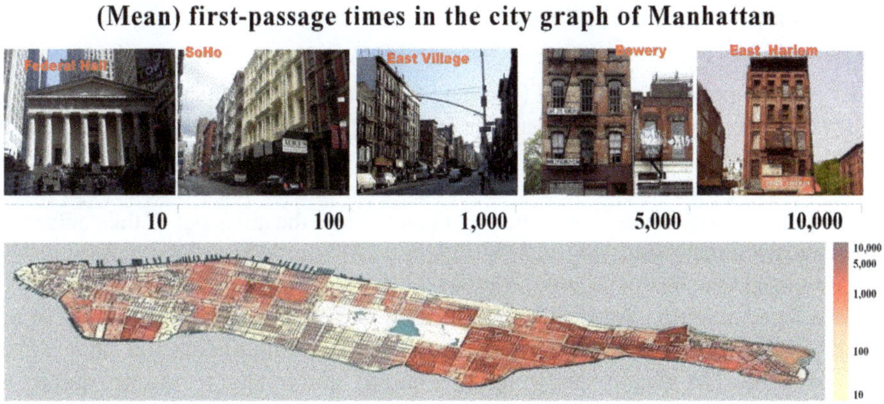

Fig. 5 Isolation map of Manhattan. Isolation is measured by first-passage times to the places in Manhattan calculated from the randomly chosen place with respect to the stationary distribution of random walks. Darker color corresponds to longer first-passage times. The first-passage times in the borough of Manhattan, NYC: (1) the Federal Hall National Memorial ~10 steps; (2) the Times square ~100 steps; (3) the SoHo neighborhood, in Lower Manhattan ~500 steps; (4) the East Village neighborhood, lying east of Greenwich Village, south of Gramercy and Stuyvesant Town ~1,000 steps; (5) the Bowery neighborhood, in the southern portion of the New York City borough of Manhattan ~5,000 steps; (6) the East Harlem (Spanish Harlem, El Barrio), a section of Harlem located in the northeastern extremity of the borough of Manhattan ~10,000 steps

tems and processes, random walks on the spatial grid of Manhattan is predominantly an ephemeral process, as the random path can hardly be predicted from the almost regular structure of urban grid, and each navigation step has almost no consequence for the future trajectory of the random walk (Fig. 4d).

The notion of isolation acquires a statistical interpretation by means of random walks. The recurrence time of random walks to a node in a undirected graph depends upon the connectivity of the node, the number of its nearest neighbors—the local structural property of the node in the graph. Times are easy to reach by whatever origin–destination route while very many random steps would be required in order to get into a statistically isolated site. Being a global characteristic of a node in the graph, the first-passage time assigns absolute scores to all nodes based on the probability of paths they provide for random walkers.

The first-passage time can, therefore, be considered as a natural statistical centrality measure of the node within the graph (Blanchard and Volchenkov 2009). A visual pattern displayed on Fig. 5 represents the pattern of structural isolation (quantified by the first-passage times) in Manhattan (darker color corresponds to longer first-passage times). It is interesting to note that the spatial distribution of isolation in the urban pattern of Manhattan (Fig. 5) shows a qualitative agreement with the map of the tax assessment value of the land in Manhattan reported by Rankin (2016) in the framework of the RADICAL CARTOGRAPHY project.

5.3 German Organic Cities

Older cities appear to be mingled together, without a rigorous plan. This quality is a legacy of earlier unplanned or organic development, and is often perceived by today's tourists to be picturesque. They usually have a hub, or a focus of several directional lines, or spokes which link center to edge, and sometime there is a rims of edge lines. Most of the trading centers are at the city center, while the areas out of the center are the more residential ones.

The spatial structure of organic cities was shaped in response to the socioeconomic activities maximizing ease of navigation in the areas, which are most likely to be visited by different people from inside and outside, but minimizes the same when it is undesired. The diagram showing the relations between the recurrence and first-passage times to the same spaces of movement in the spatial graph of Bielefeld is presented on Fig. 6b. The values of first-passage times by random walks in the spatial graph of Bielefeld just slightly exceed recurrence times to the same places, so that random walkers are not trapped in the urban fabric of organic cities, in contrast to labyrinths.

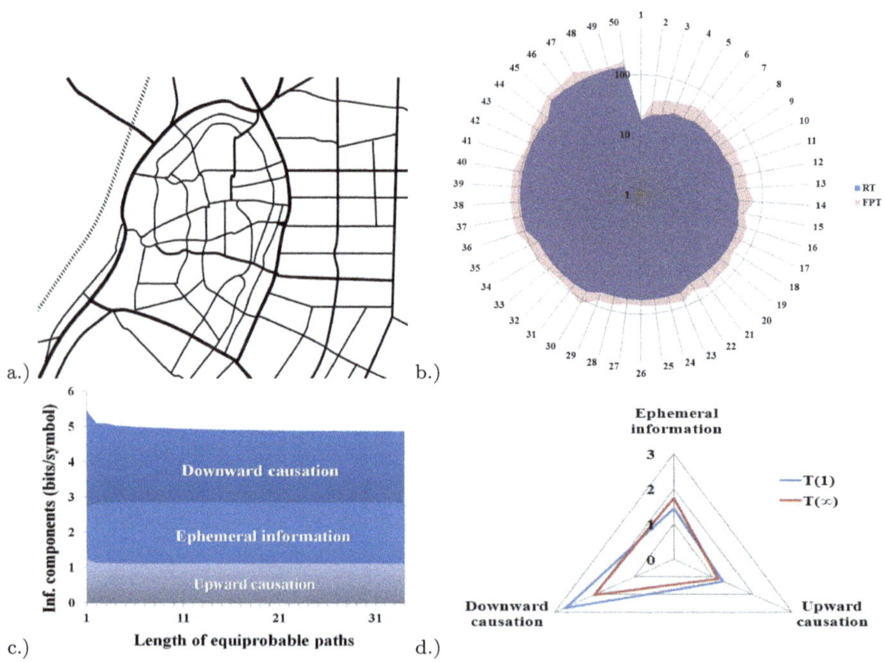

Fig. 6 **a** The "hidden" city of Bielefeld, North Rhine-Westphalia (Germany) is an example of an organic city. **b** The relations between the recurrence and first-passage times to the same spaces of movement in the spatial graph of Bielefeld. **c** and **d** Shannon's entropy quantifying uncertainty of a place for the random walks $\mathbf{T}(t)$, $t \geq 1$, on the spatial graph of Bielefeld

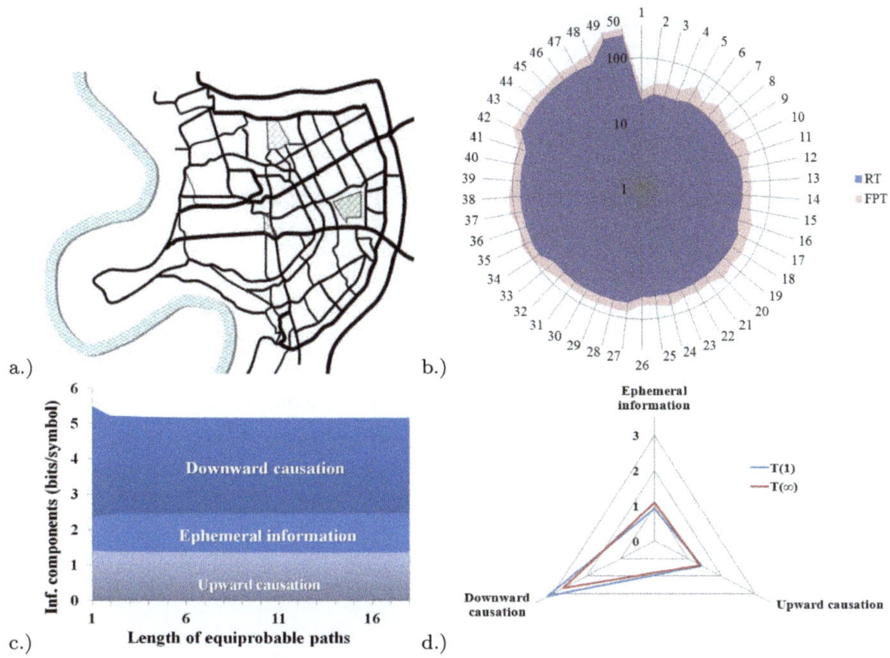

Fig. 7 **a** Another example of an organic city: Rothenburg ob der Tauber, Bavaria (Germany). **b** The relation between the recurrence and first-passage times to the same spaces of movement in the spatial graph of Rothenburg. **c** and **d** The information decomposition of Shannon's entropy for the random walks $\mathbf{T}(t)$, $t \geq 1$, on the spatial graph of Rothenburg

Old organic cities are usually not large, so that the diversity of the first-passage and recurrence times over these cities does not exceed one order of magnitude (see Fig. 6b). In Fig. 6c, d we have shown that Shannon's entropy quantifying uncertainty of a place for the random walks $\mathbf{T}(t)$, $t \geq 1$, on the spatial graph of Bielefeld consists of three well-balanced information components associated to downward and upward causation as well as ephemeral information, with the major role of downward causation revealing the importance of structural correlation for walks in the organic city.

Other organic cities may show a radial structure in which main roads converge to a central point, often the effect of successive growth over long time with concentric traces of town walls (clearly visible on the satellite image of the medieval Bavarian city, Rothenburg ob der Tauber, Fig. 7a) and citadels usually recently supplemented by ring-roads that take traffic around the edge of a town.

The diagram showing the relation between the recurrence and first-passage times to the same spaces of movement in the spatial graph of Rothenburg presented on Fig. 7b is essentially similar to that one for Bielefeld (Fig. 6b). The values of first-passage times by random walks in the spatial graph of organic cities slightly exceed recurrence times to the same places, and the maximal difference of the first-passage and recurrence times over the city graph is not very large. In Fig. 7c, d we have shown the information decomposition of Shannon's entropy for the random walks $\mathbf{T}(t), t \geq 1$, which is also similar to those diagrams obtained for the city of Bielefeld.

5.4 *The Diamond Canal Network of Amsterdam*

The central diamond within a walled city was thought to be a good design for defense. Many Dutch cities have been structured this way: a central square surrounded by a concentric canals. The city of Amsterdam (see Fig. 8a) is located on the banks of the rivers Amstel and Schinkel, and the bay IJ. It was founded in the late twelfth century as a small fishing village, but the concentric canals were largely built during the Dutch Golden Age, in the seventeenth century. Amsterdam is famous for its canals, grachten. The principal canals are three similar waterways, with their ends resting on the IJ, extending in the form of crescents nearly parallel to each other and to the outer canal. Each of these canals marks the line of the city walls and moat at different

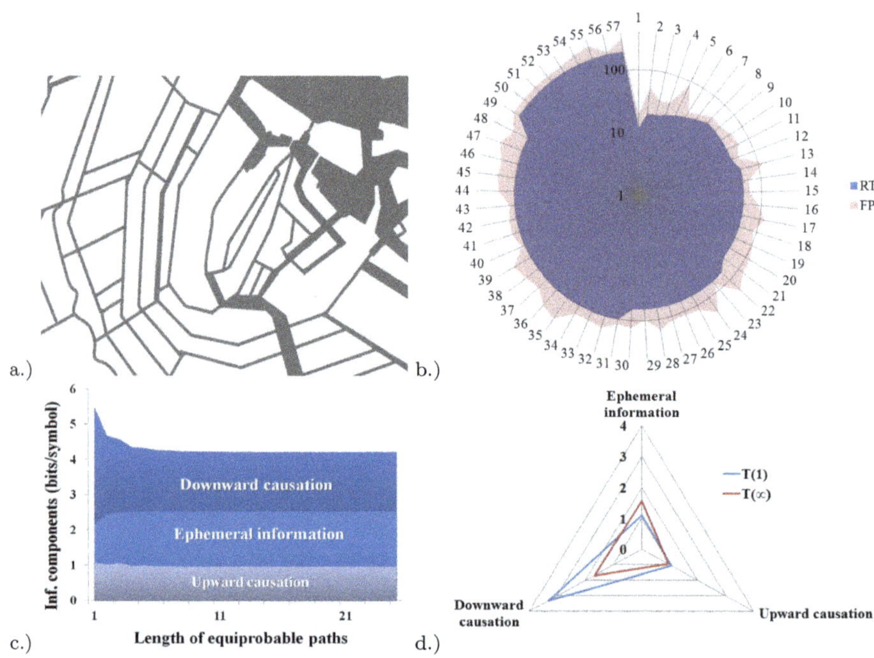

Fig. 8 **a** Amsterdam, the capital city of the Netherlands. **b** The relation between the recurrence and first-passage times to the same canals in the canal network of Amsterdam. **c** and **d** The information decomposition of Shannon's entropy for the random walks $\mathbf{T}(t)$, $t \geq 1$, over the canal network of Amsterdam

periods. Lesser canals intersect the others radially, dividing the city into a number of islands.

The diagram showing the relations between the recurrence and first-passage times to canals in the canal network of Amsterdam is presented on Fig. 8b. The values of first-passage times by random walks on the canal network of Amsterdam are very close although slightly exceeding recurrence times to the same canals. Similarly to the spatial graphs of old organic cities, the networks of urban canals are usually not large, and the diversity of the first-passage and recurrence times over these networks is either not large.

The information decomposition of Shannon's entropy for the random walks $\mathbf{T}(t)$, $t \geq 1$, over the canal network of Amsterdam is shown on (Fig. 8c, d). While the information decomposition for the locally isotropic random walks $\mathbf{T}(\infty)$ are dominated by the structural correlation of the diamond-like city canal network (downward causation), the information decomposition for the locally anisotropic random walks $\mathbf{T}(\infty)$ based on the infinitely long walks are well balanced between all three independent components of Shannon's entropy.

5.5 *The Canal Network of Venice*

The structural characteristics of the canal network in Venice revealed by random walks are quite similar to those of the canal network in Amsterdam.

The values of first-passage times by random walks on the canal network of Venice are also very close although slightly exceeding recurrence times to the same canals (see Fig. 9b). Two main canals featuring the transportation system of Venice and providing the major water-traffic corridors in the city (the Grand Canal and the Giudecca Canal) are characterized by the very short first-passage times that amplifies the diversity in accessibility to the different canals in the city canal network.

The information decomposition of Shannon's entropy for the random walks $\mathbf{T}(t)$, $t \geq 1$, over the canal network of Venice is displayed on Fig. 9c, d. While the information decomposition for the locally isotropic random walks $\mathbf{T}(\infty)$ are dominated by the structural correlation (downward causation), the major information component for the locally anisotropic random walks $\mathbf{T}(\infty)$ based on the infinitely long walks is due to ephemeral information.

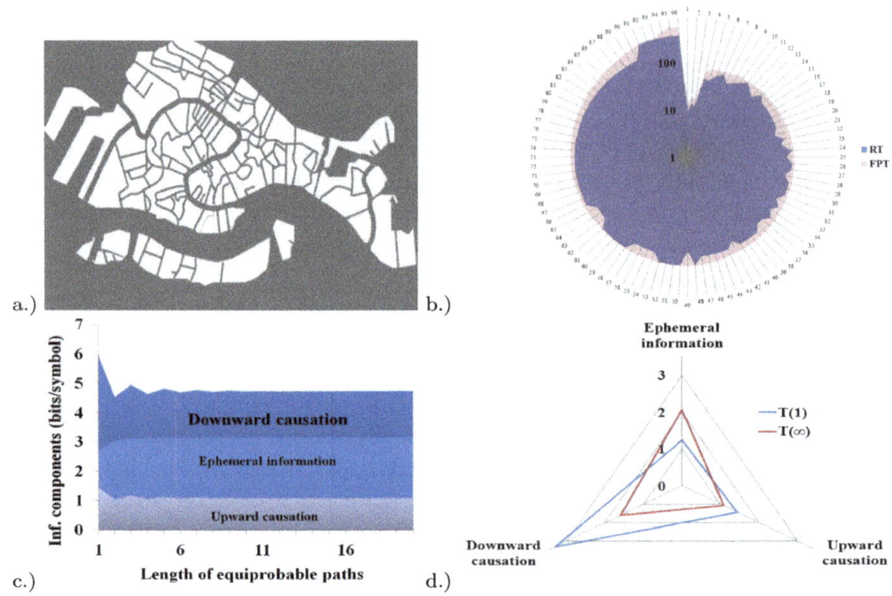

Fig. 9 **a** The scheme of the contemporary canal network of Venice. **b** The relation between the recurrence and first-passage times to the same canals in the canal network of Venice. **c** and **d** The information decomposition of Shannon's entropy for the random walks $\mathbf{T}(t)$, $t \geq 1$, over the canal network of Venice

6 Conclusion

We have presented the mathematical method for measuring structural complexity of urban spatial patterns and neighborhood's inaccessibility. The level of accessibility of nodes and subgraphs of undirected graphs can be estimated precisely in connection with random walks introduced on them. We have applied this method to the structural analysis of a maze and different cities.

References

Blanchard, Ph., Volchenkov, D., *Mathematical Analysis of Urban Spatial Networks*, Springer Series *Understanding Complex Systems*, Berlin / Heidelberg. ISBN 978-3-540-87828-5, 181 pages (2009)

Chung, F.R.K., *Lecture notes on spectral graph theory*, AMS Publications Providence (1997)

Churchill, W., Famous Quotations/Stories of Winston Churchill at http://www.winstonchurchill.org (2019)

Hanson, J.M., *Order and Structure in Urban Space: A Morphological History of London*, Ph.D. Thesis, University College London (1989)

Hillier, B., Hanson, J., *The Social Logic of Space* (1993, reprint, paperback edition ed.). Cambridge: Cambridge University Press (1984)
Iida, S., Hillier, B., "Network and psychological effects in urban movement", in A.G. Cohn, D.M. Mark (eds) *Proc. of Int. Conf. in Spatial Information Theory: COSIT 2005* published in *Lecture Notes in Computer Science* **3693**, 475-490, Springer-Verlag (2005)
Kac, M., "On the Notion of Recurrence in Discrete Stochastic Processes", *Bull. Am. Math. Soc.* **53**, 1002 (1947) [Reprinted in M. Kac *Probability, Number Theory, and Statistical Physics: Selected Papers*, K. Baclawski, M.D. Donsker (eds.), Cambridge, Mass.: MIT Press, Series: *Mathematicians of our time* Vol. **14**, 231 (1979)]
Lovász, L., "Random Walks On Graphs: A Survey." *Bolyai Society Mathematical Studies* **2**: *Combinatorics, Paul Erdös is Eighty*, 1 Keszthely (Hungary) (1993)
Orford, S., Dorling, D., Mitchell, R., Shaw M. & Davey-Smith, G., "Life and death of the people of London: a historical GIS of Charles Booth's inquiry." *Health and Place* **8** (1), 25-35 (2002)
Ortega-Andeane, P., Jiménez-Rosas, E., Mercado-Doménech, S., & Estrada-Rodrýguez, C., "Space syntax as a determinant of spatial orientation perception." *Int. J. of Psychology*, **40** (1), 11-18 (2005)
Rankin, B., *After the Map: Cartography, Navigation, and the Transformation of Territory in the Twentieth Century*. Chicago: University of Chicago Press (2016)
Shlesinger, M.F., "First encounters" *Nature* **450**(1) 40 (2007)
Vaughan, L., "The relationship between physical segregation and social marginalization in the urban environment." *World Architecture*, **185**, pp. 88-96 (2005)
Vaughan, L., Chatford, D., & Sahbaz, O., *Space and Exclusion: The Relationship between physical segregation. economic marginalization and poverty in the city*, Paper presented to Fifth Intern. Space Syntax Symposium, Delft, Holland (2005)
Volchenkov, D., "*Grammar Of Complexity: From Mathematics to a Sustainable World*", in World Scientific Series Nonlinear Physical Science (2018)
Volchenkov, D., Blanchard, Ph., "Intelligibility and first passage times in complex urban networks", *Proc. R. Soc. A* **464** 2153-2167; https://doi.org/10.1098/rspa.2007.0329 (2008a)
Volchenkov, D., Blanchard, Ph., "Random Walks Along the Streets and Channels in Compact Cities : Spectral analysis, Dynamical Modularity, Information, and Statistical Mechanics", *Phys. Rev. E* **75**, 026104 (2007)
Volchenkov, D., Blanchard, Ph., Scaling and Universality in City Space Syntax: between Zipf and Matthew. *Physica A* **387**(10), 2353 (2008b)
Wirth, L., *The Ghetto* (edition 1988) Studies in Ethnicity, transaction Publishers, New Brunswick (USA), London (UK) (1928)

Space Syntax: A Network-Based Configurational Approach to Studying Urban Morphology

Mahbub Rashid

Abstract In this chapter, I describe *space syntax*, a recent configurational approach to urban morphological studies that shows some conceptual similarities to the *network analysis* of structural sociology. First, I describe the basic concepts, methods, and measures of this configurational approach, focusing on the axial and segment map analyses—the two most commonly used methods of space syntax related to urban morphological studies. After this, I describe some of the recent mathematical developments of space syntax methods and measures. These mathematical developments include various normalization techniques for integration and choice—the two most important measures of space syntax; various clarifications on the relationships between the metric, geometric, and topological measures of axial and segment maps; various universal properties including scaling laws observed in axial and segment maps; various ways to reduce and/or eliminate the effects of boundary on space syntax measures; various ways to take into account the shape as well as the 3D of the built environment within space syntax; and, finally, various ways to integrate space syntax with GIS. Following mathematical developments, I describe various applications of space syntax methods and measures in urban morphological studies. Here, I discuss how space syntax methods and measures have been applied to describe the syntactic types and cores, and the whole and the parts relationships of spatial configurations; the processes of spatial production and reproduction of social relations, functions, and knowledge; the generative functions of spatial configurations; the relationships between spatial configurations and social capital; and the experiential aspects of urban morphological history. I conclude the chapter by highlighting the fact urban morphological studies, though an important part of the space syntax corpus, are only one of many areas of research where space syntax methods and measures have been applied in recent years.

M. Rashid (✉)
School of Architecture and Design, University of Kansas, Lawrence, KS, USA
e-mail: mrashid@ku.edu

L. D'Acci (ed.), *The Mathematics of Urban Morphology*,
Modeling and Simulation in Science, Engineering and Technology,
https://doi.org/10.1007/978-3-030-12381-9_10

1 Introduction

The city is a complex entity. Many views on this issue have been presented in the social and natural sciences literature. For my purposes here in this chapter, I will lend myself to one of these views, which suggests that the city is a dual complex system with four features (Portugali 2016). First, every city has physical and human components. The city of physical components is "the physical city", which includes every material artifacts that we, as agents, ever build, use and signify in the city; and the city of human components is "the functional city", which describes how we use the physical city for instrumental and symbolic purposes (Hillier 1996). Second, the physical city and the functional city are dialectically related to each other. The physical city emerges out of the functional behaviors and activities of agents. Once emerged, it affects the behaviors and activities of agents that created it in the first place giving rise to conditions for the physical city to change in circular causality. Third, the dialectical processes involve external representations that reside in the physical city, and internal representations that originate and reside in individual human agents in the form of ideas, intentions, memories, and thoughts and in institutional agents is the form of history, tradition, and culture. Fourth, defined by numerous complex networks of space, form, and function the physical city is complex, and so are its human and institutional agents.

The above view makes it clear that both the physical city and the functional city are important for us to understand the city and its morphology. Here, I would like to contend that the traditional concern of urban morphological studies as a part of urban geography has been the physical city (Conzen 1960; Whitehand 1981, 1987, 2001, 2007). These studies have generally described the physical city by its relatively permanent structures. They include *streets* and *open spaces*; *land division patterns* showing functionally differentiated and legally protected ownership; *land use patterns* showing the proportion, size, shape, and location of specialized use of plots; and *building fabric* showing the type, quality, and quantity of the physical structures needed for specialized use of plots. However, our interests in the dialectical processes between the physical city and the functional city, involving external representations that reside in the physical city and internal representations that originate and reside in agents, are rather recent. Lynch (1960) was probably the first person, who had discussed difficulties involved in defining how human agents conjure up the mental image of the physical city. According to Lynch, the image of a city has three components—identity, structure, and meaning. While he described the identity of the image using five elements—landmarks, nodes, paths, edges, and districts, he admitted that even these elements are not easily definable. They shift meaning depending on the purpose, time, and scale of use. He also admitted that it is even more difficult to define the structures of these elements, to define how users attribute meanings to these structures, and to define how structures affect the meaning of its constituent elements. In essence, as always, our cognition and perception of the physical city remain difficult to define (Portugali 2011). While many rigorous readings of the physical city have occurred in recent years focusing on history, society, culture, and economy, any rigorous readings of the functional city as a multilayered human cognitive and perceptual phenomenon have occurred only recently. In this chapter, I would like to argue that "space syntax" has a significant role to play in some of these more recent readings of the physical city in relation to the functional city.

2 Space Syntax: Standard Concepts, Methods, and Measures

Since the 1970s, a substantial body of research has used the "space syntax" approach to study the morphology of the physical city and its relations to the functional city. Developed by Bill Hillier and his colleagues, space syntax includes a set of theories, methods and techniques that use physical space as a central descriptive category of the society (for early theoretical and methodological introduction to space syntax, see (Hillier 1996; Hillier and Hanson 1984)). Though developed independently, the analytic foundations of space syntax are somewhat similar to that of the social network analysis of structural sociology. Whereas space syntax studies networks of physical spaces as represented in the plan of a given environment, structural sociology studies the networks of people in a society. However, both use comparable metrics to define network properties (for measures in structural sociology, see (Freeman 1977, 1979; Freeman et al. 1991); for a comparison between formal definitions of space syntax measures and network measures, see (Hillier and Hanson 1984; Hillier and Iida 2005; Jiang and Claramunt 2002; Rashid 2016)). The usefulness of space syntax theories and techniques in studies on the relationships between the physical city and the functional city depends on a homology suggesting that the centrality of a spatial unit in the network of spaces may indicate the unit's degree of access to or control over other spaces in the network, just as the centrality of an individual in a social network may indicate the individual's degree of access to or control over other individuals in the network.

In recent years, network analysis techniques have been used to study a wide range of social phenomena. They usually use various graph-theoretic measures describing how the units of a network are related to each other and to the whole. These relations can be unidirectional (where B influences A, but A does not influence B); bidirectional symmetric (where A and B influence each other in the same magnitude); bidirectional asymmetric (where A and B influence each other in different magnitudes); weighted when relations are assigned relative strengths; and non-weighted when relations are not assigned relative strengths. In social networks, these relations can have different contents based on business, friendship, authorship, email, citation, and many other forms of network that are often found in a society. Likewise, in spatial networks, the relations of spaces can have different contents based on enclosure, subdivision, overlap, intersection, adjacency, and permeability, which are generally included in space syntax studies as different configurations of spatial networks.

2.1 Configurations

In space syntax, the term "configuration" is often used to reflect the fact that, while a network of spaces as shown in the layout of an environment represents all possible ways how spatial units are related to one another, a configuration represents a state of the network describing the relations of a kind among individual spatial units. Therefore, a configuration describing adjacency may be very different from that describing permeability of the spatial units of a layout. Likewise, a configuration of weighted relations may become very different from a configuration of unweighted relations of the spatial units of a layout. Naturally, then, a network of spaces in a layout can have multiple configurations depending on the relations being explored. Indeed, the number of different configurations of a layout is limited only to how one wants to describe the layout taking into account significant relations.

Space syntax assumes that a configuration of a layout of an environment, when defined using perceptually meaningful 1D or 2D spatial units, can carry social meaning. For example, a configuration defined using lines or segments may be important because human tends to move in straight lines and along routes with few changes in directions to minimize effort and time and to maximize movement economy. A configuration of convex spaces may be important for functional as well as social reasons. For example, face-to-face interaction works better when everyone in a group can see each other in a space. People feel secured in a convex space where nothing is hidden from sight. More importantly, different spatial relations are easily described and compared when an environment is partitioned as convex spaces. Likewise, a configuration of visual fields is important because objects are often grouped based on how they are seen from different locations, and people develop awareness of each other based on who sees whom. As we will find out below, space syntax's assumption that space contains social meaning via human attribution has been proven correct by numerous studies that use space syntax methods and measures.

Space syntax also assumes that the spatial network of an environment can be described as different configurations for analytic purposes, but these configurations coexist in social reality. As Hillier writes,

> All these ways of looking at space can be seen as layers of spatial structuring, co-existing within the same plan, each with its own contribution to intelligibility and function. A spatial layout can thus be seen as offering different functional potentials. What is it like to move around in it? Does it have potential to generate interaction? Can strangers understand it? and so on. All these questions are about the relationship of space as formal potentials to different aspects of function. A layout can thus be represented as a different kind of spatial system according to what aspects of function we are interested in (Hillier 1996, p. 116)

Humans unconsciously make use of these configurations depending on the nature of everyday practice. In other words, even though configurations can be defined bottom-up for analytic purposes, they are emergent top-down in everyday practice. In this sense, spatial configurations are the emergent "structuring structures" as opposed to the relatively permanent "structured structures," to use Pierre Bourdieu's terms (Bourdieu 1992), of the environment that can modify everyday practice and eventually the society in a dialectical process where the outcome of a process can change the process that produces it. Space syntax's concern for different configurations of an environment can also be compared with J. J. Gibson's concern for environmental affordances explored in his book *The Ecological Approach to Visual Perception* (Gibson 1979). According to Gibson, environmental affordances refer to perceptually available structures of information serving one or more generic purposes for a society of animals. Therefore, an essential part of socialization is to learn how to perceive environmental affordances. By learning to use affordances, humans enter into the shared practices of their society with abilities to modify them over time to serve their purposes better.

As we have noted in the introduction of this chapter, while our interests in the relatively permanent "structured structures" of the environment, i.e., the physical city, had been longstanding in the field of urban morphology, space syntax's interests in the emergent "structuring structures" of the environment providing affordances are relatively new phenomena. Space syntax explores these structuring structures using methods and measures developed based on modern mathematics and powerful computers. These methods and measures allow space syntax to describe, analyze, and visualize different configurations of the spatial network of the physical city at different scales, and to study how different configurations are related to each other as well as to individuals and society through everyday practices.

2.2 Axial and Segment Maps

For its purpose, space syntax uses the configurations of perceptual primitives, definable as simple 1D or 2D spatial units within an environment. These units include *axial lines* representing straight lines of movement and visibility, *segments* representing the parts of axial lines broken at the points of intersections with other axial lines, *convex spaces* representing 2D shapes within each of which every point is visible from every other points, and *visual fields* representing areas seen from individual points within a spatial layout. Space syntax provides rigorous techniques showing how to represent a layout as a configuration using each of these units, which are called the axial map, the segment map, the convex map and the visual fields of the layout. It also provides techniques to analyze the relational patterns among the units within a map using such graph-theoretic (or, syntactic) measures as closeness, betweenness, centrality, complexity and entropy (Al-Sayed 2014), some of which are defined below. Additionally, space syntax allows users to color the map based on the syntactic values of individual units. The color generally ranges from red representing higher values to blue representing lower values of a space syntax measure (Fig. 1a–d). Different software programs such as *Depthmap, Confeego,* and *Spatialist* (Al-Sayed 2014; Turner 2010; Gil et al. 2007; Peponis and Spatialist 1997) are available for performing different space syntax analysis of spatial configurations.

Most studies of urban areas and cities in the space syntax literature use the techniques of the axial map analysis and, more recently, the techniques of the segment map analysis. As noted above, they are drawn based on a rigorous set of mathemat-

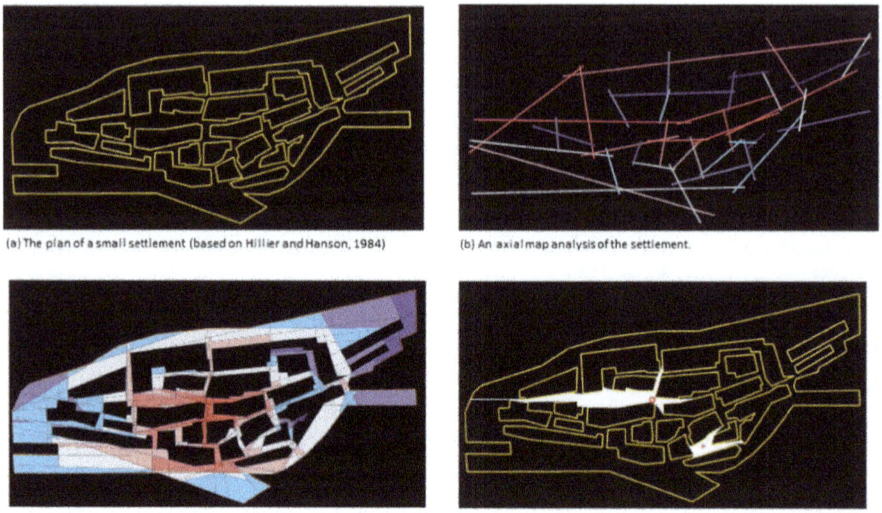

(a) The plan of a small settlement (based on Hillier and Hanson, 1984)

(b) An axial map analysis of the settlement.

(c) A convex map analysis of the settlement. The convex partition in this case are defined using the extension lines of all the surfaces of urban blocks. See Peponis et al (1997) for details.

(d) The visual fields of two locations within the settlements.

Fig. 1 Different types of syntactic analysis using space syntax tools and techniques. These analyses were performed by Spatialist (Peponis et al. 1997)

ical rules. For example, the axial map of a spatial layout is defined by the minimal set of longest axial lines, such that any point in the layout is visible from an axial line of the map, and such that any pair of disjoint lines in the map that can be linked by a third line are so linked. Algorithms for generating such a map are provided in the space syntax literature (Turner et al. 2005; Peponis et al. 1998). Therefore, the axial and segment maps make it possible to compare streets and street networks with different characters. Such comparative configurational studies are necessary because streets serve as linkages to and from various spaces and streets while providing in them spaces for potential activity. As a result, they establish a variety of interface conditions with other urban components. Streets are also related to each other in a contiguous and hierarchical manner enhancing or inhabiting access, control, and continuity. This hierarchy, which is different from the hierarchy assigned to streets based on the amount of traffic they may carry, is defined by the streets' network properties—how connected they are with other streets in a network of streets. As Schumacher writes, "Although the notion of spatial progression or promenade in architecture and urban design is one of modern theory's most belabored abstractions, the idea of 'getting there,' not as prescribed '*promenade architecturale*' but as a choice system of understood multiple routes to multiple goals, is apparently essential" (Schumacher 1986, p. 145).

The axial and segment maps also make it possible to compare cities or city areas with very different patterns of figure and ground or built and unbuilt spaces. In an attempt to describe such differences, Ellis (1986) distinguished *the structure of spaces*, defined by contiguous building patterns commonly found in traditional cities, from *the structure of solids*, where buildings exist as objects on a plane. Ellis argues that these conceptions have a historical as well as a typological significance, and they are important models that guide our actual design. In a structure of spaces, streets and buildings are inseparable. As a result, streets work as places as well as links, symbolizing the collective interests and values of the community they serve. In contrast, in a structure of solids streets lack volume. The functions of streets as links and places become separated. They are reduced to roads carrying traffic, and people are driven off the streets not only by a huge increase in vehicular traffic but also by a lack of functions along their sides. While important, for a long time these concepts have remained referential only, because there was a lack of techniques sensitive enough to describe the variations in built and unbuilt spaces commonly found between these two extreme conditions. A structure of solids can evolve into a structure of spaces owing to growth by infill rather than expansion reflecting a need to remain compact for improved communication and to preserve resources. In contrast, a structure of spaces can be modified to a structure of solids by taking out or tearing down contiguous buildings to reduce density and overcrowding. The axial maps and segment maps are useful in this regard. Using various network (or syntactic) measures, they can easily describe changes in the relationships between built and unbuilt spaces.

2.3 Measures

Among several syntactic measures, *integration* has been the most important measure when applied to the axial and segment maps of space syntax. The integration value of a line in the axial or segment map is defined in terms of the shortest routes between the line and all of the others in the network (defining "shortest" in terms of the fewest changes in direction) (Hillier 1996). As such, integration is similar to the normalized closeness centrality index or the closeness index defined in the early 1950s by structural sociologists and reviewed by Freeman in the late 1970s. *Choice* has been another important centrality measure of space syntax. While integration is closeness, choice is betweenness for structural sociologists. Unlike integration, choice gives the degree to which a line lies on simplest paths from one line to another line in the network. The choice value of a given axial line is determined by dividing the number of the shortest paths between any two lines in the axial map containing the given line divided by all the shortest paths between any two lines in the map. In contrast, the choice value of a segment is calculated by replacing shortest paths with paths that have the lowest angular cost for each possible origin and destination pair of segments on the given segment (Turner 2007). The mathematical definitions of these two as well as some other important measures of the axial and segment maps are provided here.

1. ***Total number of axial lines*** describes the quantity of axial lines in the axial map of a spatial system.

2. ***Mean axial line length*** describes the typical granularity of the axial lines of a spatial system.
3. ***Axial articulation***, based on (Hillier and Hanson 1984), at the level of urban block is defined as the number axial lines per urban block. It indicates the granularity of a spatial system in relation to urban blocks. A lower axial articulation value indicates a higher degree of axiality, and a higher value a greater breakup of axiality.
4. ***Grid axiality***, or the degree of axial deformation of the grid, is measured by comparing the number of axial lines in a spatial system with the number that could exist in a regular grid with the same number of islands or blocks. The expression for this, as given in (Hillier and Hanson 1984), is as follows: $Grid\ Axiality = \frac{2\sqrt{I}+2}{L}$, where I is the number of islands or blocks and L is the number of axial lines. The result is a number between zero and one. A higher grid axiality (*GA*) value indicates a stronger approximation to a grid, and a low value a greater degree of grid axial deformation. According to (Hillier and Hanson 1984), in general values of 0.25 and above indicate a grid-like system, while values of 0.15 and below denote a more axially deformed system.
5. ***Axial ringiness*** (l) is the number of rings in the axial map of a spatial system as a proportion of the maximum possible planar rings for that number of axial lines. It is expressed as $l = \frac{2L-5}{I}$, where L is the number of axial lines and I is the number of rings, blocks, or islands. This value may exceed 1, since the axial map is nonplanar. However, in practice values greater than 1 are unusual (Hillier and Hanson 1984). *Axial ringiness* is a measure of restrictions indicating how freely one is able to move in an area.
6. ***Axial connectivity*** of a line is simply the number of other lines incident on it. Put another way, it is the "degree" of a particular node—that is, the number of edges incident on a node—in a graph. It is, therefore, expressed as $C_i = \sum_{j=1}^{n} a_{ij}$, where C_i is the connectivity of a line i and a_{ij} is the entry of the ith row and jth column of the connectivity matrix A of the graph of an axial map.
   Mean axial connectivity of an axial map is the sum connectivity (C_i) values of all individual axial lines divided by the number of axial lines (L), which can be expressed as $C = \sum_{i=1}^{m} C_i/L$. It indicates how connected the lines are at the local level of a map. It is the typical value for the number of options available to move from any one line to another in the map.
7. ***Axial controllability*** is a measure related to *control* that describes the extent to which a given line controls the access to the lines that are adjacent (immediately connected by an edge) to it. If a line has n immediate neighbors, then the line gives $1/n$ of its control to each of its immediate neighbors, and these values are then summed for each receiving line to give the control value of that line (Hillier and Hanson 1984). Therefore, the control of space or line is inversely proportional to the connectivity of the adjacent lines. The formula for this measure is $ctrl_i = \sum_{j=1}^{n} a_{ij} \frac{1}{c_{ij}}$. In other words, the control for the ith line can be computed by multiplying its adjacency vectors—the row of zeros and ones in the adjacency matrix A—by the reciprocal of connectivity values and

summing the products. The products for the spaces that are directly connected will equal the connectivity reciprocals, while they will be zero for those spaces that are not connected. The sum of the products is, therefore, the sum of the connectivity reciprocals of the connected spaces.

Being a local measure, control has an obvious limitation. For example, if a line with a high control value is connected to the lines with very low control values, then its ability to control may indeed be very little. Conversely, if a line with a low control value is connected to the lines with very high control values, then its ability to control may be greater. In order to overcome this limitation, Turner (2001) proposes *controllability*. For an axial line, it is defined simply as the ratio of the total number of nodes (or lines) up to radius 2 to its connectivity (i.e., the total number of nodes at radius 1). One can use controllability as a measure of freedom or choice in movement.

Mean axial controllability of an axial map, therefore, indicates the typical controllability of the axial lines in a map. It describes the lack of freedom one may experience to move within an area. Higher mean controllability values indicate greater overall lack of freedom or choice in movement, and lower values indicate lower overall lack of freedom or choice in movement.

8. ***Axial integration*** is the reciprocal of the real relative asymmetry (*RRA*) value of an axial line, which is a function of the mean depth (*MD*) value of the line. The *RRA* value of a line *i* is shown as $RRA_i = \frac{RA_i}{D_n}$, where relative asymmetry $RA_i = \frac{2(MD_i-1)}{(n-2)}$, $D_n = \frac{2(n(log_2((n+2)/3)-1)+1)}{(n-1)(n-2)}$, and $MD_i = \frac{\sum_{j=1}^{n} d_{ij}}{(n-1)}$, where n is the number of lines, d_{ij} is the shortest distance between two lines i and j in a graph G, and $\sum_{j=1}^{n} d_{ij}$ is the total depth of the ith line (Hillier and Hanson 1984).

   A higher *RRA* value indicates a greater segregation, and a lower value indicates a higher integration of a node with the other nodes in the entire graph. Therefore, a higher integration value, which is the reciprocal of an *RRA* value, indicates greater accessibility, and a lower value indicates lower accessibility of a node with respect to the other nodes of a graph.

   Mean axial integration of an axial map, therefore, indicates the typical accessibility of the axial lines in a map. Higher mean integration values indicate greater overall accessibility, and lower values indicate lower overall accessibility of the map. It describes the amount of ease one may experience to move within an area.

9. ***Axial betweenness*, or *choice***, as it is called in the space syntax community, of an axial line or a segment is calculated by generating the shortest paths between all segments within the system that lie on the segment (Hillier and Iida 2005). The mathematical formula to calculate betweenness or choice is $C_B(P_i) = \sum_j \sum_k g_{jk}(p_i)/g_{jk}(j < k)$, where $g_{jk}(p_i)$ is the number of geodesics (the shortest path) between nodes p_j and p_k that contain node p_i, and g_{jk} the number of all geodesics between p_j and p_k (Turner 2001).

   Unlike integration, choice gives the degree to which a line lies on the simplest paths from one line to another line in the network. In simple words, integration

measures how easy is it to go from one line to all other lines of a network, thus indicating the potential of a line for to-movement. In contrast, choice measures how likely is it for a line to be chosen on paths from one line to another in a network, indicating its potential for through-movement (Hillier 2005).
Mean axial choice of an axial map, therefore, indicates the overall potential for through-movement of the lines in a map. Higher mean choice values indicate greater overall potential for through-movement, and lower values indicate lower overall potential for through-movement of the map. It describes the amount of ease one seeks while moving from one line to another within an area.

10. ***Intelligibility*** of an urban area describes how difficult it is to predict the global network structure (e.g., *integration*) based on the local network information (e.g., *connectivity*) of an environment (Hillier 2007). An urban area is considered intelligible when the local network measure explains a large amount of variance of the global network measure. Statistically, the intelligibility of an urban area is given by the coefficient of determination R^2, with a value ranging from zero to one. Intelligibility therefore will be used as a measure of the complexity of axial maps, indicating how difficult it is to comprehend the global network structure based on the local network information of an environment.
11. ***Entropy and relativized entropy*** measure the evenness in the frequency of distribution of nodes from each other in terms of distance. Entropy occurs in many fields, including informatics, and is proposed for use in space syntax by Hiller et al. (1987). The entropy of a node, s_i, can be expressed using Shannon's formula of uncertainty as shown here: $s_i = \sum_{d=1}^{d_{max}} -p_d \log p_d$, where d_{max} is the maximum depth from a node or a vertex (v_i), and p_d is the frequency of distance d from the vertex. According to the expression, if nodes are unevenly distributed in relation to a given node, the entropy is low for the node. If nodes are evenly distributed, the entropy is higher for the node.
Relativized entropy takes account of the *expected* distribution from the node. That is, in most cases, one expects the number of nodes encountered as one moves through the graph to increase up to the mean depth and then decline afterward. Turner (2001) provides the following formula of relativized entropy: $r_i = \sum_{d=1}^{d_{max}} -p_d \log \frac{p_d}{q_d}$, where $q_d = \frac{L_i^d}{d!} e^{-L_i}$ and L_i is the mean path length defined as the average distance or number of steps needed to reach any other node in the graph using the shortest distance or number of steps possible in each case.
Since entropy and relativized entropy measure evenness in distance and not the distance itself, the entropy with uneven distribution is always low regardless of the distances of nodes from each other. Therefore, ***mean relativized entropy*** can be used for measuring the complexity, and not the accessibility, of axial maps.

As described above, in a segment map the axial lines of an axial map are broken into segments at their intersections. For the angular segment analysis (ASA), then, the angles turned from a starting segment to any other segment within the map are recorded and used (Dalton 2001; Turner 2001). As opposed to the axial map analysis, where the number of turns from a starting point to an end point is treated as the cost of a journey, in the ASA the angular sum from a starting point to an end point is treated as the "cost" of a journey through the graph. More recently, many studies have demonstrated that there are excellent correlations between various ASA measures and traffic flows in urban areas (Turner 2007, 2009; Hillier and Iida 2005; Hillier et al. 2010). Some of the measures of a segment map include

12. ***Total number of segments*** describes the quantity of segments in a segment map.
13. ***Mean segment length*** describes the typical granularity of the segments of a spatial system.
14. ***Segment integration/segment angular closeness*** of a segment is the mean of all the angles of all the shortest paths on the segment (Turner 2007), which can be shown as: $C_\theta(x) = \frac{1}{n}\sum_{i=1}^{n} d_\theta(x, i)$, where n is the number of segments in a graph, and d_θ is the angle between any two segments on the shortest path on segment x. Angular mean depth, when weighted by the length of segment, can be shown as: $C_\theta^l(x) = \frac{\sum_{i=1}^{n} d_\theta(x,i)l(i)}{\sum_{i=1}^{n} l(i)}$, where l is the length of segment.
    Mean segment integration of an axial map, therefore, indicates the typical accessibility of the segments in a map. Higher mean integration values indicate greater overall accessibility, and lower values indicate lower overall accessibility of the map. It describes the amount of ease one may experience while moving within an area.
15. ***Segment choice/segment angular betweenness*** of a segment is calculated by replacing the shortest paths with paths that have the lowest angular cost for each possible origin and destination pair of segments on the given segment (Turner 2007). Thus the angular betweenness value for a segment x in a graph of n segments is defined as follows: $B_\theta(x) = \frac{\sum_{i=1}^{n}\sum_{j=1}^{n} \sigma(i,x,j)}{(n-1)(n-2)}$ such that $i \neq x \neq j$, where $\sigma(i, x, j) = 1$ if the shortest path from i to j passes through x, and $= 0$ otherwise.
    Since longer segments are likely to lead to more journeys simply because more possible origins and destinations may be fitted along them, a weight constructed by multiplying the length of the origin segment by the length of the destination segment is assigned to each segment on the shortest path to calculate a weighted choice measure. The origin and destination of the path themselves are given half this weight, since on average one would start and conclude a journey at the middle of each segment. The weighted choice measure, thus, is defined as $B_\theta^l(x) = \sum_{i=1}^{n}\sum_{j=1}^{n} \sigma^l(i, x, j)$ such that $i \neq j$. The weighted sigma function σ^l used by the B_θ^l is slightly more complicated than the standard one: if the shortest path from i to j passes through x, it is simply $l(i)l(j)$ (the length of segment i times the length of segment j); if x is the origin i, then σ^l is $l(x)l(j)/2$; and if x is the destination j, it is $l(i)l(j)/2$; otherwise, if x is not on the shortest path between i and j, and if the origin or destination of the shortest path is not from i or j, then σ^l is zero.
    Mean segment choice of a segment map, therefore, indicates the typical potential for through-movement of the segments in a map. Higher mean choice values

indicate greater typical potential for through-movement, and lower values indicate lower typical potential for through-movement of the segments in the map. It describes the amount of ease one seeks while moving from one line to another within an area.

3 Mathematical Developments in Space Syntax

Over the last two to three decades, several mathematical developments have occurred in space syntax methods and measures. Some of the most significant developments affecting urban morphological studies include those investigating (1) how to normalize integration and choice—the two most important space syntax measures; (2) how space syntax measures are related to metric measures; (3) if spatial configurations as described using space syntax methods and measures follow scaling laws; (4) how to minimize boundary effects on space syntax measures; (5) how to enhance space syntax methods and measures for shape-sensitive 2D descriptions; (6) how to enhance space syntax methods and measure for 3D descriptions; and (7) how to integrate spaces syntax methods and measures with GIS. I briefly discuss these mathematical developments next.

3.1 *Normalization of Integration and Choice*

Without normalization, important space syntax measures, such as integration and choice, change as the size of a spatial network changes. Therefore, it is important to normalize these measures in relation to size for comparative studies involving spatial networks of different sizes. Besides facilitating comparative studies, normalized measures are also needed for comparative studies among various stages of growth of a spatial network. Additionally, they are needed for comparative studies among different subareas of a network at the same scale or at different scales. Hillier and Hanson (1984) already introduced relativization, standardization, and/or normalization techniques for integration, allowing researchers to compare spatial networks of different sizes using this measure. Since then, several other normalization techniques for integration have been proposed. For various reasons described below, no normalization technique for choice was suggested until 2012. I turn to some of these normalization techniques next.

3.1.1 Normalization of Integration

Over the years, two generic kinds of normalization techniques have been suggested for integration. The first kind of normalization techniques assumes that the mean depths (MDs) of the nodes of an axial graph are uniformly distributed for any system with the same number of spaces for the purpose of *RA* and *RRA* transformations. In contrast, the second kind makes no such assumptions, because MDs for all non-isomorphic systems with the same number of nodes result in a positively skewed distribution (Livesey and Donegan 2003). The first kind of normalization techniques are provided by Hillier and Hanson (1984), Krüger (1989), and Teklenburg et al. (1993), while the second kind of techniques are provided by Livesay and Donegan (2003), and Krüger and Vieira (2012).

The earliest normalization technique of the first kind for integration was provided by (Hillier and Hanson 1984). Recall that integration, a measure of relative asymmetry (*RA*), is a normalized measure of the mean depth (MD = the total depth of a node divided by the number of nodes in the graph), and is given by $RA = \frac{2(MD-1)}{n-2}$, where n is the number of nodes in the graph. The normalization technique of Hillier and Hanson (1984) involves dividing the *RA* value of a node of a graph with n nodes with the *RA* value of the root node of a diamond shape graph (also called D-value) that has n nodes at the middle level, ½ n nodes at one level above and below the middle level, ¼ n nodes at the levels above and below the ½ n level, 1/8 n nodes at the levels above and below the ¼ n level, and so on until there is one node at the deepest level and one node (the standardized line) at the root of the graph. Known as the "real relative asymmetry" or the *RRA* of the node, it varies between 0 and D^{-1} (which can be greater than 1). In contrast, *RA* varies between 0 and 1.

Krüger (1989) suggests another normalization technique of the first kind, where he divides the *RA* value of a node of a graph with n nodes with the *RA* value of the root node located at the corner of a grid-like standardized graph with n nodes at the middle level, $n-1$ nodes at one level above and below the middle level, $n-2$ nodes at the levels above and below the $n-1$ level, $n-3$ nodes at the levels above and below the $n-2$ level, and so on until there is one node at the deepest level and one node (the standardized line) at the root of the graph. As with Hillier and Hanson's *RRA* values, Krüger's transformed *RRA* values are bounded below by 0 and above by the inverse of the standardized RA values.

Teklenburg et al. (1993) suggest yet another normalization technique of the first kind, where they give an integration score (as opposed to the *RA* value) to a line of an axial map with n nodes based on the comparison of the total depth of the line in the axial map with the mean total depth of a standardized axial map with n nodes. The standardized axial map in their case is a complete bipartite graph. They define the integration score (*I*) as: $I = ln\left(\frac{L-2}{2}\right)/ln(TD - L + 1)$, where L is the number of axial line in the axial map, and TD is the total depth of a given line. In a simulation study, they show that in any given urban area the distribution of integration scores corresponds perfectly with the distribution of the *RRA* values of Hillier and Hanson (1984) and Krüger (1989), and that individual integration scores vary less than the RRA values. Since the integration scores cannot be calculated for an axial map containing a line with total depth less than or equal to the total number of lines, Teklenburg et al. (1993) advise to use the integration score only for urban areas and very large buildings.

Recognizing the fact that each configuration of a given set of spaces is unique although *RRA* values may be repeated for some of the configurations, Livesay and

Donegan (2003) suggest a normalization technique of the second kind, incorporating the distributions of *RA* values generated from all non-isomorphic configurations of a given number of spaces. To normalize the mean depth of a given system, Livesay and Donegan (2003) take the sum of the mean depths generated by all the non-isomorphic systems with a given number of spaces minus the mean depth of the given system, and divide it by the sum of the mean depths generated by all the non-isomorphic systems with the given number of spaces.

Also to take into account the fact that the distribution of depth in an axial map is not uniform, Krüger and Vieira (2012) provide the most recent technique of the second kind to normalize *RA*. They not only consider the mean values given by relative asymmetry (*RA*), but also the corresponding standard deviations of the lines. As they write, "it makes a great deal of difference whether the distribution is spread out over a broad range or clustered closely around the mean" (page 195). The authors calculate the standardized total depth of an axial line (STD_i) using the standard deviation of depth for the axial line as the root node of an axial graph, and the standard deviation of depth for the root node of a diamond shape graph with the same number of nodes as the axial graph. They then use the standardized total depth to calculate the scaled mean depth ($SMD_i = \frac{STD_i}{n-1}$ where n is the number of nodes), and use the scaled mean depth to calculate the scaled relative asymmetry of the axial line (SRA_i) using the following standard formula: $SRA_i = (SMD_i - SMD_{imin}/SMD_{max} - SMD_{imin})$.

This brief survey of different normalization techniques of integration makes it clear that researchers remain interested in the problem. Given the fact that normalization can be done in different ways producing different results, Conroy-Dalton and Dalton (2007) use a decay function to calculate depth values that do not require relativization. While in the standard mean depth calculation the farther a node is from the root node the more value it adds to the mean depth, in the technique proposed by Conroy-Dalton and Dalton (2007) this is reversed: farther nodes add less value while closer nodes add more value to the mean depth calculation. In other words, by introducing the reciprocal of the distance between a node and the root node in the calculation of mean distance of the root node from all other nodes, they are effectively turning the mean depth calculation into a distance-decay model. In this model, one could simulate integration at different radii by simply changing the decay functions. For example, a decay function of 6.5 is able to simulate integration at radius-3 without relativization, because the number of nodes remains constant for all decay functions. In yet another proposed technique, Park (2005) replaces the D-value of Hillier and Hanson (1984) with maximum entropy arguing that the technique is mathematically more elegant, that it makes integration definable for any order, and that it changes the nature of integration from a dimensionless quantity into a physical quantity with a clearly defined dimension. That is because a diamond-shaped graph cannot be defined for all orders, and a *RA* value is not necessary to define integration (Park 2005).

Despite the fact that several techniques are available to normalize integration, it should be noted here that the technique based on a diamond-shaped graph proposed by Hillier and Hanson (Hillier and Hanson 1984) remains robust in practice to this day. As Hillier et al. (2012) writes, "The normalization was not beautiful, but it was effective. A good alternative was proposed by Teklenburg et al. (1993) which is implemented in Depthmap, but by and large, the D-value normalization has proved robust and has continued to be used successfully in the axial analysis of cities (see for example Hillier 2002) and for predicting movement at the design stage of projects." (Hillier et al. 2012, p. 156)

3.1.2 Normalization of Angular Choice and Integration

Unlike integration, no normalization technique existed for choice, which is another important space syntax measure, until the ones suggested by Hillier et al. (2012). The reason for not having a normalization technique for choice was simple. In axial analysis using topological distance, choice was not as powerful as integration for postdiction of traffic movements. This, however, changed after the techniques of segment angular analysis was introduced in the early 2000s. Choice calculated based on angular distance proved as good a measure as integration in postdiction if not better, and a better measure than integration in prediction (Hillier and Iida 2005). Therefore, a need for normalized angular choice was felt, but the normalization technique used in axial integration could not be applied to angular choice and integration. That is because a standardized axial graph in the form of a diamond or grid, or a bipartite graph does not exist for angular distance.

As proposed by Hillier et al. (2012), the normalization techniques for angular choice and integration turn out to be much simpler than those techniques used for axial integration. For normalized angular choice, they simply divide total choice by total angular depth for each segment in the system. That is because the greater the depth, the more the choice value will be reduced by being divided by a higher total depth number. They however define the normalized angular choice (*NACH*$_r$) at radius r using the following formula: $NACH_r = \log(ACH_r + 1)/\log ATD_r + 3)$, where ACH_r indicates angular choice and ATD_r indicates total depth at radius r. They use $(ACH_r + 1)$, because for a segment with no choice the logarithm of 0 will be meaningless. For the same segment, ATD_r will be -1. In this case, adding 1 gives 0 and adding 2 gives 1. Since taking the reciprocal of the logarithm of 0 or 1 is also meaningless, they use $(ATD_r + 3)$. They then show that *NACH* does not have any correlation with the size of the segment map in segment numbers and that it better postdicts actual movement rates than choice in order to validate the normalization technique.

In contrast, for normalized angular integration (*NAIN*) Hillier et al. (2012) use their study finding that indicates angular total depth of a segment (*TD*) increases at a rate of $NC^{1.2}$ where *NC* is node count of the map. Therefore, they use $TD/NC^{1.2}$ to normalize *TD*, and then use the reciprocal of $TD/NC^{1.2}$ to normalize angular integration in relation to the size of the segment map described in segment numbers. Like *NACH*, they then show that *NAIN* does not have any correlation with the size of the segment map in segment numbers and that it better postdicts actual movement rates than angular integration in order to validate the normalization technique.

3.2 Geometric and Topological Measures Versus Metric Measures in Space Syntax

The standard graph-theoretic techniques of space syntax, where spatial units are treated as vertices and their connections as edges, are not sensitive to metric distances. As a result, these techniques are not able to capture many important metrically defined aspects of buildings and cities. Therefore, how to describe spatial configuration taking into account metric distance has been a matter of persistent interest among many space syntax researchers (Hillier 1996, 2012; Hillier and Iida 2005; Rashid 2016; Hillier et al. 2010; Carvalho and Penn 2004; Peponis et al. 2008; Shpuza 2014; Wagner 2008). Already in 1996, Hillier (1996) set out a partitioning theory explaining the interactions among metric distance, topological distance, and shape. Using theoretical examples, he showed that, when building coverage and traveling distance are kept constant in free space, a grid with smaller blocks at the center and larger blocks at the edges has lower mean trip length from all cells to all cells than a regular grid. In contrast, a grid with larger blocks placed at the center and smaller blocks at the edges has higher mean trip length from all cells to all cells than a regular grid. Hillier (2012) also provides several other examples of such interactions.

As we have noted earlier, the two most useful measures of space syntax have been integration and choice in empirical studies on traffic flows and their outcomes. These measures use topological distance and/or angular distance to assess the cost involved in trips, without taking into account metric distance, land use and other demographic and socioeconomic factors. This is different from traditional traffic modeling techniques that use metric distance, land use, and other demographic and socioeconomic factors to predict travel behaviors. This has raised the question if space syntax would be better served if its measures were weighted by metric distance or if the dual graphs of space syntax were replaced with primal graphs and metric distance (Ratti 2004; Steadman 2004; Porta 2006; Porta et al. 2006; Crucitti et al. 2006). We now have several space syntax studies clarifying the differences between topological and angular distances used in space syntax and metric distance used in the other traffic models (Hillier 1996; Hillier and Iida 2005; Hillier et al. 2010; Peponis et al. 2008, 2007).

Some of these studies show that, in the dual structure of the city, for example, the foreground structure is identified by integration and choice measures defined based on topological and angular distances. In contrast, metric measures help partition the background structure into a patchwork of semi-discrete areas as a function of urban block geometry (Hillier et al. 2010). These studies also show that integration and choice, computed based on topological angular distances in the axial and segment maps, explain the global structure-functions relations such as natural movement (Hillier and Iida 2005). In contrast, metric distance explains the local structure-function relations as exemplified by the distance-decay effects, and the phenomenon of grid intensification required to minimize trip distances in areas of intensive use (Fotheringham 1981; Hillier 1999a, b; 2002; Siksna 1998). Additionally, they show that the substitution of metric for topological and angular distances in integration and choice may have catastrophic effects on the ability of these measures to account for the structure-function relations at the global scale (Hillier and Iida 2005; Hillier et al. 2010).

Moreover, these studies show that integration computed using topological or angular distance, and integration computed using topological or angular distance and weighted by lengths show statistically significant, very strong, positive correlations, indicating that they may be statistically similar (Hillier et al. 2010). In contrast, when each relation in an axial map or a segment is weighted by metric lengths for computing integration, integration produces a concentric pattern from center to edge trivializing its ability to differentiate spatial structures (Hillier et al. 2010; Porta 2006). Therefore, it has been suggested that, instead of weighting axial lines or segments by metric length, a better way to consider metric length in configurational analysis may be to consider integration computed at different radiuses. If putting no radius restriction on integration is equivalent to considering the total street length in the analysis, then putting a radius restriction may be equivalent to considering the length up to certain distance defined metrically, topologically or angularly. For example, the angular integration at angular radius 4 (up to four right angles from each root segment) for London and the total street length from each segment within the radius has a very strong correlation with an r^2 of 0.97, indicating that at this radius one has effectively taken into account the total street length of London in computing integration.

The fact that we can increase the correlation between the total street length and integration by increasing radius up to a certain level can be explained by the *averaging* and *overlapping* effects. According to the averaging effect, as we increase the radius, differences in segment lengths average themselves out, so that the total segment length becomes a very close function of segment counts. This is indicated by the fact that segment count is strongest predictor of integration with restricted radius (Dalton 2005). According to the overlapping effects, as we increase the radius, overlaps among segments that can be reached from individual root segments increase. The overlaps among the segments within the reach of all segments are complete when no radius restriction is imposed (Hillier et al. 2010).

The relationship between integration and street length suggested above has been formalized as "directional reach" by Peponis et al. (2007, 2008): While "reach" is the aggregate street length that can be accessed from the midpoint of each road segment subject to a limitation of distance, a related concept "directional distance" is the average number of direction changes needed in order to access all the spaces within reach. Their studies show that these measures discriminate well between different morphologies of street networks; that they are well correlated with other standard space syntax measures such as integration and choice; and that these measures can be used to model the effects of spatial configuration upon movement in ways that compare favorably to standard space syntax measures.

However, it is noteworthy that while researchers in general did not see any benefit of the weighting of integration by metric length for the structure-function relations, Zheng et al. (2010) show that, there is a negative correlation between the logarithm of road width and the total depth of the axial maps for a set of Chinese cities, and when integration is weighted by street width the distribution of integration become more realistic for the cities under investigation. Since integration is an inverse function of the total depth, the negative regression between the logarithm of road width and the total depth of the axial maps may suggest that a positive regression may exist between integration weighted by street width and traffic and pedestrian flows. However, studies on the effects of integration weighted by street width on the structure-function relationship have not been conducted to verify the claim. It is also noteworthy that Rashid (2016) has found that several of the metric, angular and topological properties of the urban layouts of the 2 mile by 2 mile downtown areas of more than 100 large cities from different countries around the world follow scaling laws. These studies therefore suggest that more studies are required to fully understand the relationships among the metric, topological and angular properties of the layouts of cities (see below for additional information on this topic).

3.3 Scaling and Universality in the Axial and Segment Maps

Scaling and allometric relationships in buildings and cities are important because they show us the way the size and location of buildings and cities are constrained by geometry, and help us detect the underlying structure of cities as patterns of urban self-organization (Rashid 2016). Studies involving the structure of axial maps have revealed several universal properties including scaling laws (Carvalho and Penn 2004; Shpuza 2014; Wagner 2008; Crucitti et al. 2006; Figueiredo and Amorim 2005; Jiang 2007; Volchenkov and Blanchard 2008). One universal property found in these studies is that axial maps generally have a small number of long lines and a large number of small lines regardless of the size of the city (Rashid 2016; Carvalho and Penn 2004; Hillier 2002). In their study, for example, Carvalho and Penn (2004) find that the length of axial lines of axial maps of a sample of 36 cities from 24 different countries are largely independent of city size; that the length shows self-similarity across morphologically relevant ranges of scales of two orders of magnitude; that the sample of cities represent three different classes as per the exponents of power relations—one with an exponent close to 2, one with an exponent close to 3, and the last one does not follow a single curve; that the cities with an exponent close to 2 have a dominant global structure as opposed to the dominant local geometric structures of the cities with an exponent close to 3; and that the dominant global geometric structures of a city seem to have appeared at a late stage of urban growth, and change the structure drastically using long axial transportation routes.

Another universal property found in these studies is that, patterns formed by longer and shorter lines axial lines follow consistent geometric laws, again, regardless of the size of the city. For example, longer lines generally connect with other longer lines at their ends by nearly straight intersections in order to create the dominant foreground structure with multi-directional sequences to facilitate flow within the city (Hillier 2010, 2012; Figueiredo and Amorim 2005). In contrast, shorter lines, in spite of generally having highly variable geometric patterns, tend to form clusters near longer near-straight lines in form of grid-like patterns. These grid-like patterns of the background structure contain lines either intersecting each other or ending on each other at nearly right angles (Hillier et al. 2010).

Rashid (2016) studied the rank-size distributions of all the measures defined earlier in this paper. These measures include the 11 measures of the axial maps, the 4 measures of the segment maps, and the two measures related to both the axial and segment maps of the 2 miles by 2 miles downtown areas of more than 100 big cities from different parts of the world. The findings of his statistical analysis show that, out of these 17 measures, 11 have dominant logarithmic distributions with less dominant linear and power distributions, 5 have dominant linear distributions with less dominant logarithmic and power distributions, and 1 has dominant power distribution with less dominant logarithmic and linear distributions. Interestingly, the two less dominant functions in each case also describe the rank-size distribution quite well, indicating that smaller families of less dominant distributions may be nested within a larger family of a more dominant distribution of each of these measures.

Whether dominant or not, in all the power relations of the geometric measures of axial and segment maps the values of parameter α are unequal and less than one, indicating inconsistent and slow rates of change for these measures with changes in rank order. Among these, the ratio of mean segment length to mean axial line length has the lowest rate of change ($\alpha = -0.186$), while the mean segment choice has the highest rate of change ($\alpha = -0.727$). A pairwise comparison of the rank-size distributions of the measures of axial maps and segments maps shows somewhat similar logarithmic patterns for the number or count, length, integration, and choice of axial lines and segments, indicating faster rates of change for cities in higher ranks and slower rates of change for cities in lower ranks. Similarities in the rank-size distributions of the measures of the axial and segments maps are also shown by the fact that the rank-size plots of the ratio of segment and axial line lengths and that of the ratio of segment and axial line numbers indicate that the rate of change in the number of segments in relation to the number of axial lines is somewhat similar to the rate of change in the length of segments in relation to the length of axial lines.

The rank-size distributions of the other measures of the axial maps in Rashid's study (Rashid 2016) show that axial articulation, grid axiality, axial ringiness, and mean axial connectivity are best described by logarithmic functions, indicating variable rates of change with rank order. In contrast, mean axial controllability and intelligibility are best described by linear functions, indicating a somewhat constant amount of change with rank order. Among all the measures, only the mean relativized entropy of the axial map shows a dominant power distribution, indicating a constant rate of change with rank order.

3.4 Boundary Effects on Space Syntax Measures

Street network analysis using space syntax often excludes elements beyond an artificially defined boundary. Such omissions can affect the results of space syntax analysis, and can raise questions regarding the reliability of space syntax analysis (Ratti 2004). For example, integration (closeness centrality), which calculates the average distance from each node to all other nodes on the network, is sensitive to the spatial distribution and density of the nodes. Any changes to the network boundary not only include and/or exclude new nodes, but these nodes are likely to change the overall relational patterns among all the nodes affecting their integration. Choice (betweenness centrality), which calculates the frequency that certain nodes are used in the shortest paths between all pairs of nodes on the network, is sensitive to the connections of routes. That is because changes to the network boundary can change routes, thus changing the distribution of shortest paths through the network. These limitations caused by the boundaries of a study area are also significant because it is difficult to predict the impact of any newly included or excluded nodes on the results of an analysis.

A common way to solve the problem has been to extend the boundary of a study area to urban edges or to existing natural and/or artificial edge conditions beyond which the city cannot grow. In today's urban context, however, such natural and/or artificial limits of the city hardly exist, or exist only at the metropolitan or regional scale. Since it is unwise and inefficient to include a metropolitan region in studies involving local areas, a reasonable way to define the boundary of study areas is needed. Several space syntax publications deal with this problem (Hillier and Iida 2005; Turner 2007; Peponis et al. 2008; Chiaradia et al. 2012; Gil 2015, 2016; Hillier 1993; Krafta 1994; Park 2009; Penn 1998).

In studies involving local areas, according to Hillier et al. (1993), the most effective way to eliminate the boundary effects may be to define a cut-off distance for analysis based on the maximum depth of any line from any other line within the study area. Therefore, if the maximum depth of any line from any other line within the study area is 7, then extend the study area in a way that will include all lines that lie within in 7 steps from any line within the study area. They call this area "the catchment area of the catchment area," and use it in many studies already cited above. This method however does not tell us the best way to define an area for determining the global syntactic measures of the area, where maximum depth from any line to all the other lines may vary from one line to another. In a situation where no natural and/or artificial limits exist to define the global whole, the process may force us to consider an unnecessarily large study area to eliminate boundary effects.

Gil (2015, 2016) studies the edge effects on the street network centrality of 100 study areas randomly selected from the Randstad region in the Netherlands. The author computes the centrality measures of circular areas of 800 m, 1.5, 3, 5 and 10 km radius at each location, where the area with the 800 m radius boundary represents the neighborhood scale and the area with 10 km radius boundary represents the metropolitan scale. The author also computes the centrality measures of four additional areas around each locations defined by shifting a 10-km radius boundary by 5 km from the center of these locations toward the North, South, East and West. He then compares these measures with the centrality measures computed at the global scale for the same 100 locations within the entire metropolitan region as the edge effect free benchmarks. The results of his study show that that as the network boundary increases in size, the correlations between the local and the regional area measures increase and the results for the various locations become more consistent with a smaller spread, converging with the results of the regional analysis. Nevertheless, the behavior of the results for the different centrality measures and types of distance is not the same. The results also show that a street network model using an analysis buffer, as suggested by Hillier et al. (1993), has a positive impact on the reliability of the centrality measures of street networks.

Based on the study, Gil points out, while the results clearly indicate that centrality measures are likely to vary with changes in study area boundaries, they do not indicate if centrality measures computed using any one boundary are better than those computed using the other boundaries for empirical studies. He also points out that as much as one would like to have a standardized solution to consistently deal with edge effects, many analysis parameters and boundary design options are at stake here. Different measures of network centrality and types of distance capture and describe different characteristics of the model. One should choose them based on relevance to the research problem, and use different ones to obtain a more complete description of the study area (Gil 2016). As long as there is a reasonable sound explanation, the boundary of a study area can be an important aspect of space syntax analysis (Park 2009).

3.5 Shape and Space in Space Syntax

While the standard convex maps, axial maps, and visibility graphs of space syntax have quite successfully served their purposes, many studies have shown concerns regarding how these maps and graphs are defined (Turner et al. 2005; Peponis et al. 1998a, b; Ratti 2004; Larkham 1988; Batty 2004; Batty and Rana 2002; Peponis 1997; Psarra 2003; Psarra and Grajewski 2001; Yoon 2009; Rashid 1998, 2011, 2012a, b; Carvalho and Batty 2004, 2005; Batty and Rana 2004; Turner et al. 2001). Some of the problems identified in these studies include: (1) the standard definition of a convex map as the fewest number of fattest convex spaces that cover a spatial system (Hillier and Hanson 1984) is not mathematically well defined; (2) the standard definition of an axial map as the fewest number of longest axial lines that cover a spatial system (Hillier and Hanson 1984) is not mathematically well defined as well; (3) the standard visibility graph analysis of space syntax do not include mathematically well-defined techniques to specify a sufficient set of isovists needed to describe an environment; and (4) the methods to generate convex maps, axial maps, and isovists are not explicitly integrated into a single framework, hence remain unrelated to each other even though they represent different aspects of the same spatial network.

Therefore, many alternative ways have been suggested defining rigorously the convex maps, axial maps, and visibility graphs of space syntax using different elements of the shape of a plan (Peponis et al. 1997, 1998a, b; Rashid 1998, 2012). For example, since defining a minimum convex partition is an intractable problem, Hillier (1996) provides a method to identify a set of overlapping convex spaces instead of a unique convex partition. The method extends all the extendible surfaces within a plan, which, of course, produces a very large number of overlapping convex spaces. For his overlapping set, Hillier selects only those spaces where each side has at least one wall surface partially or fully covering the side (Peponis et al. 1997). In contrast, Peponis et al. (1997) provide at least three different ways to define convex partitions. One of these is called a minimum partition because it uses a minimum number of convex spaces to cover a plan. This partition, however, may not always contain the fattest possible convex space as required by the early definition. In addition, a minimum partition is not uniquely defined because a plan may have more than one such partitions (Rashid 2012).

Likewise, Batty and Rana (2002, 2004) use isovists to unambiguously generate axial maps. They argue that axial lines can be approximated by the maximum diameters of isovists, and propose a generic algorithm for sorting isovists by their diameters and using the axial map as a summary of the extent to which isovists overlap (intersect) and are accessible to one another. They demonstrate their techniques for the small French town of Gassin used originally by Hillier and Hanson (1984). Though the method provides an unambiguous way to generate an axial representation based on isovists, sometimes it generates axial lines that are not connected in the way they would be in a traditional axial map ensuring continuity in movement. Additionally, the method identifies several well-defined processes to determine the scale of resolution at which isovists must be drawn, but these processes involve rules unrelated to the environment it describes.

In another example, Peponis et al. (1998a) generate linear representations of plans based on the s-partitions (defined by the extensions of extendible walls or surfaces) and e-partitions (defined by the extensions of extendible walls and diagonals). The traditional axial map, which is one of the three different linear representations presented by these authors, is defined as the minimum set of lines needed to cross all s-lines (as opposed to "cover all convex spaces" of the traditional definition) and to complete all movement rings of a configuration. The techniques used by Peponis and his colleagues to generate linear representations are somewhat related to an "all-lines map" used by Hillier (1996). Hillier's all-lines map includes all diagonals, e-lines and s-lines (Peponis et al. 1997; Rashid 1998, 2012), but does not provide any method to extract an economic linear representation as is done by Peponis et al. (1998a). Later, some of the ideas presented by Peponis et al. were used to automate the generation of axial representations by Turner et al. (2005).

Finally, it is noteworthy that Peponis et al. (1998b) and Rashid (1998, 2011, 2012) provide different ways to rigorously describe visibility within plans, as potential alternatives to the standard visibility graph analysis of space syntax (Turner et al. 2001). Taken together, the systems of configurational analysis developed by Peponis et al. (1997, 1998a, b) and Rashid (Rashid 1998, 2011, 2012a, b) not only help redefine convex maps, axial maps, and visibility graphs, but also help establish systematic relationships between spatial configurations and shape. Therefore, they help solve a longstanding problem of space syntax that treated these maps and graphs mathematically unrelated to each other. However, any cognitive importance of a shape-based description of spatial configurations as provided by Peponis et al. and Rashid is yet to be established firmly, which is discussed elaborately in Rashid (2012a, pp. 748–751).

3.6 3D Descriptions in Space Syntax

Height becomes important for cities due to topography and buildings. Land surfaces rise and fall and buildings vary in height. The standard techniques of space syntax have been limited to describing the configuration of the two-dimensional (or 2D) space based on the argument that we are able to read many three-dimensional (or 3D) qualities of space from its 2D representations. Yet, topographical changes and building heights have indelible impact on urban form, structure, and processes. Finding ways to include the 3D qualities of space in space syntax techniques, either as building heights or as topographical changes, should allow researchers model the environment more accurately for studies on urban form, structure, and processes.

In recent days, various ways to study the 3D qualities of the environment within the space syntax environment have been proposed by extending or complementing its standard methods and techniques (Asami et al. 2003; Ratti 2005; Schroder et al. 2007; Suleiman et al. 2013; Wang et al. 2007). These methods can be put into several categories. First, there are those that modify the standard techniques of space syntax, making them more sensitive to topographical changes. Then, there are those that suggest ways to associate external representations to the standard space syntax techniques making them more sensitive to the 3D qualities of buildings. Finally, there are those techniques that are different from the standard space syntax techniques but are able to describe the 3D qualities of space in addition to some of the same things that the standard space syntax techniques can do, as has been argued.

Examples of the first category may include the techniques suggested by Asami et al. (2003) and Schroder et al. (2007). Schroder et al. (2007) quantifies urban visibility using the 3D axial lines generated based on algorithms identified by both Peponis et al. (1998a) and Turner et al. (2005) for the creation of "all line axial maps" in 2D space. Using the techniques, Schroder et al. provide computerized tools that are able to identify areas of low and high visibility in 3D spaces. The authors hope that incorporating 3D into space syntax and representing the space in the way that the individual sees it will allow researchers to gain a better understanding of 3D spaces and improve how social processes operate within these 3D spaces.

In contrast, Asami et al. (2003) suggest two techniques to consider height changes in topography. In one of these techniques, "extended axial lines" are used to take into account changes in height of the surface of streets. Since axial maps are drawn based on 2D maps, the endpoints of an axial line may not remain mutually visible due to changes in topography between these points. In order to avoid any such case, an extended axial line is introduced as a composite of line segments, whereas the endpoints of every segment remain mutually visible at the eye-level (i.e., 1.5 m above the surface of the street). In other words, to follow the curvature of street, the axial line is allowed to break into segments as long as the endpoints of each segment remain mutually visible. Mathematically, the process considers an axial line as multiple segments without changing the line itself, and computes the syntactic values of these segments. In the other technique, Asami et al. (2003) uses a weighting function to take into account the fact that any two neighboring segments of an extended axial line may remain partly visible to each other due to height changes. They determine the weighting function based on the angle of incident formed by any two neighboring lines. In this case, any computation of syntactic values occurs only after the weighting values are applied to the edges between any two nodes representing axial segments. While syntactic values computed using the first technique of Asami et al. (2003) showed strong associations with buildings and commercial activities along streets, it failed to identify traditional local centers. In contrast, syntactic values computed using the second technique was not able to improve the postdictive abilities of the standard space syntax techniques.

Examples of the second category may include the techniques suggested by Wang et al. (2007). In this case, attempts are made to integrate the image of the city with space syntax techniques by weighting the integration values of spaces based on the importance of these spaces as image points. For their purpose, the authors divide an urban environment as a set of cells, and characterize each cell as an image point. They consider two kinds of image points. Functional image points are important for functions they serve. Formal image points are important for their strong visual characteristics and symbolic contents. A cell could serve as both image points. The authors measure each image point based on such characters as prominence, visibility, typicality and special meaning, and compute the general distinctiveness (*GD*) of each cell as an image point. A cell with lower GD values (<20%) becomes part of the ambience of the environment, while a cell with higher value become detached. The authors then use the *GD* value to calculate the weighted integration of a space using the formula shown here: $I_{wi} = \delta I_i$, where I_{wi} is weighted integration, I_i is integration computed using the standard space syntax method described previously, δ is the weight coefficient given by: $\delta = GD \times \gamma$, and γ is the correction value as given in the table below

Correction value	Functional image point	Formal image point
γ	$1 + \left(P_{av} + \sum_{k=1}^{n} \frac{P_k}{n}\right)\%$	$1 + \left(P_{av} + \sum_{k=1}^{n} \frac{P_k}{n}\right)\%$

Where P_{av} is the weight value for EV (user evaluation), P_k is the weight value for such variables as PV (prominence value), VL (visibility and localization values), LT (categorization and level of typicality), UM (unique membership of the category), and SM (specific meaning). Based on their study findings, Wang et al. argue that weighted integration may describe and predict the distribution of people's behavior better than standard integration.

Examples of the third category may include the techniques suggested by Ratti (2005). In this study, the author presents a number of ways the urban digital elevation model (DEM) could provide complementary information to space syntax. As noted by the author, initially defined and used in the geosciences, the DEM is a compact way of storing 3D information of the surface of the Earth using a 2D matrix of elevation values, where each pixel represents the height of the surface. Nowadays, the DEM is increasingly being used in urban areas, where the pixels also store information on building heights. Using an urban DEM, the author presents how to calculate (1) the line of sight, i.e., the axial line, in the street network; (2) the height-to-width ratio (also known as H/W or aspect ratio; (3) the viewshed or isovist (also see (Suleiman 2013) on this issue); (4) the accumulated metric and topological distance (i.e., integration) from a location to all other locations; (5) the generalized distance in terms of time, friction, etc.; and (6) the accessibility of buildings from given locations. Although none of these techniques shows how the standard techniques of space syntax, such as the axial and segment map analysis or the visual field analysis, can be enhanced taking into account the 3D qualities of the environment, the fact that each of them can theoretically be performed on a DEM may indicate that systematic relationships among them can be established.

3.7 Space Syntax and GIS

Since the late 1990s, the members of the space syntax community have been trying to integrate space syntax with GIS for many reasons. First, they want to combine space syntax's emphasis on the configurations of space with the geographer's emphasis on socioeconomic processes and her more particular notions of "place" (Jiang et al. 2003; Jones et al. 2009). Second, they want to get access to the more widely available GIS data, since cost associated with data collection has been an important limiting factor in space syntax research and its commercial acceptance (Dhanani 2012; Marcus 2005; Reis et al. 2005, 2007; Stahle 2007; Ståhle et al. 2005; Topcu and Kubat 2009; Vaughan 2005, 2007). Third, they want to take advantage of the powerful modeling and visualization techniques of GIS (Gil 2007).

The use of GIS data in space syntax studies however has not been straightforward. For this, many technical problems needed to be solved: how to convert the centerlines maps of GIS to the axial or segment map of space syntax (Dalton 2003); how to assign spatially distributed data (plots, building, land use, poverty, crime, etc.) to individual streets (Vaughan 2005, 2007); or how best to implement space syntax routines within GIS software programs or GIS capabilities within space syntax software programs for statistical analysis and visualization purposes. In solving these technical problems, researchers often needed to solve basic theoretical questions involving the differences between a primal graph of street networks used in GIS and a dual graph of street networks of space syntax (Porta 2006; Batty 2004; Jiang and Claramunt 2004), the mathematical differences between a centerline map and an axial or segment map (Turner 2005, 2007), the usefulness of axial maps over centerline maps for configurational analysis (Jiang and Claramunt 2004), and the definition of accessibility and how to define accessibility at fine spatial scales of streets and buildings (Jiang 1999).

In several papers, Jiang and colleagues (Jiang et al. 2000, 2003, 2004; Jiang 1999; Jiang and Claramunt 1999, 2000) introduced space syntax to the GIS community and proposed several computational techniques to integrate space syntax with GIS. In one of these papers (Jiang 1999), the author reviewed geographic and geometric accessibility, defining accessibility in space syntax as a special case of geometric accessibility. Accessibility, which is the relative nearness of one place i to other places j, can be defined in the following generalized terms: $A_i = \sum_j f\left(W_j d_{ij}\right)$. where W_j is some index of the attraction of j and d_{ij} is a measure of impedance, typically the distance or travel time of moving from I to j. In land use-transportation modeling, accessibility defined based on a gravitational equation therefore can be considered a special type of accessibility, which is expressed in the following form: $A_i = K \sum_j f\left(P_j d_{ij}^{-\alpha}\right)$, where P_j is the population at j, K is the gravitational or scaling constant, and α is a friction of distance parameter. However, at smaller scales where accessibility between buildings and streets are considered, the attraction may not be meaningful. In such cases, setting the attraction $W_j = 1$, we can define geometric accessibility in the following manner: $A_i = \sum_j A_{ij} = \sum_j f\left(d_{ij}\right)$.

In their paper, Jiang et al. (1999) argued that while most proprietary GIS had the measurement of accessibility at geographic or thematic level in many of their standard functions and extensions, they did not have tools for defining accessibility measures at fine scales involving the geometry of urban structure in terms of streets and buildings. Therefore, they implemented the measures of geometric accessibility of space syntax through a software extension within the desktop GIS ArcView, showing how to input street systems within the software through new drawing tools, generate accessibility measures through new computational functions, and present these visually using standard ArcView outputs.

Several years later, in an attempt to integrate geographic and geometric accessibility, Stahle et al. (2005, 2007) introduce a new GIS-based application, called the *Place Syntax Tool* (*PST*). In *PST*, attraction can be defined in terms of population density, pedestrian density, land use density, and any other place variables, and distance can be metric or topological. To do so, they define d_{ij} in the following manner: $d_{ij} = h\left(\tau\left(x_{ij}^m, y_{ij}^m, t_{ij}^m, e_{ij}^m, \ldots\right), \theta^z\right)$, where $\tau(.)$ is a representation of space, θ^z is a vector characterizing the preference of individuals, and x, y, t, e, etc., are different distance variables, where x is walking distance, y is bird's flight, t is travel time, and e is topological distance measured as axial steps. Using the above definition of distance, they provide the full *PST* formula for a "geographical accessibility measure with axial lines" in the following form: $A_i = \sum_j f(g\left(W_j \theta^\alpha\right), h\left(\tau\left(e_{ij}^m\right)\theta^z\right))$. According to the authors, even though the purpose of *PST* is well described by this formula, *PST* is also able to calculate x_{ij}^m, y_{ij}^m, and a combination of e_{ij}^m and x_{ij}^m. According to the studies reported by Stahle et al., accessibility to population and floor areas within axial lines computed using *PST* predicts pedestrian movement as good as the standard integration of space syntax does. *PST* measures are also better predictors of pedestrian movement than walking distance and straight-line distance (bird's flight).

Besides introducing space syntax as a special case of accessibility within GIS, Jiang and Claramunt (2004) also introduced techniques for a structural representation of a street network using graph principles, where vertices represent named streets and links represent street intersections. Based on this representation, so-called connectivity graph, they provided local and global measures to qualify the status of each individual vertex within the graph. They also provided two street selection algorithms based on these structural measures, and applied the approach to a middle-sized Swedish city validating its findings against the findings of the standard space syntax techniques.

Since the road centerline maps of cities are easily available GIS data, several studies examine how best to analyze these maps using space syntax techniques. In one of these studies, Dalton et al. (2003) examine how to convert the road centerline maps of the TIGER (Topologically Integrated Geographic Encoding and Referencing) files of United States Census Bureau based on the concept of angular fractional depth. Turner (2005) also examines the same issue using the angular change between the segments (i.e., the segmental angular depth) of the centerline maps. Peponis et al. (2007, 2008) provide parametric definitions of two measures of connectivity—reach and directional distance, and implement their computation using a new software program that runs on standard GIS representations of street centerlines. Liu and Jiang (2012) develop an automatic solution to generate axial lines from street centerlines using the Gestalt principle of good continuity.

Space syntax researchers often use the syntactic variables to explain the variations in census data, land use data, traffic data, crime data, topographical data, street maps, and satellite images, which are some of the easily available data sets in GIS for cities around the world. Reis et al. (2005, 2007) examine security and crime in a Brazilian city using GIS and Space syntax. Omer and Gabay (2007) examine the correlation between the syntactic variables of space syntax and the homogeneity/ heterogeneity of income distribution in six towns in Israel using detailed GIS social census. Using GIS and space syntax, Topcu and Kubat (2009) study urban and spatial factors that affect the urban land values in residential areas of Istanbul. van Nes et al. (2012) and Ye and van Nes (2014) use GIS to study the relationships between the metrics generated by various space analysis tools and socioeconomic data. The space analysis tools they use include (1) "space matrix" that focuses on various types of density on the urban block such as FSI, GSI, OSR (or spaciousness), network density (N) and the number of floors (L); (2) space syntax measuring accessibility in terms of topological and geographical distance; and (3) the Function Mix model measuring the degree of mix of functions in terms of dwellings, work places and amenities. Vaughan et al. (2005, 2007) and Carpenter and Peponis (2010) study the spatial distribution of poverty and social segregations using GIS data. Peponis et al. (2007) and Jiang et al. (2009) study the relationship between accessibility and traffic distribution in cities using GIS data.

Given the importance of GIS for space syntax studies, efforts to integrate GIS and space syntax tools and techniques remain unabated. Developed by Gil et al. (2007), *Confeego* is a plug-in for the MapInfo Professional GIS. It offers the standard space syntax tools, and it can import results from different space syntax software. It includes functionality for querying, displaying and analyzing multiple layers of information. It may be used to perform a multitude of analysis involving large data sets. It may however not be ideal for beginners with no GIS experience. First developed by Turner (2010), *Depthmap* is a popular software program among space syntax researchers, with functionalities of a traditional GIS system. As we have already noted above, several other tools also exist to run space syntax routines within the GIS environment. Besides new software programs and plug-ins, Dhanani et al. (2012) examined if freely available Volunteered Geographic Information (VGI) such as the Open Street Map (OSM) could be used in place of the costly hand-drawn axial maps of space syntax. Beyhan (2011) examined the potential of Free and Open Source Software for GIS (FOSS4GIS) for the extraction of adjusted graphs and calculation of basic space syntax and social network analysis parameters, thus opening up avenues for new research involving these three different areas.

4 Research Applications of Space Syntax

Space syntax concepts, methods, and measures have been applied to different environments including cities, districts, neighborhoods, and individual buildings, characterizing the syntactic structures of the spatial configurations of these environments. They have been also applied to study the spatial distribution and organization of objects, people, and functions explaining the social and cultural meanings of space in these environments. Additionally, they have been applied to study how various syntactic properties of spatial configurations affect such everyday practices as pedestrian and vehicular flows, face-to-face interactions, navigation, crime, and urban liveliness. Studies are also being conducted using space syntax methods and measures to understand the effects of spatial configurations on social capital and to understand the history of cities and societies in relation to space. While we review some of these research applications next, the reader is encouraged to consult the following sources for more on research using space syntax: *Journal of Space Syntax*, *Environment and Planning B*, and *Space Syntax Symposium Proceedings.*

4.1 The Syntactic Cores and Types of Spatial Configurations

In space syntax, the syntactic core of an environment is generally defined by a subset of spatial units in an axial map, a segment map, or a convex map, whose values for a syntactic measure are generally ranked in the top 10% percentile. For example, the 10% most integrated axial lines of an axial map form the integration core of the axial map. A syntactic type of the axial maps refers to a common shape formed by the integration cores of the axial maps of different cities, urban areas or buildings (Hillier 1996; Hillier and Hanson 1984; Hillier and Vaughan 2007; Peponis 1989; Hillier et al. 1987).

Fig. 2 Different types of syntactic cores. This analysis was performed by *Depthmap* (Turner 2007)

In space syntax studies, the syntactic cores of the spatial configurations of cities and urban areas often display different syntactic types. One syntactic type is a tree-like pattern where the most integrated lines branch out like a tree into different areas of the city (Fig. 2a). Another syntactic type is a grid or deformed grid pattern, where the most integrated lines form a fine-grain mesh with cells no bigger than a few urban blocks (Fig. 2b). Yet another syntactic type is a deformed wheel, where the most integrated lines form a pattern of a ring with spikes (Fig. 2c). Still, another syntactic type is a super grid, where the most integrated lines form a net-like pattern with very large cells, each containing numerous urban blocks (Fig. 2d). In many other cases, some combinations of different types of syntactic cores may be found. It is worth noting here that the geographical reach or span of a syntactic core may vary from one city to another. In some cases, the syntactic core of the city may remain confined within a limited area. In some other cases, the core may span across the city covering much of its areas.

Both the shape and span of syntactic cores provide different configurational opportunities affecting such generic functions as movement, interaction, and navigation at different scales of the city. For example, a tree-like syntactic core may suggest a hierarchy in the structure of the city, where accessibility may diminish from the center to the periphery of the core. In contrast, accessibility remains relatively stable when an area has a syntactic core in the form of a grid with a fine-grain mesh, emphasizing accessibility over mobility. In comparison, a super grid pattern generally serves a much larger area than a grid defined by a fine-grain mesh, emphasizing mobility over accessibility. They are often found in cities that have grown very rapidly over a vast area using a street network containing very large cells. In this type, very large cells often contain different street configurations due to speculative developments. In space syntax studies, the shape and span of syntactic cores are often used to explain the economic, social and cultural functions of cities (for example, see (Hillier 1996, 2005; Hillier and Vaughan 2007; Peponis 1989; Hillier et al. 1987; Ahmed et al. 2014; Hanson 1989; Karimi 1997, 2012; Shpuza 2009; Omer and Zafrir-Reuven 2010; Topcu and Kubat 2012, 2007; Rashid and Alobaydi 2015; Rashid and Shateh 2012)).

4.2 *The Whole and the Parts Relationships of Spatial Configurations*

It is possible to study the whole and the parts relationships of the spatial configuration of an environment in different ways using space syntax. One common way to do this is to compute the syntactic values of the spatial units of a spatial configuration at a restricted radius (e.g., integration at radius-3), identify the local syntactic cores of the configuration using the values, and then study how the local cores are distributed, and how they are related to each other and to the syntactic core of the whole. It is also possible to analyze statistically local areas as independent systems or as systems embedded in a larger system revealing the whole and the parts relationships. Concerning this, different scenarios can be imagined. In one scenario, a part of the city may show good correlations between its local and global syntactic values both as an independent system and as a system embedded in the larger system, indicating that the part possesses identity while preserving its relationship with the larger entity. In another scenario, a part of the city may show good correlations between local and global syntactic values as an independent system but not as a system embedded in the larger system, indicating that the part possesses identity without having any strong relationship with the larger entity. In yet another scenario, a part of the city may show poor correlations between local and global syntactic values as an independent system but may show good correlations between the same as a system embedded in the larger system. As a result, the part may lack identity as an independent system but it may form an integral part of the larger system. Finally, a part of the city may show poor correlations between local and global syntactic values as an independent system as well as a system embedded in the larger system, indicating that the part lacks identity as an independent system and as a system embedded in the larger system. Using some of these scenarios, in one of many such studies, Hillier (1996) shows how some modern housing estates fall apart as local areas and how some old commercial areas preserve identity as well integrity within the global system of the City of London.

The whole and the parts relationships of the physical city is probably best exemplified in Hillier's concept of the dual city (Hillier 2002, 2005, 2006; Hillier and Vaughan 2007). According to this concept, all cities are created by a dual process with each part of the process exploiting the relationships between movement and spatial networks in different ways. One part uses space in the form of a *foreground* network to generate life in the city, and the other part uses space in the form of a *background* network to conserve life that already exists in different parts of the city. According to Hillier, the generative role of the foreground network in the city is primarily determined by public processes including microeconomic factors. These processes tend to follow similar logic in cities and manifest themselves in similar global structures of spatial networks in cities. Most often, these structures take the shape of a deformed wheel in an attempt to bring the center and the periphery of the city closer. In contrast, the conservative role of the background network is primarily determined by local processes including cultural factors. These processes vary not only from one city to another but also from one part to another part of a city, and manifest themselves in different local structures. According to Hillier, every city is unique in the way it finds a balance between the two parts of the dual process in its syntactic structures. We illustrate this fact using the City of Baghdad as an example (Figs. 3, 4 and 5). These figures show the dual structures of the city for three different measures of space syntax—axial integration, segment integration, and segment choice.

To illustrate the dual processes and structures, Hillier often uses Nicosia in Cyprus as an example (Hillier 2002, 2005, 2006; Hillier and Vaughan 2007). The axial map analysis of Nicosia picks up a global deformed-wheel spatial structure reflecting the more consistent nature of economic processes. It also picks up more nuanced local spatial structures indicating differences in local processes. Interestingly, these local structures match well with the territories better known as the Greek and Turkish quarters. About the spatial structures of these quarters, Hillier writes, "Their line geometry is different. In the Turkish quarter, lines are shorter, their angles of incidence have a different range, and there is much less tendency for lines to pass through each other … Syntactically, the Turkish area is much less integrated than the Greek area. We can also show that it is less intelligible, and has less synergy between the local and global aspects of space" (Hillier and Vaughan 2007, p. 221). This and many other studies (Chiaradia 2012; Vaughan 2005, 2007; Rashid and Alobaydi 2015; Vaughan et al. 2005) make it clear that it may be possible to map different territories of a city onto its structures of spatial networks described using space syntax to understand the whole and the parts relationships in terms of social processes.

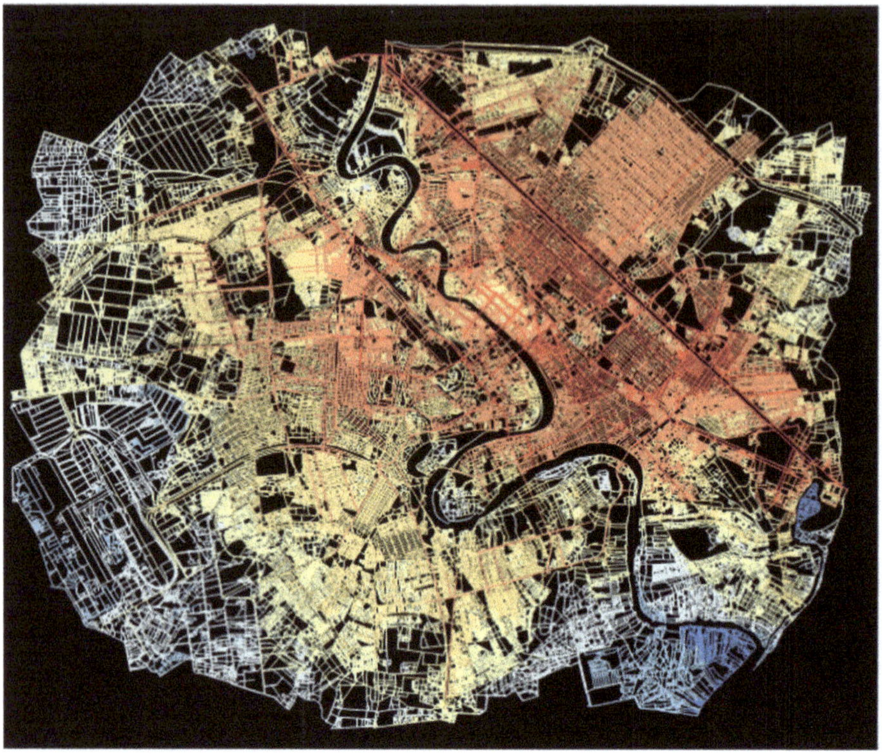

Fig. 3 The dual structure in the axial map of Baghdad colored using Integration-Rn. This analysis was performed by *Depthmap* (Turner 2007)

4.3 The Re/Production of Society and Culture Using Spatial Configurations

The processes by which spatial configurations produce and reproduce social relations, functions, and knowledge are often non-discursive and unconscious. Space syntax often uses different concepts to bring these processes to the level of consciousness. For example, Hillier uses the concepts of *ordinary* and *strange* towns to explain the processes for the city (Hillier 1996). These are somewhat similar to the concepts of long and short models, and weak and strong programs describing genotypical structures, which he and others use to explain how space produces and reproduces social relations, functions, and knowledge (Hillier 1987; Bafna 2001, 2012; Hanson 2003; Orhun et al. 1995; Rashid et al. 2014; Markus 1987, 1993). Hillier uses a ritual to illustrate the long model whose purpose is to eliminate randomness, and a party to illustrate the short model whose purpose is to maximize the randomness of encounter through spatial proximity and movement. Therefore, short models encourage the

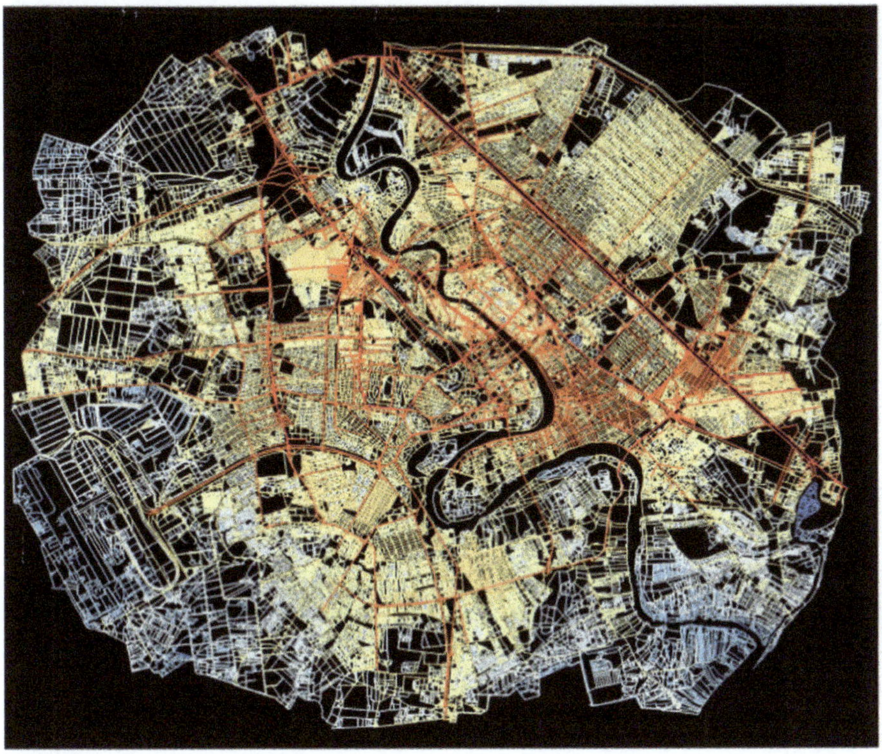

Fig. 4 The dual structure in the segment map of Baghdad colored using Integration-Rn. This analysis was performed by *Depthmap* (Turner 2007)

production of social relations, functions and knowledge, and long models encourage the conservation or reproduction of the same (Hillier 1996).

If short and long models are a bit too abstract, weak and strong programs give short and long models clear spatial dimensions in terms of interfaces between *inhabitants* and *visitors*. Inhabitants are those who have some degree of control of space; and visitors are those who lack control of space. Thus, teachers, doctors, priests, and householders are inhabitants, while pupils, patients, congregations, and guests are visitors (Hillier 1996). In a strong program building, like a courthouse, spaces are inscribed with a long model to maintain old relations by eliminating unnecessary interfaces. In contrast, in a weak program building, like a shopping mall or a clubhouse, spaces are inscribed with a short model to generate new relations by maximizing interfaces (Markus 1993).

According to Hillier, both the ordinary and strange towns participate in social production and reproduction. However, the ordinary town is concerned more with social production than reproduction. In contrast, a strange town is concerned more with social reproduction than production. In the ordinary city, like the City of London, streets are primarily used to facilitate flow, which is essential for this city to act as

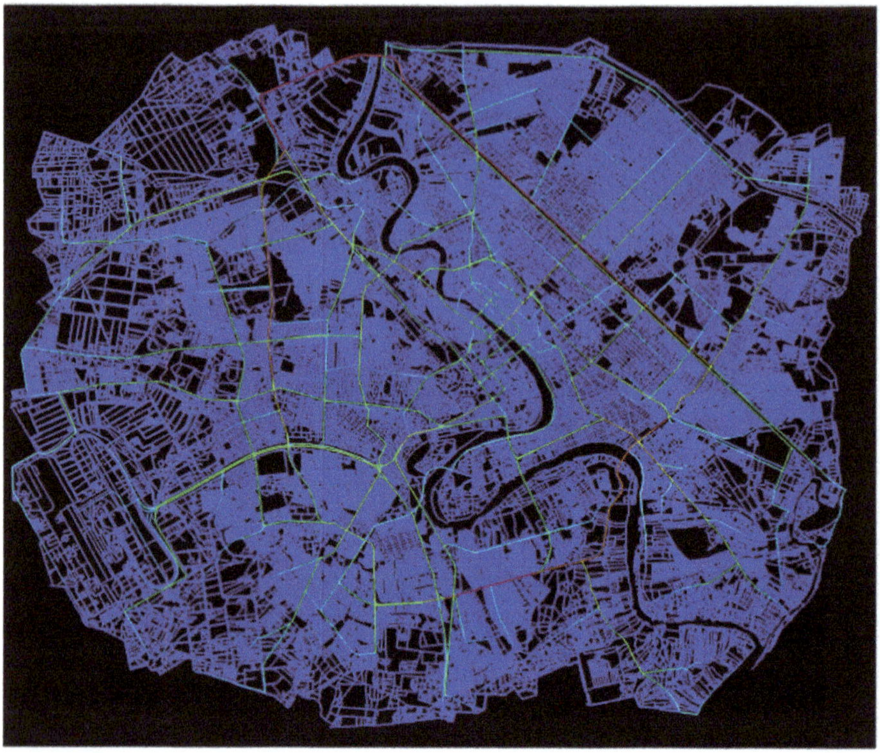

Fig. 5 The dual structure in the segment map of Baghdad colored using Choice-Rn. This analysis was performed by *Depthmap* (Turner 2007)

a center of economic and social transactions. Therefore, it minimizes the effects of major buildings on the flow by placing them at less disruptive angles in relation to streets. In an axial map, this is often represented as axial continuity, where long lines often intersect with other long lines at wide angles reducing sharp turns or stops and improving flow (Hillier 1996, 2005, 2006, 2007, 2012). In this sense, in the ordinary city physical space is more instrumental than symbolic, though its symbolic purpose is never fully eliminated.

In contrast, in the strange town, like the towns that act as centers for governing institutions, regulating bureaucracies and dominant religious ceremonies, the relationships among important buildings are more important than flow. Here, important buildings are placed in a way that maximizes obstruction to flow. Put another way, here streets may stop right at important buildings. The strange city generally has one or more major axes, each may end at important buildings and each may be expanded laterally and more or less uniformly so that important buildings are always in sight. In an axial map, this is often represented as axial discontinuity, where long lines end abruptly at important buildings disrupting flow (Hillier 1996, 2005, 2006, 2007,

2012). In contrast to the physical space of the ordinary city, therefore, the physical space of the strange town is more symbolic than instrumental.

4.4 The Generative Functions of Spatial Configurations

Despite some criticisms and limitations (Ratti 2004; Rashid 2012; Jiang and Liu 2009; Griffiths 2011; Hillier and Netto 2002; Hillier 2004), rarely does a set of theory, methods and modeling techniques explore as many generative functions of spatial networks as does space syntax. Some of these functions include movement (or traffic flow), co-presence, co-awareness, encounter, and avoidance. Among these, traffic flow is arguably the most important urban function that the syntactic structure of spatial configuration is able to describe, explain, and predict. Some of the highly referenced articles and books that report studies on the density and distribution of traffic flow using space syntax theories, techniques and measures include (Hillier 1996, 2005; Hillier et al. 1993; Penn et al. 1998; Peponis et al. 1997). Other recent space syntax studies explaining the flow of pedestrians include (Baran et al. 2008; Chiaradia et al. 2005; Law et al. 2012; Parvin et al. 2007); explaining the flow of bicycles include (McCahil and Garrick 2008; Raford 2007); explaining the flow of cars include (Jiang and Liu 2009; Barros 2007; Barros et al. 2014; Scoppa et al. 2009); and explaining pedestrian flow in multilevel systems include (Parvin et al. 2007; Chang and Penn 1998; Rashid 1997). These studies cover wide-ranging issues related to traffic flows. Many other recent studies also use space syntax theories, techniques, and measures to explain outcomes related to traffic flows, such as travel time (Barros et al. 2007; Paul 2009); transit ridership (Chiaradia et al. 2005; Ozbil et al. 2009); pedestrian safety (Raford and Ragland et al. 2004); walkability (Choi and Sayyar et al. 2012); legibility, wayfinding, and spatial decision-making (Hillier 2003, 2012; Hölscher et al. 2006; Kim 2001; Lee and Ryu 2007; Long 2007; Long and Baran 2006; Penn 2003; Dalton et al. 2012; Hölscher and Brösamle 2007; Kim and Penn 2004; Peponis et al. 1990); antisocial behavior (Friedrich et al. 2009), crime, fear, and safety (Hillier 2004; Listerborn 1999; Nubani and Wineman 2005); air pollution (Croxford 1996); and issues related to environmental perception (Ortega-Andeane et al. 2005).

In general, studies using space syntax methods and techniques report strong correlations between vehicular and pedestrian flows along axial lines and the integration values of the lines in urban areas, even in cases where user differences, metric properties of the study areas, and land use and density that could potentially affect movement are not considered explicitly. Strong correlations found between integration and vehicular and pedestrian flows eventually led space syntax researchers to propose what is now known as "the theory of natural movement" (Hillier et al. 1993). According to this theory, spaces that are more connected to all the other spaces in buildings, urban areas and cities will attract more movements than those that are less connected. On the one hand, the theory has led to several postulations related to traffic flow and several aspect of urban life such as urban buzz, safety, and multifunctionality. On the other hand, any observed deviations from this theory and its related postulations have allowed researchers to seek morphological reasons explaining these deviations. Interestingly enough, most often than not, they have been able to describe these deviations based on the nature of syntactic cores and the whole and the parts relationships in the case of urban areas and cities; and based on the nature of genotypical structures (long and short models, strong and weak programs) in the case of buildings. Since our interest here is in urban morphology, a summary of the findings of studies in urban areas are provided below:

- Traffic flows, both vehicular and pedestrian, are affected more by the geometrical and topological properties than by the metric properties of street configurations (Hillier and Iida 2005; Hillier et al. 2010).
- The importance of syntactic structures for the distribution of traffic flows is indicated by the fact that flow densities are higher in cities where local integration cores overlap with global integration cores and where integration cores are not locally confined, but are spread over the city and its different parts (Peponis et al. 1989; Min 1993; Hillier 1987).
- In areas with better correlations between integration and connectivity, traffic flows, and integration show better correlations. This is taken to indicate that the distribution of traffic flows in an area may be affected by the relationships between local and global syntactic properties of the area (Peponis et al. 1989, 1997; Hillier 1987).
- The correlations between integration and traffic flows always seem to increase when local areas are treated as parts of the larger systems, even in cases where local areas themselves are well connected. This may indicate that traffic flows may be more than a local phenomenon, and may be affected by how parts are related to the whole (Peponis et al. 1997, 1989; Hillier 1987).
- Vehicular flows show better correlations with global integration for primary streets, but they show better correlation with local integration for secondary streets. This may indicate that vehicular flows are globally oriented on primary streets, but are locally oriented on secondary streets (Penn et al. 1998; Rashid and Bindajam 2015).
- Vehicular flows are better correlated with street capacity for primary streets than they are for secondary streets. This finding, in conjunction with the above finding, may indicate that the volume of traffic on a primary street may depend on its integration as well as width (Penn et al. 1998).
- In suburban areas where inconsistent correlations between integration and pedestrian flows are observed, the integration cores do not seem to cover the area. Instead, they remain biased towards major streets with shops. This is taken to indicate that syntactic structures play an important role in the distribution of flows (Peponis et al. 1997; Hillier 1987).
- In some housing estates, integration and pedestrian flow show no correlations when integration is computed for the estates only. However, correlations get stronger when integration is computed with the estates considered as parts of the larger urban systems. This is because the local areas are poorly connected within them-

selves, but are well connected to the larger systems within which they are located. This is also because pedestrian flows in these estates are oriented towards the larger spatial systems, as if there is an urge to get out of these "unsafe" areas (Hillier 1987, 1988).

4.5 Spatial Configurations and Social Capital

Bourdieu (1972) defines the location, position, and disposition of agents in social space using the volume and structure of economic and cultural capital. He also observes that to define the location, position and disposition of agents in social space, it is necessary to analyze the relations between the structures of social space and those of physical space (Bourdieu 1996; Bourdieu 2000). However, he lacked a robust way to describe physical space in term of location, position, and disposition. Therefore, he was unable to study the homologies that might exist between the physical space and social space of cities. Using space syntax methods and techniques it is possible to show how physical space may be associated with Bourdieu's social capital. For example, since visibility serves different social functions in different settings, profits associated with visibility may vary in these settings. In workplaces, higher visibility has been associated with increased social interaction among workers (Rashid et al. 2006, 2009, 2014, 2016). In museums and stores, higher visibility has been associated with increased traffic at displays (Choi et al. 1999; Lu and Peponis 2014; Peponis et al. 2004). In urban areas, higher visibility has been associated with increased traffic leading to more economic activities, reduced crime and improved safety (Hillier 2004; Nubani and Wineman 2005; López and Van Nes 2007; López 2005).

Bourdieu identified different forms of profits with the division and control of space as a scarce resource. These include "profits of localization," "profits of position or rank," and "profits of occupation." Bourdieu's profits of localization (i.e., the proximity of a socioeconomic class or social institution to limited, desirable resources, such as financial centers, hospitals, or symbolic landmarks) can be quantified in terms of the syntactic relationships of those resources with respect to the spaces monopolized by a class or an institution at multiple overlapping scales. In this regard, it is interesting to note that in recent studies using space syntax, spatial accessibility has been able to provide a good account of many issues that may be directly or indirectly related to profits of localization, such as density and diversity (Peponis et al. 2007; Marcus 2007, 2010), land use (Kasemsook 2003; Mora Vega 2003; Mulders-Kusumo 2005; Narvaez et al. 2012), income distribution (Omer and Gabay 2007; Narvaez et al. 2012; Mora Vega 2003; Omer and Goldblatt 2012), social equity (Lima 2001), and urban segregation (Vaughan 2005, 2007; Legeby 2010, 2011, 2013, Vaughan et al. 2005).

Bourdieu's profits of position or rank occur when one can derive economic and/or symbolic profits from the distinction that come along with occupying a place in physical space. Again, space syntax can help define any such distinction in spatial terms. For example, the position of every space can be defined in relation to every other space in a configuration to determine how central a space is in the configuration. Some spaces may occupy a more central location defined in terms of accessibility related to "diffusion potential" (i.e., the potential of a space to be used when agents move from any space to any other space in the configuration). Characterized by high *integration* values, these spaces may help promote social interaction, increase diversity and improve social integration by facilitating flow from everywhere to everywhere. In contrast, some spaces may occupy a more central location in terms of accessibility related to "channeling potential" (i.e., the potential of a space to be used when agents move from one given space to another given space). Characterized by high *choice* values, such spaces may channel more flow between an origin and a destination, thus affecting regular social transactions. In this regard, it is interesting to note that in recent studies using space syntax, spatial accessibility has been able to provide a good account of residential property values (Topcu and Kubat 2009; Narvaez et al. 2012; Chiaradia et al. 2009, 2013; Matthews and Turnbull 2007; Wang and Tsai 2009), and office rent patterns (Enström and Netzell 2008).

Bourdieu's profits of occupation occur when physical space can help maintain distance and exclude undesirable intrusion. In cities that are built upon the modern city planning principle of single land use zones, Bourdieu's profits of occupation may be more directly related to these land use zones and less to syntactic structures. In contrast, in traditional cities that had developed over a long period, different land uses often comingle and show interdependency taking advantage of the profits of localization and of position using accessibility and visibility structures. Therefore, in these traditional cities profits of occupation may be more directly related to syntactic structures and less to land use zones. In this regard, Marcus (Marcus 1999, 2001, 2007, 2010; Marcus and Legeby 2012) provides some hints regarding how space syntax can be used to study profits of occupation. Several space syntax studies on urban segregation (Vaughan 2005, 2007; Legeby 2010, 2013; Legeby and Marcus 2011; Vaughan et al. 2005) also cover issues related to Bourdieu's profits of occupation.

4.6 Spatial Configurations and Urban Historical Studies

The application of space syntax methods and techniques to study historical maps can provide experientially relevant morphological descriptions that are often inaccessible to urban and architectural historians. Therefore, it is not surprising that space syntax has a long tradition of research in urban and architectural history beginning with the work of Hanson (1989). Using space syntax, many studies look at the evolution of settlement morphologies within broad national and international historical contexts (Shpuza 2009; Medeiros and de Holanda 2005, 2007; Medeiros et al. 2003, 2009; Pinho and Oliveira 2009). Many others studies look at the syntactical morphological histories in relation to movement and land use (Rashid and Bindajam 2015; Azimzadeh 2003; Azimzadeh and Bjur 2001, 2005, 2007, Karimi 2000; Read 2000), often with a focus on historical urban areas (Topcu and Kubat 2007, 2012; Rashid and Shateh 2012; Perdikogianni et al. 2003; Trigueiro and Medeiros 2007; Dai and Dong 2005; Haynie and Peponis 2009; Kubat 1997). Still, others have explored the extent to which urban form serves to materialize and mediate socioeconomic differentiation and status in different historical periods (Vaughan et al. 2005; de Holanda 2000; Zhu 2004).

Yet, there is a lack of interest in studying the diachronic processes of change, adaptation and decay of cities and urban areas using space syntax. With the exception of a few, many historical studies using space syntax methods take a synchronic view of the city undermining its temporal nature. As a result, they risk separating physical space from the historically embedded events that make the space socially and culturally meaningful. For space syntax to engage more directly with the history of the physical space of cities, it is necessary to acknowledge the temporality of space and of society using multiple layers of description that Geertz (1973) called "thick description."

Within space syntax, a multiplicity of description needed for historical studies can be achieved in several ways (Rashid and Shateh 2012; Griffiths 2005, 2008, 2009, 2011, 2013; Griffiths et al. 2010). Among these, the most obvious one involves a periodization of the morphological history of the physical space of a city into morphological phases—a technique familiar to historical geographers for representing changing spatial relations over time (e.g., (Conzen 1960)), and then describe and analyze these morphological phases using any one or more techniques of space syntax (Rashid and Shateh 2012). In this case, the selection of the morphological phases needs to be historically contingent for a meaningful understanding of the relationship between the physical city and the functional city. For example, though it is expected that cities with fewer changes in political history would show fewer morphological changes than cities with more frequent changes in political history, this may not always be the case. Another way to achieve multiplicity in the historical studies using space syntax may be to use its different descriptive techniques emphasizing different experiential qualities of space, and then study the correlations between these qualities and relevant social phenomena to gain historical insights on how spatial configuration might have affected social change and vice versa.

It should however be noted here that it is always easier to conduct synchronic than diachronic studies on the relationships between physical space and social phenomena using space syntax. That is because, the production of space are often the result of a long process of deliberation and negotiation among multiple agencies. A rigorous analysis of such a process would require a model able to handle the complex dynamics arising from the interactions of these agencies. Space syntax does not provide this model, even though its analytic methods and techniques provide the basis for evaluating the configurations of the city in relation to negotiation processes that might have occurred during various stages of development and growth. However, such evaluations may involve balancing the particular demands of historical narratives with the generalizing descriptions of the configurations of the city (Griffiths 2011).

5 Summary and Conclusions

This chapter provided an account of the mathematical developments and research applications of space syntax concepts, methods and measures relevant to urban morphological studies. Due to its focus on urban morphological studies, several other important areas of research applications were not discussed. They include numerous studies on environmental psychology and behavior including spatial cognition and wayfinding in different environments, mostly within buildings, using space syntax methods and measures (Hölscher et al. 2006; Kim 2001; Penn 2003; Hölscher and Brösamle 2007; Kim and Penn 2004; Peponis et al. 1990; Peponis and Wineman 2002; Emo 2012; 2014; Emo et al. 2012; Franz et al. 2005; Franz and Wiener 2005, 2008; Haq 2003; Haq and Girotto 2003; Haq 2005; Haq and Zimring 2003; Li and Klippel 2010; Montello 2007; Mora Vega 2009; Yun and Kim 2007). In addition, studies describing design processes using space syntax techniques (Al-Sayed et al. 2010; Dursun 2007; 2009; Arnold 2011; El-Khouly and Penn 2014; El-Khouly et al. 2012; Jeong and Ban 2011a, b, c; Kasemsook and Paksukcharern 2005; Koch 2010; Kubat et al. 2003, 2012; Vaughan et al. 2007) were also not included in this chapter. Despite limitations, the account provided in this chapter makes it clear that space syntax offers many advantages for the configurational studies of urban morphology. First, by focusing on the perceptually significant dimensions of the physical city, space syntax places humans squarely in the forefront of the analysis of space. Second, space syntax provides mathematical rigor without being overly rigorous. As a result, it is able to describe many psychologically and sociologically important qualitative properties of the physical city that had defied any proper description for a very long time. Many have criticized space syntax for this reason without appreciating the fact that human spatial perception and cognition or human social processes may not always follow rigorous mathematical rules and structures. Therefore, any methods and techniques for describing and explaining human processes must be appropriately flexible, and they cannot be too specific to spatial types, scales, and functions. Third, space syntax provides measures on continuous rather than discrete scales, which make intuitive geometric as well as psychological and sociological sense. Finally, using its methods and measures, space syntax is able to study everyday practices in relation to the physical city at any scale. It is can also explain how the everyday practices are related not only to the local space but also to the larger physical space that contains the local space in the physical city.

In summary, studies using space syntax methods and techniques have been able to make significant contributions to our understanding of the relationships between the physical city and the functional city, between spatial configurations and society in various ways.

- Enhancing our understanding of the city as a complex system, these studies have been able to characterize the syntactic structures, as well as the whole and the parts relationships, of the spatial configurations of cities and urban areas.
- Enhancing our understanding of the role of spatial configurations in the production and reproduction of society and culture, these studies have been able to describe

and/or explain many social, economic, and cultural phenomena using such genotypical structures as weak and strong programs, short and long models, and ordinary and strange towns.

- Enhancing our understanding of the generative functions of spatial configurations, these studies have been able to describe, explain and/or predict traffic flows (or movement) and other associated behavioral outcomes such as crime, safety, legibility, and wayfinding.
- Though space syntax studies on the relationships between social capital and spatial configurations and between history and spatial configurations have been limited, there are many reasons to believe that it may be possible to enhance our understanding on such relationships in the near future using space syntax methods and techniques.

To conclude, it should be noted that how humans perceive, conceive, use and inhabit the geometry of the physical city remains limitless in the functional city. By showing us how to describe the emergent structuring structures or experiential structures of the physical city, space syntax has opened up a fertile area of investigation in spatial sciences related to the functional city. So far, the space syntax literature has explored only some of these configurations and their meanings at different scales for different functions in different contexts. Many more of these configurations are yet to be identified, described and investigated. Therefore, the space syntax community still has much work left to do.

References

Ahmed, B., R. Hasan, and K. Maniruzzaman, *Urban morphological change analysis of Dhaka city, Bangladesh, using space syntax.* ISPRS International Journal of Geo-Information, 2014. **3**(4): p. 1412–1444.

Al-Sayed, K., R.C. Dalton, and C. Hölscher, *Discursive design thinking: The role of explicit knowledge in creative architectural design reasoning.* AI EDAM, 2010. **24**(2): p. 211–230.

Al-Sayed, K., et al., *Space Syntax Methodology*. 2014 (4th Edition), Bartlett School of Architecture, UCL: London, UK.

Arnold, T., *Using space syntax to design an architecture of visual relations.* The Journal of Space Syntax, 2011. **2**(2): p. 201–222.

Asami, Y., et al. *Introducing the third dimension on Space Syntax: Application on the historical Istanbul*. In *Proceedings of the 4th International Space Syntax Symposium*. 2003. London, UK: University College London.

Azimzadeh, M. and B. Klarqvist. *Metamorphosis and Evolution in Cities*. In *Proceedings of the 3rd International Space Syntax Symposium*. 2001. Atlanta, GA: University of Michigan, Ann Arbor.

Azimzadeh, M. and H. Bjur. *The urban palimpsest: the interplay between historical generated layers in the urban spatial system and urban life*. In *Proceedings of the 6th International Space Syntax Symposium*. 2007. Istanbul, Turkey: Istanbul Technical University Faculty of Architecture.

Azimzadeh, M. and H. Bjur. *Transforming Cities. The role of the configuration of the network of public spaces in urban life*. In *Proceedings of the 5th International Space Syntax Symposium*. 2005. Delft, NL: TU Delft Faculty of Architecture.

Azimzadeh, M. *Survival of bazaars: global spatial impact and local self-organising processes*. In *Proceedings of the 4th International Space Syntax Symposium*. 2003. London: University College London.

Bafna, S. *Geometrical intuitions of genotypes*. In *Proceedings of the 3rd International Space Syntax Symposium*. 2001. Atlanta, USA: University of Michigan, Ann Arbor.

Bafna, S., *Rethinking genotype: Comments on the sources of type in architecture.* The Journal of Space Syntax, 2012. **3**(1): p. 69–80.

Baran, P.K., D.A. Rodríguez, and A.J. Khattak, *Space syntax and walking in a new urbanist and suburban neighbourhoods.* Journal of Urban Design, 2008. **13**(1): p. 5–28.

Barros, A.P.B.G., P.C.M. da Silva, and F.R.B. de Holanda. *Exploratory study of space syntax as a traffic assignment tool*. In *Proceedings of the 6th International Space Syntax Symposium*. 2007. Istanbul, Turkey: Istanbul Technical University Faculty of Architecture.

Barros, A.P.B.G., et al., *Analysis of trip generating developments by space syntax: a case study of Brasília, Brazil.* Journal of Transport Literature, 2014. **8**: p. 07–36.

Batty, M., *A new theory of space syntax*, in *CASA Working Papers*. 2004, Centre for Advanced Spatial Analysis, University College London: London, UK.

Batty, M. and S. Rana, *Reformulating space syntax: the automatic definition and generation of axial lines and axial maps*, in *UCL Working Paper Series*. 2002, UCL Centre for Advanced Spatial Analysis: London, UK. p. 1–44.

Batty, M. and S. Rana, *The automatic definition and generation of axial lines and axial maps.* Environment and Planning B: Planning and Design, 2004. **31**(4): p. 615–640.

Beyhan, B. *Developing space syntax tools for free and open source software for GIS*. In *19th International Conference on Geoinformatics*. 2011. IEEE.

Bourdieu, P., *Outline of a Theory of Practice. Translated by Richard Nice*. Vol. 16. [1972] 1977, Cambridge, UK: Cambridge university press.

Bourdieu, P., *Physical Space, Social Space and Habitus*, in *Vilhelm Aubert Memorial Lecture. Rapport 10*. 1996, Institutt for Sosiologi og samfunnsgeografi, Universitetet i Oslo: Oslo.

Bourdieu, P., *The logic of practice. Translated by Richard Nice*. 1992, Stanford, CA: Stanford University Press.

Bourdieu, P., *The weight of the world: Social suffering in contemporary society. Translated by Translated by Priscilla Parkhurst Ferguson, Susan Emanuel, Joe Johnson and Shoggy T. Waryn* 2000, Stanford, CA: Stanford University Press.

Carpenter, A. and J. Peponis, *Poverty and connectivity.* The Journal of Space Syntax, 2010. **1**(1): p. 108–120.

Carvalho, R. and A. Penn, *Scaling and universality in the micro-structure of urban space.* Physica A: Statistical Mechanics and its Applications, 2004. **332**: p. 539–547.

Carvalho, R. and M. Batty, *Automatic extraction of hierarchical urban networks: A micro-spatial approach.* Computational Science-ICCS 2004, 2004: p. 1109–1116.

Carvalho, R. and M. Batty, *Encoding geometric information in road networks extracted from binary images.* Environment and Planning B: Planning and Design, 2005. **32**(2): p. 179–190.

Chang, D. and A. Penn, *Integrated multilevel circulation in dense urban areas: the effect of multiple interacting constraints on the use of complex urban areas.* Environment and Planning B: Planning and Design, 1998. **25**(4): p. 507–538.

Chiaradia, A., E. Moreau, and N. Raford. *Configurational exploration of public transport movement networks: a case study, the London Underground*. In *Proceedings of the 5th International Space Syntax Symposium*. 2005. Delft, NL: TU Delft Faculty of Architecture.

Chiaradia, A., et al., *Compositional and urban form effects on centres in Greater London.* Proceedings of the Institution of Civil Engineers-Urban Design and Planning, 2012. **165**(1): p. 21–42.

Chiaradia, A., et al., *Compositional and urban form effects on residential property value patterns in Greater London.* Proceedings of the Institution of Civil Engineers-Urban Design and Planning, 2013. **166**(3): p. 176–199.

Chiaradia, A., et al. *Residential property value patterns in London: space syntax spatial analysis*. In *Proceedings of the 7th International Space Syntax Symposium*. 2009. Stockholm, Sweden: KTH Royal Institute of Technology.
Choi, E. and S. Sayyar. *Urban Diversity and Pedestrian Behavior: Refining the concept of land-use mix for walkability*. In *Proceedings of the 8th International Space Syntax Symposium* 2012. Santiago de Cile: PUC.
Choi, Y.K., *The morphology of exploration and encounter in museum layouts*. Environment and Planning B: Planning and Design, 1999. **26**(2): p. 241–250.
Conroy-Dalton, R. and N. Dalton. *Applying depth decay functions to space syntax network graphs*. In *Proceedings of the 6th International Space Syntax Symposium*. 2007. İstanbul, Turkey: Istanbul Technical University Faculty of Architecture.
Conzen, M.R.G., *Alnwick, Northumberland: a study in town-plan analysis*. Transactions and Papers. 1960, London: Institute of British Geographers. iii-122.
Croxford, B., A. Penn, and B. Hillier, *Spatial distribution of urban pollution: civilizing urban traffic*. Science of the Total Environment, 1996. **189**: p. 3–9.
Crucitti, P., V. Latora, and S. Porta, *Centrality measures in spatial networks of urban streets*. Physical Review E, 2006. **73**(3): p. 036125.
Dai, X. and Q. Dong. *Centrality process in orthogonal grid–case study Suzhou*. In *Proceedings of the 5th International Space Syntax Symposium Proceedings*. 2005. Delft, NL: TU Delft Faculty of Architecture.
Dalton, N. *Fractional configurational analysis and a solution to the Manhattan problem*. in *Space Syntax 3rd International Symposium*. 2001. Georgia Institute of Technology, Atlanta: University of Michigan.
Dalton, N., J. Peponis, and R. Dalton. *To tame a TIGER one has to know its nature: extending weighted angular integration analysis to the description of GIS road-centerline data for large scale urban analysis*. In *Proceedings of the 4th International Space Syntax Symposium*. 2003. London, UK: University College London.
Dalton, N. *New measures for local fractional angular integration or towards general relativisation in space syntax*. In *Proceedings of the 5th International Space Syntax Symposium* 2005. Delft, NL: TU Delft Faculty of Architecture.
Dalton, R.C., C. Hölscher, and A. Turner, *Understanding Space: The Nascent Synthesis of Cognition and the Syntax of Spatial Morphologies*. Environment and Planning B: Planning and Design, 2012. **39**(1): p. 7–11.
de Holanda, F., *Class footprints in the landscape*. Urban Design International, 2000. **5**: p. 189–198.
Dhanani, A., et al. *From the axial line to the walked line: Evaluating the utility of commercial and user-generated street network datasets in space syntax analysis*. In *Proceedings of the 8th International Space Syntax Symposium* 2012. Santiago de Cile: PUC.
Dursun, P. *Architects are Talking About Space*. In *Proceedings of 7th International Space Syntax Symposium* 2009. Stockholm, Sweden: KTH Royal Institute of Technology.
Dursun, P. *Space syntax in architectural design*. In *Proceedings of the 6th International Space Syntax Symposium*. 2007. Istanbul, Turkey: Istanbul Technical University Faculty of Architecture.
El-Khouly, T. and A. Penn, *On an integrated analytical approach to describe quality design process in light of deterministic information theory*, in *Design Computing and Cognition'12*. 2014, Springer. p. 451–470.
El-Khouly, T.A.I. and A. Penn. *Order, structure and disorder in space syntax and linkography: intelligibility, entropy, and complexity measures*. In *Proceedings of the 8th International Space Syntax Symposium* 2012. Santiago De Chile: Pontifica Universidad Católica.
Ellis, W.C., *The spatial structure of streets*, in *On streets*, S. Anderson, Editor. 1986, MIT Press. p. 133–149.
Emo, B., *Seeing the axial line: evidence from wayfinding experiments*. Behavioral Sciences, 2014. **4**(3): p. 167–180.
Emo, B., *Wayfinding in real cities: Experiments at street corners*. Spatial Cognition VIII, 2012: p. 461–477.

Emo, B., et al. *Wayfinding and Spatial Configuration: evidence from street corners*. In *Proceedings of the 8th International Space Syntax Symposium*. 2012. Santiago de Chile: PUC.
Enström, R. and O. Netzell, *Can space syntax help us in understanding the intraurban office rent pattern? Accessibility and rents in downtown Stockholm.* The Journal of Real Estate Finance and Economics, 2008. **36**(3): p. 289–305.
Figueiredo, L. and L. Amorim. *Continuity lines in the axial system*. In *Proceedings of the 5th International Space Syntax Symposium*. 2005. Delft, NL: TU Delft Faculty of Architecture.
Fotheringham, A.S., *Spatial structure and distance-decay parameters.* Annals of the Association of American Geographers, 1981. **71**(3): p. 425–436.
Franz, G. and J.M. Wiener. *Exploring isovist-based correlates of spatial behavior and experience.* In *Proceedings of the 5th International Space Syntax Symposium*. 2005. Delft, NL: TU Delft Faculty of Architecture.
Franz, G. and J.M. Wiener, *From space syntax to space semantics: a behaviorally and perceptually oriented methodology for the efficient description of the geometry and topology of environments.* Environment and Planning B: Planning and Design, 2008. **35**(4): p. 574–592.
Franz, G., et al. *Graph-based models of space in architecture and cognitive science-a comparative analysis*. In *17th International Conference on Systems Research, Informatics and Cybernetics*. 2005.
Freeman, L.C., *A set of measures of centrality based on betweenness.* Sociometry, 1977: p. 35–41.
Freeman, L.C., *Centrality in social networks conceptual clarification.* Social networks, 1979. **1**(3): p. 215–239.
Freeman, L.C., S.P. Borgatti, and D.R. White, *Centrality in valued graphs: A measure of betweenness based on network flow.* Social networks, 1991. **13**(2): p. 141–154.
Friedrich, E., B. Hillier, and A. Chiaradia. *Anti-social behaviour and urban configuration using space syntax to understand spatial patterns of socio-environmental disorder*. In *Proceedings of the 7th International Space Syntax Symposium*. 2009. Stockholm, Sweden: KTH Royal Institute of Technology.
Geertz, C., *The interpretation of cultures: Selected essays*. 1973: Basic books.
Gibson, J.J., *The Ecological Approach to Visual Perception: Classic Edition*. [1979] 2014, New York: Taylor & Francis.
Gil, J., C. Stutz, and A. Chiaradia. *Confeego: tool set for spatial configuration studies*. In *Proceedings of the 6th International Space Syntax Symposium*. 2007. Istanbul, Turkey: Istanbul Technical University Faculty of Architecture.
Gil, J. *Examining 'edge effects': Sensitivity of spatial network centrality analysis to boundary conditions*. In *Proceedings of the 10th International Space Syntax Symposium*. 2015. London, UK: Uniersity College London.
Gil, J., *Street network analysis "edge effects": Examining the sensitivity of centrality measures to boundary conditions.* Environment and Planning B: Planning and Design, 2016.
Griffiths, S., *GIS and Research Into Historical "Spaces of Practice": Overcoming the Epistemological Barriers*, in *History and GIS*. 2013, Springer. p. 153–171.
Griffiths, S., *Historical space and the interpretation of urban transformation : the spatiality of social and cultural change in Sheffield c. 1770–1910*, in *Bartlett School of Graduate Studies*. 2008, University of London: London.
Griffiths, S. *Historical space and the practice of "spatial history": the spatio-functional transformation of Sheffield 1770–1850*. In *Proceedings of the 5th International Space Syntax Symposium*. 2005. Delft, NL: TU Delft Faculty of Architecture.
Griffiths, S. *Persistence and Change in the Spatio-temporal Description of Sheffield Parish 1770–1910*. In *Proceedings of the 7th International Space Syntax Symposium* 2009. Stockholm, Sweden: KTH Royal Institute of Technology.
Griffiths, S., *Temporality in Hillier and Hanson's theory of spatial description: some implications of historical research for Space Syntax.* The Journal of Space Syntax, 2011. **2**(1): p. 73–96.
Griffiths, S., et al., *The persistence of suburban centres in Greater London: combining Conzenian and space syntax approaches.* Urban Morphology, 2010. **14**(2): p. 85–99.

Hanson, J., *Decoding homes and houses*. 2003, Cambridge, UK: Cambridge University Press.

Hanson, J., *Order and structure in urban design: the plans for the rebuilding of London after the Great Fire of 1666*. Ekistics, 1989. **56**(334/335): p. 22–42.

Haq, S. and C. Zimring, *Just down the road a piece the development of topological knowledge of building layouts*. Environment and behavior, 2003. **35**(1): p. 132–160.

Haq, S., G. Hill, and A. Pramanik. *Comparison of configurational, wayfinding and cognitive correlates in real and virtual settings*. In *Proceedings of the 5th International Space Syntax Symposium*. 2005. Delft, NL: TU Delft Faculty of Architecture.

Haq, S., *Investigating the syntax line: configurational properties and cognitive correlates*. Environment and Planning B: Planning and Design, 2003. **30**(6): p. 841–863.

Haq, S. and S. Girotto. *Ability and intelligibility: Wayfinding and environmental cognition in the designed environment*. In *Proceedings of the 4th international space syntax symposium*. 2003. London, UK: University College London.

Haynie, D. and J. Peponis. *Atlanta: A Morphological History*. In *Proceedings of the 7th International Space Syntax Symposium*. 2009. Stockholm, Sweden: KTH Royal Institute of Technology.

Hillier, B. and A. Penn, *Rejoinder to Carlo Ratti*. Environment and Planning B: Planning and design, 2004. **31**(4): p. 501–511.

Hillier, B., *A theory of the city as object: or, how spatial laws mediate the social construction of urban space*. Urban Design International, 2002. **7**(3–4): p. 153-179.

Hillier, B., *Against enclosure*, in *Rehumanizing housing*, N. Teymur, T. Markus, and T. Wooley, Editors. 1988, Butterworth: London. p. 63–88.

Hillier, B., *Can streets be made safe?* Urban Design International, 2004. **9**(1): p. 31–45.

Hillier, B., *Centrality as a process: accounting for attraction inequalities in deformed grids*. Urban Design International, 1999a. **4**(3–4): p. 107-127.

Hillier, B., J. Hanson, and H. Graham, *Ideas are in things: an application of the space syntax method to discovering house genotypes*. Environment and Planning B: planning and design, 1987. **14**(4): p. 363–385.

Hillier, B., J. Hanson, and J. Peponis, *Syntactic analysis of settlements*. Architecture et comportement/Architecture and Behaviour, 1987. **3**(3): p. 217–231.

Hillier, B. and J. Hanson, *The Social Logic of Space*. 1984, Cambridge, UK: Cambridge University Press.

Hillier, B. and L. Vaughan, *The city as one thing*. Progress in Planning, 2007. **67**(3): p. 205–230.

Hillier, B. and S. Iida. *Network and psychological effects in urban movement*. In *International Conference on Spatial Information Theory*. 2005. Springer.

Hillier, B. and S. Iida, *Network and psychological effects in urban movement*, in *Spatial information theory*. 2005, Springer. p. 475–490.

Hillier, B., *Space is the machine: a configurational theory of architecture*. 2007, Cambridge, UK: Cambridge University Press.

Hillier, B., *Space is the machine: a configurational theory of architecture*. [1996] 2007, London, UK: Space Syntax.

Hillier, B. *Studying cities to learn about minds: How geometric intuitions shape urban space and make it work*. In *Space Syntax and Spatial Cognition Workshop Proceedings. Available at* http://www.space.bartlett.ucl.ac.uk/sc06/proceedings. 2006. Bremen: Transregional Collaborative Research Center SFB/TR 8.

Hillier, B., *Studying cities to learn about minds: Some possible implications of space syntax for spatial cognition*. Environment and Planning B: Planning and Design, 2012. **39**(1): p. 12–32.

Hillier, B., T. Yang, and A. Turner, *Normalising least angle choice in Depthmap and how it opens up new perspectives on the global and local analysis of city space*. Journal of Space syntax, 2012. **3**(2): p. 155–193.

Hillier, B. *The Architectures of Seeing and Going: are Cities Shaped by Bodies or Minds? And is there a Syntax of Spatial Cognition?* In *Proceedings of the 4th International Space Syntax Symposium*. 2003. London: University College London.

Hillier, B., *The art of place and the science of space*. World Architecture, 2005. **185**: p. 96–102.

Hillier, B., *The hidden geometry of deformed grids: or, why space syntax works, when it looks as though it shouldn't.* Environment and Planning B: planning and Design, 1999b. **26**(2): p. 169–191.
Hillier, B. and V. Netto, *Society Seen Through the Prism of Space: Outline of a Theory of Society and Space.* Urban Design International, 2002. **7**: p. 181–203.
Hillier, B., et al., *Metric and topo-geometric properties of urban street networks: some convergences, divergences and new results.* The Journal of Space Syntax, 2010. **1**(2): p. 258–279.
Hillier, B., et al., *Natural movement: or, configuration and attraction in urban pedestrian movement.* Environment and Planning B: planning and design, 1993. **20**(1): p. 29–66.
Hillier, B., et al., *Creating life: Or, does architecture determine anything?* Architecture and Comportment/Architecture and Behavior, 1987. **3**(3): p. 233–250.
Hölscher, C. and M. Brösamle. *Capturing indoor wayfinding strategies and differences in spatial knowledge with space syntax.* In *Proceedings of the 6th International Space Syntax Symposium.* 2007.
Hölscher, C., R. Conroy Dalton, and A. Turner. *Space Syntax and Spatial Cognition* in *Space Syntax and Spatial Cognition Workshop Proceedings. Available at* http://www.space.bartlett.ucl.ac.uk/sc06/proceedings. 2006. Bremen: Transregional Collaborative Research Center SFB/TR 8.
Jeong, S.K. and Y.U. Ban, *Computational algorithms to evaluate design solutions using Space Syntax.* Computer-Aided Design, 2011a. **43**(6): p. 664–676.
Jeong, S.-K. and Y.-U. Ban, *Developing a topological information extraction model for space syntax analysis.* Building and Environment, 2011b. **46**(12): p. 2442–2453.
Jeong, S.-K. and Y.-U. Ban, *Methods to Recognize and Manage Spatial Shapes for Space Syntax Analysis.* KIEAE Journal, 2011c. **11**(6): p. 95–100.
Jiang, B., C. Claramunt, and M. Batty, *Geometric accessibility and geographic information: extending desktop GIS to space syntax.* Computers, Environment and Urban Systems, 1999. **23**(2): p. 127–146.
Jiang, B., *A topological pattern of urban street networks: universality and peculiarity.* Physica A: Statistical Mechanics and its Applications, 2007. **384**(2): p. 647–655.
Jiang, B., B. Huang, and V. Vasek, *Geovisualisation for planning support systems*, in *Planning Support Systems in Practice*. 2003. p. 177–191.
Jiang, B. and C. Claramunt. *A Comparison study on space syntax as a computer model of space*. In *2nd International Symposium on Space Syntax* 1999. Brasilia: Universidade de Brasilia.
Jiang, B. and C. Claramunt, *A Structural Approach to the Model Generalization of an Urban Street Network.* GeoInformatica, 2004. **8**(2): p. 157–171.
Jiang, B., C. Claramunt, and B. Klarqvist, *Integration of space syntax into GIS for modelling urban spaces.* International Journal of Applied Earth Observation and Geoinformation, 2000. **2**(3–4): p. 161-171.
Jiang, B. and C. Claramunt. *Extending space syntax towards an alternative model of space within GIS.* In *3rd European Agile Conference on Geographic Information Science*. 2000. Helsinki, Finland.
Jiang, B. and C. Claramunt, *Integration of space syntax into GIS: new perspectives for urban morphology.* Transactions in GIS, 2002. **6**(3): p. 295–309.
Jiang, B. and C. Claramunt, *Topological analysis of urban street networks.* Environment and Planning B: Planning and design, 2004. **31**(1): p. 151–162.
Jiang, B. and C. Liu, *Street-based topological representations and analyses for predicting traffic flow in GIS.* International Journal of Geographical Information Science, 2009. **23**(9): p. 1119–1137.
Jones, C., et al. *A multi-disciplinary perspective on the built environment: Space Syntax and cartography–the communication challenge.* In *Proceedings of the 7th International Space Syntax Symposium.* 2009. Stockholm, Sweden: KTH Royal Institute of Technology.
Karimi, K., *A reflection on 'Order and Structure in Urban Desig'.* The Journal of Space Syntax, 2012. **3**(1): p. 38–48.
Karimi, K., *The spatial logic of organic cities in Iran and the United Kingdom.* Proc. Space Syntax 1st Intl. Sym. London. pp, 1997: p. 1–17.

Karimi, K., *Urban conservation and spatial transformation: preserving the fragments or maintaining the 'spatial spirit'*. Urban Design International, 2000. **5**: p. 221–231.

Kasemsook, A. and K. Paksukcharern. *Planning a deep island: introducing space syntax to an urban planning process for Phuket, Thailand*. In *Proceedings of the 5th International Space Syntax Symposium Proceedings*. 2005. Delft, NL: TU Delft Faculty of Architecture.

Kasemsook, A. *Spatial and functional differentiation: A symbiotic and systematic relationship*. In *Proceedings of the 4th International Space Syntax Symposium* 2003. London: University College London.

Kim, Y.O. and A. Penn, *Linking the spatial syntax of cognitive maps to the spatial syntax of the environment.* Environment and behavior, 2004. **36**(4): p. 483–504.

Kim, Y.O. *The role of spatial configuration in spatial cognition*. In *Proceedings of the 3rd International Space Syntax Symposium*. 2001. Geogia Tech, Atlanta: University of Michigan, Ann Arbor.

Koch, D., *Architecture re-configured.* The Journal of Space Syntax, 2010. **1**(1): p. 1–16.

Krafta, R., *Modelling intraurban configurational development.* Environment and Planning B: Planning and Design, 1994. **21**(1): p. 67–82.

Krüger, M. and A.P. Vieira, *Scaling relative asymmetry in space syntax analysis.* The Journal of Space Syntax, 2012. **3**(2): p. 194–203.

Krüger, M., *On node and axial maps: distance measures and related topics*, in *European Conference on the Representation and Management of Urban Change*. 1989, University of Cambridge: Cambridge.

KUBAT, A., et al. *Application of Space Syntax in Developing A Regeneration Framework for Sharjah's Heritage Area*. In *Proceedings of the 8th International Space Syntax Symposium*. 2012. Santiago De Chile: PUC.

Kubat, A.S., *The morphological characteristics of Anatolian fortified towns.* Environment and Planning B: Planning and Design, 1997. **24**(1): p. 95–123.

Kubat, A.S., *The morphological history of Istanbul.* Urban Morphology, 1999. **3**: p. 28–40.

Kubat, A.S., et al. *An urban redevelopment proposal for Istanbul's Galata district*. In *Proceedings of the 4th International Space Syntax Symposium*. 2003.

Larkham, P.J., *Agents and types of change in the conserved townscape.* Transactions of the Institute of British Geographers, 1988: p. 148–164.

Law, S., A. Chiaradia, and C. Schwander. *Towards a multimodal space syntax analysis: A case study of the London street and underground network*. In *Proceedings of the 8th International Space Syntax Symposium*. 2012. Santiago de Chile: PUC.

Lee, S. and S. Ryu, *Multiple path-finding models using Kalman filtering and space syntax techniques.* Transportation Research Record: Journal of the Transportation Research Board, 2007(2029): p. 87–95.

Legeby, A. *Configuration and co-presence: the underpinnings of job opportunities*. In *Proceedings of the 9th International Space Syntax Symposium* 2013. Seoul, Korea: Sejong University.

Legeby, A., *From housing segregation to integration in public space.* The Journal of Space Syntax, 2010. **1**(1): p. 92–107.

Legeby, A. and L. Marcus, *Does the urban structure of Swedish cities inhibit the sharing of public space?* Built Environment, 2011. **37**(2): p. 155–169.

Li, R. and A. Klippel. *Using space syntax to understand knowledge acquisition and wayfinding in indoor environments*. In *Cognitive Informatics (ICCI), 2010 9th IEEE International Conference*. 2010. IEEE.

Lima, J. *Compensatory urban form*. In *Proceedings of the 3rd International Space Syntax Symposium*. 2001. Georgia Tech, Atlanta: University of Michigan, Ann Arbor.

Listerborn, C. *Women's fear and space configurations*. In *2nd International Symposium on Space Syntax*. 1999. Brasilia, Brazil: University of Brasilia.

Liu, X. and B. Jiang, *Defining and generating axial lines from street center lines for better understanding of urban morphologies.* International Journal of Geographical Information Science, 2012. **26**(8): p. 1521–1532.

Livesey, G.E. and A. Donegan. *Addressing normalisation in the pursuit of comparable integration*. In *Proceedings of the 4th International Space Syntax Symposium* 2003. London, UK: University College London.

Long, Y. and P. Baran. *Methodology for analyzing the relationships between objective and subjective evaluations of urban environment: space syntax, cognitive maps, and urban legibility*. In *Space Syntax and Spatial Cognition Workshop Proceedings. Available at* http://www.space.bartlett.ucl.ac.uk/sc06/proceedings. 2006. Bremen: Transregional Collaborative Research Center SFB/TR 8.

Long, Y., *The relationships between objective and subjective evaluations of the urban environment: space syntax, cognitive maps, and urban legibility*, in *Design*. 2007, North Carolina State University.

López, M.J. and A. Van Nes. *Space and crime in Dutch built environments: macro and micro scale spatial conditions for residential burglaries and thefts from cars*. In *Proceedings of the 6th International Space Syntax Symposium*. 2007. Istanbul: Istanbul Technical University Faculty of Architecture.

López, M.J.J. *The spatial behavior of residential burglars*. In *Proceedings of the 5th International Space Syntax Symposium*. 2005. Delft, NL: TU Delft Faculty of Architecture.

Lu, Y. and J. Peponis, *Exhibition visitors are sensitive to patterns of display covisibility*. Environment and Planning B: Planning and Design, 2014. **41**(1): p. 53–68.

Lynch, K., *The image of the city*. 1960, Cambridge, MA: MIT Press.

Marcus, L. and A. Legeby. *The need for co-presence in urban complexity: Measuring social capital using space syntax*. In *Proceedings of the 8th International Space Syntax Symposium*. 2012. Santiago de Chile: PUC.

Marcus, L. *Physical planning for economical growth: A study of urban areas*. In *Proceedings of the 2nd International Space Syntax Symposium* 1999. Brasilia, Brazil: University of Brasilia.

Marcus, L. *Plot Syntax: A configurational approach to urban diversity*. In *Proceedings of the 5th international space syntax symposium*. 2005. Delft, NL: TU Delft Faculty of Architecture.

Marcus, L. *Spatial capital and how to measure it: an outline of an analytical theory of the social performativity of urban form*. In *Proceedings of the 6th international space syntax symposium*. 2007. Istanbul: Istanbul Technical University Faculty of Architecture.

Marcus, L., *Spatial capital*. The Journal of Space Syntax, 2010. **1**(1): p. 30–40.

Marcus, L. *The impact of land-division on long-term occupation: The possibility of such a thing as natural occupation*. In *Proceedings of the 3rd International Space Syntax Symposium*. 2001. Atlanta, GA: University of Michigan, Ann Arbor.

Markus, T.A., *Buildings as classifying devices*. Environment and Planning B: Planning and Design, 1987. **14**(4): p. 467–484.

Markus, T.A., *Buildings & power: Freedom and control in the origin of modern building types*. 1993: Psychology Press.

Matthews, J.W. and G.K. Turnbull, *Neighborhood street layout and property value: The interaction of accessibility and land use mix*. The Journal of Real Estate Finance and Economics, 2007. **35**(2): p. 111–141.

McCahil, C. and N. Garrick, *The applicability of space syntax to bicycle facility planning*. Transportation Research Record: Journal of the Transportation Research Board, 2008(2074): p. 46–51.

Medeiros, V., F. de Holanda, and E. Trigueiro. *From Compact Colonial Villages to Sparse Metropolis*. In *Proceedings of the 4th International Space Syntax Symposium*. 2003. London: University College London.

Medeiros, V. and F. de Holanda. *Structure and size: Brazilian cities in an urban configurational world scenario*. In *Proceedings of the 6th International Space Syntax Symposium*. 2007. Istanbul: Istanbul Technical University Faculty of Architecture.

Medeiros, V. and F. de Holanda. *Urbis Brasiliae: investigating topological and geometrical features in Brazilian Cities*. In *Proceedings of the 5th International Space Syntax Symposium*. 2005. Delft: TU Delft, Faculty of Architecture.

Medieros, V., F. de Holanda, and A.P. Barros. *The Myth of the Intention: The Portuguese Urban Heritage Overseas*. In *Proceedings of the 7th International Space Syntax Symposium*. 2009. Stockholm, Sweden: KTH Royal Institute of Technology.

Min, Y., *Housing layout design-Neighborhood morphology, pedestrian movement and strategic choices*. Nordic Journal of Architectural Research, 1993. **2**: p. 75–95.

Montello, D.R. *The contribution of space syntax to a comprehensive theory of environmental psychology*. In *Proceedings of the 6th International Space Syntax Symposium*. 2007. Istanbul, Turkey: Istanbul Technical University Faculty of Architecture.

Mora Vega, R.I. *Informal economy and spatial configuration in Quilicura, Santiago*. In *Proceedings of the 4th International Space Syntax Symposium*. 2003. London: University College London.

Mora Vega, R.I. *Land uses in the context of a perfect grid: the case of Barcelona*. In *Proceedings of the 4th International Space Syntax Symposium*. 2003. London: University College London.

Mora Vega, R.I., *The cognitive roots of space syntax*. 2009, University College London: London, UK.

Mulders-Kusumo, C. *Is a railway station a central urban place? Spatial configuration study of retail distribution pattern around railway stations*. In *Proceedings of the 5th International Space Syntax Symposium*. 2005. TU Delft: TU Delft, Faculty of Architecture.

Narvaez, L., A. Penn, and S. Griffiths. *Space syntax economics: decoding accessibility using property value and housing price in Cardiff, Wales*. In *Proceedings of the 8th International Space Syntax Symposium*. 2012. Santiago de Chile: PUC.

Nubani, L. and J. Wineman. *The role of space syntax in identifying the relationship between space and crime*. In *Proceedings of the 5th International Space Syntax Symposium*. 2005. TU Delft: TU Delft Faculty of Architecture.

Omer, I. and O. Zafrir-Reuven, *Street patterns and spatial integration of Israeli cities*. The Journal of Space Syntax, 2010. **1**(2): p. 280–295.

Omer, I. and R. Goldblatt, *Urban spatial configuration and socio-economic residential differentiation: The case of Tel Aviv*. Computers, Environment and Urban Systems, 2012. **36**(2): p. 177–185.

Omer, I. and R.a. Gabay. *Social homogeneity and space syntax of towns in Israel*. In *Proceedings of the 6th International Space Syntax Symposium*. 2007. Istanbul: Istanbul Technical University Faculty of Architecture.

Orhun, D., B. Hillier, and J. Hanson, *Spatial types in traditional Turkish houses*. Environment and planning B: Planning and design, 1995. **22**(4): p. 475–498.

Ortega-Andeane, P., et al., *Space syntax as a determinant of spatial orientation perception*. International Journal of Psychology, 2005. **40**(1): p. 11–18.

Ozbil, A., J. Peponis, and S. Bafna. *The effects of street configuration on transit ridership*. In *Proceedings of the 7th International Space Syntax Symposium*. 2009. Stockholm, Sweden: KTH Royal Institute of Technology.

Park, H.-T. *Before integration: A critical review of integration measure in space syntax*. In *Proceedings of the 5th International Space Syntax Symposium*. 2005. TU Delft: TU Delft Faculty of Architecture.

Park, H.-T. *Boundary effects on the intelligibility and predictability of spatial systems*. In *Proceedings of the 7th International Space Syntax Symposium*. 2009. Stockholm, Sweden: KTH Royal Institute of Technology.

Parvin, A., A.M. Ye, and B. Jia. *Multilevel pedestrian movement*. In *Proceedings of the 6th International Space Syntax Symposium*. 2007. Istanbul: Istanbul Technical University Faculty of Architecture.

Paul, A., *An integrated approach to modeling vehicular movement networks: Trip assignment and space syntax*. 2009, Texas Tech University.

Penn, A., *Space syntax and spatial cognition or why the axial line?* Environment and Behavior, 2003. **35**(1): p. 30–65.

Penn, A., et al., *Configurational modelling of urban movement networks*. Environment and Planning B: Planning and Design, 1998. **25**(1): p. 59–84.

Peponis, J., C. Ross, and M. Rashid, *The structure of urban space, movement and co-presence: the case of Atlanta.* Geoforum, 1997. **28**(3): p. 341–358.

Peponis, J., C. Zimring, and Y.K. Choi, *Finding the building in wayfinding.* Environment and behavior, 1990. **22**(5): p. 555–590.

Peponis, J. and J. Wineman, *Spatial structure of environment and behavior*, in *Handbook of environmental psychology*, R.B. Bechtel and A. Churchman, Editors. 2002, John Wiley & Sons, Inc.: New York. p. 271–291.

Peponis, J., S. Bafna, and Z. Zhang, *The connectivity of streets: reach and directional distance.* Environment and Planning B: Planning and Design, 2008. **35**(5): p. 881–901.

Peponis, J., et al., *Describing plan configuration according to the covisibility of surfaces.* Environment and Planning B: Planning and Design, 1998b. **25**(5): p. 693–708.

Peponis, J., et al. *Measuring the configuration of street networks.* In *Proceedings of the 6th international space syntax symposium.* 2007. Istanbul, Turkey: Istanbul Technical University Faculty of Architecture. Liu, X. and B. Jiang, *Defining and generating axial lines from street center lines for better understanding of urban morphologies.* International Journal of Geographical Information Science, 2012. **26**(8): p. 1521–1532.

Peponis, J., et al., *Measuring the effects of layout upon visitors' spatial behaviors in open plan exhibition settings.* Environment and Planning B: Planning and design, 2004. **31**(3): p. 453–473.

Peponis, J., et al., *On the description of shape and spatial configuration inside buildings: convex partitions and their local properties.* Environment and Planning B: Planning and Design, 1997. **24**(5): p. 761–781.

Peponis, J., et al., *On the generation of linear representations of spatial configuration.* Environment and Planning B: planning and design, 1998a. **25**(4): p. 559–576.

Peponis, J., et al. *Street connectivity and urban density.* In *Proceedings of the 6th International Space Syntax Symposium.* 2007. Istanbul, Turkey: Istanbul Technical University Faculty of Architecture.

Peponis, J., et al., *The spatial core of urban culture.* Ekistics, 1989. **56**(334/335): p. 43–55.

Peponis, J., et al., *Spatialist.* 1997, Georgia Tech Reserarch Corporation: Atlanta, GA.

Perdikogianni, I. *Heraklion and Chania: a study of the evolution of their spatial and functional patterns.* In *Proceedings of the 4th International Space Syntax Symposium.* 2003. London: University College London.

Pinho, P. and V. Oliveira. *Combining Different Methodological Approaches to Analyze the Oporto Metropolitan Area.* In *Proceedings of the 7th International Space Syntax Symposium.* 2009. Stockholm, Sweden: KTH Royal Institute of Technology.

Porta, S., P. Crucitti, and V. Latora, *The network analysis of urban streets: a dual approach.* Physica A: Statistical Mechanics and its Applications, 2006. **369**(2): p. 853–866.

Porta, S., P. Crucitti, and V. Latora, *The network analysis of urban streets: a primal approach.* Environment and Planning B: planning and design, 2006. **33**(5): p. 705–725.

Portugali, J., *Complexity, cognition and the city.* 2011, Heidelberg: Springer Science & Business Media.

Portugali, J., *What Makes Cities Complex?*, in *Complexity, Cognition, Urban Planning and Design: Post-Proceedings of the 2nd Delft International Conference*, J. Portugali and E. Stolk, Editors. 2016, Springer International Publishing: Cham. p. 3–19.

Psarra, S. and T. Grajewski. *Describing shape and shape complexity using local properties.* In *Proceedings of the 3rd International Space Syntax Symposium.* 2001. Atlanta, GA: University of Michigan, Ann Arbor.

Psarra, S. *Top-down and bottom-up characterisations of shape and space.* In *Proceedings of the 4th International Space Syntax Symposium.* 2003. London, UK: University College London.

Raford, N., A. Chiaradia, and J. Gil, *Space syntax: The role of urban form in cyclist route choice in central London. Available at:* http://escholarship.org/uc/item/8qz8m4fz. 2007, Safe Transportation Research & Education Center: UC Berkeley.

Raford, N. and D. Ragland, *Space syntax: Innovative pedestrian volume modeling tool for pedestrian safety.* Transportation Research Record: Journal of the Transportation Research Board, 2004(1878): p. 66–74.

Rashid, M. and A.A.A. Bindajam, *Space, movement and heritage planning of the historic cities in Islamic societies: Learning from the Old City of Jeddah, Saudi Arabia.* URBAN DESIGN International, 2015. **20**(2): p. 107–129.

Rashid, M. and D. Alobaydi, *Territory, politics of power, and physical spatial networks: The case of Baghdad, Iraq.* Habitat International, 2015. **50**: p. 180–194.

Rashid, M., D. Boyle, and M. Crosser, *Network of spaces and interaction-related behaviors in adult intensive care units.* Behavioral Science, 2014. **4**: p. 487–510.

Rashid, M. and H. Shateh, *The Dialectics of Functional and Historical Morphology in the Evolution of a City: The Case of the Stone Town of Zanzibar.* Journal of Architecture, 2012. **17**(6): p. 889–924.

Rashid, M., J. Wineman, and C. Zimring, *Space, behavior, and environmental perception in open-plan offices: a prospective study.* Environment and Planning B: Planning and Design, 2009. **36**(3): p. 432–449.

Rashid, M., *Mutual visibility of points in building plans.* The Journal of Architecture, 2011. **16**(2): p. 231–266.

Rashid, M., N. Khan, and B. Jones, *Physical and Visual Accessibilities in Intensive Care Units: A Comparative Study of Open-Plan and Racetrack Units.* Critical Care Nursing Quarterly, 2016. **39**(4): p. 313–334.

Rashid, M., *On Space Syntax as a Configurational Theory of Architecture from a Situated Observer's Viewpoint.* Environment and Planning B: Planning and Design, 2012a. **39**(4): p. 732–754.

Rashid, M., *On the configurational studies of building plans from the viewpoint of a situated observer*, in *Architecture*. 1998, Georgia Institute of Technology: Atlanta, GA.

Rashid, M. *Revisiting John Portman's Peachtree Center Complex in Atlanta.* In *Proceedings of the 1st International Space Syntax Symposium*. 1997. London, UK: University College London.

Rashid, M., *Shape-sensitive configurational descriptions of building plans.* International Journal of Architectural Computing, 2012b. **10**(1): p. 33–52.

Rashid, M., *The Geometry of Urban Layouts: A Global Comparative Study*. 2016: Springer International Publishing.

Rashid, M., et al., *Spatial layout and face-to-face interaction in offices—a study of the mechanisms of spatial effects on face-to-face interaction.* Environment and Planning B: Planning and Design, 2006. **33**(6): p. 825–844.

Ratti, C., *Space syntax: some inconsistencies.* Environment and Planning B: Planning and Design, 2004. **31**(4): p. 487–499.

Ratti, C., *The lineage of the line: space syntax parameters from the analysis of urban DEMs.* Environment and Planning B: Planning and Design, 2005. **32**(4): p. 547–566.

Read, S., *The grain of space in time: the spatial/functional inheritance of Amsterdam's centre.* Urban Design International, 2000. **5**(3–4).

Reis, A., et al. *Perception of security, visual connections, children and adolescents behaviour in open spaces*. In *Proceedings of the 5th international Space syntax symposium*. 2005. TU Delft: TU Delft Faculty of Architecture.

Reis, A.T., L. Vedana, and C. Dittmar. *An analysis of street robbery and residential burglary through integration of axial lines, segments connectivity and GIS*. In *Proceedings of the 6th international space syntax symposium* 2007. Istanbul, Turkey: Istanbul Technical University Faculty of Architecture.

Schroder, C., W. Mackaness, and F. Reitsma. *Quantifying urban visibility using 3D space syntax*. In *Proceedings of GISRUK*. 2007. National University of Ireland.

Schumacher, T., *Buildings and streets: notes on configuration and use*, in *On streets*, S. Anderson, Editor. 1986, MIT Press. p. 133–149.

Scoppa, M., S. French, and J. Peponis. *The effects of street connectivity upon the distribution of local vehicular traffic in metropolitan Atlanta*. In *Proceedings of the 7th International Space Syntax Symposium*. 2009. Stockholm, Sweden: KTH Royal Institute of Technology.

Shpuza, E., *Allometry in the syntax of street networks: evolution of Adriatic and Ionian coastal cities 1800–2010.* Environment and Planning B: Planning and Design, 2014. **41**(3): p. 450–471.

Shpuza, E. *Evolution of Street Networks in Adriatic and Ionian Coastal Cities 1769–2007*. In *Proceedings of the 7th International Space Syntax Symposium*. 2009. Stockholm, Sweden: KTH Royal Institute of Technology.

Siksna, A., *City centre blocks and their evolution: a comparative study of eight American and Australian CBDs*. Journal of Urban design, 1998. **3**(3): p. 253–283.

Stahle, A., L. Marcus, and A. Karlström, *Place Syntax Tool—GIS Software for Analysing Geographic Accessibility with Axial Lines*, in *New developments in space syntax software*, A. Turner, Editor. 2007, ITU Faculty of Architecture: Istanbul, Turkey. p. 35–42.

Ståhle, A., L. Marcus, and A. Karlström. *Place Syntax: Geographic accessibility with axial lines in GIS*. In *Proceedings of the 5th international space syntax symposium*. 2005. Delft, NL: TU Delft Faculty of Architecture.

Steadman, P., *Developments in space syntax*. Environment and Planning B: Planning and Design, 2004. **31**(4): p. 483–486.

Suleiman, W., T. Joliveau, and E. Favier, *A new algorithm for 3D isovists*, in *Advances in Spatial Data Handling*. 2013, Springer. p. 157–173.

Teklenburg, J., H. Timmermans, and A. Van Wagenberg, *Space syntax: standardised integration measures and some simulations*. Environment and Planning B: Planning and Design, 1993. **20**(3): p. 347–357.

Topçu, M. and A.S. Kubat. *Morphological comparison of two historical Anatolian towns*. In *Proceedings of the 6th International Space Syntax Symposium*. 2007. Istanbul, Turkey: Istanbul Technical University Faculty of Architecture.

Topcu, M. and A.S. Kubat. *Old and new city: morphological analysis of Antakya*. In *Proceedings of the 8th International Space Syntax Symposium*. 2012. Santiago de Chile: PUC.

Topcu, M. and A.S. Kubat. *The analysis of urban features that affect land values in residential areas*. In *Proceedings of the 7th International Space Syntax Symposium* 2009. Stockholm, Sweden: KTH Royal Institute of Technology.

Trigueiro, E. and V. Medeiros. *The Bridge, the market, a centrality forever lost and some hope*. In *Proceedings of the 6th International Space Syntax Symposium*. 2007. Istanbul, Turkey: Istanbul Technical University Faculty of Architecture.

Turner, A., *UCL Depthmap: Spatial Network Analysis Software*. 2010, London.

Turner, A., A. Penn, and B. Hillier, *An algorithmic definition of the axial map*. Environment and Planning B: planning and design, 2005. **32**(3): p. 425–444.

Turner, A. *A program to perform visibility graph analysis*. In *Proceedings of the 3rd Space Syntax Symposium*. 2001. Atlanta, GA: University of Michigan, Ann Arbor.

Turner, A. *Angular analysis*. In *Proceedings of the 3rd international symposium on space syntax*. 2001. Georgia Institute of Technology: University of Michigan.

Turner, A. *Could a road-centre line be an axial line in disguise*. In *Proceedings of the 5th International Space Syntax Symposium*. 2005. TU Delft: TU Delft Faculty of Architecture.

Turner, A., *From axial to road-centre lines: a new representation for space syntax and a new model of route choice for transport network analysis*. Environment and Planning B: Planning and Design, 2007. **34**(3): p. 539–555.

Turner, A., *The role of angularity in route choice*, in *Spatial information theory*. 2009, Springer. p. 489–504.

Turner, A., et al., *From isovists to visibility graphs: a methodology for the analysis of architectural space*. Environment and Planning B: Planning and design, 2001. **28**(1): p. 103–121.

Van Nes, A., M. Berghauser Pont, and B. Mashhoodi. *Combination of Space syntax with spacematrix and the mixed use index: The Rotterdam South test case*. In *Proceedings of the 8th International Space Syntax Symposium*. 2012. Santiago de Chile: PUC.

Vaughan, L., J. Hanson, and R. Conroy-Dalton. *Teaching Space Syntax Through Reflective Practice*. In *Proceedings of the 6th International Space Syntax Symposium*. 2007. Istanbul, Turkey: Istanbul Technical University Faculty of Architecture.

Vaughan, L., *The relationship between physical segregation and social marginalisation in the urban environment*. World Architecture, 2005. **185**(185): p. 88–96.

Vaughan, L., *The spatial syntax of urban segregation.* Progress in Planning, 2007. **67**(3): p. 205–294.
Vaughan, L., et al., *Space and exclusion: does urban morphology play a part in social deprivation?* Area, 2005. **37**(4): p. 402–412.
Volchenkov, D. and P. Blanchard, *Scaling and universality in city space syntax: between Zipf and Matthew.* Physica A: Statistical Mechanics and its Applications, 2008. **387**(10): p. 2353–2364.
Wagner, R., *On the metric, topological and functional structures of urban networks.* Physica A: Statistical Mechanics and its Applications, 2008. **387**(8): p. 2120–2132.
Wang, J., Q. Zhu, and Q. Mao. *The three-dimensional extension of space syntax.* In *Proceedings of the 6th International Space Syntax Symposium.* 2007. Istanbul, Turkey: Istanbul Technical University Faculty of Architecture.
Wang, W.-C.J. and H.-W. Tsai. *Natural movement versus land value.* In *Proceedings of the 7th International Space Syntax Symposium.* 2009. Stockholm, Sweden: KTH Royal Institute of Technology.
Whitehand, J.W.R., *Urban morphology.* Historical geography: progress and prospect, 1987: p. 250–76.
Whitehand, J.W.R., *British urban morphology: the Conzenion tradition.* Urban Morphology, 2001. **5**(2): p. 103–109.
Whitehand, J.W.R. *Conzenian urban morphology and urban landscapes.* In *Proceedings of the 6th International Space Syntax Symposium.* 2007. Istanbul: Istanbul Technical University Faculty of Architecture.
Whitehand, J.W.R., ed. *The Urban Landscape: Historical Development and Management Papers by M. R. G. Conzen* Institute of British Geographers Special Publication 13. 1981, Academic Press.
Ye, Y. and A. van Nes, *Quantitative tools in urban morphology: Combining space syntax, space-matrix and mixed-use index in a GIS framework.* Urban morphology, 2014. **18**(2): p. 97–118.
Yoon, C. *Alternative Geometry for Space Syntax.* In *Proceedings of the 7th International Space Syntax Symposium.* 2009. Stockholm, Sweden: KTH Royal Institute of Technology.
Yun, Y.W. and Y.O. Kim. *The effect of depth and distance in spatial cognition.* In *Proceedings of the 6th International Space Syntax Symposium.* 2007. Istanbul, Turkey: Istanbul Technical University Faculty of Architecture.
Zheng, X., et al., *An extension method of space syntax and application.* Advancing Computing, Communication, Control and Management, 2010: p. 7–14.
Zhu, J., *Chinese spatial strategies: imperial Beijing, 1420–1911.* 2004, London: Routledge Curzon.

Applied Mathematics on Urban Space

A. van Nes

Abstract The aim of this chapter is to explain what Space Syntax is. Firstly, the types of spatial elements used in Space Syntax is discussed, secondly, the various mathematical formulas of various space syntax methods are elaborated, and finally Space Syntax' contribution to theory building on built environments are discussed.

1 Introduction

Research and methodological development on built environments are located on the intersection between natural, human and sociological sciences. During the Greek and Roman period, as well as in the Renaissance period, mathematics was applied in the geometry of buildings as well as in urban planning. During the twentieth century, various approaches from the social and human sciences were applied for understanding how cities work in relation to society. Due to large social and economic changes during the Industrial Revolution, the role of health, population growth and economic development had to be treated together with uncontrolled urban growth. There exist several writings on how we understand cities and how we should design them. However, concrete methods on analysing urban form and urban space are lacking in most of these writings.

In the past three decades the space syntax method, developed by Bill Hillier and his colleagues at the University College London, has been applied to urban studies. This method consists of calculating configurative spatial relationships in built environments.

According to Hillier, space syntax is four things. First, space syntax is operating with a concise definition of urban space. Second, it is a family of techniques for analysing cities as the networks of space formed by the placing, grouping and orientation of buildings. Third, it is a set of techniques for observing how these networks

---

A. van Nes (✉)
Western Norway University of Applied Sciences, Bergen, Norway
e-mail: avn@hvl.no; a.vannes@tudelft.nl

Faculty of Architecture, TU-Delft, The Netherlands

L. D'Acci (ed.), *The Mathematics of Urban Morphology*,
Modeling and Simulation in Science, Engineering and Technology,
https://doi.org/10.1007/978-3-030-12381-9_11

of space relate to functional patterns such as movement, land use, area differentiation, migration patterns and even social well-being and malaise. Fourth, based on the empirical results from the two first things, space syntax has made it possible to make a set of theories about how urban space networks relate in general to the social, economic and cognitive factors which shape them and are affected by them. The techniques have been applied to a large number of cities in different parts of the world. In this way, a substantial database now exists of cities which have been studied at some level using space syntax (Hillier et al. 2007).

What space syntax calculates is the two primary all-to-all (all street segments to all others) relations. On the one hand, it measures the *to-movement*, or accessibility, potential, of each street segment with respect to all others. On the other hand, it measures the *through-movement* potential of each street segment with respect to all pairs of others. Each of these two types of relational pattern can be weighted by three different definitions of distance. The metric distance measures the city's street and road net as a system of shortest paths, while the topological distance, calculate the city's street and road net as a system of fewest turns paths. Finally, the geometrical distance gives a picture of the city's street and road net as a system of least angle change paths. Each type of relation can be calculated at different radii from each street segment, defining radius again either in terms of shortest, fewest turns or least angle paths (Hillier and Iida 2005, pp. 557–558).

Space syntax is under constant development. Its contribution to theories on built environments and methodology develop at the intersection of natural, social and technical sciences. So far, research projects range from anthropology or cognitive sciences to applied mathematics and informatics and touch upon philosophical issues. The evolution of space syntax asks for communication not just between various cultural contexts, but likewise between different scientific domains. Space syntax research is placed in the overlap between the applied mathematics from the natural sciences, cognition and orientation issues from the human sciences and human behaviour (social and economic related activities) from the social sciences.

2 Definition of Urban Space

Hillier distinguishes between *extrinsic* and *intrinsic* properties of space. Extrinsic ones determine the way in which spatial units relate to one another. Here settlements are regarded as sets of spaces defined by surrounding physical objects. Volumes, textures and size are not taken into consideration. Here, spaces are shape-free. It is just their inter-relational aspect or structure that is taken into consideration. Each space has one or more functions either in terms of occupation or with regard to movement (Hillier 1999, p. 1). Extrinsic properties of space determine both built form and its possible function.

Intrinsic properties of space relate to visible aspects of things we can see, i.e. shape, size, volumes and texture of physical objects or built mass. They present themselves mostly through geometrical properties. They account for the articulation of social meaning via a built form (Hillier 1999, p. 1). We have many words for describing the extrinsic properties of space. Words like “a narrow street, a large square, a massive building etc” make it possible to describe the artefacts of a city.

Describing the extrinsic properties of space, thus, requires to represent parts of the complex urban reality in models. Therefore, spatial modelling of built environments simplifies reality. In the case of space syntax, the fewest and longest sight lines axial map is seen as a representation of publicly accessible spaces. It models or represents the required correspondence between world and model. Naturally, the model is simpler than the thing modelled.

3 The Method of Calculations

Recent software development has made it possible to calculate and combine geometrical relationships (in terms of angular weighting between axial lines) and metric relationships with topological spatial relationships.

A global axial integration analysis implies to calculate how spatial integrated a street axe is in terms of the total number of direction changes to all others in a town or city. The fewer changes of direction to all other streets, the higher global integration values. Conversely, streets with many direction changes to all others tend to have low global integration values. Hence, they are spatially segregated.

Figure 1 left shows an axial map of a simple settlement, named town X, consisting of the main street with some side streets and some smaller back streets. The right upper corner in Fig. 1 shows an integration analysis of the settlement carried out by the computer program Depthmap. Various integration values are represented by colours. The red axes are the most integrated ones, while the blue ones are the most segregated ones. As shown in the lower part in Fig. 1, the justified graph shows how the system can be experienced from the most integrated main street and the segregated back street. In the case of the fewest line axial map, the lines are represented as nodes and the intersection of lines as connections between the nodes. The colours of the circles are the same as in the axial integration analyses. The justified graph shows how a whole system is connected to one another in an abstract manner.

The more integrated a street is, the shorter topological distance it has to all other streets. Other way around, the more segregated a street is, the longer topological distance to all other streets. When comparison the justified graphs of the back and the main street, the graph’s structure looks more like a “tree” in the back street case, while it looks more like a “bush” in the case of the main street. The higher number of spaces in a shorter topological distance from space, the more likely this space will have a high integration value. The “bush” shaped justified graph is thus topological shallow. If most of the spaces are located many topological steps away, the more likely this space implies low integration values. In this case, it is a highly segregated space, and the “tree” shaped justified graph is topologically deep.

Since town X has a simple spatial system consisting of a few streets, it can be calculated manually. The computer programs Axwoman (Jiang et al. 2000) and Depthmap

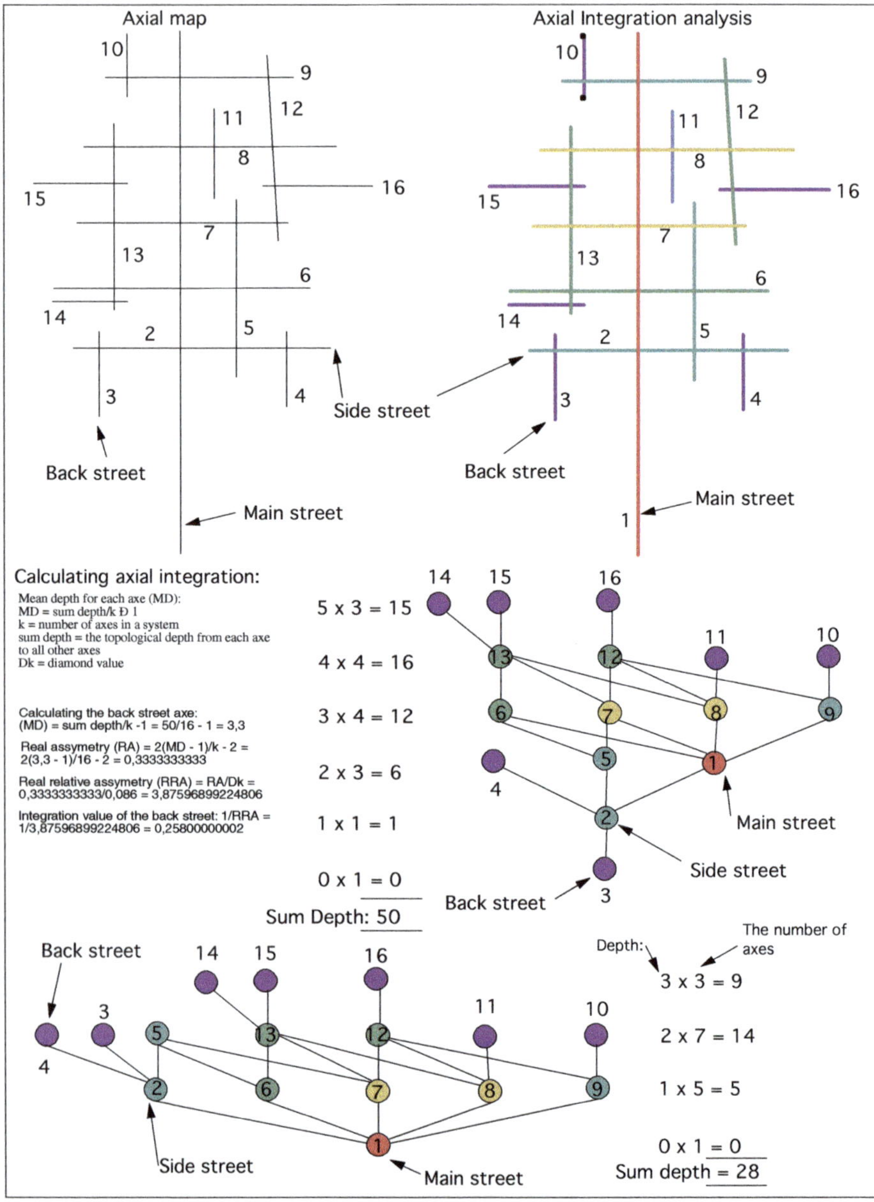

Fig. 1 A simple example on calculating spatial relationships

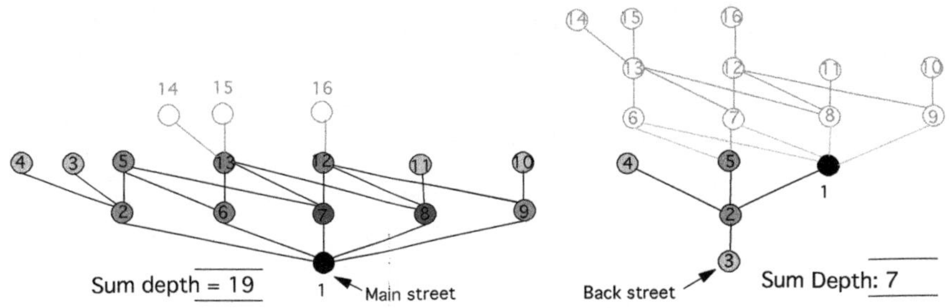

Calculating the main street axe:
(MD) = sum local depth/klocal -1 = 19/13 - 1 = 1,583333333

Real asymmetry (RA) = 2(MD - 1)/k - 2 =
2(1,583333333 - 1)/13 - 2 = 0,10606060606

Real relative asymmetry (RRA) = RA/Dk =
0,10606060606/0,276 = 0,38427755819

Local integration value of the main street: (1/RRA) =
1/0,38427755819 = **2,6022857143**

Calculating the back street axe:
(MD) = sum local depth/k -1 = 7/5 - 1 = 1,75

Real asymmetry (RA) = 2(MD - 1)/k - 2 =
2(1,75 - 1)/5 - 2 = 0,5

Real relative asymmetry (RRA) = RA/Dk =
0,5/0,352 = 1,4204545455

Local integration value of the back street: (1/RRA) =
1/ 1,4204545455 = **0,7039999998**

Fig. 2 The principle of calculating local integration

(Turner 2007) are able to calculate large cities with thousands of axes inter-related with one another.

As research shows, commercial activities take place in the most global integrated streets (Hillier 1996, p. 175; Hillier et al. 1993, pp. 31, 36 and 61; van Nes 2002, pp. 287–303). Dwelling areas are mostly located in the segregated areas (Hillier 1996, pp. 175–179; Hillier and Hanson 1984, p. 140).

A local axial integration analysis contributes to reducing the edge effect from global integration analyses. Figure 2 illustrates what local integration with a radius like 3 is like in town X. It calculates the value of the axe in a topological radius like 3 from each street. When calculating the value of the main street, the axes number 14, 15 and 16 are not taken into account. In the case of the back street, only 5 axes are included in the local integration analysis. The system in the left shows a local integration analysis of every street in town X.

Like global axial integration, local axial integration can measure the spatial impacts of the whole city before and after urban interventions. As research has shown, the flow rates of pedestrians through cities correlate with local integration values while vehicle flow rates correspond with global integration ones (Hillier et al. 1998, pp. 59, 84). Moreover, local axial integration gives an indication of local shopping areas in a city.

Figure 3 shows a global and local integration analysis of Haarlem. The locally most integrated streets, coloured in red, are where the most vital pedestrian-friendly shopping streets are. Along these streets, there is a mixture of individual shops, chain stores and cafés. Along the globally integrated streets, Haarlem's main city centre is located with a high number of shops, large chain stores, the municipality hall and the cinema.

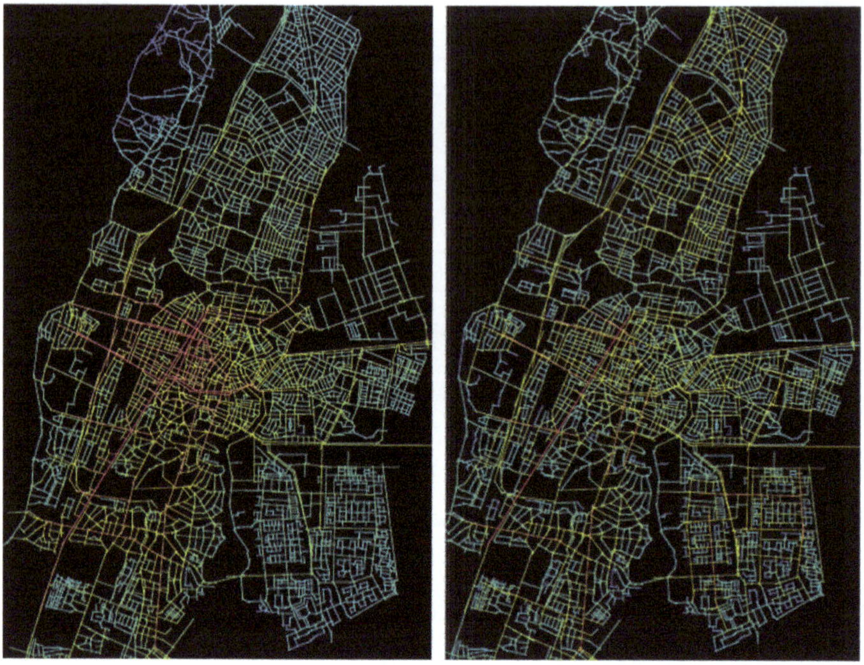

Fig. 3 Global (left) and local (right) axial integration of Haarlem

Applying global and local integration analyses on Dutch cities tend to show weak results. Axial analyses count each change of direction as one topological step, even though the angle is close to 180°. In this way, many centres in cities with curved streets tend to get a broken up street net consisting of many short axial lines. As follows, the local axial integration values will get low in these kinds of areas, which does not always correspond with the location pattern of shops.

The angular analysis is essentially an extension of visibility graph analysis and axial analysis (Turner 2001, p. 30.1). What the angular analysis adds to the various integration analyses is that each axial line is weighted by the angle of their connections to other axial lines. As shown in Fig. 4, two axes which are almost 180° has a shallow angle of incidence while two axes with almost 90° have a sharp angle of incidence (Dalton 2001, p. 26.7). For making angular analyses, the axial map is broken up into segments. For example, a long axis crossing several other axes consist of several segments. Now it is possible to make integration analyses of the segments as well as taking the segments' angular relationship into account in the spatial analyses.

As research has shown, streets' angular relationships play a role in the way people orientate themselves through built environments. People tend to conserve linearity through their routes, with minimal angular deviation (Conroy Dalton 2001, p. 47.8). By changing direction, people tend to choose an angle close to 90° or to 180°. Urban blocks with rare angles, like those of 30 and 60°, tend to make people get lost.

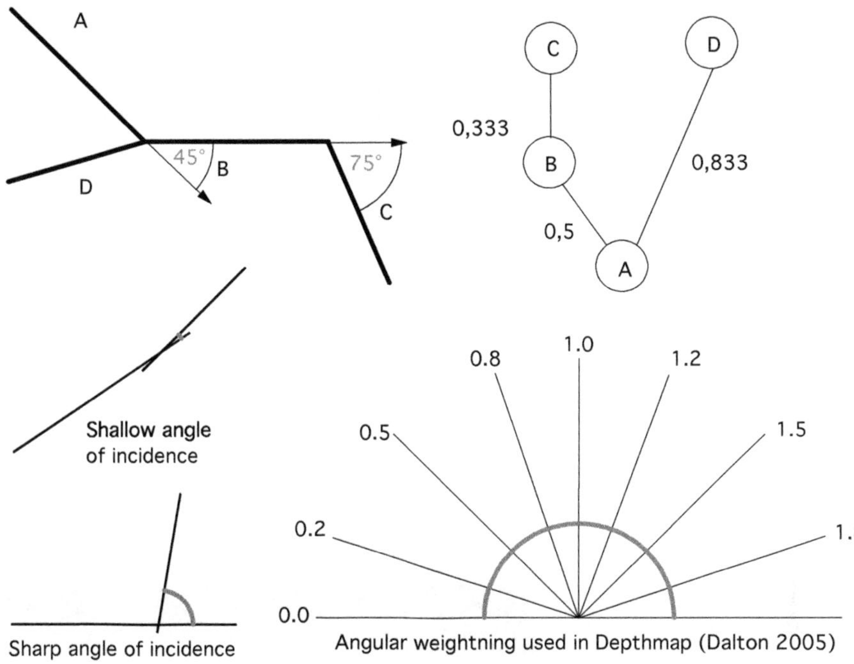

Fig. 4 Angular weighting of axial lines

Moreover, people tend to choose the longest street with the shortest angle towards their aimed direction. With other words, people choose the straightest possible routes in order to avoid the complexity for way-finding through urban street grids (Conroy Dalton 2001, p. 47.11).

In order to calculate the angular segment analyses of a street segment, the total angular turn from one segment to another segment is calculated. Values are given on different angles. For example, an angle with 45° has the value like 0,5, and angle with 90° has the value like 1, and an angle with 180° has a value like 2 (Turner 2005, p. 148). Consider four street segments connected to one another in different angles, as illustrated in Fig. 4. The depth from segment A to B is 0.5, since one makes a turn of 45°. The depth from segment A to C is 0.833 due to that one has to make a turn of 45° first to segment B and then a turn of 30° to segment B. The values are shown in the justified graph on the upper right in Fig. 4.

When calculating the angular mean depth, or the local angular integration, the case shown in Fig. 4 is used as an example. The angular mean depth from street segment A, is calculated as follows:

$$angular\ mean\ depth\ of\ (A) = \frac{(B)0.5 + (C)0.833 + (D)0.833}{(3)} = \mathbf{0.722} \qquad (1)$$

The angular mean depth analysis highlights the main routes network through cities and regions. Moreover, the edge effect from the axial analysis is reduced. The least angle analysis seems to be the best predictor of movement, followed closely by the fewest turns (the results from the global and local integration analyses). Metric distance comes far behind the two first ones (Hillier et al. 2007).

The software Depthmap is able to make a segment map from an axial map. This map is the base for the angular analysis. At every node, the angular relationship between the connected lines can now be taken into account. When trying out various radii, the following results are obtained: the lower the radius, the more the main routes in the small local centres of a city are highlighted, the higher the radius, the more the main routes through the city are highlighted. As shown in Fig. 5, a local angular integration analysis highlights the most vital shopping streets in Haarlem as well as the streets where most of the larger companies and municipality hall are located.

One of the criticisms of space syntax has been that it does not take into account metrical properties in analyses of the mobility network (Ratti 2004, p. 501). As

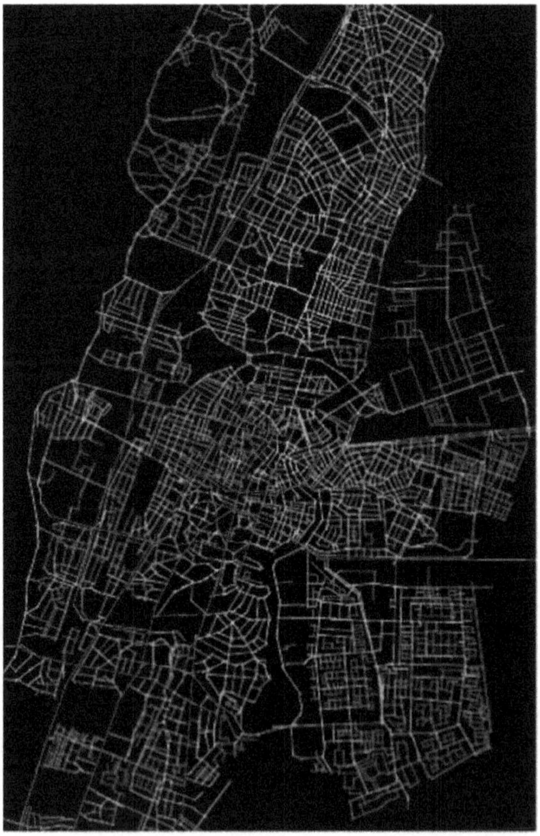

Fig. 5 Angular analyses of Haarlem centre with R = 3

mentioned earlier, metrical distances show the least correlation between pedestrian and car traffic flow rates and the spatial analyses in comparison with the geometrical and topological distances. However, there is a difference between metrical distance and metrical radii.

Figure 6 shows some examples of how various types of radii affect various types of street networks. Town Y has on the left side of its main street an orthogonal street network and on the right side an organic street structure. When applying a two-step analysis from the main route axis, almost all streets in the strict orthogonal grid can be reached within two direction changes. Conversely, the local catchment area in the organic street network is rather poor. Likewise, when applying the two-step analysis to a street segment (below left in the figure), the local catchment area reduces slightly for both types of street networks. When applying the metrical radius from a segment, the strict orthogonal street network as well as the organic street network has more or less the same catchment area. The next step is to show what metrical radiuses add to the analyses of topological and geometrical distances.

Figure 7 shows the most used space syntax analyses of town Y. In the global axial and segment analyses, the main street is the highest integrated street. When splitting the axial map into segments, and running a segment integration analyses, the middle part of the main route is the most integrated on both global and local levels. When analysing the angular choice on global and local levels, the main route shows and a side street shows the largest through movement potentials. The angular segment integration analysis measures the "to-movement" potentials whereas the angular choice analysis measures the "through-movement" potentials on various scale levels. The first one highlights the potentials for urban centres, whereas the latter for the potentials for movement flow between various urban areas.

In order to demonstrate what a metrical radius implies for urban centrality when combining it with topological and geometrical distances, an example of a new and an old town is used. The Dutch new town Zoetermeer is 50 years old and has around 123.500 inhabitants. Originally, it was a small village until the late 1960s with 7000 inhabitants. The new town was constructed with the intention to catch up the population growth in the Hague City. The car traffic routes and pedestrian and bicycle routes are separated. Figure 8 shows angular choice segment analyses with a low and a high metrical radius for Zoetermeer's mobility network. All routes are included in the analyses. As can be seen in the figure on the left, the streets where the children's playgrounds are located are highlighted in red. Conversely, as shown in the figure on the right, the main routes between the various local areas are highlighted in red. These routes are trafficked by only vehicles. Zoetermeer's main mega shopping mall is located in the middle, where the density of the integrated main routes is the highest. The degree of orientability is low for visitors in Zoetermeer, due to its labyrinthly structured local street network and that the main route network is located far outside the various local residential areas.

When applying the same analyses to the old Dutch town Haarlem, a different structure can be seen. Haarlem was founded in 1245, and today it has around 155.500 inhabitants. Figure 9 shows an analysis of Haarlem's street and road network with a low radius. The town's local shopping streets are highlighted in red. Likewise, all

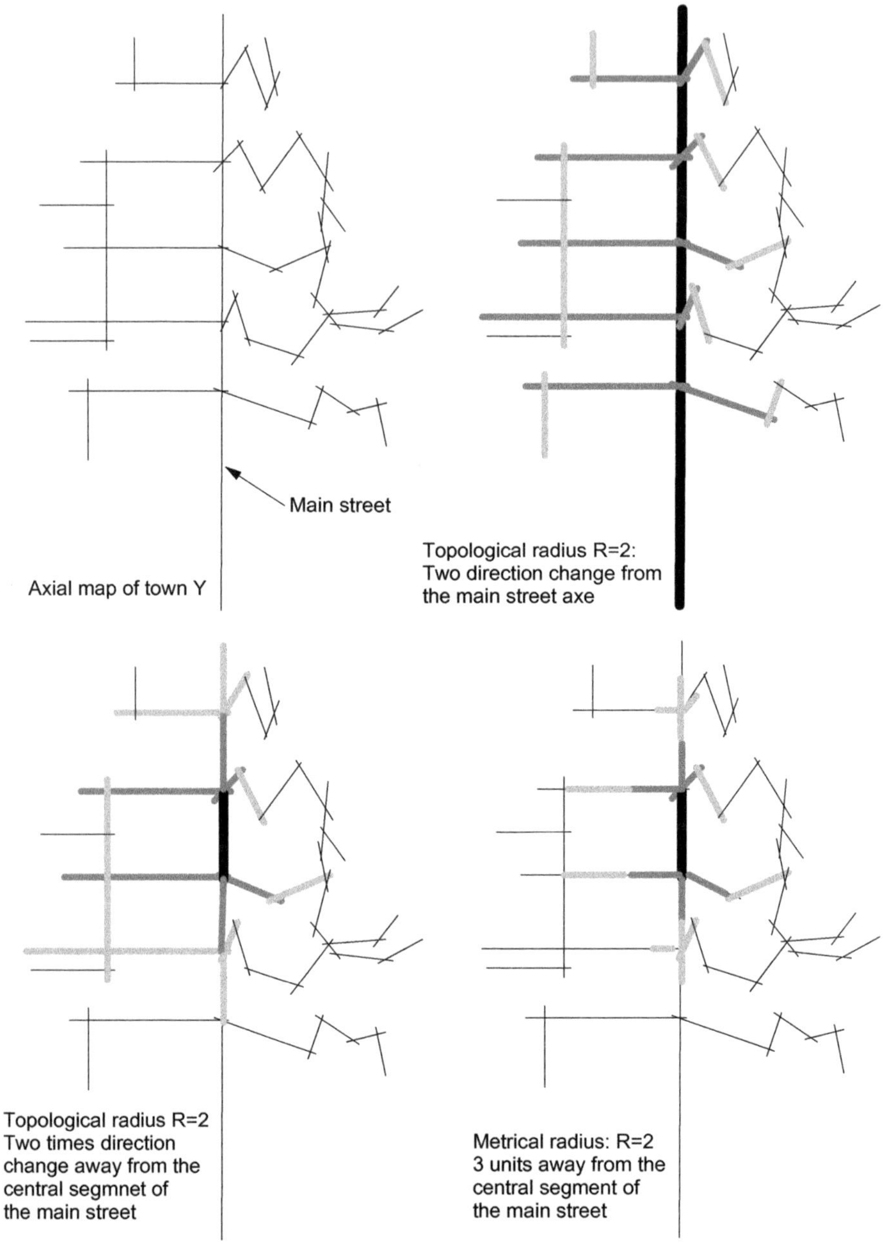

Fig. 6 The difference between a topological radius and a metrical radius

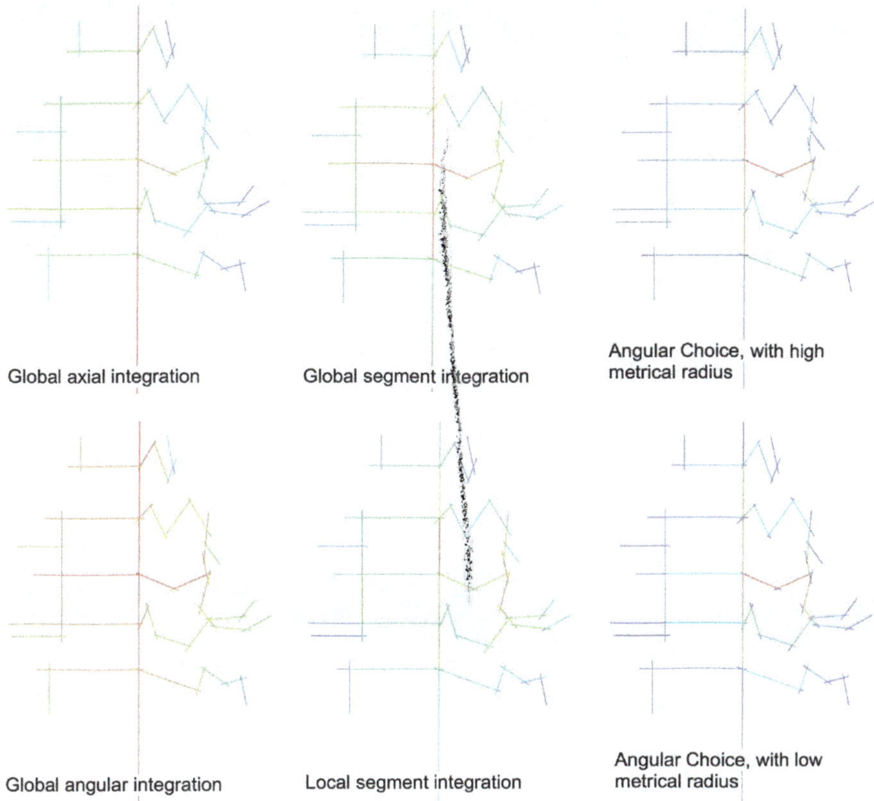

Fig. 7 Various integration analyses of Town Y

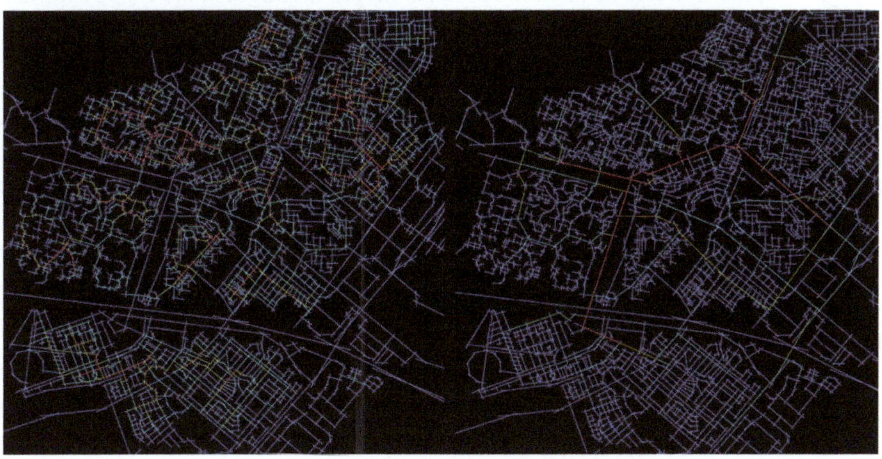

Fig. 8 Angular choice analyses with a low (left) and high (right) metrical radius for Zoetermeer

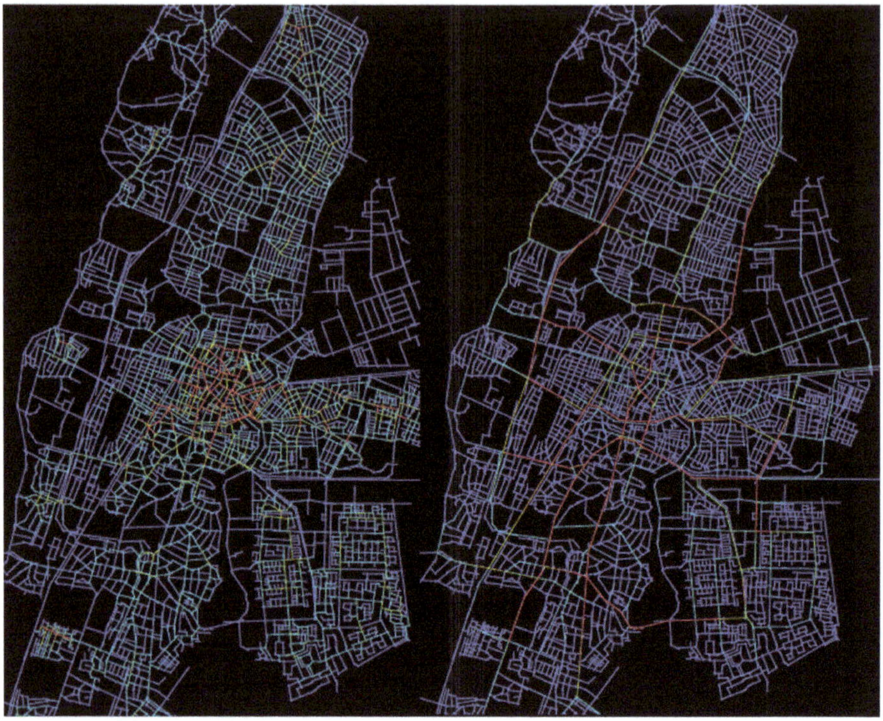

Fig. 9 Angular choice analyses with a low (left) and high (right) metrical radius for Haarlem

its central shopping streets are highlighted. When applying a high radius, as shown in Fig. 9, the main routes through the urban areas are highlighted in red. These main routes cross the local centres and the main centre highlighted in the small radius analysis.

As presumed, an optimal location for shops tends to be in streets that are accessible within a short metrical radius to a neighbourhood's dwellers as well as to a place that catches the through travellers. Therefore, the variation of shops in local shopping areas tends to be higher in old towns than in new towns. Even though the density of inhabitants can be high in a post-war neighbourhood, these areas generally offer a local supermarket. The supermarket is mostly used by the dwellers of the neighbourhood. In pre-war neighbourhoods, the variation of shops tends to be much higher than in post-war areas, due to their strategic location in the mobility network. These streets can easily be reached from the main route network.

A criticism often directed at new towns tends to be that they are "sleeping towns" with a lack of street life (van Casteren 2008). The use of the metrical radii can shed some light on this discussion. The main route network in post-war neighbourhoods is separated from the local centres whereas it is integrated with the local centres in pre-war neighbourhoods.

All cities are made up of a very large number of short streets and a very small number of long streets and roads. This can be seen on all scale levels in which gives the city street networks a clearly fractal structure. The foreground network is largely made up of longer streets or roads whose ends are linked by highly obtuse, nearly straight connections. The longer the line, the more likely it is to end with a nearly straight connection. The main routes through cities on all scale levels tend to consist of a set of longer lines connected to each other with almost 180° angles. The main routes net through urban areas it the armature for linking the city's edges to its centre. The natural interface of co-presence through movement from centres to edges is made efficient and possible, and affects the location of economic activities. The foreground network is the general component of the city (Hillier et al. 2007, pp. 2–4).

Conversely, the background network is largely made up of shorter streets, which tend to intersect and end at a near right angle. The shorter the street, the more likely it is to end at a right angle. Most silent dwelling streets tend to be metrically short. The background network reproduces the cultural pattern, and is the conservative component of the city. Various cultures have different local radius measures (Hillier et al. 2007, pp. 2–4).

In many ways, the generative spatial laws are the spatial parameter for human cognition. What space syntax measures are various degrees of inter-visibility i.e. how many people can see each other, and various degree of accessibility i.e. how a city's edges can reach its centre. The latter one concern the through movement pattern of "betweenness". When built environments grow in a natural way, extensions occur through the way avoiding blocking longer streets or roads (Hillier and Iida 2005, pp. 557–557).

Experimenting with mathematical formulas contributes to new discoveries (Hacking 1991). In many ways, space syntax is developed through "trial and error" research. New phenomena are created through experiments and tested out on a wide range of built environments. The challenges have been to give names on these various discovered phenomena and to present it to an audience outside the research community. At least experiments contribute to new development or improvement of existing methods for analysing urban space and to provide understandings on how built environments works.

All these spatial measurements can be correlated with a variety of other numerical data expressing social activities such as flows of human movement through the street net, land use pattern, land values, and distribution of crime. Thus, spatial and social factors can be correlated with one another. Hence, it is possible to study one urban area in correlation with its whole city.

So far, in urban research where space syntax has been applied, the following results were obtained. First of all, the degree of spatial integration is a strong predictor of pedestrian and traffic flow rates (Hillier et al. 1998, pp. 80–84). Pedestrian flows tend to follow various local integration values, while vehicle movement tends to follow the global integration values. As research has shown, applying the space syntax method on the configuration of the street net is better able to explain how the spatial set up of built environments influences the flow of traffic than the modelling techniques of the road engineers (Penn 2003, p. 33). Moreover, correlations between various degrees of integration and distribution of various types of crime have been found in research (Hillier and Sahbaz 2005; López and van Nes 2007; Hillier and Shu 2000). Moreover, a strong correlation was also found between streets' degree of integration and land values (Desyllas 2000), building density (Van Nes et al. 2012), land use diversity (Ye et al. 2014), degree of ethnic mixture (van Nes and Aghabeick 2015), degree of gender mixture (van Nes and Rooij 2015) and location pattern of shops (van Nes 2002).

4 Space Syntax' Contribution to Theory Building and Understanding on How to Build Environments Works

Building systematic theories on the built environment is still in a beginning phase. The reason is that most writings on built environment tend to have a *normative* approach (Hillier and Hanson 1984, p. 5), where most authors describe *how* to make a good city. What is lacking is a description of what a good city *is* or how a city *function* spatially and in relation to society. Conversely, writings in the field of urban sociology lack concise definitions of space or understandings on the physical framework on where various social interactions take place.

The use of space syntax has contributed to an understanding of the spatial structure of the city as an object shaped by a society on the one hand and on the other hand how it can generate or affect certain socio-economical processes in a society. To some extent, space syntax is able to predict some types of economic processes as an effect on urban interventions. Likewise, space syntax provides understandings on the spatial possibilities for certain social activities such as crime, social segregation and anti-social behaviour. It is all about how spatial integration and segregation conditions social integration and segregation.

The biggest challenge at this moment is to build descriptive theories on how cities work. It has to be done from three different perspectives. The first approach is on the relation between society and space. Here the focus is to get understandings on how activities in a society influence the shape, pattern and structure of a built environment. Research in the field of social anthropology and archaeology belongs under this approach. The aim is to gain an understanding of various cultures based on their built form. A hermeneutic approach is used here, and therefore clear explanations or theory building between cause (the society) and effect (the built form) from the natural science tradition cannot be done (von Wright 1971).

The second approach focuses only on the spatial relationships of the built environment. In line with the positivistic tradition, it is obvious that Hillier's theories on spatial laws or combinatorics (Hillier 1996, Chap. 8) have a strong link between cause and effect:

The principle of centrality: A central placed object increases the topological depth more than one placed at the edge.

The principle of extension: Partitioning a longer line increases the topological depth than a short one.

The principle of contiguity: Contiguous blocks increase topological depth more than separate ones.

The principle of compactness: Straight lines increase topological depth more than "curled" lines.

Intentions and human rationalities are not taken into account here.

The third approach is on the relationship between space and society. This approach has both a hermeneutic and positivistic approach. It is about how built form affect activities in society. Here again, human rationality has to be taken into account. Where the human intentions are unambiguous, it is possible to predict the effects on society as an effect on spatial changes of built form. Marked rationality is an unambigious rationality, where it is about profit maximising. Therefore, various space syntax researches have contributed to the *theory of the natural movement economic process*. The spatial structure of the street network influences the movement rates through an urban street net and where economic activities take place. Attractors, such as shops, retail and large firms tend to locate themselves along the most integrated streets (Hillier et al. 1993, pp. 31 and 61).

Figure 10 shows the relationship between configuration, attraction (the location of shops) and movement. It explains how a built environment function independent on planning processes regard the location pattern of economic activities, human movement through the urban network and the configuration of the street grid. Movement and attractors influence each other. The more people in a street, the more it attracts shops to locate along these streets. The more shops locating along a street, the more they attract people into this street. It gives a multiple effect process. After all, movement and attractors do not influence the configuration of the street net.

Likewise, the recently proposed *theory of the natural urban transformation process* (Ye et al. 2014) is rooted in marked rationality. The more integrated or accessible the street network is, the higher density of the buildings and the higher degree of

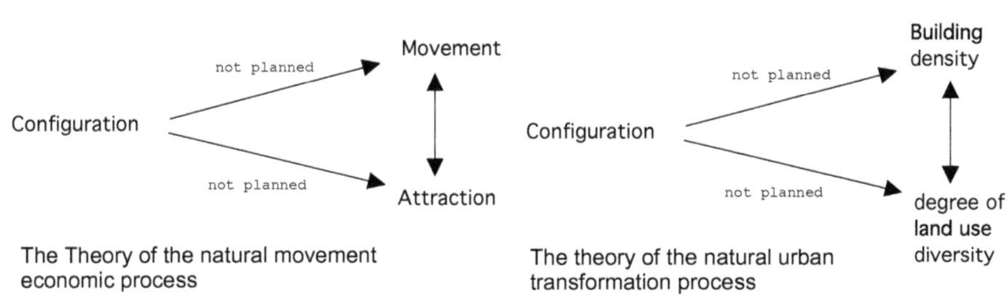

Fig. 10 The Theory of the natural movement economic process and the Theory of the natural urban transformation process

land use diversity. The spatial structure of the street network is the underlying driver for urban transformation processes.

Other kinds of human rationality, such as the occurrence of crime, anti-social behaviour, the location of various ethnical groups in cities and the occurrence of fear in built environments cannot be predicted. Therefore, there exist no theories on for example the relationship between space and anti-social behaviour.

Political forces and organisatoric constraints can overrun the spatial forces in built environments. A strong planning system on different levels and organisatoric constraints can block a natural location of economic activities at strategic optimal locations. Likewise, ethnic conflicts can contribute to that people avoid central integrated areas. Therefore one has to be aware of a country's planning system, political forces or ethnic conflicts on the one hand, and on the other hand the generative power of the street and road net.

In essence, subsequent considerations distinguish between a theory able to offer an *explanation* of phenomena and a theory proposing an *understanding* thereof. As concluded, the theories on spatial combinatorics, the natural movement economic process, and natural urban transformation process can offer an explanation of changes in a built environment in terms of cause and effect, while research related to social rationality, archaeology or historical research, space and crime or anti-social behaviour, cognitive aspects aims at an understanding of the culture or meaning associated with the causes at issue. Moreover, research concerning how activities in society affects urban space requires a hermeneutic approach, whereas research concerning how a spatial layout can affect activities in society requires both a positivistic as well as a hermeneutic approach. Seemingly, human behaviour as an effect on spatial structure depends on the type of rationality of human intentions and behaviour the research is focusing on. Marked rationality can use positivistic explanation models, whereas other kinds of rationality rely on hermeneutic ones.

What does space syntax add to studies on built environment? At least it offers concise spatial tools to measure spatial changes in the built environment, independent on context related situations where cultural aspect must be taken into account. In this way, space syntax is able finding some spatial evidence on presumptions and observations. Even though Norberg-Schulz criticises a quantitative approach in studies on built environment (Norberg-Schulz 1967, p. 202), a space syntax approach can at least provide some exact evidence on how some spatial components of built environments create lively or quiet places based on applying some simple mathematic calculations on spatial relationships.

References

J. van Casteren. Lelystad. Prometheus, Amsterdam, 2008.

R. Conroy Dalton. The secret is to follow your nose. In J. Peponis, J. Wineman, and S. Bafna, editors, Proceedings Space Syntax. 3rd International Symposium, Atlanta, 2001.

N. Dalton. Fractional configurational analysis and a solution to the manhattan problem. In J. Peponis, J. Wineman, and S. Bafna, editors, Proceedings Space Syntax. 3rd International Symposium, Atlanta, 2001.

J. Desyllas. The relationship between urban street configuration and office rent patterns in Berlin. PhD thesis, The Bartlett, University College London, 2000.

I. Hacking. Experimentation and scientific realism. In The Philosophy of science, pages 247–260. Boyd, Gasper and Trout ed., Mass, 1991.

B. Hillier. Space is the Machine. Cambridge University Press, 1996.

B. Hillier. Space as paradigm for describing emergent structure in strongly relational systems. Bartlett School of Architecture, UCL, 1999.

B. Hillier and J. Hanson. The Social Logic of Space. Cambridge University Press, 1984.

B. Hillier and S. Iida. Network effects and psychological effects: a theory of urban movement. In A. van Nes, editor, Proceedings Space Syntax. 5th International Symposium, Delft, 2005.

B. Hillier and O. Sahbaz. High resolution analysis of crime patterns in urban street networks: an initial statistical sketch from an ongoing study of a london borough. In A. van Nes, editor, Proceedings Space Syntax. 5th International Symposium, Delft, 2005.

B. Hillier and C. F. Shu. Crime and urban layout: the need for evidence. In S. Ballintyne, K. Pease, and V. McLaren, editors, Secure Foundations. Key issues in crime prevention, crime reduction and community safety. Institute for Public Policy research, London, 2000.

B. Hillier, A. Penn, J. Hanson, et al. Natural movement: or, configuration and attraction in urban pedestrian movement. Environment and Planning B: Planning and Design, 20:29–66, 1993.

B. Hillier, A. Penn, D. Banister, and J. Xu. Configurational modelling of urban movement network. Environment and Planning B: Planning and Design, 25:59–84, 1998.

B. Hillier, A. Turner, T. Yang, and H. T. Park. Metric and topo-geometric properties of urban street networks. In A. S. Kubat, editor, Proceedings Space Syntax. 6th Interna- tional Symposium, Istanbul, 2007.

B. Jiang, C. Claramunt, and B. Klarqvist. An integration of space syntax into GIS for modelling urban spaces. International Journal of Applied Earth Observation and Geoinformation, 2:161–171, 2000.

M. López and A. van Nes. Space and crime in dutch built environments. macro and micro scale spatial conditions for residential burglaries and thefts from cars. In A. S. Kubat, editor, Proceedings Space Syntax. 6th International Symposium, Istanbul, 2007.

C. Norberg-Schulz. Intensjoner i arkitekturen. Universitetsforlaget, Oslo, 1967.

A. Penn. Space Syntax and Spatial Cognition Or Why the Axial Line. Environment and Behaviour, 35:30–65, 2003.

C. Ratti. Space syntax: some inconsistencies. Environment and Planning B: Planning and Design, 31:501–511, 2004.

A. Turner. Angular analysis. In J. Peponis, J. Wineman, and S. Bafna, editors, Proceedings Space Syntax. 3rd International Symposium, Atlanta, 2001.

A. Turner. Could a road-centre line be an axial line in disguise? In A. van Nes, editor, Proceedings Space Syntax. 5th International Symposium, Delft, 2005.

A. Turner. UCL depthmap 7: From isovist analysis to generic spatial net- work analysis. In A. Turner, editor, New Developments in Space Syntax Software, Istanbul, 2007.

A. van Nes. Road Building and Urban Change. The effect of ring roads on the dispersal of functions in Western European towns and cities. PhD thesis, Agricultural University of Norway, 2002.

A. van Nes & Aghabeik, L 2015, Ethnic groups and spatial behaviour in Rotterdam's neighbourhoods. In K Karimi, L Vaughan, K Sailer, G Palaiologou & T Bolton (Eds.), Proceedings of the 10th international space syntax symposium (pp. 102:1–102:17). London: Space Syntax Laboratory, The Bartlett School of Architecture, UCL.

A. van Nes & Rooij, L de 2015, The perceived safety and spatial behaviour in three different neighbourhoods in Rotterdam. In: K Karimi, L Vaughan, K Sailer, G Palaiologou & T Bolton (Eds.), Proceedings of the 10th international space syntax symposium (pp. 139:1–139:19). London: Space Syntax Laboratory, The Bartlett School of Architecture, UCL.

A. van Nes, Berghauser Pont, Meta and Mashhoodi, Bardia, 2012, Combination of Space Syntax with Spacematrix and the Mixed Use Index. The Rotterdam South test case. in Greene, M. (ed.), 2012, "Proceedings Space Syntax. 8th International Symposium," Pontificia Universidad Católica de Chile, Santiago, Chile.
G. H. von Wright. Explanation and Understanding. International Library of Philosophy and Scientific Method, Lowe & Brydone Ltd., London, 1971.
Ye, Yu, and Van Nes, Akkelies 2014, Quantitative tools in urban morphology: Combining space syntax, spacematrix and mixed-use index in a GIS framework. in: Journal of Urban Morphology, Vol 18, number 2, pp 97–118.

The Morphology and Circuity of Walkable and Drivable Street Networks

Geoff Boeing

Abstract Circuity, the ratio of network distances to straight-line distances, is an important measure of urban street network structure and transportation efficiency. Circuity results from a circulation network's configuration, planning, and underlying terrain. In turn, it impacts how humans use urban space for settlement and travel. Although past research has examined overall street network circuity, researchers have not studied the relative circuity of walkable versus drivable circulation networks. This study uses OpenStreetMap data to explore relative network circuity. We download walkable and drivable networks for 40 US cities using the OSMnx software, which we then use to simulate four million routes and analyze circuity to characterize network structure. We find that walking networks tend to allow for more direct routes than driving networks do in most cities: average driving circuity exceeds average walking circuity in all but four of the cities that exhibit statistically significant differences between network types. We discuss various reasons for this phenomenon, illustrated with case studies. Network circuity also varies substantially between different types of places. These findings underscore the value of using network-based distances and times rather than straight-line when studying urban travel and access. They also suggest the importance of differentiating between walkable and drivable circulation networks when modeling and characterizing urban street networks: although different modes' networks overlap in any given city, their relative structure and performance vary in most cities.

1 Introduction

Street networks organize and structure human spatial dynamics and flows in a city. They underlie commutes, discretionary trips, and the location decisions of households and firms. Accordingly, substantial research has been conducted in recent years to better characterize the topological and geometric characteristics of urban

G. Boeing (✉)
School of Public Policy and Urban Affairs, Northeastern University, Boston, MA, USA
e-mail: g.boeing@northeastern.edu

L. D'Acci (ed.), *The Mathematics of Urban Morphology*,
Modeling and Simulation in Science, Engineering and Technology,
https://doi.org/10.1007/978-3-030-12381-9_12

street networks (Gastner and Newman 2006; Batty 2005; Fischer 2004). Topological character describes the configuration of the network and includes measures of connectivity, centrality, and clustering. Geometric character describes the network's distances, areas, and densities. Both intermingle to define the network's structure, efficiency, and performance.

Street network analysis has become an important bridge between graph theory and urban morphology and planning (Barthelemy 2017; Barthelemy 2008; Cardillo et al. 2006; Masucci et al. 2009, 2013; Alexander 1965). Recent studies have found that cities can be clustered and classified according to their network's structural characteristics (Strano et al. 2013; Porta et al. 2006a, b; Crucitti et al. 2006; Ravulaparthy and Goulias 2014). In other words, cities' circulation networks exhibit spatial signatures that can be quantified and operationalized to differentiate types of places. One branch of this research literature augments urban morphology studies with graph-theoretic topological analyses of street networks (Porta et al. 2006b; Crucitti et al. 2006; Jiang and Claramunt 2004; Boeing 2018a; Barthelemy et al. 2013). Another branch considers the implications of network configuration and geometry on transportation and circulation (Zhong et al. 2014, 2017; Parthasarathi 2011, 2013; Jiang 2009).

Circuity, the ratio of network distances to straight-line distances, is an important measure of network structure and transportation efficiency (Huang and Levinson 2015). Interest in circuity is not new. In 1929, the modernist polemic Le Corbusier (2007) wrote: "The circulation of traffic demands the straight line; it is the proper thing for the heart of a city. The curve is ruinous, difficult and dangerous; it is a paralyzing thing… The winding road is the Pack-Donkey's Way, the straight road is man's way." In particular, Corbusier argued that planners must eradicate walkable, self-organized streets and paths from traditional cities to enable the development of deliberate, rational, straight-line roads for cars. Similar plans were enacted throughout the twentieth century, including Robert Moses's "meat ax" carving up New York's neighborhoods to make room for new expressways (Caro 1974; Hall 1996; Boeing 2017b). Over time, evolving street network pattern and design standards that impact circuity have been bureaucratized by the ITE, the Federal Housing Administration, and the Urban Land Institute (Southworth and Ben-Joseph 1995, 1997; Jackson 1985).

Circuity results from a network's planning, configuration, and underlying terrain (Boeing 2019). In turn, it impacts how humans use urban space for settlement and travel. Levinson and El-Geneidy (2007, 2009) demonstrate that households tend to select residential locations that offer less circuitous work commutes. Giacomin and Levinson (2015) show that average US metropolitan circuities rose between 1990 and 2000, as recent peripheral suburban development has featured more circuitous street network designs. O'Sullivan and Morrall (1996) find that walking trip circuity (counterintuitively) increases in Calgary during winter months, as pedestrians avoid shortcuts that have been rendered impassible by ice. Most circuity research has treated a city's street network as a single entity. However, multiple circulation networks—that may be disambiguated by mode (e.g., walking, driving, and biking)—overlap to constitute the city's complete multimodal circulation network.

Past research has measured the circuity of car (Giacomin and Levinson 2015; Ballou et al. 2002), bike (Boisjoly and El-Geneidy 2016; Dill 2004), and transit (Huang and Levinson 2015) networks. However, the relative circuity of walkable networks versus drivable networks has been underexplored. On one hand, car-only routes like freeways can provide straight paths to link opposite sides of the city by cutting across grids and winding surface streets. On the other hand, such routes may have circuitous elements such as cloverleaf interchanges and are engineered to optimize trip time—a function of distance and speed—rather than distance itself (cf. Levine et al. 2012). Drivable networks may also include one-way streets, but pedestrians can traverse them bidirectionally. Moreover, walking networks provide paths across parks, mid-block cut-throughs, passageways between buildings, and other shortcuts that driving networks lack. The relative circuity of drivable versus walkable networks depends on the magnitude of the effect that each's unique features have on enabling trips to approximate straight-line travel. This effect likely varies by city, as a function of topography, planning history, and circulation network design (Ballou et al. 2002; Qureshi et al. 2002).

Although past research has examined overall street network circuity in various cities (Huang and Levinson 2015; Levinson and El-Geneidy 2007, 2009; Giacomin and Levinson 2015; O'Sullivan and Morrall 1996; Ballou et al. 2002; Boisjoly and El-Geneidy 2016; Dill 2004), researchers have not studied the relative circuity of walking versus driving networks across multiple cities. This study uses open-source software and open data to explore relative network circuity. We download OpenStreetMap data for walkable and drivable circulation networks in 40 US cities using the OSMnx software, which we then use to simulate four million routes and analyze each's circuity. We find statistically significant differences between driving circuity and walking circuity in all but two cities. Driving circuity exceeds that of walking in all but four of the cities with statistically significant differences. Moreover, circuity varies in different kinds of places. These findings suggest the importance of using network-based distances to study urban access as well as the importance of differentiating between circulation network types. Although walkable and drivable networks overlap in every city, their relative structure and performance differ in most cities.

The following section briefly introduces the mathematics of graph theory and the data and tools used to study street networks. Then we present the methods used to analyze relative circuity and the results of this analysis. Finally, we discuss these results in the context of urban morphology and planning.

2 Analytical Background

Street network analysis uses the mathematics of graph theory (Newman et al. 2010; Derrible and Kennedy 2009; Barthelemy et al. 2011). A graph $G = (V, E)$ is composed of a set V of vertices connected to one another by a set E of edges (Trudeau 1994). An undirected graph's edges point mutually in both directions, but a directed graph's edges are one-way, from an origin vertex to a destination vertex. Multigraphs allow multiple edges between a pair of graph vertices. A planar graph can be represented in two dimensions with edges only intersecting at vertices—otherwise, it is nonplanar. Street networks are nonplanar graphs due to bridges and underpasses. Primal representations model street segments as edges and intersections and dead-ends as vertices (Porta et al. 2006; Ratti et al. 2004). This study models drivable street networks as primal, nonplanar, directed, multigraphs weighted by length, and walkable networks as the same but undirected. Street networks are spatial graphs. Their vertices and edges are embedded in space and, in turn, have geometric characteristics such as circuity that rely on lengths and areas—alongside the standard topological traits of all graphs (O'Sullivan 2014; Boeing 2018b).

Street network data traditionally come from various sources, including disparate municipal and state repositories, expensive commercial data sets, and (in the US) the census bureau's TIGER/Line roads shapefiles. Many studies rely on the latter because of its accessibility and comprehensiveness. However, TIGER/Line suffers from substantial spatial inaccuracies, broad classifiers that lump multiple path types together, and the misrepresentation of traffic-diverting bollards as through-streets (Frizzelle et al. 2009). The latter in particular biases routing results. OpenStreetMap offers an alternative data source. It is an online collaborative mapping project that covers the entire world (Corcoran et al. 2013; Jokar Arsanjani et al. 2015). As of 2017, OpenStreetMap contained over 4.4 billion geospatial objects in its database, along with over 1.5 billion tags which describe these features. The objects comprise streets, trails, building footprints, land parcels, rivers, power lines, points of interest, and many other features.

These data are added to the OpenStreetMap database in typically one of two ways. The first is through large-scale imports of publicly available data sources, such as census TIGER/Line data or municipalities' shapefiles. The second is through the many individual additions and edits performed on an ongoing basis by OpenStreetMap's users and contributors. OpenStreetMap's data are largely high quality (Barron et al. 2014; Girres and Touya 2010; Haklay 2010; Maron 2015; Zielstra and Hochmair 2011; Neis et al. 2007; Basiri et al. 2016). In 2007, it imported the TIGER/Line roads as a foundation, and since then, the user community has added additional features, richer attribute data, and spatial corrections (Zielstra et al. 2013). Of particular relevance to the present study, OpenStreetMap data go far beyond those available in TIGER/Line as they include pedestrian paths, park trails, passages between buildings and through blocks, and finer grained codes classifying various path and street types.

Researchers typically acquire street network data from OpenStreetMap in one of three ways. First, Overpass provides an API that allows users to query for geospatial features. However, its query language is somewhat difficult to use directly. Second (and accordingly), a handful of commercial services have sprung up as middlemen, downloading data extracts for certain areas or bounding boxes and then providing them to users. However, these services are often expensive, slow, and not customizable. While they may work well for studying the street network within a single bounding box, they are inconvenient for acquiring data in multiple precisely bounded study sites. Further, they provide data as geometric shapefiles, which do not lend themselves immediately to nonplanar, graph-theoretic network analysis.

A third method for acquiring OpenStreetMap network data is OSMnx. OSMnx is a free, open-source Python package for downloading and analyzing street networks from OpenStreetMap (Boeing 2017a). It can query by bounding box, address plus network distance, polygon (e.g., from a shapefile), or by place name (which resolves to a polygon representing the place's borders) such as cities, boroughs, or counties. OSMnx can download drivable, walkable, or bikeable networks, as well as other infrastructures such as power lines or subway systems. Walkable and drivable paths are identified by OpenStreetMap metadata. Once the network has been downloaded, OSMnx automatically assembles it into a nonplanar directed multigraph and corrects its topology to retain vertices only at true intersections and dead-ends. This simplification process faithfully retains the true geometry and length of each street segment. Finally, OSMnx can analyze these street networks in various ways, including shortest path calculations, topological measures such as centrality and clustering, and geometric measures such as intersection density and circuity (Boeing 2018a).

3 Methods

This study uses OSMnx to calculate shortest path distances along walkable and drivable street networks in various cities. It builds on the research designs of (Levinson and El-Geneidy 2007, 2009; Giacomin and Levinson 2015) by simulating 100,000 routes in each city and calculating circuities as a function of network distance and straight-line distance. Unlike some previous morphological studies, we use nonplanar networks as a superior representation of topology (Boeing 2018b; Karduni 2016) and we utilize the fine-grained classifier codes in OpenStreetMap data to create separate (directed) driving and (undirected) walking networks that are more detailed than those of most prior studies. To study the difference in circuity between walkable and drivable networks, we examine 40 US cities. We select cities across the breadth of the nation, including most of the largest cities as well as several medium cities, small towns, and suburbs for contrast. For each city, we draw a convex hull around the municipal borders, then download the street network within this hull.

This technique offers two advantages. First, it allows us to focus on municipalities—the scale of urban planning jurisdiction and decision-making—and their immediate vicinities without including the suburbs, exurbs, and urban fringe at the periphery of the broader metropolitan area. Second, it adjusts for substantial quirks in the shapes of municipal borders. Some cities' borders snake along a narrow linear feature to connect two disparate hemispheres. Others exhibit concavity and bend around large elbow curves. These quirks would cause inflated circuity scores as trips are forced to route around city borders instead of taking shorter and more direct routes that briefly cross through a neighboring town. Convex hulls solve this problem while still constraining the analysis to a city and its immediate environs. They also help us mostly avoid large bodies of water around which some metropolitan areas wrap—-such as the San Francisco Bay and the Puget Sound—and which significantly impact metropolitan-scale circuity.

To acquire the street network data, we use OSMnx to download each city's walkable and drivable street networks—constrained to the convex hull—from OpenStreetMap. OSMnx uses OpenStreetMap's fine-grained tags to identify walkable and drivable streets and paths. In case of a disconnected network, we retain only the largest connected component. Then, for each city and network type, we simulate 50,000 random routes. This number of simulations was arrived at after a sensitivity analysis revealed that the means typically converge at stable values around this number. These randomized origin–destination pairs need not reflect the spatial distribution and weighting of real-world trips, as our goal is instead to characterize the structure of each network as a whole rather than just its most well-worn paths.

For each simulation, we randomly select two vertices, calculate the shortest path between them using Dijkstra's algorithm (1959), and then calculate the great-circle distance between these two vertices. The great-circle distance ζ_{gc} is calculated as

$$\zeta_{gc} = r \arccos(\sin(\Phi_1)\sin(\Phi_2) + \cos(\Phi_1)\cos(\Phi_2)\cos(|\lambda_1 - \lambda_2|))$$

where r represents the Earth's radius of approximately 6371 km and Φ_1, λ_1, Φ_2, and λ_2 represent, in radians, the geographical latitude and longitude of two points. The great-circle distance ζ_{gc} thus represents the shortest distance along the curved surface of the earth, and is more accurate than the Euclidean distance. We calculate each walking or driving route's circuity as

$$\psi = \frac{\zeta_{net}}{\zeta_{gc}}$$

where, for each route, ψ represents circuity, ζ_{net} represents the shortest path network distance between the origin and destination vertices, and ζ_{gc} represents the great-circle distance between these vertices. Thus, a route's circuity is the ratio of the shortest path network distance to the great-circle distance between the origin and destination. Figure 1 illustrates the difference between a shortest path route and a straight line between two vertices in Manhattan's driving network, accounting for one-way streets.

We hypothesize that the average circuity of a city's walkable circulation network differs from that of its drivable circulation network. To test this hypothesis, for each city, we conduct a t-test to ascertain the statistical significance of the difference between the simulated walking routes' average circuity and that of the simulated driving routes, to see if we may reject the null hypothesis H_0:

$$H_0 : \mu_w = \mu_d$$
$$H_1 : \mu_w \neq \mu_d$$

where, for each city, μ_w is the mean ψ of the routes along its walking network and μ_d is the mean ψ of the routes along its driving network. We conduct two-sided t-tests because it is not known *a priori* if μ_w is expected to be greater than μ_d or vice versa.

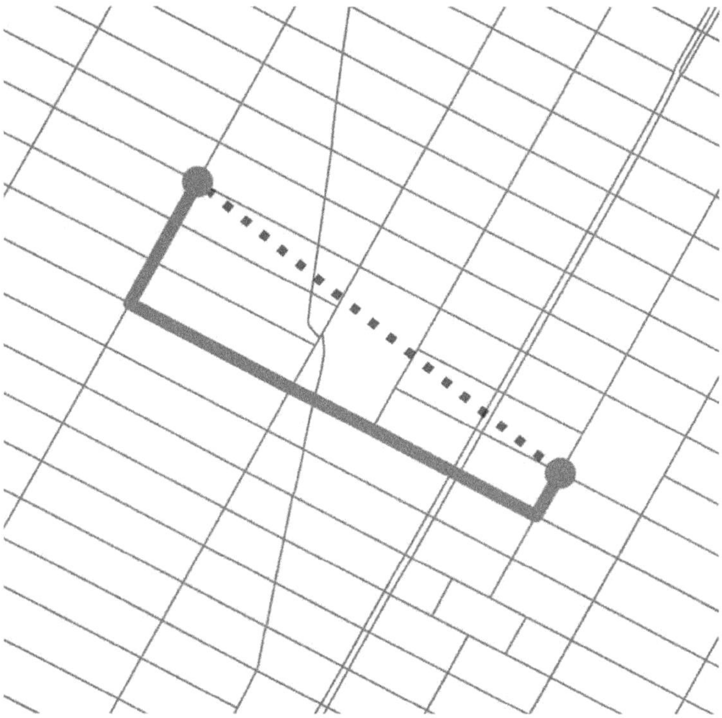

Fig. 1 A section of Manhattan's drivable network, showing the shortest path (thick solid line) between two vertices, accounting for one-way streets, and the great-circle path (thick dashed line)

For instance, if the effects of pedestrian pathways and cut-throughs exceed the effect of freeways on minimizing trip circuity, we would expect μ_d to be greater than μ_w (that is, walking allows for more direct routes). Conversely, if the effect of freeways exceeds drawbacks of one-way streets and the effects of pedestrian pathways and cut-throughs on minimizing trip circuity, we would expect μ_w to be greater than μ_d (that is, driving allows for more direct routes). Next, for each city, we calculate an indicator φ to represent the ratio of average driving circuity to average walking circuity:

$$\varphi = \frac{(\mu_d - 1)}{(\mu_w - 1)}$$

We subtract 1 from each term in the φ ratio because the minimum possible circuity is 1 (cf. 34). The following section presents the results of these route simulations and the resulting statistical analyses.

4 Results

Table 1 presents the average circuity of walking routes (μ_w) and driving routes (μ_d) in each city, with significance levels denoted by asterisks in the table. It also presents the mean distance of routes along the driving network (δ_d), the mean distance of routes along the walking network (δ_w), and the φ ratio expressed as a percentage. We find that the average circuities of driving routes and walking routes differ by a statistically significant margin in 38 out of the 40 cities studied. Moreover, we find that the average driving circuity exceeds that of walking in 34 out of the 38 cities that have a statistically significant difference.

The mean distance of routes along the driving network, δ_d, and the mean distance of routes along the walking network, δ_w, demonstrate how much distance can be saved, on an average trip, by taking the more direct mode of travel. For instance, in Manhattan, the average walking route is 0.7 km shorter than the average driving route. However, the values of δ_w and δ_d are correlated with street network size, so cities with larger spatial extents demonstrate higher trip distances on average (cf. Levinson et al. (2012)).

To adjust for this scaling effect, μ_d and μ_w represent circuity as the mean of the simulated ψ values (each route's ratio of shortest path network distance to great-circle distance). Cities with orthogonal street grids or robust radial freeway systems tend to have the least circuitous average driving routes: in Philadelphia, Chicago, Detroit, Dallas, and Los Angeles the average driving route is only 18 to 20% more circuitous than the straight-line distance from the origin to destination. More recently developed suburbs with curvilinear residential street networks have the most circuitous average driving routes: in Sugar Land and Redmond, the average driving route is over 52% more circuitous than the straight-line distance. Similar effects are seen in the walking networks. The average walking route in Manhattan, Chicago, and Detroit is less than 18% more circuitous than the straight-line distance from origin to destination. However, the average walking route in Sugar Land and Redmond are 40% more circuitous than the straight-line distance.

The primary focus of this study is on the relationship between driving network circuity and walking network circuity (φ in Table 1). Manhattan and San Francisco exhibit the greatest values of φ. In each of these cities, the average driving route is over 46% more circuitous than the average walking route. In contrast, in San Diego and Kansas City, the average driving route is 13% and 34% less circuitous, respectively, than the average walking route.

5 Discussion

These statistical results suggest that in most cities, the average circuity of the walking routes differs from that of the driving routes. In 38 of the 40 cities studied, this difference is statistically significant and the average circuity of driving routes exceeds

Table 1 Average Circuity of Walking and Driving Routes

	μ_d	μ_w	δ_d (km)	δ_w (km)	φ (%)	
Atlanta, GA	1.243	1.226	12.49	11.68	7.3	***
Baltimore, MD	1.232	1.221	8.68	8.24	4.8	***
Boston, MA	1.255	1.191	9.14	7.89	33.9	***
Charlotte, NC	1.267	1.248	19.71	19.14	7.7	***
Chicago, IL	1.194	1.178	18.79	18.40	8.8	***
Cincinnati, OH	1.341	1.332	12.35	11.51	2.6	***
Cleveland, OH	1.213	1.208	12.26	11.53	2.3	***
Dallas, TX	1.177	1.180	24.23	24.33	−1.5	***
Denver, CO	1.242	1.209	15.19	13.23	15.3	***
Detroit, MI	1.190	1.178	13.03	12.36	6.3	***
Fargo, ND	1.336	1.291	7.64	7.00	15.6	***
Gary, IN	1.324	1.285	7.90	6.93	13.6	***
Kansas City, MO	1.223	1.339	21.51	23.08	−34.2	***
Las Vegas, NV	1.272	1.268	13.72	14.02	1.3	**
Los Angeles, CA	1.198	1.196	28.47	27.93	0.6	
Louisville, KY	1.274	1.254	18.85	17.47	7.7	***
Manhattan, NY	1.209	1.142	7.57	6.87	47.6	***
Miami, FL	1.259	1.246	7.83	7.93	5.5	***
Minneapolis, MN	1.241	1.223	8.24	7.40	8.4	***
Orlando, FL	1.306	1.300	13.60	13.79	2.0	***
Philadelphia, PA	1.200	1.184	12.85	12.79	8.9	***
Phoenix, AZ	1.256	1.224	25.53	24.41	14.0	***
Pittsburgh, PA	1.345	1.328	9.58	9.18	5.3	***
Portland, ME	1.389	1.347	6.09	6.15	12.2	***
Portland, OR	1.289	1.264	12.05	11.97	9.5	***
Redmond, WA	1.522	1.396	5.79	5.18	31.8	***
Riverside, CA	1.312	1.289	10.78	11.30	7.9	***
Salem, MA	1.505	1.487	4.83	4.66	3.8	***
San Antonio, TX	1.218	1.199	22.43	21.77	9.9	***
San Diego, CA	1.307	1.354	25.90	25.91	−13.3	***
San Francisco, CA	1.308	1.210	11.46	6.96	46.2	***

(continued)

Table 1 (continued)

	μ_d	μ_w	δ_d (km)	δ_w (km)	φ (%)	
Scranton, PA	1.376	1.349	5.09	5.08	7.7	***
Seattle, WA	1.289	1.251	12.29	11.08	15.1	***
St Augustine, FL	1.373	1.331	4.39	3.88	12.9	***
St Louis, MO	1.204	1.193	8.55	7.73	5.4	***
Stamford, CT	1.340	1.340	6.93	6.63	0.1	
Sugar Land, TX	1.523	1.405	7.92	7.15	29.2	***
Tampa, FL	1.281	1.267	14.85	13.99	5.3	***
Vicksburg, MS	1.363	1.379	6.74	6.16	−4.1	***
Walnut Creek, CA	1.470	1.392	5.60	5.29	19.9	***

Note 1 km = 0.62 mi. μ_w = mean circuity of the simulated routes along the walking network, μ_d = mean circuity of the simulated routes along the driving network, δ_d = mean distance (km) of routes along the driving network, δ_w = mean distance (km) of routes along the walking network, and φ represents how much μ_d exceeds μ_w expressed as a percentage. Finally, **indicates a statistically significant difference between μ_w and μ_d at the 0.01 level and ***indicates significance at the 0.001 level

that of walking routes in all but four of the cities with statistically significant results. In other words, on average, driving routes tend to be more circuitous than walking routes in most cities.

To interpret these network circuity findings, we use Manhattan and San Diego as illustrative cases. Manhattan's average driving route is 48% more circuitous than its average walking route. Figure 2 shows Manhattan's drivable street network on the left and its walkable street network on the right. We can immediately see that the walking network is much denser than the driving network. While the walking network contains 11,857 vertices and 1,331 km of streets/paths (n.b. physical streets are equivalent to edges in an undirected graph), the driving network contains only 4,889 vertices and 1,064 km of streets. Although it excludes expressways around the periphery of the island, the walking network provides numerous mid-block passages, pedestrian walkways, and frequent juncture points that in aggregate allow for more direct routes. In particular, Central Park poses an obstacle to straight-line driving, but the park's dense mesh of walking paths provides many cross-cutting and diagonal routes for pedestrians whose origins and destinations lie on either side of it. Manhattan also has many one-way streets that pedestrians may traverse bidirectionally, improving walking efficiency. In this case, the findings suggest that the effects of pedestrian pathways, cut-throughs, and bidirectionality exceed the effect of car-only motorways on trip directness.

As a contrasting example—unlike Manhattan and most of the cities studied—San Diego's φ value is both negative and statistically significant. Its average driving route is 13% less circuitous than its average walking route. Figure 3 shows San Diego's drivable street network on the left and its walkable street network on the right. As in Manhattan, we can see that the walking network is denser than the driving network.

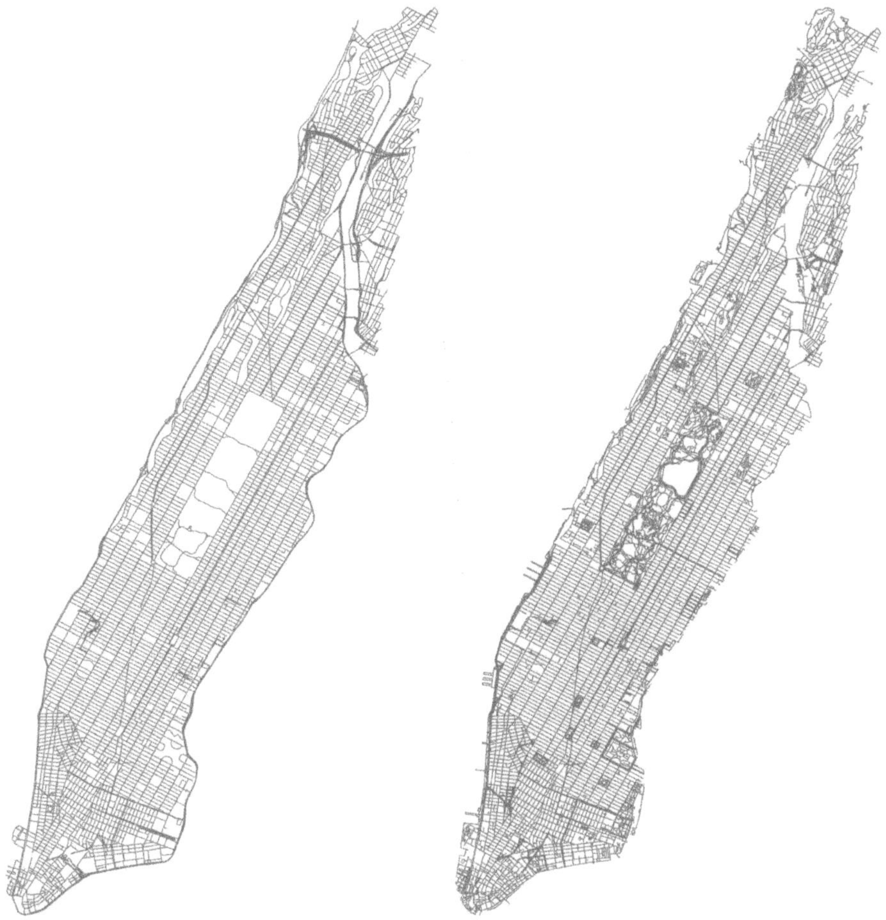

Fig. 2 Manhattan's driving network (left) and walking network (right). Note that its convex hull includes a sliver of the Bronx

While the walking network has 94,118 vertices and 14,293 km of streets/paths, the driving network has only 51,544 vertices and 11,283 km of streets.

There seem to be three primary reasons why San Diego has a negative φ value. First, its highway system runs at cross-cutting diagonals, providing direct routes across the city that local streets cannot match. Second, its hills and canyons create substantial open space, natural preserves, and in turn network thresholds. The walking network provides dendritic, circuitous access into these spaces without offering efficient links across them. Third, Coronado Island (technically a peninsula connected to the mainland by a tombolo) connects to downtown San Diego via the San Diego–Coronado Bridge for reasonably direct driving routes. However, this bridge is automobile-only, so walking routes between Coronado and the majority of the city must route along the peninsula and around the entire San Diego Bay (similarly,

Fig. 3 San Diego's driving network (left) and walking network (right)

the Kansas City network has 7 driving bridges over the Missouri River but only 1 walkable bridge). In other words, due to topography and design, San Diego relies particularly heavily on its freeway system for shortest path connectivity. Accordingly, in this case, the findings suggest that the effect of car-only motorways exceeds the effects of pedestrian paths and passageways on route directness. Nevertheless, we find that this effect is the exception rather than the rule across our study sites.

Empirical studies in the urban economics, planning, and transportation literature today often rely on Euclidean or great-circle distances to model accessibility. These findings add to the growing body of evidence that such straight-line measures are both inaccurate and inconsistent. Average network distances are at least 14% and sometimes over 50% longer than the straight-line distance from origin to destination. The magnitude of this phenomenon varies substantially between, for instance, Chicago and Sugar Land. This variation results from different topographies, planning eras, transportation technologies, and design paradigms. Thus, researchers and practitioners should use network-based distances and travel times to prevent biased distance measures in different kinds of places.

Moreover, we find that route circuity along walkable versus drivable networks differs significantly in most cities. A city's circulation network cannot be accurately modeled as a single monolithic entity. Rather, it comprises an interwoven set of overlapping circulation networks available to different modes of travel. Some one-way streets in the driving network may be accessible in both directions to pedestrians in the walking network. Similarly, pedestrian paths and mid-block passages may be unavailable to drivers, while expressways and certain bridges may be unavailable to pedestrians. Moreover, driving networks are more likely to be engineered to minimize travel time at high speeds rather than to minimize distance: distance itself is more important for modes with lower travel speeds. Measuring circuity depends on carefully defining what the circulation network includes and which modes of travel are of interest to the study. Thus, urban street network structure and performance cannot be sufficiently assessed without specifying network types and travel modes.

6 Conclusion

This study examined the relative circuity of walkable and drivable urban circulation networks by simulating four million routes using OpenStreetMap data and the OSMnx software. It found that, in most cities, driving networks tend to produce more circuitous routes than walking networks do. Specifically, average driving circuity exceeds average walking circuity in all but four of the cities that have statistically significant differences. Old, dense cities like Manhattan and San Francisco saw the greatest effects, with average driving routes over 46% more circuitous than average walking routes. Network circuity also varies substantially between different types of places. These findings underscore the value of using network-based distances and times rather than straight-line when studying urban travel and access. They also suggest the importance of differentiating between walkable and drivable circulation networks: although these networks overlap, their relative structure and performance vary in most cities.

This study used simulated route distances as an indicator of circuity and network efficiency. Travel time is another important measure of efficiency. Future research can weigh walking and driving routes by travel time to compare how simulated trip times vary by mode in different places as a function of network structure. Simulated trips could also be weighted by the likelihood of each being a real-world trip, based on travel survey data. This study randomized routes to examine overall network structure, but using actual trips would shed light on real-world travel behavior circuity and reduce bias from random sampling when neighborhood trips may be more common. Travel surveys could also provide information about average trip distances by mode. Beyond simple measures of network access, incorporating impedances based on grade, streetscape, traffic, and other data would provide a superior representation of routes. Moreover, it would be useful to explore how common planar models affect the results of route analysis (Boeing 2018b). Finally, this study focused on US cities, but future work could use OSMnx and OpenStreetMap's worldwide data to compare networks in other countries to investigate the structure and performance of older European cities or rapidly growing African and Asian cities.

References

Alexander, C. A City Is Not a Tree. *Architectural Forum*, Vol. 122, No. 1, 1965, pp. 58–62.

Ballou, R. H., H. Rahardja, and N. Sakai. Selected Country Circuity Factors for Road Travel Distance Estimation. *Transportation Research Part A: Policy and Practice*, Vol. 36, No. 9, 2002, pp. 843–848. https://doi.org/10.1016/S0965-8564(01)00044-1.

Barron, C., P. Neis, and A. Zipf. A Comprehensive Framework for Intrinsic OpenStreetMap Quality Analysis. *Transactions in GIS*, Vol. 18, No. 6, 2014, pp. 877–895. https://doi.org/10.1111/tgis.12073.

Barthelemy, M., and A. Flammini. Modeling Urban Street Patterns. *Physical Review Letters*, Vol. 100, No. 13, 2008. https://doi.org/10.1103/PhysRevLett.100.138702.

Barthelemy, M., P. Bordin, H. Berestycki, and M. Gribaudi. Self-Organization versus Top-down Planning in the Evolution of a City. *Scientific Reports*, Vol. 3, 2013. https://doi.org/10.1038/srep02153.

Barthelemy, M. *The Structure and Dynamics of Cities: Urban Data Analysis and Theoretical Modeling*. Cambridge University Press, Cambridge, England, 2017.

Barthelemy, M. Spatial Networks. *Physics Reports*, Vol. 499, No. 1–3, 2011, pp. 1–101. https://doi.org/10.1016/j.physrep.2010.11.002.

Basiri, A., M. Jackson, P. Amirian, A. Pourabdollah, M. Sester, A. Winstanley, T. Moore, and L. Zhang. Quality Assessment of OpenStreetMap Data Using Trajectory Mining. *Geospatial Information Science*, Vol. 19, No. 1, 2016, pp. 56–68. https://doi.org/10.1080/10095020.2016.1151213.

Batty, M. Network Geography: Relations, Interactions, Scaling and Spatial Processes in GIS. In *Re-Presenting GIS* (D. J. Unwin and P. Fisher, eds.), John Wiley & Sons, Chichester, England, 2005, pp. 149–170.

Boeing, G. A Multi-Scale Analysis of 27,000 Urban Street Networks: Every US City, Town, Urbanized Area, and Zillow Neighborhood. *Environment and Planning B: Urban Analytics and City Science*, online first, 2018a. https://doi.org/10.1177/2399808318784595.

Boeing, G. OSMnx: New Methods for Acquiring, Constructing, Analyzing, and Visualizing Complex Street Networks. *Computers, Environment and Urban Systems*, Vol. 65, No. 126–139, 2017a. https://doi.org/10.1016/j.compenvurbsys.2017.05.004.

Boeing, G. Planarity and Street Network Representation in Urban Form Analysis. *Environment and Planning B: Urban Analytics and City Science*, online first, 2018b. https://doi.org/10.1177/2399808318802941.

Boeing, G. *We Live in a Motorized Civilization: Robert Moses Replies to Robert Caro*. University of California, Berkeley, 2017b. https://doi.org/10.2139/ssrn.2934079.

Boeing, G. Urban Spatial Order: Street Network Orientation, Configuration, and Entropy. arXiv: 1808.00600, 2019. https://arxiv.org/pdf/1808.00600.

Boisjoly, G., and A. El-Geneidy. Are We Connected? Assessing Bicycle Network Performance through Directness and Connectivity Measures, a Montreal, Canada Case Study. Presented at the Transportation Research Board 95th Annual Meeting, Washington, DC, 2016.

Cardillo, A., S. Scellato, V. Latora, and S. Porta. Structural Properties of Planar Graphs of Urban Street Patterns. *Physical Review E*, Vol. 73, No. 6, 2006. https://doi.org/10.1103/PhysRevE.73.066107.

Caro, R. A. *The Power Broker: Robert Moses and the Fall of New York*. Alfred A. Knopf, New York, NY, 1974.

Corbusier, L. "The Pack-Donkey's Way and Man's Way" and "A Contemporary City." In *The Urban Design Reader* (M. Larice and E. Macdonald, eds.), Routledge, New York, NY, 2007, pp. 66–75.

Corcoran, P., P. Mooney, and M. Bertolotto. Analysing the Growth of OpenStreetMap Networks. *Spatial Statistics*, Vol. 3, 2013, pp. 21–32. https://doi.org/10.1016/j.spasta.2013.01.002.

Crucitti, P., V. Latora, and S. Porta. Centrality Measures in Spatial Networks of Urban Streets. *Physical Review E*, Vol. 73, No. 3, 2006, p. 036125.

Derrible, S., and C. Kennedy. Network Analysis of World Subway Systems Using Updated Graph Theory. *Transportation Research Record: Journal of the Transportation Research Board*, Vol. 2112, 2009, pp. 17–25. https://doi.org/10.3141/2112-03.

Dijkstra, E. W. A Note on Two Problems in Connexion with Graphs. *Numerische Mathematik*, Vol. 1, No. 1, 1959, pp. 269–271. https://doi.org/10.1007/BF01386390.

Dill, J. Measuring Network Connectivity for Bicycling and Walking. Presented at the Transportation Research Board 83rd Annual Meeting, Washington, DC, 2004.

Fischer, M. M., D. A. Henscher, K. J. Button, K. E. Haynes, and P. R. Stopher. GIS and Network Analysis. In *Handbook of Transport Geography and Spatial Systems*, Pergamon Press, Oxford, England, 2004, pp. 391–408.

Frizzelle, B., K. Evenson, D. Rodriguez, and B. Laraia. The Importance of Accurate Road Data for Spatial Applications in Public Health: Customizing a Road Network. *International Journal of Health Geographics*, Vol. 8, No. 24, 2009. https://doi.org/10.1186/1476-072X-8-24.

Gastner, M. T., and M. E. J. Newman. The Spatial Structure of Networks. *The European Physical Journal B: Condensed Matter and Complex Systems*, Vol. 49, No. 2, 2006, pp. 247–252. https://doi.org/10.1140/epjb/e2006-00046-8.

Giacomin, D. J., and D. M. Levinson. Road Network Circuity in Metropolitan Areas. *Environment and Planning B: Planning and Design*, Vol. 42, No. 6, 2015, pp. 1040–1053. https://doi.org/10.1068/b130131p.

Girres, J.-F., and G. Touya. Quality Assessment of the French OpenStreetMap Dataset. *Transactions in GIS*, Vol. 14, No. 4, 2010, pp. 435–459. https://doi.org/10.1111/j.1467-9671.2010.01203.x.

Haklay, M. How Good Is Volunteered Geographical Information? A Comparative Study of OpenStreetMap and Ordnance Survey Datasets. *Environment and Planning B: Planning and Design*, Vol. 37, No. 4, 2010, pp. 682–703. https://doi.org/10.1068/b35097.

Hall, P. *Cities of Tomorrow: An Intellectual History of Urban Planning and Design in the Twentieth Century*. Blackwell Publishers, Malden, MA, 1996.

Huang, J., and D. M. Levinson. Circuity in Urban Transit Networks. *Journal of Transport Geography*, Vol. 48, 2015, pp. 145–153. https://doi.org/10.1016/j.jtrangeo.2015.09.004.

Jackson, K. T. *Crabgrass Frontier: The Suburbanization of the United States*. Oxford University Press, New York, NY, 1985.

Jiang, B., and C. Claramunt. Topological Analysis of Urban Street Networks. *Environment and Planning B: Planning and Design*, Vol. 31, No. 1, 2004, pp. 151–162. https://doi.org/10.1068/b306.

Jiang, B. Ranking Spaces for Predicting Human Movement in an Urban Environment. *International Journal of Geographical Information Science*, Vol. 23, No. 7, 2009, pp. 823–837. https://doi.org/10.1080/13658810802022822.

Jokar Arsanjani, J., A. Zipf, P. Mooney, and M. Helbich, Eds. *OpenStreetMap in GIScience*. Springer International, Cham, Switzerland, 2015.

Karduni, A., A. Kermanshah, and S. Derrible. A Protocol to Convert Spatial Polyline Data to Network Formats and Applications to World Urban Road Networks. *Scientific Data*, Vol. 3, 2016, p. 160046. https://doi.org/10.1038/sdata.2016.46.

Levine, J., J. Grengs, Q. Shen, and Q. Shen. Does Accessibility Require Density or Speed? *Journal of the American Planning Association*, Vol. 78, No. 2, 2012, pp. 157–172. https://doi.org/10.1080/01944363.2012.677119.

Levinson, D., and A. El-Geneidy. Network Circuity and the Location of Home and Work. Presented at the University Transport Study Group Conference, Harrowgate, England, Jan 3–5, 2007.

Levinson, D., and A. El-Geneidy. The Minimum Circuity Frontier and the Journey to Work. *Regional Science and Urban Economics*, Vol. 39, No. 6, 2009, pp. 732–738. https://doi.org/10.1016/j.regsciurbeco.2009.07.003.

Levinson, D. Network Structure and City Size. *PLoS ONE*, Vol. 7, No. 1, 2012, p. e29721. https://doi.org/10.1371/journal.pone.0029721.

Maron, M. How Complete Is OpenStreetMap? Mapbox, Nov 19, 2015. https://www.mapbox.com/blog/how-complete-is-openstreetmap/.

Masucci, A. P., D. Smith, A. Crooks, and M. Batty. Random Planar Graphs and the London Street Network. *The European Physical Journal B: Condensed Matter and Complex Systems*, Vol. 71, No. 2, 2009, pp. 259–271. https://doi.org/10.1140/epjb/e2009-00290-4.

Masucci, A. P., K. Stanilov, and M. Batty. Limited Urban Growth: London's Street Network Dynamics since the 18th Century. *PLoS ONE*, Vol. 8, No. 8, 2013, p. e69469. https://doi.org/10.1371/journal.pone.0069469.

Neis, P., D. Zielstra, and A. Zipf. The Street Network Evolution of Crowdsourced Maps: OpenStreetMap in Germany 2007–2011. *Future Internet*, Vol. 4, No. 1, 2011, pp. 1–21. https://doi.org/10.3390/fi4010001.

Newman, M. E. J. *Networks: An Introduction*. Oxford University Press, Oxford, England, 2010.

O'Sullivan, D. Spatial Network Analysis. In *Handbook of Regional Science* (M. M. Fischer and P. Nijkamp, eds.), Springer-Verlag, Berlin, Germany, 2014, pp. 1253–1273.

O'Sullivan, S., and J. Morrall. Walking Distances to and from Light-Rail Transit Stations. *Transportation Research Record: Journal of the Transportation Research Board*, No. 1538, 1996, pp. 19–26. https://doi.org/10.3141/1538-03.

Parthasarathi, P., D. Levinson, and H. Hochmair. Network Structure and Travel Time Perception. *PLoS ONE*, Vol. 8, No. 10, 2013, p. e77718. https://doi.org/10.1371/journal.pone.0077718.

Parthasarathi, P. *Network Structure and Travel*. Dissertation. University of Minnesota, Minneapolis, MN, 2011.

Porta, S., P. Crucitti, and V. Latora. The Network Analysis of Urban Streets: A Dual Approach. *Physica A: Statistical Mechanics and its Applications*, Vol. 369, No. 2, 2006b, pp. 853–866. https://doi.org/10.1016/j.physa.2005.12.063.

Porta, S., P. Crucitti, and V. Latora. The Network Analysis of Urban Streets: A Primal Approach. *Environment and Planning B: Planning and Design*, Vol. 33, No. 5, 2006a, pp. 705–725. https://doi.org/10.1068/b32045.

Qureshi, M., H.-L. Hwang, and S.-M. Chin. Comparison of Distance Estimates for Commodity Flow Survey: Great Circle Distances versus Network-Based Distances. *Transportation Research Record: Journal of the Transportation Research Board*, No. 1804, 2002, pp. 212–216. https://doi.org/10.3141/1804-28.

Ratti, C. Space Syntax: Some Inconsistencies. *Environment and Planning B: Planning and Design*, Vol. 31, No. 4, 2004, pp. 487–499. https://doi.org/10.1068/b3019.

Ravulaparthy, S., and K. Goulias. Characterizing the Composition of Economic Activities in Central Locations: Graph-Theoretic Approach to Urban Network Analysis. *Transportation Research Record: Journal of the Transportation Research Board*, Vol. 2430, 2014, pp. 95–104. https://doi.org/10.3141/2430-10.

Southworth, M., and E. Ben-Joseph. Street Standards and the Shaping of Suburbia. *Journal of the American Planning Association*, Vol. 61, No. 1, 1995, pp. 65–81.

Southworth, M., and E. Ben-Joseph. *Streets and the Shaping of Towns and Cities*. McGraw-Hill, New York, NY, 1997.

Strano, E., M. Viana, L. da Fontoura Costa, A. Cardillo, S. Porta, and V. Latora. Urban Street Networks, a Comparative Analysis of Ten European Cities. *Environment and Planning B: Planning and Design*, Vol. 40, No. 6, 2013, pp. 1071–1086. https://doi.org/10.1068/b38216.

Trudeau, R. J. *Introduction to Graph Theory*. Dover Publications, New York, NY, 1994.

Zhong, C., M. Schläpfer, S. M. Arisona, M. Batty, C. Ratti, and G. Schmitt. Revealing Centrality in the Spatial Structure of Cities from Human Activity Patterns. *Urban Studies*, Vol. 54, No. 2, 2017, pp. 437–455. https://doi.org/10.1177/0042098015601599.

Zhong, C., S. M. Arisona, X. Huang, M. Batty, and G. Schmitt. Detecting the Dynamics of Urban Structure through Spatial Network Analysis. *International Journal of Geographical Information Science*, Vol. 28, No. 11, 2014, pp. 2178–2199. https://doi.org/10.1080/13658816.2014.914521.
Zielstra, D., H. H. Hochmair, and P. Neis. Assessing the Effect of Data Imports on the Completeness of OpenStreetMap – A United States Case Study. *Transactions in GIS*, Vol. 17, No. 3, 2013, pp. 315–334. https://doi.org/10.1111/tgis.12037.
Zielstra, D., and H. Hochmair. Comparative Study of Pedestrian Accessibility to Transit Stations Using Free and Proprietary Network Data. *Transportation Research Record: Journal of the Transportation Research Board*, Vol. 2217, 2011, pp. 145–152. https://doi.org/10.3141/2217-18.

Part IV
Complexity

Emergence of Complexity in Urban Morphology

Segun Goh, Keumsook Lee and M. Y. Choi

Abstract We analyze the building distribution data of Seoul City and build a model to understand urban morphology. Performing coarse-graining of the data and employing a lattice description, we find that the urbanized area forms a fractal in the proper range of the length scale. In particular, the building is classified into three categories: 'Residential', 'Work', and 'Commercial', and two-state variables are assigned to each lattice site, which elucidates the functionally organized pattern as well as the form of the urbanized area. Based on empirical observations, we finally propose a layered Ising-type model with frustrated inter-layer coupling, where boundary conditions and external fields are defined appropriately to simulate formation of the patterns. This model, reproducing successfully the heterogeneous spatial patterns, provides a clue for understanding the origin of characteristic urban morphology.

1 Introduction

Cities are highly organized spatial structures that have been developed by human beings. As a result of the organization, each city forms a distinctive pattern depending on its own topographical, historical, economical, and political environments. Nevertheless, substructures composing urban areas should be organized in a proper way for the cities to function properly. Specifically, each city should provide the bases and infrastructures for urban dwellers including shelters, job opportunities, transportation, and so on. To elucidate the pattern of the urbanized area called urban

S. Goh
Institut für Theoretische Physik II: Weiche Materie, Heinrich-Heine-Universität Düsseldorf, 40225 Düsseldorf, Germany

K. Lee
Department of Geography, Sungshin Women's University, Seoul 02844, Korea

M. Y. Choi (✉)
Department of Physics and Astronomy and Center for Theoretical Physics, Seoul National University, Seoul 08826, Korea
e-mail: mychoi@snu.ac.kr

L. D'Acci (ed.), *The Mathematics of Urban Morphology*,
Modeling and Simulation in Science, Engineering and Technology,
https://doi.org/10.1007/978-3-030-12381-9_13

morphology, various studies have been conducted (White and Engelen 1993; Makse et al. 1998; Batty 2001; Andersson et al. 2002; Bitner et al. 2009; Frasco et al. 2014; Masucci et al. 2015).

Meanwhile, the city is one of the most representative complex phenomena among social systems. Recently, the complex system perspectives begin to find general acceptance in studies of social systems (Bouchaud 2008; Castellano et al. 2009; Bettencourt 2013). Many social phenomena are now regarded as emergent and collective properties, exhibiting such characteristics as criticality and scaling (Bettencourt 2013; Goh et al. 2014), and have been widely probed by means of statistical mechanics. For instance, studies of urban growth (Chen 2012; Louf and Barthelemy 2013) and urban structures including street networks (Gudmundsson and Mohajeri 2013; Masucci et al. 2015) and public transportation systems (Goh et al. 2014; Louf et al. 2014) have been conducted extensively from this viewpoint. Still current understanding as to the fundamental rules governing the formation/emergence of cities remains in its infancy, due to the inherent complex nature of the urban system. Specifically to the case of urban morphology, most existing studies, focusing on the properties of individuals, are descriptive rather than theoretical and there still lacks an integrated framework for urban morphology.

Furthermore, insufficiency of real data has also prevented proper understanding of the urban formation and the land use pattern. In principle, most detailed measurement of a social system including a city would be possible if the locations and movements of every individual are specified. Indeed individual dynamics has proved to be traceable mostly by analyzing the mobile phone data (Candia et al. 2008; Kang et al. 2012); nevertheless, complete enumeration of a whole urban system remains yet to be achieved. Moreover, a city involves various scales, ranging from individuals (Candia et al. 2008) to urban substructures such as the central business district (CBD) (Bitner et al. 2009; Louf and Barthelemy 2013). In addition, there might exist strong fluctuations when individual dynamics is taken into account in detail. Therefore, the most relevant measure should be determined carefully.

Related to the land use pattern and urban morphology, the distribution of buildings provides an appropriate measure (Batty et al. 2008; Schläpfer et al. 2015). The building distribution typically characterizes unique features of each city and provides an accurate description of the detailed configuration of the urban area. Undoubtedly, buildings are not distributed randomly in space; rather, they tend to aggregate in a city, accompanying economic activities. To measure the tendency toward aggregation of urban economic activities, distance-based methods have been developed (Duranton and Overman 2005; Jensen 2006; Marcon and Puech 2009). Movements in a city take place under the constraint guided by such inhomogeneous building distribution, and the city operates accordingly. Dealing with the building distribution data of the urbanized area rather than individual dynamics, one can also bypass the technical difficulty and make use of the relatively static nature of the building.

Here we present a review of our previous work on the building distribution (Goh et al. 2016) together with some new developments. We first examine the building distribution of Seoul City, which includes information on the location, area, and functional characteristics of *every* building. Following the viewpoint of statistical mechanics and complex systems, we regard urban morphology as a collective behavior of constituents coupled to each other. To capture distinguished features of emergent phenomena, we adopt a coarse-grained description of Seoul City, placing its urban areas on a lattice. We not only provide a description of the observed phenomena but also infer the underlying principles of urban land use patterns, constructing a model Hamiltonian with relevant components: The frustrated inter-layer coupling is proposed as a fundamental ingredient for the formation of urban morphology and the criticality as a necessary condition for the emergence of complex land use patterns. The role of heterogeneous fields is also addressed.

2 Fractal Nature of Urbanized Regions in Seoul

Before proceeding further, we describe briefly the Seoul building data set (Building Register Data 2011). As addressed already, it contains quantitative information on *every* building located in the administrative area of Seoul City in the year of 2011. The attributes include the number of floors, structure, gross floor area, position/address, zip code, use and type of the roof of the building. There were 678,594 buildings in Seoul at the time of the data gathering, the total area of which amounted to 429.34 km^2 (cf. note the area of Seoul, 605.25 km^2). In particular, uses of the buildings are classified into 29 types, e.g., apartments, offices, neighborhood facilities, and so on. As the use of the building represents the land use directly, it serves as an indicator of the urban morphology which should comprise differentiated patterns as well as shapes of the urbanized areas. Representative attribute types and their ratios in terms of the gross floor area are displayed in Table 1. Among various attributes of buildings, longitude/latitude coordinates and gross floor areas turn out to be relevant information for specifying the configurations of coarse-grained areas, as discussed in this and the next sections. (According to the Korean Enforcement Decree of the Building Act, the gross floor area of a building is defined to be the sum of the floor areas of all the floors of the building.)

We now extract the urbanized area of Seoul City from the building data set. It is well known that the urbanized areas constitute fractals (Fotheringham 1989; Batty and Xie 1999), as to which the diffusion-limited aggregation model was studied extensively (Witten and Sander 1981; Fotheringham 1989; Batty and Xie 1999; Chen 2012). We also verify the fractal nature of the urbanized area and compute the fractal dimension from the building distribution data, which, unlike the data analyzed in previous studies, provide direct measurement. In particular, the dimension needs to be specified accurately since it plays a crucial role in the collective behavior.

However, there exist severe drawbacks in the bare description where the use/location of a certain building is fixed at a particular attribute/value. For example, shopping facilities around a residential area, categorized as commercial in the bare description, should be considered to be a residential-commercial complex. Moreover, although the surface areas occupied by buildings are different from one another, each building is put on an equal footing if regarded as a point element. In particular,

Table 1 Classification of buildings and the ratio in terms of the gross floor area of each attribute. The attribute of each building is provided in the raw data set. We further classify the use of building into three categories: residential (R), work place (W), and commercial (C) buildings. There are 58,079 buildings (6.56% in terms of the area) the uses of which are not categorized. Among them, 55,146 buildings (5.8%) have no attribute in the raw data, while attributes such as farming/fishing houses, transportation facilities, hazardous material storage facilities or vehicle maintenance facilities in the data (2933 buildings, 0.68%) are, in consideration of uncertainty, not categorized

Attribute	Ratio	Category
Apartment	0.3136	R
Multifamily house	0.1496	R
Multiplex house	0.0681	R
Detached house	0.0488	R
Row house	0.0181	R
Employee apartment	0.0012	R
Office	0.0828	W
Education/Research facility	0.0450	W
Manufacturing facility	0.0240	W
Neighborhood facility	0.1286	C
Large store facility	0.0194	C
Religious facility	0.0114	C
Hotel	0.0070	C
Accommodation	0.0045	C
Cultural/Broadcast facility	0.0035	C
Sports facility	0.0031	C
Food sanitation facility	0.0031	C
Market place	0.0009	C
Medical facility	0.0009	C
Public sanitation facility	0.0005	C
Amusement facility	0.0005	C
Inn	0.0004	C
Ritual facility	0.0003	C
Recreational facility	0.0001	C

such a fine-grained description on a length scale smaller than the typical size of a single building is obviously superfluous, for it would generate fictitious empty spaces, which are in reality occupied by buildings, and in consequence introduce heterogeneity, which is an artifact. This may make an undesirable alteration on the behavior of system, unless there exist infinitely many buildings so that the size can be ignored.

Here we follow the process suggested in the recent study (Goh et al. 2014) and perform coarse-graining of the urbanized areas into small boxes, to circumvent the redundancy in the fine-grained description. Constructing a coarse-grained lattice by putting a site on the face of each box, we can also utilize methods developed for lattice systems. The lattice constant and the total number of boxes or sites on the

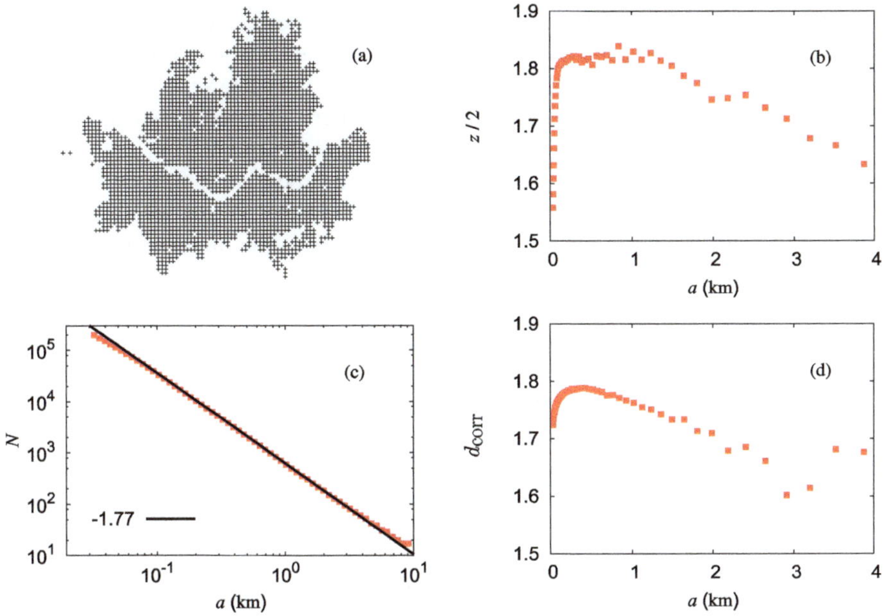

Fig. 1 Lattice structure of the urbanized area in Seoul City and its fractal dimension. **a** A coarse-grained lattice with lattice constant $a = 432\,\text{m}$. **b** Half the coordination number $z/2$ versus lattice constant a, which gives the dimension $d = 1.82 \pm 0.01$ in the range $100\,\text{m} < a < 1500\,\text{m}$. Note the strong finite-size effects for $a < 100\,\text{m}$ and strong boundary effects for $a > 1500\,\text{m}$. **c** Number N of sites versus lattice constant a, disclosing the scaling relation between them. Fitted to the solid line of slope -1.77, it leads to the fractal dimension $d = 1.77 \pm 0.01$. **d** Correlation dimension d_{corr} versus lattice constant a, displaying a pattern similar to (**b**). In the range $100\,\text{m} < a < 1000\,\text{m}$, we obtain $d = 1.78 \pm 0.01$

lattice are denoted by a and N, respectively. A typical example of the coarse-grained lattice is shown in Fig. 1a.

To compute the fractal dimension of the lattice, we first examine the coordination number, which stands for the number of the nearest neighbors, and observe that its value varies from site to site. Labeling the site dependence, we write z_i to denote the coordination number of site i. In fact, the presence of empty spaces and consequent variations in the number of the nearest-neighboring sites apparently reflect the fractal nature of the lattice. Namely, whereas $z_i = z = 2d$ at every site on a d-dimensional regular lattice, in the case of a fractal lattice, the coordination number has a noninteger mean value z and there should thus exist empty sites. In Fig. 1b, which exhibits half the (mean) coordination number depending on the lattice constant, we obtain directly the fractal dimension of the coarse-grained lattice of Seoul City: $d = 1.82 \pm 0.01$.

The presence of empty spaces, however, does not guarantee that the lattice is a fractal. For example, when the sites are removed randomly, the average value of z becomes noninteger; nevertheless, the dimension of the lattice remains the same as that of the original lattice. Further, the average value of z could also be reduced when the sites are removed orderly; such a reduction, which is exemplified by the

Han (river) in the case of Seoul, brings on the boundary effects rather than the fractal nature. Therefore, the fractal character of the lattice should be confirmed in a more concrete way. Examining how the number of sites changes with the lattice constant, we verify the scaling relation between them: $N \sim a^{-d}$, which confirms a fractal with dimension $d = 1.77 \pm 0.01$ (see Fig. 1c). In addition, one can also consider the correlation sum $C(r)$ (Grassberger and Procaccia 1983), and probe its scaling relation. Computing $C(r)$, we indeed observe the scaling relation $C(r) \sim r^{d_{\text{corr}}}$ and plot in Fig. 1d the obtained values of the correlation dimension d_{corr} as the lattice constant is varied. The fractal character of the lattice is manifested again, with dimension $d = d_{\text{corr}} = 1.78 \pm 0.01$.

Note here the appropriate length scale for the coarse-grained description. First, we observe sharp decrease in the fractal dimension for small values of the lattice constant, $a < 100$ m, including the correlation dimension computed directly from the point element description ($d_{\text{corr}} \approx 1.67$). This has its origin in the artificial heterogeneity introduced in the fine-grained description. Second, in the opposite case of large values, $a > 1000$ m, boundary effects are evident. Gathering the results from three different approaches, we finally conclude that the urbanized area of Seoul City is indeed a fractal of the dimension $d \approx 1.8$, which is consistent with the results in previous studies (Batty and Xie 1999; Shen 2002). In the following, we restrict our analysis to the coarse-grained system satisfying these conditions.

3 State Vector Description of Urban Areas

In this section, we specify the state of each site on the fractal lattice, constructed of the coarse-grained areas in Seoul City, and analyze the distribution of configurations of the whole system. For simplicity, one may define the state of a site according to, e.g., whether the area of the site is active or inactive. As remarked, however, cities are highly patterned systems consisting of functionally organized modules, such as CBDs, industrial complexes, and residential areas. Even the actual distribution of modules differs from one city to another, and coexistence of them is a key ingredient which defines and enables cities, especially metropolises such as Seoul, to work as self-sufficient spatial structures for human life. It is therefore desirable to categorize the state of each site in detail, so as to reflect the land use. In previous studies (White and Engelen 1993; Decraene et al. 2013a, b), the urban land use has been categorized as three distinct allocations. Also the urban life has been classified into three representative activities, based on the eigenmode analysis of mobile phone-usage data (Park et al. 2010). Following these, we classify the building into three distinct categories: residential, work place, and commercial buildings (denoted by r-, w- and c-categories hereafter), as summarized in Table 1. Note that one may subdivide the categories in a more detailed way, appropriate for the specific purpose of the study (Lee and Holme 2015; Daggitt et al. 2016).

Finally, we consider the total gross floor area of each category in an area, which is denoted by $\sigma_{i,x}^{\text{raw}}$ with i and x ($= r, w, c$) being lattice site and category indices,

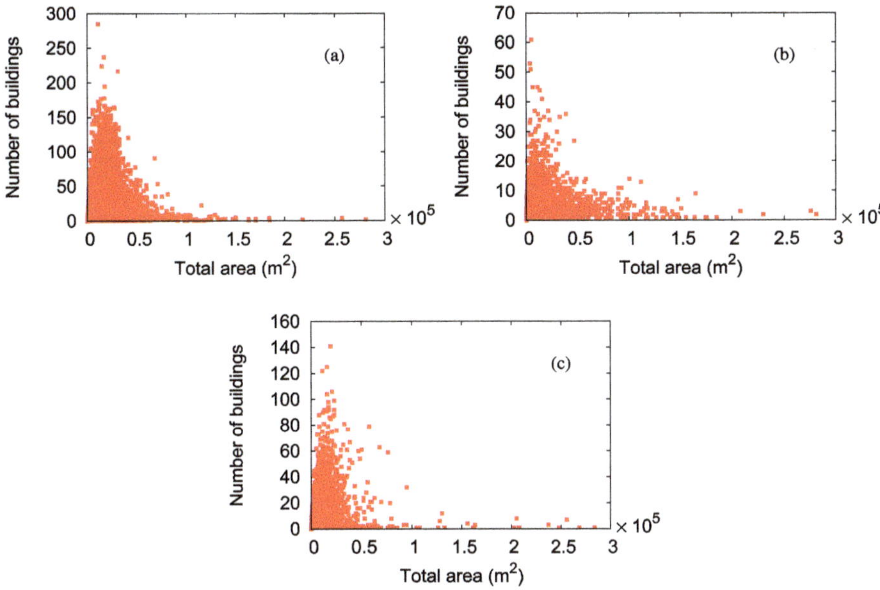

Fig. 2 Numbers of buildings versus the total areas of buildings in **a** residential, **b** work place, and **c** commercial categories. The length scale is given by $a = 138\,\mathrm{m}$

respectively, and regard it as a component of the state vector defined in coarse-grained area i. Namely, the state of each site is described by the state vector, specifying the total areas of the buildings of three categories: $\vec{\sigma}_i^{\mathrm{raw}} \equiv (\sigma_{i,r}^{\mathrm{raw}}, \sigma_{i,w}^{\mathrm{raw}}, \sigma_{i,c}^{\mathrm{raw}})$. In other words, the total gross floor areas and uses of buildings in each coarse-grained area quantify the amplitude and the direction of the state vector at the site. It is also conceivable to adopt the number of buildings as a measure of the amplitude. However, Fig. 2 discloses that there are no correlations between the numbers and the gross floor areas of buildings due to the large variations in the size of buildings. Therefore, the analysis based on the number of buildings could give an inaccurate description of the reality.

Until this stage, the state of each site is represented by continuous variables $\sigma_{i,x}^{\mathrm{raw}}$. Henceforth, for further simplicity, we adopt a two-state description, characterizing the total gross floor area of each category as a binary variable. Setting an appropriate threshold area, we assign $\sigma_{i,x} = \pm 1$ as the x-component of the state of lattice site i if the sum of the gross floor areas of buildings of category x in the corresponding area is larger/smaller than the threshold. The state of site i is then specified by $\vec{\sigma}_i \equiv (\sigma_{i,r}, \sigma_{i,w}, \sigma_{i,c})$, where each component $\sigma_{i,x}$ can be regarded as an Ising spin. Accordingly, each site can take eight states; $\vec{\sigma}_i = (-1, -1, -1)$, $(1, -1, -1)$, $(-1, 1, -1)$, $(-1, -1, 1)$, $(1, 1, -1)$, $(1, -1, 1)$, $(-1, 1, 1)$, and $(1, 1, 1)$, which are, for convenience, abbreviated as 0-, r-, w-, c-, rw-, cr-, wc-, and rwc-states, respectively.

Then the remaining issue is how to determine the threshold of the total gross floor area. In search of the appropriate value, we examine the rank distribution of

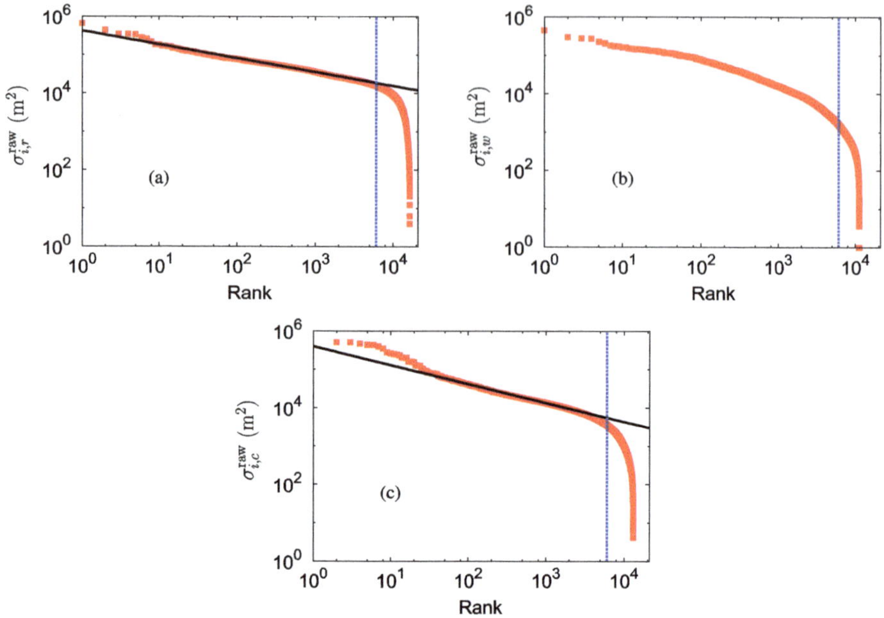

Fig. 3 Rank distributions (red squares) of total gross floor areas for **a** residential, **b** work place, and **c** commercial buildings. The lattice constant is taken to be $a = 138$ m. Power-law behaviors are observed in (**a**) and (**c**), down to the 30 percentile indicated by vertical blue lines (6093 among the total of 20,309 boxes in this coarse-grained scale). The black solid lines describe the power-law distributions with exponents (**a**) 0.36 and (**c**) 0.49

the total gross floor areas and plot the results in Fig. 3. Interestingly, residential and commercial building distributions follow power-law distributions in the high-rank regimes. Specifically, the power-law behavior persists approximately down to the top 30 percentile. We therefore consider this value as a lower limit for the threshold. In contrast, the rank distribution of work place buildings decays exponentially rather than algebraically. Furthermore, the number of boxes in which at least a single work place building is located, is smaller than those of residential and commercial buildings. In the case of lattice constant $a = 138$ m, for example, only 55% of boxes contain work place buildings and we adopt it as an upper limit. Note that these behaviors are consistently observed irrespective of the lattice constant. One may then take any percentile value for the threshold, and assign $\sigma_{i,x} = \pm 1$ to site i if the total gross floor area of category x is larger/smaller than the specific percentile value. It is pleasing that threshold values taken from criteria between 30 and 55% turn out to give consistent results, and in this study we show only the results of the median (50%) criterion for the threshold.

The site classified accordingly in the three-category description, shown in Fig. 4, manifests characteristic nonuniform distributions of sites in different states. First of all, sites in the r-, w-, and c-states are primarily located close to the boundary, as shown in Fig. 4a. On the other hand, rwc-state sites exhibiting mixed land use

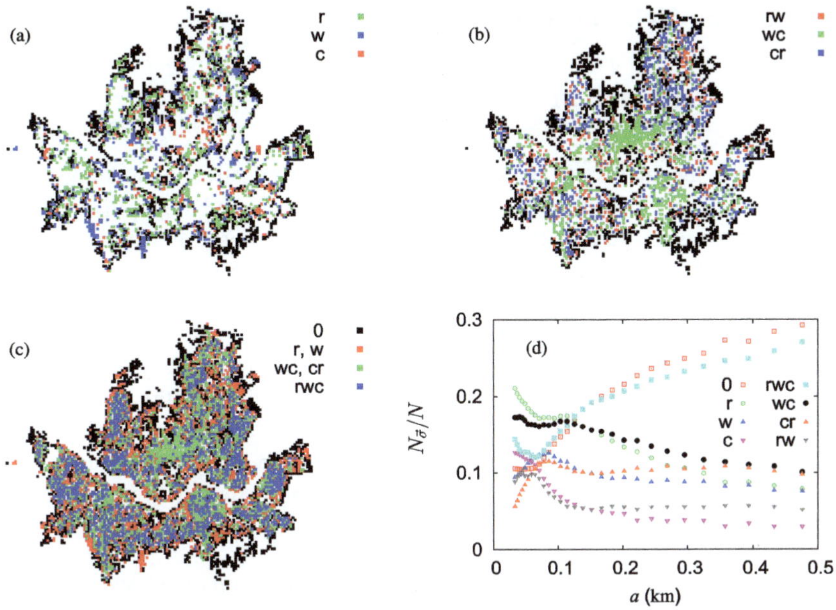

Fig. 4 Three-category Ising-type description of Seoul City under the median criterion. **a** Sites in the r-, w-, and c-states are depicted. It is observed that sites near the boundary of Seoul City are mostly in these states. **b** Sites in the rw-, wc-, and cr-states are displayed. Note that downtown areas are populated mainly by the wc-state sites; in effect buildings of the residential category are expelled from the downtown, giving rise to an urban doughnut pattern. **c** Shown are sites in the rwc-state as well as in the 0-state. In contrast to r-, w-, and c-state sites, rwc-state sites are located away from the boundary. **d** Fraction $N_{\vec{\sigma}}/N$ of sites in state $\vec{\sigma}$ are presented versus the lattice constant a. Notably, the fractions of c- and rw-state sites are smaller than those of other state sites

are relatively centered, forming clusters from place to place. It is also observed in Fig. 4b that the CBD areas of Seoul are represented mostly by the wc-state. One can further confirm that sites in the r- or w-state are located mainly near the boundary and easily distinguished from the mixed use sites (rwc-state) or the CBD (wc-state) at the center. These discriminative features reflect the spatially modularized pattern of the urbanized area. Finally, we compute the fraction of states $N_{\vec{\sigma}}/N$, where $N_{\vec{\sigma}}$ denotes the number of lattice sites in state $\vec{\sigma}$, and plot in Fig. 4d the results as the lattice constant a is varied. With the finite-size effects for $a \lesssim 0.1$ km excepted, there exist fewer c-state sites than r- and w-state ones and fewer rw-state sites than wc- and cr-state ones. These characteristics are expected to originate from the coupling between the buildings of different categories. In particular, the commercial category appears to play a distinctive role in the pattern formation, subordinate to the principal components of the human life, habitation, and production.

4 Modeling Urban Morphology

4.1 Model Description

Equipped with the analysis in the previous section, we construct a model describing the major characteristics of the land use pattern. Based on the Ising-spin description, we introduce three layers of the lattice, each of which hosts the state component corresponding to each category. Specifically, we place the Ising spin $\sigma_{i,x}$ at site i on layer x, describing the x-category state of the coarse-grained area i. The Hamiltonian of the urban morphology system is then proposed in the following way: We first assume that there exists a tendency toward aggregation (Batty and Xie 1994; Schweitzer 2003; Iannone et al. 2011) between the buildings of the same category, which may be taken into account by the "ferromagnetic" nearest-neighbor interactions on a layer, i.e., between the same category components of the state vector. The Hamiltonian describing such intra-layer coupling or interactions between the buildings of category x then reads

$$\mathcal{H}_x(\{\sigma_{i,x}\}) = -\sum_i h_{i,x}\sigma_{i,x} - J_x \sum_{\langle i,j\rangle} \sigma_{i,x}\sigma_{j,x}, \quad (1)$$

where the temperature has been absorbed in defining the Hamiltonian and the nonuniform external field $h_{i,x}$ has been introduced to measure advantages and disadvantages in constructing the building of category x in area i. For example, if the slope of area i is steep, we may assign a negative value of the field $[h_{i,x} < 0]$ to mimic the difficulty to construct buildings in that area. History of the city, infrastructures such as the road/subway networks, or even the price of land can also be good candidates for the field. We may also take into account the urban development planning by assigning positive external fields, and probe the consequence of the planning.

Leaving the detailed verification for further studies, we here assign negative fields only at the boundary as a simple but significant example. Such negative fields reflect the minimal constraint, imposed mostly by the mountains and rivers located at the boundary of Seoul City. When specifying the strength of the field at the boundary, we consider the following two factors: (1) If a lattice site is surrounded by many outer points, the negative field should be strong. (2) If the nearest neighbor coupling is strong, the effects of the boundary should also be strong. In addition, we also apply weak fields on the whole lattice to simulate the efforts of urban settlers to maintain the city. Accordingly, the external fields are specified by

$$h_{i,x} = \begin{cases} h_x - z_i J_x & \text{at the boundary } \Omega, \\ h_x & \text{elsewhere.} \end{cases}$$

Now the strength of h_x controls the activity of the city. For example, if the city is in its developmental phase, h_x should have large values. To model a sustaining city like Seoul, we assume that the total field strength vanishes:

$$\sum_i h_{i,x} = -\sum_{i\in\Omega} z_i J_x + N h_x = 0, \tag{2}$$

which yields the uniform field

$$h_x = \frac{1}{N}\sum_{i\in\Omega} z_i J_x, \tag{3}$$

assigned on the whole lattice

Next, we turn to the interactions between the buildings of distinct categories in an area. Such on-site inter-layer coupling terms are written in the most general form:

$$\begin{aligned}\mathcal{H}_{\text{int}} = &- K_{rw}\sum_i \sigma_{i,r}\sigma_{i,w} - K_{wc}\sum_i \sigma_{i,w}\sigma_{i,c} \\ &- K_{cr}\sum_i \sigma_{i,c}\sigma_{i,r}.\end{aligned} \tag{4}$$

According to the analysis in the previous section, both residential and work place buildings tend to attract commercial buildings. We thus assume ferromagnetic interactions between the commercial layer and other layers, i.e., $K_{wc} > 0$ and $K_{cr} > 0$. This assumption is indeed consistent with the observed phenomena that commercial land uses tend to locate near residential or working areas having high level of floating populations (Mushinski and Weiler 2002). Because profits generated by commercial activities can stem from distributing products and services, one of the most important factors for the location of the commercial property is the geographical proximity to its demands, consuming products or services. On the other hand, residential and work place buildings appear mutually exclusive, implying that the coupling between residential and work place layers is antiferromagnetic: $K_{rw} < 0$. As well known, residences tend to avoid locating near industrial uses since industrial facilities generate negative externality, causing noises, pollution, and traffic jams. Thus, even though there exist a certain level of housing demands (e.g., factory workers may prefer to live close to their working places), people are hesitant to reside near these facilities in general due to their negative impact on the environment (Burnell 1985). The schematic diagram of the on-site inter-layer couplings is given in Fig. 5a. It is remarkable that the inter-layer coupling is frustrated; otherwise (i.e., if all couplings are ferromagnetic), the system should consist mostly of the rwc-state (at the center) and the 0-state (at the boundary). Then the modular structure of the urban area could not be explained.

Finally, we arrive at the total Hamiltonian in the form

$$\mathcal{H}(\{\sigma_{i,x}\}) = \mathcal{H}_r + \mathcal{H}_w + \mathcal{H}_c + \mathcal{H}_{\text{int}}, \tag{5}$$

which, with the simplifying assumption $-K_{rw} = K_{wc} = K_{cr} \equiv K$ and $J_x \equiv J$, takes the simple form

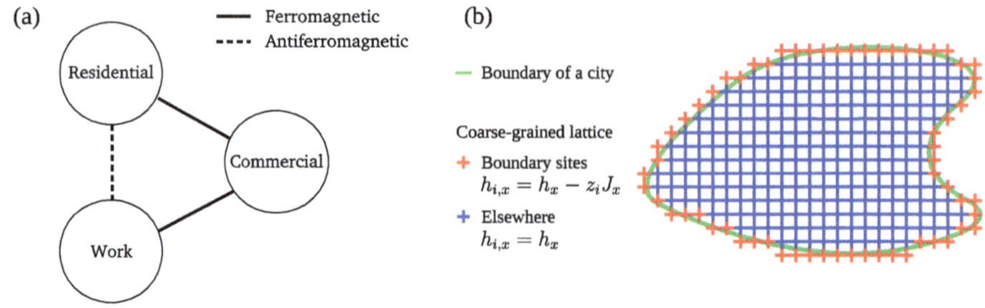

Fig. 5 Model description. **a** A schematic diagram of the inter-layer coupling, displaying frustration due to the antiferromagnetic interaction between the residential layer and the work place layer. **b** External fields applied in the city and boundary conditions imposed at the boundary

$$\begin{aligned}\mathcal{H}(\{\sigma_{i,x}\}) = & -h\sum_i \left(\sigma_{i,r} + \sigma_{i,w} + \sigma_{i,c}\right) \\ & - J\sum_{\langle i,j\rangle}\left(\sigma_{i,r}\sigma_{j,r} + \sigma_{i,w}\sigma_{j,w} + \sigma_{i,c}\sigma_{j,c}\right) \\ & - K\sum_i\left(-\sigma_{i,r}\sigma_{i,w} + \sigma_{i,w}\sigma_{i,c} + \sigma_{i,c}\sigma_{i,r}\right). \end{aligned} \tag{6}$$

Here the effects of the external field at the boundary have been considered by fixing boundary conditions: $\sigma_{i,x} = -1$ at the boundary, as illustrated in Fig. 5b. Then Eq. (3) determines completely the uniform field h, and only the coupling constants J and K remain as control parameters.

Remarkably, the frustrated inter-layer coupling explains directly the rarity of c- and rw-states observed in the building data (see Fig. 4d). For convenience of notation, we define $\tilde{\sigma}^{\pm}_{i,x} \equiv (1 \pm \sigma_{i,x})/2$, which takes values 0 and 1 depending on the state of the site, and compute the total number of sites of state $\vec{\sigma}$ through the use of the relation $N_{\vec{\sigma}} = \sum_i \tilde{\sigma}^{\pm}_{i,r}\tilde{\sigma}^{\pm}_{i,w}\tilde{\sigma}^{\pm}_{i,c}$ (where the signs are chosen in accord with the state $\vec{\sigma}$). Expressing the numbers of rw- and c-states in terms of the spin states, we then have

$$\begin{aligned} N_{rw} + N_c &= \sum_i \left(\tilde{\sigma}^{+}_{i,r}\tilde{\sigma}^{+}_{i,w}\tilde{\sigma}^{-}_{i,c} + \tilde{\sigma}^{-}_{i,r}\tilde{\sigma}^{-}_{i,w}\tilde{\sigma}^{+}_{i,c}\right) \\ &= \frac{N}{4} + \frac{1}{4}\sum_i\left(\sigma_{i,r}\sigma_{i,w} - \sigma_{i,w}\sigma_{i,c} - \sigma_{i,c}\sigma_{i,r}\right). \end{aligned} \tag{7}$$

Accordingly, the third term, depending on the inter-layer coupling constant K, of the Hamiltonian in Eq. (6) can be rewritten in the form $4K(N_{rw} + N_c)$, with the constant term $4KN$ neglected. This manifests that the coupling constant K controls the fraction of c- and rw-states: in particular, $N_{rw}, N_c \to 0$ as $K \to \infty$.

Note here that the eight states of the system are connected to each other via symmetries of the system. In the thermodynamic limit ($N \to \infty$), the boundary effects become negligible, and the uniform field h also becomes vanishingly small. In consequence the Hamiltonian given by Eq. (6) carries $Z_2 \times Z_3$ symmetry: The three states (rwc-, r-, and w-states) as well as the three (0-, wc-, and cr-states) are connected by Z_3 symmetry while the two states in each of the three pairs (rwc- and 0-states, r- and wc-states, w- and cr-states) are connected by Z_2 symmetry. Depending on the values of J and K, the system may thus exhibit ordered/disordered states together with the intermediate critical state.

4.2 Monte Carlo Simulations and Mean-Field Approximation

To probe the emergent phases of the model, we first carry out Monte Carlo simulations under the fixed boundary conditions and weak external fields. Fractal lattices with the lattice constant $a = 138, 183, 244, 325$ and $432\,\text{m}$ are used; 10^5 Monte Carlo steps are performed to equilibrate the system and another 10^5 steps to compute the ensemble averages.

To make use of the $Z_2 \times Z_3$ symmetry inherent in the Hamiltonian, we introduce q-fold symmetry order parameters $m_q (q = 2, 3, 6)$ and their fluctuations χ_q. In view of the symmetry among the three categories ($x = r$, w, and c), we take into account the state $\vec{\sigma}_i = (1, 1, 1)$ and its symmetric counterparts, and define the order parameters as

$$m_2 \equiv \left\langle \left| \frac{1}{N} \sum_i \left(\tilde{\sigma}^+_{i,r} \tilde{\sigma}^+_{i,w} \tilde{\sigma}^+_{i,c} - \tilde{\sigma}^-_{i,r} \tilde{\sigma}^-_{i,w} \tilde{\sigma}^-_{i,c} \right) \right| \right\rangle, \tag{8}$$

$$m_3 \equiv \left\langle \left| \frac{1}{N} \sum_i \left(\tilde{\sigma}^+_{i,r} \tilde{\sigma}^+_{i,w} \tilde{\sigma}^+_{i,c} + e^{2\pi i/3} \tilde{\sigma}^-_{i,r} \tilde{\sigma}^+_{i,w} \tilde{\sigma}^-_{i,c} \right. \right. \right.$$
$$\left. \left. \left. + e^{4\pi i/3} \tilde{\sigma}^+_{i,r} \tilde{\sigma}^-_{i,w} \tilde{\sigma}^-_{i,c} \right) \right| \right\rangle, \tag{9}$$

$$m_6 \equiv \left\langle \left| \frac{1}{N} \sum_i \left(\tilde{\sigma}^+_{i,r} \tilde{\sigma}^+_{i,w} \tilde{\sigma}^+_{i,c} + e^{\pi i/3} \tilde{\sigma}^-_{i,r} \tilde{\sigma}^+_{i,w} \tilde{\sigma}^+_{i,c} \right. \right. \right.$$
$$+ e^{2\pi i/3} \tilde{\sigma}^-_{i,r} \tilde{\sigma}^+_{i,w} \tilde{\sigma}^-_{i,c} + e^{\pi i} \tilde{\sigma}^-_{i,r} \tilde{\sigma}^-_{i,w} \tilde{\sigma}^-_{i,c}$$
$$\left. \left. \left. + e^{4\pi i/3} \tilde{\sigma}^+_{i,r} \tilde{\sigma}^-_{i,w} \tilde{\sigma}^-_{i,c} + e^{5\pi i/3} \tilde{\sigma}^+_{i,r} \tilde{\sigma}^-_{i,w} \tilde{\sigma}^+_{i,c} \right) \right| \right\rangle, \tag{10}$$

where $\langle \cdots \rangle$ stands for the ensemble average. Then their fluctuations are measured by

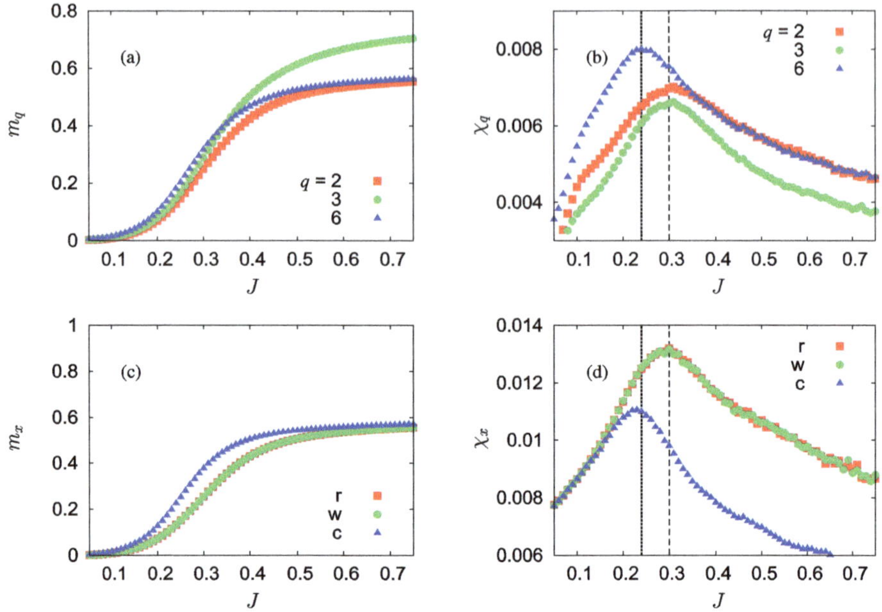

Fig. 6 Order parameters **a** m_q and **c** m_x and their fluctuations **b** χ_q and **d** χ_x. The dotted and dashed vertical lines in (**b**) and (**d**) indicate the values of J at which χ_q reaches the peak: $J = 0.24$ for $q = 6$ and $J = 0.30$ for $q = 2, 3$. It is evident that m_r and m_w (similarly, χ_r and χ_w) exhibit the same behavior while χ_c peaks at a smaller value of J than χ_r and χ_w. The lattice constant and the inter-layer coupling constant are taken to be $a = 138\,\text{m}$ and $K = 0.2$, respectively

$$\chi_2 \equiv \left\langle \left| \frac{1}{N} \sum_i \left(\tilde{\sigma}_{i,r}^{+} \tilde{\sigma}_{i,w}^{+} \tilde{\sigma}_{i,c}^{+} - \tilde{\sigma}_{i,r}^{-} \tilde{\sigma}_{i,w}^{-} \tilde{\sigma}_{i,c}^{-} \right) \right|^2 \right\rangle - \left\langle \left| \frac{1}{N} \sum_i \left(\tilde{\sigma}_{i,r}^{+} \tilde{\sigma}_{i,w}^{+} \tilde{\sigma}_{i,c}^{+} - \tilde{\sigma}_{i,r}^{-} \tilde{\sigma}_{i,w}^{-} \tilde{\sigma}_{i,c}^{-} \right) \right| \right\rangle^2 \tag{11}$$

and by χ_3 and χ_6 defined in the same manner. Figure 6a, b display the obtained results of m_q and χ_q as the intra-layer coupling constant J is varied. Interestingly, we observe two distinctive phase transitions at $J = 0.24$ and at $J = 0.30$ when the inter-layer coupling is taken to be $K = 0.2$. We have also performed simulations of the two-dimensional regular lattice with periodic boundary conditions, varying K, up to the linear size of $L = 1024$, and obtained similar behaviors (results not shown). Rather than computing the rigorous phase diagram employing the finite-size scaling, we focus on the accompanying phenomena applicable to urban morphology, computing the order parameter m_x and its fluctuations χ_x:

$$
m_x = \left\langle \frac{1}{N} \sum_i \sigma_{i,x} \right\rangle,
$$

$$
\chi_x = \left\langle \left(\frac{1}{N} \sum_i \sigma_{i,x} \right)^2 \right\rangle - \left\langle \frac{1}{N} \sum_i \sigma_{i,x} \right\rangle^2. \tag{12}
$$

As shown in Fig. 6c, d, layer c corresponding to the commercial category displays rather distinctive behavior compared with layers r (residential category) and w (work place category): The symmetry is broken earlier at a lower value of J, giving rise to larger values of m_c than m_r and m_w in the ordered phase. As expected, fluctuations exhibit peaks at the critical points estimated from the behavior of χ_q.

To elucidate this observation, we take the mean-field approximation: Neglecting fluctuations around the mean value m_x, we write the mean-field Hamiltonian in the form

$$
\begin{aligned}
\mathcal{H}_{\rm MF} = &- \left(J m_r - \frac{K}{2} m_w + \frac{K}{2} m_c + h \right) \sum_i \sigma_{i,r} \\
&- \left(J m_w - \frac{K}{2} m_r + \frac{K}{2} m_c + h \right) \sum_i \sigma_{i,w} \\
&- \left(J m_c + \frac{K}{2} m_r + \frac{K}{2} m_w + h \right) \sum_i \sigma_{i,c}.
\end{aligned} \tag{13}
$$

It is then easy to obtain the self-consistent equations for the order parameters:

$$
\begin{aligned}
m_r &= \tanh \left(2dJ m_r - \frac{K}{2} m_w + \frac{K}{2} m_c + h \right), \\
m_w &= \tanh \left(2dJ m_w - \frac{K}{2} m_r + \frac{K}{2} m_c + h \right), \\
m_c &= \tanh \left(2dJ m_c + \frac{K}{2} m_r + \frac{K}{2} m_w + h \right).
\end{aligned} \tag{14}
$$

In the thermodynamic limit, we set $h = 0$ and take only positive solutions, i.e., $m_r, m_w, m_c \geq 0$, observable in simulations. In the ordered phase, we find $m_r = m_w \equiv m_s$ and $m_c \equiv m_l$ with

$$
\begin{aligned}
m_s &= \tanh \left[2dJ m_s + \frac{K}{2} (m_l - m_s) \right] \\
&= \tanh \left[2dJ m_s - \frac{K}{2} m_s + \frac{K}{2} m_l \right], \\
m_l &= \tanh \left[2dJ m_l + K m_s \right] \\
&= \tanh \left[2dJ m_l - \frac{K}{2} m_l + \frac{K}{2} m_l + K m_s \right],
\end{aligned} \tag{15}
$$

where $m_l > m_s$. Solving the equations numerically with the fractal dimension $d = 1.8$, we obtain the critical value $J_c = 0.250$ for $K = 0.2$, which coincides with the Z_6-symmetry-breaking point $J = 0.24$ in Monte Carlo simulations. Note, however, that the intermediate phase emerging for $0.24 < J < 0.30$ is not verified in the mean-field approximation and still remains to be investigated.

4.3 Criticality in the Urban Morphology of Seoul

Finally, in this subsection, the model predictions are compared with the results from data analysis. We first introduce the "macroscopic state" of the system described by the eight-component state vector

$$\mathbf{S} \equiv \frac{1}{N_S}(N_0, N_r, \ldots, N_{rwc}), \tag{16}$$

where $N_S \equiv [N_0^2 + \cdots + N_{rwc}^2]^{1/2}$ is the normalization factor. In other words, while the microscopic state details the state of every site and is thus characterized by N (microscopic) state vectors $\vec{\sigma}_i$ identifying the state of site i $(= 1, 2, \ldots, N)$, the macroscopic state specifies just the number of sites in each of the eight states $\vec{\sigma}$. Comparison of the macroscopic state vectors $\mathbf{S}_\mathrm{M}$ and $\mathbf{S}_\mathrm{D}$, obtained via simulations of the model and from the building data, respectively, should expose the discrepancy between the simulation results and the real data. In particular, the inner product of the two-state vectors provides a convenient measure of how faithful the model description is, and we thus define the accuracy factor $Q \equiv \mathbf{S}_\mathrm{M} \cdot \mathbf{S}_\mathrm{D}$, quantifying the accuracy of the simulation results. Here we encounter again the threshold issue. To treat the simulation results and data on an equal footing, we use the order parameters from simulations to fix the threshold when coarse-graining the building data. Specifically, the positive state is assigned to the first $\frac{m_x+1}{2}N$ sites in descending order of area.

Note that the accuracy factor also manifests the criticality of the system. As shown in Fig. 7, the accuracy factor reaches the maximum in the range $0.24 < J < 0.30$. In other words, simulations reproduce the building data best in the intermediate phase between breaking of Z_6 symmetry and that of Z_2 or Z_3 symmetry. In the critical region, the accuracy factor becomes almost unity ($Q \approx 1$). Table 2 summarizes the distribution ratios of the eight different states, obtained from the building data and from the model, for various values of the lattice constant a. The coupling constants J and K are adjusted to the values at which Q is maximum for given a. Note the remarkable agreement in the state distributions between simulations and real data.

Encouraged by the successful results of the model, we are now in a position to examine the urban morphology generated by numerical simulations. For reference, we first analyze the Seoul building data set applying the threshold value obtained from the simulation results with the lattice constant $a = 138$ m. Extracted urban morphology of Seoul City is displayed in Fig. 8a. Note that the results are consistent with those presented in Fig. 4c even though the threshold value and the lattice constant

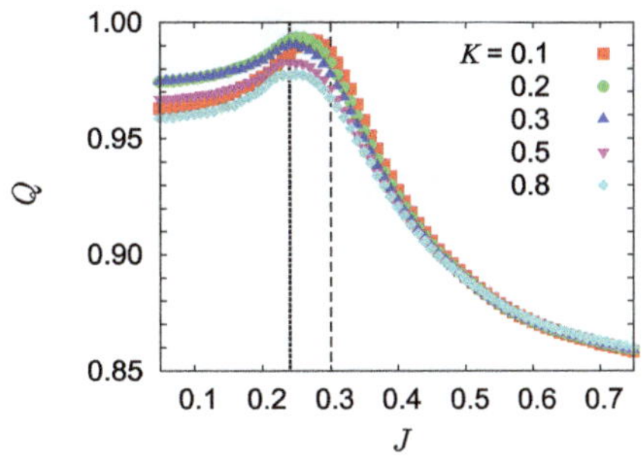

Fig. 7 Accuracy factor Q versus intra-layer coupling constant J for various values of the inter-layer coupling constant K. Q reaches the peak in the vicinity of the Z_6 symmetry breaking, revealing the criticality of the city. The lattice constant is given by $a = 138\,\text{m}$

Table 2 Fractions of states $N_{\vec{\sigma}}/N$ for several values of the lattice constant, extracted from (a) the Seoul building data and (b) numerical simulations of the model with coupling constants J and K at which Q is maximum

State $\vec{\sigma}$	138 m	183 m	244 m	325 m	432 m
(a) Building data					
$rwc(1,1,1)$	0.276	0.409	0.502	0.563	0.623
$r(1,-1,-1)$	0.126	0.092	0.070	0.054	0.042
$w(-1,1,-1)$	0.083	0.066	0.056	0.050	0.036
$0(-1,-1,-1)$	0.123	0.106	0.107	0.116	0.125
$wc(-1,1,1)$	0.154	0.121	0.095	0.071	0.056
$cr(1,-1,1)$	0.133	0.107	0.081	0.068	0.051
$c(-1,-1,1)$	0.064	0.058	0.050	0.043	0.035
$rw(1,1,-1)$	0.041	0.041	0.039	0.035	0.032
(b) Simulations					
$rwc(1,1,1)$	0.270	0.402	0.488	0.556	0.622
$r(1,-1,-1)$	0.101	0.067	0.054	0.047	0.038
$w(-1,1,-1)$	0.101	0.067	0.054	0.047	0.038
$0(-1,-1,-1)$	0.123	0.116	0.119	0.127	0.132
$wc(-1,1,1)$	0.157	0.125	0.104	0.081	0.062
$cr(1,-1,1)$	0.157	0.125	0.104	0.081	0.062
$c(-1,-1,1)$	0.043	0.042	0.032	0.026	0.020
$rw(1,1,-1)$	0.048	0.056	0.045	0.035	0.026

are different. We have also confirmed the consistency of urban morphology across the threshold values and coarse-graining scales in proper ranges.

When analyzing simulation results, on the other hand, one should ponder over social fluctuations and collective dynamics operating in the urban area because via Monte Carlo simulations we can probe only equilibrium fluctuations rather than the actual dynamics of the urban system. Approximately, one may regard individual choices as noise fluctuating on short-time scales. Observed phenomena, however, would change rather gradually exhibiting temporal correlations as we probe the macroscopic properties or the morphology of the city, originating from the interactions between individuals. As a result, macroscopic dynamics of the system is an

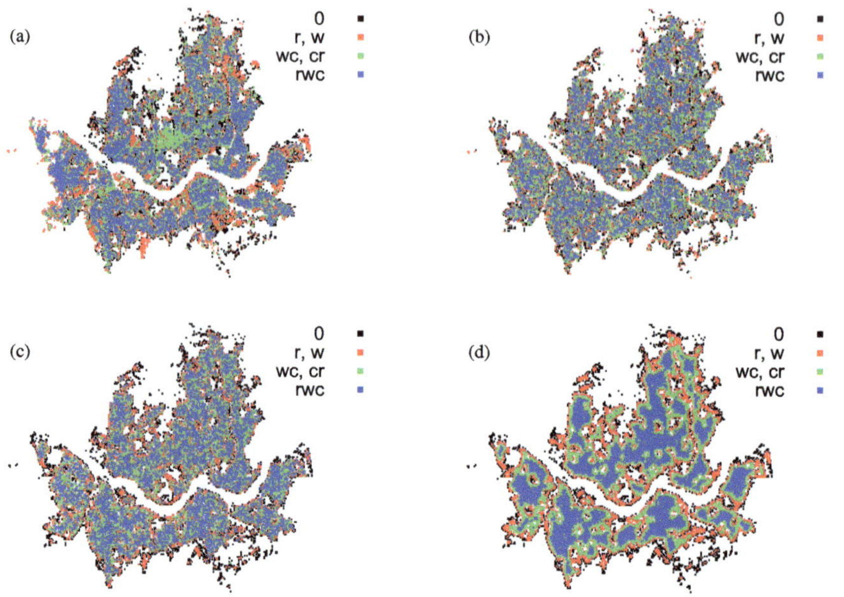

Fig. 8 Urban morphology of Seoul City, extracted from **a** the Seoul building data set and **b–d** Monte Carlo simulation results. The number of ensembles used in the averaging process is **b** 1 (snapshot, short-time limit), **c** 20 (intermediate time), and **d** 1000 (long-time limit). Fluctuations are mainly observed in the short-time limit while rather a regular pattern is extracted in the long-time limit. Both results are not quite appropriate due to the quasi-equilibrium nature of the system

emergent property which reflects the averaged effects of microscopic fluctuations. Here we assume that the urban system is in a quasi-equilibrium state and the individuals may feel mean-field-like couplings for the duration of time during which the system resides near equilibrium. Furthermore, it is obvious that there exists time delay in the administrative procedure of data gathering and planning. We, therefore, conclude that the observed urban morphology is a consequence of averages over the time associated with decision making and executing as well as observations. However, the time scale should not be infinitely long as usually assumed in equilibrium statistical mechanics. According as technology, relations to other cities, and natural environment evolve in time, the city should develop and come out of the current quasi-equilibrium state, manifesting its nonequilibrium nature. If the actual trajectory of the system should be followed in the phase space, the relevant averaging time scale could be specified. Leaving this for further study, we here take the average of the results of Monte Carlo simulations, varying the number of ensembles, and thus probe the macroscopic patterns explicitly.

The detailed procedure taken to prepare the figure is summarized as follows: We first evaluate the frequency of each state appearing in the system, and choose the lattice sites where the frequency of the given state is high. In some sites, there could be more than one state whose frequencies are high enough to be depicted while there also exist sites of which no states display high frequencies. To assign a single state to each site, we first choose the highest one among overlapped states in the former case, and leave the states of sites undetermined in the latter case. Then the above frequency comparison process is repeated only with the remaining sites whose states are not determined until the state of every site is specified. Note that we enforce the number of sites in a given state, i.e., the fraction of a state, to equal the ensemble average value computed from the simulation data displayed in Table 2. Accordingly, if the fraction of a state is once fulfilled, the state is also excluded in the repeated procedure. In this manner, we specify the most probable state in every site while constraining the fractions of states to be the ensemble average values.

Figure 8 presents the obtained results, where the number of ensembles used in the averaging process is given by (b) 1 (snapshot, short-time limit), (c) 20 (intermediate time) and (d) 1000 (long-time limit). As shown, it is successfully simulated to locate the sites of r- and w-states which are distributed mainly around the boundary, irrespectively of the averaging time. These results reveal unambiguously the interplay between boundary conditions and overall external fields. Also observed are clustering patterns, especially of the rwc-state, which become clearer as the averaging time is increased. As expected, the clustering patterns resemble the results from the data better in the intermediate time scale.

On the other hand, emergence of the CBD clusters consisting mostly of the wc-state is not observed in the simulation. This is presumably due to the fact that the CBD of Seoul involves historical formation and land price issues. To observe the emergence of the wc-state cluster, we may thus need to apply an additional field, taking into consideration such issues in the CBD area. Furthermore, working areas are apparently categorized into further details. Specifically, the working areas may be split into two representative subcategories: Most of working areas located on the edge of Seoul are either factories or warehouses. As well known, many industrial facilities tend to locate further from the center of the city, since they prefer to consume large spaces of the land at cheaper prices (Wood 1974). In contrast, office facilities tend to cluster at certain areas in a city and contribute formation of the CBD cluster, which are generated by agglomeration economies. Agglomeration economies facilitate more face-to-face interactions, sharing inputs, and integrating the labor market; these factors reintensify agglomeration economies, resulting in the clustering of office firms (O'Sullivan 2011). With the external field issues put together, the historically quenched area of the city may serve as a seed of the cluster formation.

To address quantitatively the cluster formation issue, we further probe a correlation-like quantity between nearest-neighboring areas for each category. Specifically, we define an agglomeration index as

$$E_x \equiv \frac{1}{N}\sum_{\langle i,j\rangle} \sigma_{i,x}\sigma_{j,x}, \tag{17}$$

which corresponds to the energy contribution of the intra-layer coupling per area to the full Hamiltonian $\mathcal{H}$. Having large values in case that neighboring areas are in the same states, this index quantifies the agglomeration tendency of the respective category. Figure 9a presents results extracted from the data under the median criterion.

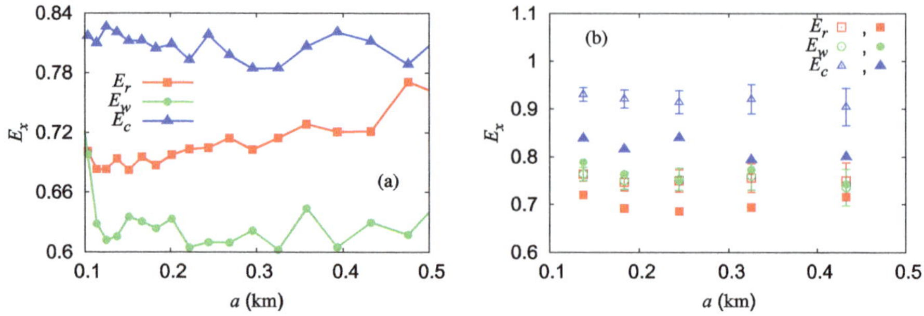

Fig. 9 Agglomeration index E_x for $x = r, w$, and c. In **a** the values of E_x extracted from the data under the median criterion are shown. Varying the lattice constant a, we observe that commercial areas exhibit stronger agglomeration than residential and work place areas. In **b** simulation results (open symbols with error bars) are compared with the data (filled symbols), with a positive state assigned to $N(m_x + 1)/2$ sites. Again, the values of E_c are larger than those of E_r and E_w, both in the data and simulation results

It is interesting to observe that the values of E_c are larger than those of E_r and E_w; such a difference might be attributed to the frustrated interactions. As discussed already in Sect. 4.2, the symmetry breaking of the commercial category is distinctive from that of other categories, although such two distinctive symmetry breakings may not be preserved in the thermodynamic limit. Here, rather than seeking rigorously for verification of the effects solely from the frustration, we interpret the results in combination with the boundary effects. Indeed we have confirmed that, for instance, states σ_r and σ_w are more frequently located near the boundary than state σ_c as a consequence of the interplay between the frustrated interactions and the boundary conditions, leading to smaller values of E_r and E_w than E_c. In other words, it is agglomeration of the commercial areas that promotes the clustering of areas with σ_{rwc}, σ_{wc}, and σ_{cr}. We also present direct comparison of the data and simulation results in Fig. 9b, where, instead of the median criterion, a positive state has been assigned to $N(m_x + 1)/2$ sites as in Fig. 8a, and observe again the tendency that E_c is larger than E_r and E_w. Note, however, that the values of E_w are in general larger than those of E_r, which contrasts with Fig. 9a. This may reflect the absence of the scale invariance of the work place category (see Fig. 3b). It is thus desirable to determine a specific scale in the case of the work place category, which is left for further study.

To conclude, properly averaged simulation results, e.g., Fig. 8c, successfully describe the overall differentiated patterns and fluctuating features (except for the formation of the CBD cluster due to the aforementioned issues): The explicitly broken translational symmetry due to the boundary conditions contributes the formation of the spatial structures while the criticality of the system accounts for the statistics and fluctuating patterns. The agglomeration index E_x, computed for each category, discloses that the commercial category plays a significant role in the cluster formation. We also expect that these simulation results may shed light on the future land use pattern of Seoul unless the system is perturbed dramatically. For example, we observe recently evidence for the reformation and population growth in downtown areas. Also the mixed land use issue attracts much attention in the context of urban planning. As shown in Fig. 8d, the clustered mixed use (rwc-state) areas are observed ubiquitous in the central regions of cities and in fact coincident with the actual downtown areas of Seoul as well.

5 Discussion

We have analyzed the building data set of Seoul City, to extract the land use pattern. Based on the interactions between buildings, we have constructed a three-layer Ising model, through which urban morphology has been studied. To be specific, we have performed coarse-graining of the system, which is described not only by the urbanized area but by three categories of the building. The resulting coarse-grained system has been probed and emergence of the distribution patterns capturing urban morphology has been attributed to the inter- and intra-layer couplings between the buildings. In particular, the coupling between residential areas and work places has turned out to be mutually repulsive or antiferromagnetic. Further, it has been confirmed that boundary conditions are the key ingredients for generating the urban spatial structure. With minimal postulates, as summarized above, we have indeed reproduced successfully the urban morphology of Seoul.

As the model requires only the social/natural boundary of the city, one can simulate morphology of other cities around the world and examine the applicability of the model. We further stress here the distinctive characteristics of commercial land uses. Related to the recent urban planning issue to improve the proximity of jobs from housing, the locations/distributions of the commercial land uses may play an important role in urban formation, minimizing the distances between the work and residential areas.

In principle, an extension of our model to the dynamics of urban growth should be straightforward. For instance, one may simply consider an initial configuration with 0-states for all areas and probe equilibration dynamics. We note, however, that a proper time scale needs to be specified quantitatively; as shown in Fig. 8, microscopic fluctuations should be averaged in this time scale. Moreover, it is also necessary to consider time-dependent external fields for a realistic model for each city.

Meanwhile, there still remain several issues to resolve or improve. First of all, we point out that the detailed phase diagram is not obtained. Even though the behavior of the model at criticality indeed resembles that observed in the real data set, we have not evaluated rigorously the critical behavior. Comparing the exponents of the model with the building data of various cities, one may probe the possibility of the universality class among different cities. Furthermore, if the correlation function, for example, could be evaluated accurately, one may predict responses of a city to external perturbations such as natural and/or social disasters.

Second, note that the fractal lattice constructed of the Seoul building data set is not fully connected in general. Most of all, the Han (river), divides the urbanized area of Seoul into two separate parts. As a consequence, each part of the lattice is decoupled and completely independent of each other if the coupling is assumed to exist between nearest-neighboring sites. In fact, it is well known that the two parts of Seoul, usually called Gangbuk (northern part) and Gangnam (southern part), indeed exhibit endemic characteristics. The Han, approximately 1 to 2 km wide estimated by the length of the bridges, is rather wider than most rivers in other cities and is therefore likely to generate boundary effects. Obviously, however, Seoul should be regarded as a single city. A simple alternative is to modify the intra-layer coupling to be long-ranged, e.g., $J(r) \sim 1/r^{\alpha}$ with appropriate exponent α. On the other hand, one may introduce disordered coupling J_{ij}, reflecting the transportation infrastructure such as the subway or road network. For example, we may assign a positive coupling constant J_{ab} between the areas a and b connected by a bridge.

Furthermore, one may assume asymmetric inter-layer couplings, e.g., $K_{wc} \neq K_{cr}$, to elucidate the difference between the distributions of r- and w-states (or wc- and cr-states). As shown in Fig. 4d, the states are not evenly distributed among six other states as well as c- and rw-states. Since the relation between the inter-layer coupling terms and the average number of a state could be obtained as in Eq. (7), one can extend the Hamiltonian to model the state distributions in more detail, according to the specific objective of the study. Such asymmetric couplings may indeed arise in the discrete spin description of the system, since the spin state $\sigma_{i,x} = 1$ does not necessarily represent the same amount of buildings or more essentially, of population, irrespective of category x. For instance, commercial buildings are generally larger in floor areas than residential buildings and expected to attract more people, which suggests that $K_{wc} > K_{rw}$. In this regard, one may consider an extended description (Louf and Barthelemy 2016) to take such heterogeneity into account.

Finally, the origin of the criticality still remains unanswered. To address this issue properly, one should clarify the concept of social temperature beforehand. From a practical point of view, various types of uncertainty with the origin ranging from variability at individual levels to large-scale perturbations due to urban planning or technological advances introduce noises or fluctuations, the strength of which is measured by "temperature". Such an approach can be justified theoretically by the maximum entropy principle (Jaynes 1957, 2003) in statistical mechanics. Turning to the criticality, we note that typically in a physical system, the temperature should be tuned in an appropriate way to achieve criticality of the corresponding system. Indeed in our model, the temperature is fine-tuned to describe the statistics of the morphology of Seoul precisely. As usual, the thermodynamic limit is an essential assumption for the criticality in a concrete manner. Here it is worthwhile to point out that such fine tuning is not applicable to a real urban system. In this regard, the framework of self-organized criticality (Bak et al. 1987; Bak 1996) may provide an adequate explanation. According to the idea of self-organized criticality, complex systems exhibit critical behaviors because critical points serve as attractors for their dynamics.

Specifically, in the case of an urban system, the coupling via transportation which enables the urban movements may play a crucial role in controlling fluctuations and/or the social temperature of the urban morphology. We speculate that if the urban morphology corresponds to an ordered state, there would be too heavy load on the transportation system. It may cause the redistribution of the buildings to minimize the transportation cost. If, on the other hand, the urban morphology remains in the disordered state, the urban system may not function properly. In consequence, critical points are expected to provide the only possible attractors for dynamics of the urban system which is successful and sustainable. To verify this hypothesis is left for further study.

Acknowledgements We thank Prof. K. Kim for his contributions at the early stage of this work and acknowledge the support by the National Research Foundation of Korea through the Basic Science Research Program (Grant No. 2016R1D1A1A09917318).

References

2011 Building Register Data, Electronic Architectural Information System, Ministry of Land, Infrastructure, and Transport (Republic of Korea, 2011).

C. Andersson, K. Lindgren, S. Rasmussen, and R. White, "Urban growth simulation from 'first principles'", Phys. Rev. E **66**, 026204 (2002).

P. Bak, *How nature works: the science of self-organized criticality* (Springer Science & Business Media, New York, 1996).

P. Bak, C. Tang, and K. Wiesenfeld, "Self-organized criticality: An explanation of the $1/f$ noise", Phys. Rev. Lett. **59**, 381 (1987).

M. Batty, "Exploring isovist fields: space and shape in architectural and urban morphology", Env. Plan. B **28**, 123–150 (2001).

M. Batty and Y. Xie, "Modelling inside GIS: Part 1. Model structures, exploratory spatial data analysis and aggregation", Int. J. Geogr. Inf. Syst. **8**, 291–307 (1994).

M. Batty and Y. Xie, "Self-organized criticality and urban development", Discrete Dyn. Nat. Soc. **3**, 109–124 (1999).

M. Batty, R. Carvalho, A. Hudson-Smith, R. Milton, D. Smith, and P. Steadman, "Scaling and allometry in the building geometries of Greater London", Eur. Phys. J. B **63**, 303–314 (2008).

L. M. Bettencourt, "The origins of scaling in cities", Science **340**, 1438–1441 (2013).

A. Bitner, R. Holyst, and M. Fialkowski, "From complex structures to complex processes: Percolation theory applied to the formation of a city", Phys. Rev. E **80**, 037102 (2009).

J.-P. Bouchaud, "Economics needs a scientific revolution", Nature **455**, 1181 (2008).

J. D. Burnell, "Industrial land use, externalities, and residential location", Urban Stud. **22**, 399–408 (1985).

J. Candia, M. C. González, P. Wang, T. Schoenharl, G. Madey, and A.-L. Barabási, "Uncovering individual and collective human dynamics from mobile phone records", J. Phys. A **41**, 224015 (2008).

C. Castellano, S. Fortunato, and V. Loreto, "Statistical physics of social dynamics", Rev. Mod. Phys. **81**, 591–646 (2009).

Y. Chen, "Fractal dimension evolution and spatial replacement dynamics of urban growth", Chaos Soliton. Fract. **45**, 115–124 (2012).

M. Daggitt, A. Noulas, B. Shaw, and C. Mascolo, "Tracking urban activity growth globally with big location data", R. Soc. Open. Sci. **3**, 150688 (2016).

J. Decraene, C. Monterola, G. K. K. Lee, and T. G. G. Hung, "A quantitative procedure for the spatial characterization of urban land use", Int. J. Mod. Phys. C **24**, 1250092 (2013a).

J. Decraene, C. Monterola, G. K. K. Lee, T. G. G. Hung, and M. Batty, "The Emergence of Urban Land Use Patterns Driven by Dispersion and Aggregation Mechanisms", PLoS ONE **8**, e80309 (2013b).

G. Duranton and H. G. Overman, "Testing for localization using micro-geographic data", Rev. Econ. Stud. **72**, 1077–1106 (2005).

S. Fotheringham, "Urban growth and form: scaling, fractal geometry, and diffusion-limited aggregation", Env. Plan. A **21**, 1447–1472 (1989).
G. F. Frasco, J. Sun, H. D. Rozenfeld, and D. Ben-Avraham, "Spatially distributed social complex networks", Phys. Rev. X **4**, 011008 (2014).
S. Goh, K. Lee, M.Y. Choi, and J.-Y. Fortin, "Emergence of criticality in the transportation passenger flow: scaling and renormalization in the seoul bus system", PLoS ONE **9**, e89980 (2014).
S. Goh, M.Y. Choi, K. Lee, and K. Kim, "How complexity emerges in urban systems: Theory of urban morphology", Phys. Rev. E **93**, 052309 (2016).
P. Grassberger and I. Procaccia, "Measuring the strangeness of strange attractors", Physica D **9**, 189–208 (1983).
A. Gudmundsson and N. Mohajeri, "Entropy and order in urban street networks", Sci. Rep. **3**, 3324 (2013).
G. Iannone, A. Troisi, C. Guarnaccia, P. D'Agostino, and J. Quartieri, "An urban growth model based on a cellular automata phenomenological framework", Int. J. Mod. Phys. C **22**, 543–561 (2011).
E. T. Jaynes, "Information theory and statistical mechanics", Phys. Rev. **106**, 620 (1957).
E. T. Jaynes, *Probability Theory: The Logic of Science*, edited by G. L. Bretthorst (Cambridge University Press, Cambridge, 2003).
P. Jensen, "Network-based predictions of retail store commercial categories and optimal locations", Phys. Rev. E **74**, 035101 (2006).
C. Kang, X. Ma, D. Tong, and Y. Liu, "Intra-urban human mobility patterns: An urban morphology perspective", Physica A **391**, 1702–1717 (2012).
M. Lee and P. Holme, "Relating land use and human intra-city mobility", PLoS ONE **10**, e0140152 (2015).
R. Louf and M. Barthelemy, "Modeling the polycentric transition of cities", Phys. Rev. Lett. **111**, 198702 (2013).
R. Louf and M. Barthelemy, "Patterns of residential segregation", PLoS ONE **11**, e0157476 (2016).
R. Louf, C. Roth, and M. Barthelemy, "Scaling in Transportation Networks", PLoS ONE **9**, e102007 (2014).
H. A. Makse, J. S. Andrade, M. Batty, S. Havlin, and H. E. Stanley, "Modeling urban growth patterns with correlated percolation", Phys. Rev. E **58**, 7054–7062 (1998).
E. Marcon and F. Puech, "Measures of the geographic concentration of industries: improving distance-based methods", J. Econ. Geogr. **10**, 745 (2009).
A. P. Masucci, E. Arcaute, E. Hatna, K. Stanilov, and M. Batty, "On the problem of boundaries and scaling for urban street networks", J. R. Soc. Interface **12**, 763 (2015).
D. Mushinski and S. Weiler, "A note on the geographic interdependencies of retail market areas", J. Regional Sci. **42**, 75–86 (2002).
A. O'Sullivan, *Urban Economics* (McGraw-Hill, New York, 2011).
J. Park, D.-S. Lee, and M. C. González, "The eigenmode analysis of human motion", J. Stat. Mech. P11021 (2010).
M. Schläpfer, J. Lee, and L. Bettencourt, "Urban skylines: building heights and shapes as measures of city size", arXiv: 1512.00946 (2015).
F. Schweitzer, *Brownian Agents and Active Particles: Collective Dynamics in the Natural and Social Sciences* (Springer-Verlag, Berlin, 2003).
G. Shen, "Fractal dimension and fractal growth of urbanized areas", Int. J. Geogr. Inf. Sci. **16**, 419–437 (2002).
R. White and G. Engelen, "Cellular automata and fractal urban form: a cellular modelling approach to the evolution of urban land-use patterns", Env. Plan. A **25**, 1175–1199 (1993).
T. A. Witten, Jr. and L. M. Sander, "Diffusion-limited aggregation, a kinetic critical phenomenon", Phys. Rev. Lett. **47**, 1400 (1981).
P. A. Wood, *Suburban Growth: Geographical Processes at the Edge of the Western City*, edited by J. H. Johnson (John Wiley, London, 1974), pp. 129–154.

On the Complex Interaction Between Mathematics and Urban Morphology

Nicola Bellomo and Pietro Terna

Abstract This chapter is devoted to some speculations on the search for possible interactions between mathematics and the dynamics of urban morphology viewed as a science, where heterogeneous human behaviors can apply an important influence over the overall dynamics. The contents first focus on the enlightening the complexity features of the class of systems under consideration, subsequently analyzes how the contribution of mathematical sciences can be developed to explore the afore mentioned dynamics. Finally, future research perspectives are brought to the attention of the reader.

1 Motivations and Plan of the Chapter

This century keeps witnessing a growing interest from the sciences where the dynamics is influenced by human behaviors, occasionally called "*soft sciences*", towards a dialogue with mathematical sciences. The aim of this scientific interaction consists in improving the predictive and explorative ability of theoretical tools typical of each specific science.

This complex dialogue is not limited to exploit the variety of technical tools that mathematics can offer, but also induces new mathematical approaches and, in some lucky cases, also new theories in the field of mathematics.

The aforementioned dialogue has constantly pervaded the interaction between the so called "*hard sciences*", typically physics and chemistry, with mathematical

N. Bellomo
Collegio Carlo Alberto, Torino, Italy
e-mail: nicola.bellomo@polito.it

N. Bellomo
Politecnico of Torino, Torino, Italy

P. Terna (✉)
University of Torino, Torino, Italy
e-mail: pietro.terna@unito.it

L. D'Acci (ed.), *The Mathematics of Urban Morphology*,
Modeling and Simulation in Science, Engineering and Technology,
https://doi.org/10.1007/978-3-030-12381-9_14

sciences. The output has generally given robustness to theories in physics and chemistry and has allowed mathematics to grow and enrich its cultural patrimonium.

Indeed, new mathematical theories have been developed which otherwise would not have been achieved in absence of the hint given by applications. This fascinating aspect of scientific growth has pervaded, already in the past century, advanced technology, where physics, chemistry, computer sciences, and mathematics have opened a scientific dialogue toward a common objective.

The last chapter of the book Bellomo et al. (2017) has critically analyzed the abuse of terms such as *hard* and *soft* sciences and it has suggested that each scientific environment can generate a "science", while the term *science of living systems* should define an environment, where different sciences can be developed accounting for the complexity features of living systems.

Mathematics is a science made of assumptions (or axioms) and subsequent rigorous analysis devoted to enlighten the consequences of the said assumptions. Some of the axioms might not have a motivation given by applications, some of them might be directly induced by them. In the former case the output has found, at least in some cases, possible applications later in time, while in the latter case, the mathematical analysis has occasionally brought to revisions of theories developed in other sciences.

A variety of research contributions developed within the aforementioned interactions is offered by the book Bellomo et al. (2017) which is specifically devoted to the *mathematical theory of active particles* to be viewed as the main theoretical tool used in our chapter. Additional bibliography is given in the next sections also referring with alternative methods.

Bearing all above in mind, let us now move to the study of the dynamics of urban morphology, where the first step consists in understanding how human behaviors produce complex, even be not predictable outputs of a dynamics of pattern formation which cannot be rapidly put into a mathematical framework. Our chapter aims at tackling this key problem, where a selection of key issues, chosen according to the author bias, are presented within a framework of concepts to be possibly formalized into mathematical structures. This formalization might be firstly focused on well defined case studies and subsequently inserted into a unified mathematical theory.

> How "urban morphology" can be properly interpreted within the broad concept of sciences of living systems?
>
> Can the concept of "mathematical models" can be defined and how models can be classified and validated?
>
> Which are the complexity features of the dynamics of "urban morphology" and how mathematics can account for them?

The answer to these three key questions is given in the sequel. In more detail: Sect. 2 refers, thanks to a certain amount of imagination, to a free speculation on the "*Metamorphosis*" by Maurits Cornelis Escher (1898–1972).

Section 3 provides a concise introduction to the concept of mathematical models, of their role, and of their validation. The classification is also related to the mathematical structure of models. Subsequently, a modeling strategy is proposed and critical analyzed also on the basis of the structural properties of the models that can be developed.

Section 4 outlines the specific features of a possible new mathematical approach somehow related to the specific features of the dynamics of urban morphology. This section proposes some speculations on possible research perspectives, mainly focused on developments of the contents of the preceding section referring to an interdisciplinary approach on the modeling and simulation of urban morphology.

The selection of the three key questions does not claim to be exhaustive, as various possible alternatives and additions can be conceived. We have been guided by own bias and the conceptual-philosophical approach we have in mind. Hopefully, our reasonings might generate further speculations.

2 Reasonings on the Metamorphosis by Cornelis Escher

Let us use our fantasy and bias towards a possible interpretation of the dynamics of urban morphology through the celebrated artistic opera known as the *Metamorphosis* by Cornelis Escher. It is a nonsymmetric representation, where the height is very small with respect to the length. The left side mixed creative geometries to mutating stylized animals, while the very right side shows the transformation of a simple geometry into the complex one of a village. It is a continuous evolution from homogeneity to heterogeneity, where the village ends with a church and a bridge linking the village to a chess plate. Moving on towards the right, the Metamorphosis reaches again the same stylized geometry of the very left side.

The reader interested to a detailed oversight of the "metamorphosis" can obtain it by visiting the WEB site: https://www.google.fr/search?q=cornelis+escher+metamorphosis where various compositions of the original piece of art are reported.

Although it is a visionary representation, we can observe that the artist modifies an existing reality. In fact, the village exists in a Mediterranean coast of south Italy. It is Atrani in the so-called Amalfi coast. Figure 1 shows the real village over a end-part of Escher's visionary interpretation immersed into some images of the real village.

Let is now focus on the *first key question* on the connection between urban morphology and dynamics of living, hence complex, systems. Indeed we have in mind that a mathematical modeling should account for the time evolution of urban morphology by taking advantage of tools delivered by the so-called *science of living systems*. The know approaches, see Bellomo et al. (2017), start with a detailed analysis of the common complexity features that are typical of all living systems and subsequently mathematical tools are derived to capture, as far as it is possible, the said features. These tools are finally specialized over each specific system under consideration.

In more detail, still following Bellomo et al. (2017), the following features can be referred, in a low or great extent, to all living systems and hence to the dynamics urban morphology:

Fig. 1 The real village and Escher's visionary interpretation

1. Ability to chase a purpose or a strategy;
2. Heterogeneous expression of the said purpose/strategy;
3. Nonlinearly additive interactions evolving due to learning;
4. Darwinist selection and time as a key variable;
5. Complexity in the interpretation of reality and chasing the black swan.

Let us leave some freedom to our fantasy and take the liberty to look at the fascinating Escher's piece of art looking for a representation inside the "Metamorphosis" of the aforementioned complexity features.

Strategy: The population, probably looking for its own wellbeing, modifies the morphology of the village looking for a more appropriate organization of the living environment for a population which might be rapidly growing. This dynamics occurs under the action of the population according to the available spaces as well as to the shape of the territory which, in this specific case, is the end of a valley between steep mountains.

Heterogeneity: The presence of an heterogeneous development and growth of the village appears with evidence in the *Metamorphosis*. The appearance appears more and more evident while we move towards the right hand side. The main factor generating this feature might arguably be the social level of the inhabitants who look for houses related to their own wealth, in addition to their culture and bias.

Nonlinearity and learning: The evolution is related to interactions, definitely multiple ones. Is nonlinearity somehow expressed? This might even reflect a multiscale dynamics. In fact, it results as the output of the action from the micro-scale of individuals to the macro-scale of the village. As mentioned, the wealth of inhabitants plays an important role in their contribution to the morphology, in the interactions with the surrounding houses. Interactions are definitely nonlocal.

Darwinist selection: The evolution is selective as shown by the transition from essential shapes to an organized village, where all available spaces are well exploited. In addition, the presence of a church, which takes an important part of the space and a somehow central position, indicates the presence of a cultural evolution.

Complexity and rare events: Suddenly the landscape changes from a village, turning into a chess plate, the only connections being a Bridge and a tower. Can this sudden change be interpreted as a so-called "black swan"? The Tower can be interpreted as an early signal that an extreme event is going to happen. The various changes in the picture can be interpreted as predictable emerging behaviors, while the last one is not a predictable event.

The fifth Item deserves some further remarks first focusing on the difficulty of the interpretation of complex systems. The village looks at the sea. If we look at the village as shown by Fig. 1, definitely the tower cannot be observed in the real village. On the other hand, if we look at it from the back of the village at a certain distance, a tower appears as it is located on the cape in front of the village as shown in Fig. 2. Leaving the fantasy towards reality, Fig. 3 shows the tower as it appears on the cape in front of the village.

Focusing now on the black swan, we know that Cornelis Escher has gone through the experience of two world wars, where a peaceful village can even be involved into a battle between the two armies of the chess plate. In this vision, the bridge can be interpreted as an early signal of a dramatic change very difficult to be foreseen, but rationally interpreted once happened.

The definition by Taleb (2007) can contribute to enlighten this complex dynamics

> "A Black Swan" is a highly improbable event with three principal characteristics: it is unpredictable; it carries a massive impact; and, after the fact, we concoct an explanation that makes it appear less random, and more predictable, than it was.

History teaches us that the battle effectively took place on the September, 8 1943 from the sea on front of the village and on the territory inside the coast.

Let us now leave our fantasy and let us return to the aim of our chapter. Indeed, we aim at showing how mathematical sciences can contribute to the derivation of models suitable to describe the dynamics of urban morphology. It is not an easy task and we cannot indicate a well settled theory although the aforementioned objective is somewhat interpreted in various pieces of research. Without claim of completeness, we refer to the collection of scientific contributions edited in Ball (2012), where it is shown that we live in a complex society operating in a complex environment. The human behaviors and urban morphology is enlightened in D'Acci (2014, 2015).

On the other hand, despite the absence of a formal theory, mathematical sciences offer tools which can be properly addressed to the aim of this paper. For instance, nonlinear interactions can be modeled by theoretical tools of game theory (Allen and Nowak 2013; Camerer 2003; Gintis 2009; Nash 1996). These tools have been applied in different field of stochastic dynamics of self-propelled entities, for instance crowd dynamics (Bellomo and Bellouquid 2015). An interesting collective learning theory has been developed in Burini et al. (2016) and already applied in various fields of social sciences (Burini et al. 2017). The concept of darwinist mutations has been introduced in Ball (2012), D'Acci (2015) referring urban morphology to the classical theory of evolution and selection (Mayr 2001). The study of *tipping points* as precursors of sudden changes has been developed in Bissell et al. (2015). Finally, let us also mention a stone guest of urban developments, namely ethical principles fighting with the search of profit (Salvi 2016). Indeed this guest appears to be definitely difficult to account for, but it cannot be forgotten.

Fig. 2 The tower from the back of the real village

Fig. 3 The tower on front of the real village

Game theory should not only account for heterogeneous and even irrational behaviors, but it needs to be properly developed to account for a complex space dynamics. Hence, we look at the contents of Sect. 4 where agents and active particles methods are introduced (Bellomo et al. 2017; Boero et al. 2015; Galam 2013; Pareschi and Toscani 2013).

3 Mathematical Models for Complex Systems

This section proposes an answer to the *second key question* by explaining in a plain and tutorial manner what a mathematical model is, how it can be validated and subsequently used. Therefore, we continue on our effort to build a bridge to bring the variety of tools that mathematics can offer towards the needs of experts in the field of urban morphology.

A sequence of definitions and concepts are brought to the attention of the reader's attention by a presentation which keeps avoiding an excess of formalization, however consistent with the conceptual rigor of mathematics. These can be viewed also as a concise handbook to be used as a users' guide towards the derivation and applications of mathematical models. In addition, some speculations on the possible way to develop a dialogue between the aforementioned research fields are proposed.

• **Mathematical models**: The dynamics of real systems can be described, at a virtual level, by mathematical models which consists in equations, generally—but not necessarily—of differential type, deemed to describe the dynamics in time and space of a variable, which is vector valued, charged to model the state of the system under consideration. If one looks at the differential system, this variable is the so-called *dependent variable* of an equation, where time and space are so-called *independent variables*. If one looks at the applications, the dependent variable, is also called *state variable* as it describe the state of the overall system.

• **Classification of models**: A possible classification of mathematical models can be related to the structure of the state variable and hence of the equation describing the dynamics. Focusing on differential models, a simple classification is as follows:

1. *Discrete state at the super-macroscopic scale*: The state variable depends on time only, while the dynamics is described by ordinary differential equations, corresponding to the components of the state variable, where its time derivative is induced by a cause, to be properly modeled, which nonlinearly depends on the state variable.
2. *Discrete state at the microscopic scale*: The overall system is constituted by several interacting entities, while the state variable, which depends on time only, is given by the whole set of all their individual states. The dynamics is modeled as in Item 1, by a large system of ordinary differential equations corresponding to all variables at the microscopic state. These models are also called *individual-based*.
3. *Continuous state—macroscopic scale*: The state variable, which can be a vector, depends on time and space being given by locally averaged quantities for a large system, viewed as a continuum, constituted by several interacting entities. The dynamics is defined by partial differential equations derived by equilibrium or conservation relations within the elementary volume of space.
4. *Stochastic models*: These models account for random fluctuations in the real system by adding a noise to the deterministic models defined in Items 1–3. The noise depends on time in the case of discrete states models and on time and space in the case of continuous models.
5. *Kinetic theory models*: The overall system is constituted by several interacting entities whose individual state is called *microscopic state*, while the state, dependent, variable is a probability distribution function over the microscopic state. The dynamics is described by a differential system derived by conservation relations within the elementary volume of the space of the microscopic sates. These models are derived in the broad framework of the so-called *kinetic theory of active particles*.

The action of the external world can be included, for all type of models, in the cause. An additional classification, which applies to all class of models, can be based on the specific aim by which models are designed and used. In more detail, we can define the following two main categories:

1. *Predictive models*: which aim at foreseeing the behavior in time of systems given appropriate initial conditions and, if needed, constraints on the solutions. These

models should reproduce, qualitatively and quantitatively, the dynamics by solution of mathematical problems.

2. *Explorative models*: which aim at exploring the behavior in time of systems under programmed variation of the parameters and of external actions. So that the scenario of a huge variety of possible outputs can be depicted.

• **Validation of models**: In general, validation of models is based on their ability to reproduce empirical data. Additional difficulty has to be tackled in the case of living systems as the validation process consists in verifying the following:

1. Ability of models to capture the complexity features of the system object of the modeling. This first step amounts to derive a mathematical structure designed for the derivation of models. Namely, the structure should show the ability to capture the complexity features that appear relevant for a certain class of systems.
2. Ability to reproduce "quantitatively" available empirical data. This step, as the next one, is operative if and only if the first step has been fully investigated. This step can be operative if and only if the requirement of the first item has been fulfilled. However, agreement with available empirical data is a necessary, but not sufficient condition for validation.
3. Ability to reproduce "qualitatively" observed collective emerging behaviors. Particularly interesting is the analysis of large deviations that might correspond to breaking of the observed patterns. A not expected case of emerging behavior is the appearance of non-predictable events with the characteristics of the previously mentioned *black swan*, see also Townsend (2013).

A key aspect of model validation is the solution of inverse problems towards the assessment of parameters which are somehow related to interactions. The effort to achieve this task can be reduced by using minimal models which include a number of parameters equal to the number of different types of interactions. However, reducing the complexity of the model means reducing also the significant number of interactions, hence this reduction generally amounts to weaken the predictive ability of models.

• **Nonlinearity and mathematical tools**: Nonlinearity appears whenever the differential system shows terms which are not linear functions of the dependent variable. Models may include different types of nonlinearity. For instance power functions of the dependent variable related to interactions (for binary interaction the power is equal to two), and parameters function of the dependent variables which appear whenever the interaction rules involve, as it naturally happens, the dependent variable.

Dealing with nonlinear models makes difficult, or even impossible, obtaining analytic solutions, hence numerical methods are required. These are strongly related to the selection of the class of equations which are deemed to support the derivation of models. Referring to the classification of mathematical models, deterministic models can be used to deal with models defined in Items 1–3, while stochastic particle methods have to be used in the case of models in Items 4–5. The latter approach might require Monte Carlo particle simulation tools (Barbante et al. 2015; Dimarco and Pareschi 2016; Pareschi and Toscani 2013). In both cases advanced computational tools are needed.

• **Modeling strategy**: Let us now define a general strategy towards the selection and application of mathematical tools to model the dynamics of urban morphology. This specific aim as well as the requirements towards their validation suggest the following sequential steps:

1. Selection the complexity features which play an important role in the dynamics of the specific systems under consideration and understanding how these features can influence the dynamics.
2. Selection of the scale and class of equations which appear appropriate towards the derivation of models, followed by the derivation of a general mathematical structure suitable to capture, as far as it is possible, the complexity features selected as in Item 1.
3. Derivation of models by implementing the said structure both with suitable models of individual-based interactions, according to a detailed interpretation of the dynamics at the micro-scale, and with models related to the interactions with the external environment.
4. Validation of models by comparison of the predicted dynamics with empirical data and development of a parameter sensitivity analysis is necessary to identify the parameters which have an effective influence on the overall dynamics.
5. Application of models to the study of the specific system of urban morphology which is object of the modeling approach focusing not only on the dynamics of the morphology, but also to the possible onset of large deviation.

It is plain that an interdisciplinary dialogue is necessary to support this strategy. In fact, Steps 1, 4, and 5 require both type of expertise, namely that of modelers and that of experts in the field of urban morphology. Once this dialogue exists, the strategy can be put in practice. Some perspective ideas are presented in the next section.

4 Towards a Mathematics for Urban Morphology

Let us now consider the *third key question* focused on the design of mathematical tools towards modeling and computation. The contents of the preceding section has enlightened the need of deriving a general structure consistent with the complexity features of each specific system and appropriate to offer the conceptual framework towards the derivation of specific models which have to be validated.

Based on these requirements, we claim that agent-based and active particles methods should be selected towards the aforementioned strategy. Accordingly, this sections presents, in the next two subsections, an introduction to both the said modeling approaches, while a third subsection enlightens the motivation of this selection and looks ahead to research perspectives.

The presentation delivers, as in the preceding sections, conceptual issues and avoids mathematical formalizations. The main objective consists in delivering to urban morphologists the basis to develop a formalized approach which, at present, does not yet provides, in the pertinent literature, exhaustive and self-consistent mathematical tools.

Let us remark that the first key question requires an assessment of the complexity features of urban morphology. The answer is that we confirm those listed in Sect. 2 and referred to Escher's Metamorphosis. We simply remark that the system is highly sensitive to external actions which can be rational and/or irrational and might be affected by economical choices and strategies. In addition different frameworks and needs of the society have to be accounted for, including the search of a sustainable urbanism (Salat 2011; Salat and Bourdic 2011).

4.1 Agent-Based Modeling Tools

Following—with modifications—the two pioneers of the Agent-Based Models (ABMs) field, Axtell and Epstein, particularly in the description of the ABMs that they introduce in Axtell and Epstein (1999), let us describe an ABM in a few steps.

1. The starting point is a population of agents, representing individuals or more generally *entities*, as the component of a generic system, that we construct using small parts of computer code operating in dedicated software environments.
   In a parallel way, we can take in consideration, as strictly related to the content of this chapter, the Batty (2009) definition of ABMs as "systems composed of individuals who act purposely in making locational/spatial decisions." This is especially relevant for cities, that "have been treated as systems for fifty year but only in the last two decades has the focus changed from aggregate equilibrium systems to more evolving systems whose structure merges from the bottom up."
2. The goal is to search for regularities at the macro level generated by the behavior of the agents (micro level, if individuals; *meso* level, if more complex entities).
3. An ABM does not introduce equations governing the effects of the agents' behavior at the macro level, but it allows us to observe the emergence of those effects, e.g., summing up the outcomes or representing them graphically.
4. Agents, on their side, can use equations to assume their decisions; equations can be very complicated, e.g., in the case of artificial neural networks. Agents can also learn from their errors, modifying the internal equations or neural networks.
5. Heterogeneity is usual in ABMs and, as a consequence of this paradigm, the agents have different equations or neural networks.

6. In any case, the suggestion is that of keeping the agents quite simple; we remember Axelrod's KISS (Keep It Simple, Stupid) principle, derived from engineers' technical jargon.
7. With ABMs we can manage boundedly rational behavior, nonequilibrium dynamics, and spatial processes.
8. Regularly, the coding techniques are based on object-oriented programming and build the agents using instances of classes, where a good class synonym is *set*.

We have many shells helping us in coping with step 8, from simple and quite intuitive, as NetLogo (https://ccl.northwestern.edu/netlogo/) to more complicated and powerful, mainly derived by Swarm (from Santa Fe Institute, in the middle of the 1990s). One example is SLAPP (https://terna.github.io/SLAPP/) described in Boero et al. (2015).

The roots of ABMs stays in the cellular automata idea that Stanislaw Ulam and John von Neumann had in the 1940s; popularized by the Game of life, created by John Conway and diffused by Martin Gardner in a celebrated article in Scientific American (Oct. 1970). Cellular automata are represented by a regular grid of cells usually with two statues (1/0; on/off); cells can change their internal value over time, as a consequence of the statuses of their neighborhood. This kind of structure very well applies to the study of urban morphology dynamics.

In a modern ABM structure, a dynamic spatial variable may be thought of as a cellular automaton, although somewhat more general. Most of all, in NetLogo we have the possibility of running the status of the patches the fill the space in which the agents (*turtes*, in jargon, due to the Logo origin) behave.

An important detail: between classical cellular automata and ABMs we can also remember the models inspired to Schelling's famous paper Schelling (1971) on segregation. In this family of models, agents strictly behave as a consequence of the presence/absence of their neighborhood. Again a structure suitable for applying to urban morphology.

Finally, ABMs fully adhere to the abilities required above for the validation of a model. From the formal point of view, they are close to a narrative of the reality, thanks to their flexibility; but most of all, they are close to a mathematical model, thanks to a rigorous representation via a computer code.

4.2 Active Particles Kinetic Modeling Tools

A natural development of agents method is offered by the so-called *kinetic theory of active particles*. Theory, mathematical tools and applications are reviewed in the book Bellomo et al. (2017), which is the main reference to this topic.

Applications have been addressed to modeling social dynamics Marsan et al. (2016), behavioral economy (Dolfin et al. 2017), learning theory (Burini et al. 2017), as well as to the modeling of systems where space dynamics can have an important role such as crowd dynamics (Bellomo and Bellouquid 2015) and vehicular traffic (Bellouquid et al. 2012; Burini et al. 2017).

An interesting historical and philosophical overview on agent methods and of their initial developments is given in the survey (Hegselmann et al. 2017). Agents methods have been recently followed by the onset of active particle methods. However, possible developments towards modeling of urban morphology are, at present, only an research plan which will be properly introduced in the next subsection.

Bearing this brief introduction in mind, let us summarize, with reference to Chaps. 3 and 4 of Bellomo et al. (2017), the main features of the method and of the modeling approach specialized for a system steadily localized in the space, but possibly expanding in space.

1. **Role of the space variable**: The urban system is distributed on a domain which can be either fixed or might evolve, generally expanding in time, starting from an initial domain, towards green areas.
2. **Representation of the system**: Which corresponds first to the assessment of the functional subsystems into which the various components of the system can be subdivided according to their functional use, for instance habitations, offices, public services, etcetera. Subsequently, to the selection of the variables, which define the so-called *state at the microscopic scale* or shortly *micro-state*, deemed to describe locally the state of the system. The state of the system is then described by a time dependent probability density over the said micro-state.
3. **Derivation of the mathematical structure**: The phenomenological interpretation contributes to derive a mathematical structure corresponding to the modeling the dynamics of urban morphology. This structure is required to have the ability to account for the complexity features presented in Sect. 2.
4. **Modeling interactions:** Interactions related to the structure mentioned in Step 3 are modeled by theoretical tools of evolutionary stochastic game theory, see Chap. 3 of Bellomo et al. (2017), where the term *stochastic* is used as interactions involve probability distributions. In addition, the modeling should be referred also to the interactions between the system and external actions which can have an important influence on the dynamics. External actions can generate mutations and selections which might end up with the strengthening or weakening of functional subsystems.
5. **Derivation of models**: Implementing these models of interactions into the structure defined in Step 3 generates the mathematical model. The derivation of models should be followed by a parameter sensitivity analysis is necessary to identify the parameters which have an effective influence on the overall dynamics as well as a validation of models by detailed analysis of their predictive ability to depict quantitatively empirical data, when available, and qualitatively observed emerging behaviors.

The validation strategy can be viewed in two steps, firstly a mathematical structure is designed with the ability of reproducing the complexity features of the class of systems under consideration and subsequently by a toning of the model by assessment of its parameters. with the aim of testing the predictive ability of the model.

In more detail, one can explore their ability to reproduce "quantitatively" available empirical data and "qualitatively" collective emerging behaviors which are observed by experiments. This strategy is independent on the specific tools used for the modeling. Hence can be referred to both approaches enlightened in Sects. 4.1 and 4.2.

In general, validation of models refers to the ability of models to reproduce the dynamics of emerging behaviors and their dependence on the variation of parameters and on the action of the external environment. In addition, validation can refer to predictive ability models which, after validation, can be used also as explorative ones. Once a model is derived simulations might put in evidence emerging behaviors not yet observed. This event should suggest further search of empirical data that might, or not, confirm the presence of the aforementioned event.

A very special case of emerging behavior is the appearance of non predictable events with the characteristics of the so-called *black swan*. A model might hopefully put in evidence this event for certain choice of the parameters of the model. This very special event is not easily predictable, however a successful model might put in evidence, as mentioned, tipping points which anticipate the arrival of this particular onset by sudden sharp changes (Bissell et al. 2015). The author of Taleb (2007) is very critical towards mathematics whenever focused on models which predict what is easily predictable. However, mathematical sciences have elaborated various visionary ideas on complex systems and complexity that nowadays are object of great attention, see as an example among various ones (Casti 1994).

The overall validation process can take advantage of the record of the dynamics before the development of the modeling approach. This very special feature might lead to the assessment of the parameters towards the effective validation based on comparison with empirical data according to the strategy which has been given above. Collection of data, at well defined time intervals, is not an easy task. In fact, these data are not always available and does not always provide data consistent, hence useful, with the modeling approach. It follows that further research perspectives cannot be simply limited to modeling, but should also be focused on the organization of databases suitable to store useful data. This topic is further expanded in the next section.

4.3 Towards Research Perspectives

A conceptual analysis on the possible applications of mathematics and computer sciences to the study of urban morphology has been presented in our chapter. Summarizing the contents: Sect. 2 has identified some of the most relevant complexity features of the overall system, where human behaviors can play an important role in the dynamics and development of the system under consideration. A possible strategy has been proposed in Sect. 3 consistently with the aim of accounting, in the modeling approach, for the said complexity features. Two different, however technically related, modeling and simulation approaches have been examined in this section as the natural materialization of the said strategy.

The overall contents of this chapter have shown that the mathematical tools reviewed in Sects. 4.1 and 4.2 appear as an interesting, however challenging, perspective. Nevertheless, these approach cannot be straightforwardly applied without appropriate developments to be planned in a well focused research program. Some reasonings on these developments are proposed in the following by selecting, among various conceivable ones and without any claim of completeness, three key topics which are, as we shall see, correlated among them.

Our reasoning account for the idea that urban morphology approaches human settlements by products that emerge over long periods by means of sequential generations of building activity. The study and critical analysis of these traces can be considered the key question of urban morphology, thus going beyond architecture and looking at the entire built landscape and its internal logic. This statement is commonly share and strongly motivates possible application of mathematics. Conversely, mathematical models are required to give information on the aforementioned traces.

• **Multivariable description, interaction of different dynamics and open systems**: The greatest part of models make use of scalar variables to define the state of agents or active particles. However, the description of the main features of urban morphology needs multi-component variables to account for the effective state of real systems, namely not only size, but also use, style, shapes of the buildings, as well as to account for additional features induced by the specific study which is developed. The theoretical solution of this problem can be rapidly formally achieved, but the practical application requires additional work not only to define the variables which play an important role, but also their hierarchy in terms of importance up to the derivation of mathematical structures deemed to account for them.

Possible examples can be obtained by the so-called *town-plan analysis* developed by the pioneer work by Conzen (1969), who suggests to account for three elements that in mathematical language would be the components that define the state of agents or active particles. Namely, *streets and their arrangement into a street-system*, *plots (or lots) and their aggregation into street-blocks*, and *buildings, in the form of the block-plans*. More recent approaches suggest to account for a variety of additional feature, nevertheless the ideas proposed in Conzen (1969) contain *in nuce* the perspective of a systems approach to urban morphology. At present, we suggest to mathematicians to look at the books by Batty (2009, 2013) which tech us also about the complexity of urban areas.

An amazing coincidence is that the cover (Batty 2009) precisely refers to the last part of the Metamorphosis introduced also in Bellomo et al. (2013) as a visual paradigm of complexity.

In addition, multi-component state variable allow to take into account also interactions of different dynamics, for instance social dynamics and morphology. Social dynamics can exert not only an influence on the growth of the urban area, but also can modify previous areas by a darwinist type selection as described in D'Acci (2014, 2015).

Unfortunately, social dynamics might also involve corruption in our society (Salvi 2016) and introduces irrational behaviors that have, however, to be included in the modeling approach. Accounting for this feature is very important towards the design of predictive models which might, ultimately, in evidence the onset of possible disasters.

Finally, let us stress that it is plain that a urban system is open to external actions which depend on the specific features of the areas to be studied. These actions generally act on the system at the individual-based scale. Therefore, the modeling approach can include them by models developed at the same scale used for the endogenous system.

• **On the use of big data**: Recently the idea of using big data to mine the future has attracted applied mathematicians and statisticians who chase prediction of future events. The so-called *big data* are stored in large databases which are everywhere although most of the applications refer to social and economical sciences, see as possible examples (McKinsey Global Institute 2001; OECD 2015). This matter belongs also to private enterprizes interested to exploit large databases towards economical initiatives.

The key feature consists in selecting a metric appropriate to define the distance between simulations stored in the database and the variables of the real case study, here we can take as an example the approach (Conzen 1969). This distance should be related to the variables selected to describe the system according to the contents of the first system as well as of studies of urban features, see for example (D'Acci 2014, 2015), where this concept is introduced and used to compare areas with different features.

Therefore the database should be closely related to specific models used for the simulations and include past history which contribute to the toning of models. These, after the assessment of parameters, can be used for predictive purposes.

• **Moving boundary problems and computational tools**: Mathematical problems generated by applications of predictive models can be stated as problems where the dependent variable is distributed in space either within space confined by fixed boundaries or with moving boundaries, which corresponds to a system expanding over nonurban areas. In the latter case the moving boundary is an unknown of the problem, but boundary conditions do not generally need to be implemented whenever the inner system is not conditioned by the external environment.

As mentioned, predictions of models are generally obtained by simulations as, due to various nonlinearities, analytic solutions are very rare and limited to very simple problems. Hence computational methods have to be developed focusing on the class of models and problems under consideration. This requirement is valid both for agent methods and for the active particles approach. The former can take advantage of efficient methods, with bounded complexity, that have been tested in different applications. The latter requires more advanced tools consisting in simulating the dynamics of very many real particles and by obtaining macroscopic variables by suitable averaging processes.

This method, called *monte carlo particle method*, has been generated by the pioneer ideas by Bird (1994) and subsequently developed by other authors Aristov (2001), Dimarco and Pareschi (2016). Recently, it has recently developed towards the simulation of particles which posses both mechanical and social features with applications to crowd dynamics Bellomo and Bellouquid (2015). Further developments are needed to make the computational tools applicable to kinetic type equations modeling the dynamics of urban morphology.

Our reasonings on the three key topics, which have been treated in this subsection, do not naively claim to be that the overall approach can be based on a straightforward application of known techniques. However, a research plan should account for all of them as it is a necessary step towards a fruitful dialogue between experts in urban morphology and mathematicians. These two different types of scientists should agree upon the various assumptions that generate a model. Subsequently, mathematicians should obtain simulations by computing for case studies suggested by experts in urban morphology, who should provide an interpretation of what is shown by simulations towards an improvement of the quality of urban areas. Indeed, this is the key feature of any interdisciplinary research.

It is an ambitious and challenging program which goes towards a systems approach to urban morphology as anticipated by visionary ideas proposed by various authors, for instance (Ball 2012; Batty 2009, 2013; D'Acci 2014). It is not an easy task, but it is worth trying also for the benefit of our society.

References

G. Ajmone Marsan, N. Bellomo, and L. Gibelli, Stochastic evolutionary differential games toward a systems theory of behavioral social dynamics, *Math. Models Methods Appl. Sci.*, **26** 1051–1093, (2016).

B. Allen and M.A. Nowak, Games on networks, *EMS Surveys Math. Sci.*, **1**, 113–151, (2013).

V.V. Aristov, Direct Methods for Solving the Boltzmann Equation and Study of Nonequilibrium Flows, Springer-Verlag, New York, (2001).

R. Axtell and J.M. Epstein, Coordination in Transient Social Networks: an Agent-Based Computational Model of the Timing of Retirement, in H.J. Aaron ed., **Behavioral dimensions of retirement economics**. Brookings Institution Press, Washington, (1999).

P. Ball, **Why Society is a Complex Matter**, Springer-Verlag, Heidelberg, (2012).

P. Barbante, A. Frezzotti, and L. Gibelli, A kinetic theory description of liquid menisci at the microscale, *Kinet. Relat. Models*, **8(2)**, 235–254, (2015).

M. Batty, Cities as complex systems: Scaling, interactions, networks, dynamics and urban morphologies, in R.A. Meyers ed., **Encyclopedia of complexity and systems science**. Springer-Verlag, Heidelberg, 1041–1071, (2009).

M. Batty, **The New Science of Cities**, MIT Press, (2013).

N. Bellomo and A. Bellouquid, On multiscale models of pedestrian crowds from mesoscopic to macroscopic, *Comm. Math. Sciences*, **13(7)**, 1649–1664, (2015).

N. Bellomo, A. Bellouquid, L. Gibelli, and N. Outada, **A Quest Towards a Mathematical Theory of Living Systems**, Birkhäuser, New York, (2017).

N. Bellomo, P. Degond, and E. Tadmor, Eds., **Active Particles, Volume 1 - Advances in Theory, Models, and Applications**, *Modeling and Simulation in Science Engineering and Technology*, Birkhäuser, New York, (2017).

N. Bellomo, D. Knopoff, and J. Soler, On the difficult interplay between life, "complexity", and mathematical sciences *Math. Models Methods Appl. Sci.*, **23**, 1861–1913, (2013).

A. Bellouquid, E. De Angelis, and L. Fermo, Towards the modeling of vehicular traffic as a complex system: A kinetic theory approach, Math. Models Methods Appl. Sci. **22**, paper n. 1140003, (2012).

G.A. Bird, "Molecular Gas Dynamics and the Direct Simulation of Gas Flows", Oxford University Press, (1994).

J.J. Bissell, C.C.S. Caiado, M. Goldstein, and B. Straughan, **Tipping Points: Modelling Social Problems & Health**, Wiley, London, (2015).

R. Boero, M. Morini, M. Sonnessa, and P. Terna, **Agent-based Models of the Economy From Theories to Applications** Palgrave Macmillan, (2015).

D. Burini, S. De Lillo, and G. Fioriti, Influence of drivers ability in a discrete vehicular traffic model *Int. J. Modern Phys. C*, **28(3)**, paper n. 1750030, (2017).

D. Burini, S. De Lillo S., and L. Gibelli, Stochastic differential "nonlinear" games modeling collective learning dynamics, *Phys. Life Rev.*, **16(1)**, 123–139, (2016).

D. Burini, L. Gibelli, and N. Outada, A kinetic theory approach to the modeling of complex living systems, in **Active Particles, Volume 1. - Advances in Theory, Models, and Applications** *Modeling and Simulation in Science Engineering and Technology*, Birkhäuser, New York, 229–258, (2017).

J. Casti, **Complexification: Explaing a Paradoxical World Through the Science of Surprise**, Harper-Collins Publisher, New York, (1994).

C.F. Camerer, **Behavioral Game Theory: Experiments in Strategic Interaction**, Princeton University Press, (2003).

M.-R.-G. Conzen **Alnwick, Northumberland: A Study in Town-plan Analysis**, Institute of British Geographers, London, (1969)

L. D'Acci, Monetary, subjective and quantitative approaches to assess urban quality of life and pleasantness in cities - Hedonic Price, Willingness-to-pay, Positional Value, life satisfaction, Isobenefit lines, *Social Indicators Research*, **115**, 531–559, (2014).

L. D'Acci, Mathematize *Urbes* by humanizing them. Cities as isobenefit landscapes: psycho-economical distances and personal isobenefit lines, *Landscape and Urban Planning*, **139**, 63–81, (2015).

G. Dimarco and L. Pareschi, Numerical methods for kinetic equations, *Acta Numerica*, **23**, 369–520, (2016).

M. Dolfin, L. Leonida, and N. Outada, Modelling human behaviour in economics and social science, *Phys. Life Rev.*, **2223**, 1–21 (2017).

J.M. Epstein and R. Axtell, **Growing Artificial Societies: Social Science from the Bottom up**, Brookings Institution Press, Washington, (1996).

S. Galam, **Sociophysics**, Springer, New York, (2013).

H. Gintis, **Game Theory Evolving**, 2nd Ed., Princeton University Press, Princeton NJ, (2009).

R. Hegselmann, Thomas C. Shelling and James M. Sakoda: The Intellectual, Technical and Social History of a Model, *J. Art. Society and Social Simul.*, **20(3)**, 15, (2017) https://doi.org/10.18564/jass.5311.

McKinsey Global Institute, Big data: the next frontier for innovation, competition and productivity, www.mckinsey.com/insights/business-technology/big-data-the-next-frontier-for-innovation, (2001).

E. Mayr, The phylosophical foundation of Darwinism, *Proceed. Am. Phylos. Society*, **145**, 448–495, (2001).

J. Nash, **Essentials of Game Theory**, Elgar, Cheltenham, (1996).

L. Pareschi and G. Toscani, **Interacting Multiagent Systems: Kinetic Equations and Monte Carlo Methods**, Oxford University Press, Oxford, (2013).

S. Salat, **Cities and Forms: On Sustainable Urbanism**, Hermann Ed., (2011).

S. Salat and L. Bourdic, Power laws for energy efficient and resilient cities, *Procedia Engineering*, **21**, 1193–1198, (2011).

S. Salvi, Corruption corrupts: Society-level rule violations affect individuals' intrinsic honesty, *Nature*, **531(7595)**, (2016), 456–457.

T.C. Schelling, Dynamic models of segregation, *J. of Mathematical Sociology*, **1(2)**, 143–186, (1971).

N.-N. Taleb, **The Black Swan: The Impact of the Highly Improbable**, Random House, New York City, (2007).

A.-M. Townsend, **Smart Cities - Big Data, Civic Hackers, and the Quest for a New Utopia**, (2013).

OECD, Organization for Economic Co-Operation and Development, Paris, France. Data Driven Innovation, Big Data for Growth and Well-Being, *OECD Publishing,* www.oecd.org/sti/ieconomy/data-driven-innovation, (2015).

A Topological Representation for Taking Cities as a Coherent Whole

Bin Jiang

It is possible that one day computer programs designed for cognition might be able to pick out these centers and rank order them by their degrees of life....
Alexander (2002–2005, Book 1, p. 365).

Abstract A city is a whole, as are all cities in a country. Within a whole, individual cities possess different degrees of wholeness, defined by Christopher Alexander as a life-giving order or simply a living structure. To characterize the wholeness and in particular to advocate for wholeness as an effective design principle, this paper developed a geographic representation that views cities as a whole. This geographic representation is topology-oriented, so fundamentally differs from existing geometry-based geographic representations. With the topological representation, all cities are abstracted as individual points and are put into different hierarchical levels, according to their sizes and based on head/tail breaks—a classification and visualization tool for data with a heavy-tailed distribution. These points of different hierarchical levels are respectively used to create Thiessen polygons. Based on polygon–polygon relationships, we set up a complex network. In this network, small polygons point to adjacent large polygons at the same hierarchical level and contained polygons point to containing polygons across two consecutive hierarchical levels. We computed the degrees of wholeness for individual cities, and subsequently found that the degrees of wholeness possess both properties of differentiation and adaptation. To demonstrate, we developed four case studies of all China and UK natural cities, as well as Beijing and London natural cities, using massive amounts of street nodes and Tweet locations. The topological representation and the kind of topological analysis in general can be applied to any design or pattern, such as carpets, Baroque architecture and artifacts, and fractals in order to assess their beauty, echoing the introductory quote from Christopher Alexander.

B. Jiang (✉)
Faculty of Engineering and Sustainable Development, Division of GIScience,
University of Gävle, SE-801 76 Gävle, Sweden
e-mail: bin.jiang@hig.se

L. D'Acci (ed.), *The Mathematics of Urban Morphology*,
Modeling and Simulation in Science, Engineering and Technology,
https://doi.org/10.1007/978-3-030-12381-9_15

Keywords Wholeness · Natural cities · Head/tail breaks · Complex networks · Scaling hierarchy · Urban design

1 Introduction

Geographic representation or representing the Earth's surface constitutes one of the main research areas in geographic information systems (GIS) and science. Existing geographic representations, such as raster and vector, are essentially geometry-based because of involved locations, directions, and sizes. These representations are mainly driven by a static map metaphor (Yuan et al. 2005; Goodchild et al. 2007) for representing and characterizing the Earth's surface. They can be used to measure things and to assess the relationship of things at local scales, such as spatial dependence. However, things are not measurable, or measurement depends on the measuring scales because of the fractal nature of geographic phenomena (e.g., Goodchild and Mark 1987; Batty and Longley 1994; Chen 2012). Spatial dependence is just one of two spatial properties. The other is spatial heterogeneity across all scales. Geometry-based representations fail to capture the spatial property of heterogeneity across all scales (Jiang et al. 2008). To better understand geographic forms or cities' structure, we must rely on topology-based representations, e.g., the topological representation of named or natural streets (Jiang and Claramunt 2004; Jiang et al. 2008). This paper develops another topological representation that views cities as a whole. We refer to topology as an instrument that enables us to see the underlying spatial heterogeneity, or scaling property of far more small things than large ones. The topology used in current GIS, however, is essentially geometry-based, referring to adjacent relationships of basic geometric elements such as points, lines, polygons, and pixels.

The geometry-based topology is a key principle for defining efficient data structure with adjacent relationships stored in databases, which can significantly reduce data storage, and subsequently speed up search and retrievals. The topology or planar topology is a rigorous method for identifying digitizing errors such as self-intersecting polygons, islands, overshoots and undershoots, and slivers (Corbett 1979; Theobald 2001). One typical example is the topologically integrated geographic encoding and referencing (TIGER), which was developed by the US Census Bureau in the 1970s. With the TIGER, adjacent geometric elements, such as points, lines, and polygons, are clearly defined and stored in the data structure. For example, two polygons can have one of eight possible relationships: Disjoint, contains, inside, equal, meet, covers, covered by, and overlap (Egenhofer and Herring 1990). Another use of topology in the GIS literature is with cartograms, in which geometric aspects such as distances and areas are dramatically distorted, but the underlying topological relationships are retained. The most well-known example of cartograms is the London Underground map, devised in 1933 by Harry Beck. Both topological data structures and cartograms still retain some geometry. In the London Underground map, the geographic locations of stations are still relatively correct, although the distances between stations are distorted. With both topological data structures and cartograms, one, however, cannot see the underlying scaling of far more small things than large ones; see Jiang et al. (2008) for details. This makes our use of topology unique—the topology among meaningful geographic features, such as streets, cities, buildings, mountains, and rivers, because the topology enables us to see the underlying scaling property or hierarchy.

We put forward a topological representation that takes cities as a coherent whole and enables us to see their underlying scaling hierarchy. This representation makes it possible for a configurational analysis based on the concept of wholeness (Alexander 2002–2005; Jiang 2015b, 2016). See Sect. 2 for an introduction to wholeness and its 15 fundamental properties. Based on the configurational analysis, we can see clearly that individual cities are differentiated from and adapt to each other to form a coherent whole. With the topological representation, we can compute the degrees of wholeness or beauty of individual cities for better understanding cities' structure. We develop some useful visualization methods (c.f., Sect. 4 for case studies) to illustrate the underlying scaling of cities in terms of their sizes and their degrees of wholeness. We further demonstrate the properties of differentiation and adaptation of cities and provide eventually analytical evidence for and insights into the structure-preserving or wholeness-extending transformations (Alexander et al. 1987, 2002–2005) as the fundamental urban design principles.

Section 2 introduces the concept of wholeness or living structure and its 15 geometric properties. Using 100 city sizes that strictly follow Zipf's law (1949), Sect. 3 illustrates the topological representation for viewing cities as a coherent whole. To further demonstrate how the topological representation adds new insights into the cities' structure, Sect. 4 reports two major properties of differentiation and adaptation from case studies by applying the representation to China and UK natural cities. Section 5 further discusses the implications of the topological representation and wholeness for planning cities with a living structure. Finally, Sect. 6 draws a conclusion and points out to future work.

2 Wholeness or Living Structure, and Its 15 Fundamental Properties

Wholeness is defined as the structure that exists in space by all various coherent entities (called centers) and how these centers are nested in, and overlap with, each other (Alexander 2002–2005). Thus, wholeness is a phenomenon that is inherent to space or matter, rather than just what is perceived as the gestalt (Köhler 1947). For the sake of simplicity, we use the Sierpinski carpet as a working example for introducing the concept of wholeness. The carpet is created iteratively, or generated, by removing one-ninth of the square of size 1 by 1 (Fig. 1a). In theory, this process continues indefinitely, but we limited it here to three scales: 1/3, 1/9, and 1/27. The different squares of the carpet are not fragmented but are a whole with a high degree of wholeness, called *living structure*. However, the carpet is not a typical living structure

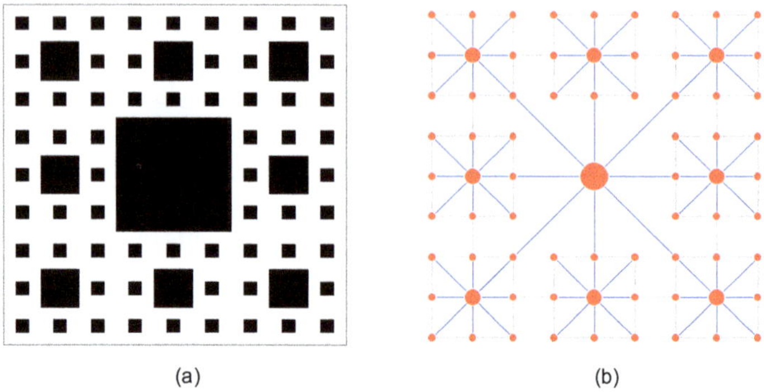

Fig. 1 The Sierpinski carpet acting or perceived as a whole (*Note* The Sierpinski carpet with the three hierarchical scales: 1/3, 1/9 and 1/27 (**a**) is represented as a complex network as a whole, consisting of many individual nodes (**b**), in which the dot sizes represent the degrees of wholeness, and the gray and blue lines indicate relationship respectively within a same scale and across two consecutive scales.)

Table 1 The 15 properties of the living structure or wholeness (Alexander 2002–2005)

Levels of scale	Good shape	Roughness
Strong centers	Local symmetries	Echoes
Thick boundaries	Deep interlock and ambiguity	The void
Alternating repetition	Contrast	Simplicity and inner calm
Positive space	Gradients	Not separateness

often seen in reality, given its strict definitions or exactitude. It is strictly defined both globally and locally in terms of the square shapes. It is strictly defined by the constant, exact 1/3 scaling ratio. This strictness or exactitude makes the carpet different from most living structures in reality, such as trees, mountains, coastlines, and cities. These real-world living structures are transformed—so-called wholeness-extending transformations—under the guidance of the 15 geometric properties, which are also called transformations (Table 1), whereas the carpet is created by following some strict mathematic rules. Despite the strictness, the carpet shares the same spirit of wholeness or living structure. A living structure is a geometrical coherence made of, or more precisely transformed by, the 15 fundamental properties. As a living structure, the carpet possesses many of these properties, although it also misses some. The carpet contains one, eight, and 64 squares respectively at three levels of scale 1/3, 1/9, and 1/27, indicating that there are far more small squares than large ones. These squares are not fragmented but constitute an interconnected whole (Fig. 1b).

The notion of far more small things than large ones refers to a recurring pattern so should be rephrased more truly as numerous smallest things, a few largest things, and some in between the smallest and the largest (Jiang and Yin 2014). The different scales form a continuum in the carpet and make it whole (Fig. 1b), so it creates life.

There are many strong centers in the carpet, creating a sensation of centeredness in a recursive manner. For example, the largest square is a strong center surrounded by the eight middle-sized squares, which constitute strong centers surrounded by the 64 smallest squares. Each square is surrounded by a boundary, but it is less obvious in the carpet. The property of alternating repetition is missing in the mathematical carpet, but it often appears in real carpets; see Alexander (1993, 2002–2005) for more examples. In the carpet, the spaces between the black squares (the white part) are positive spaces but are not as well-shaped as the black squares. In many good designs, such as the Nolli map of Rome, the ground space (the spaces between figural patterns) is as well-shaped as the figure space, forming positive spaces. The carpet is a good shape, since it consists of many squared shapes. The carpet is locally and globally symmetrical. This is true for most of real carpets. In the carpet, the local symmetries are all the same and uniform. This appearance lacks alternating repetition. However, a living structure is not necessarily globally symmetrical, but local symmetries are essential. The Alhambra plan, well studied by Alexander (2002–2005), is a typical example of local symmetries.

The property of deep interlock and ambiguity is clearly missing in the carpet, but it appears in many real carpets. The black and white contrast is obvious in the carpet. The property of gradient is less obvious with the carpet. Roughness is completely missing in the carpet because of exactitude of the shapes and scaling ratio. The property of echoes, in which the parts echo the whole, is shown in the carpet. The largest square is the void. Each of the eight middle-sized squares is surrounded by eight smallest squares, forming a kind of simplicity and inner calm, just as the eight middle-sized squares. None of the squares are separated from each other, but form a scaling hierarchy, which underlies the complex network as a whole (Fig. 1b). With the complex network, there are two kinds of links: those among the same scales that are undirected or mutually directed; and those across two consecutive scales that are directed from smaller to larger ones. The property of not separateness, or the 15 properties in general, advocate for a new worldview in which we see things in their wholeness, which underlies the topological representation we put forward in this paper. This new worldview differs fundamentally from the twentieth-century mechanistic worldview. Wholeness evokes a sense of beauty that we can experience, life comes from wholeness, and eventually order emerges from wholeness. Wholeness is therefore the source of life, or beauty or order (or good design), which all can be interchangeably referred to. The wholeness that produces life and evokes beauty in buildings and cities is "*a direct result of the physical and mathematical structure that occurs in space, something which is clear and definite, and something which can be described and understood*" (Alexander 2002–2005, Book 1, p. 62). The physical and mathematical structure can be described as a hierarchical graph (Jiang 2015b, 2016), which helps address not only why a design is beautiful, but also how much beauty the design has. The proposed topological representation aims to characterize the structure for cities as a coherent whole.

3 The Topological Representation for Taking Things as a Whole

The topological representation takes cities as a whole. To illustrate the representation, let us create 100 cities that exactly follow Zipf's law (1949): 1, 1/2, 1/3, …, and 1/100. The 100 cities are given some random locations in a two-dimensional space (Fig. 2a). According to their sizes, the cities are put into three different hierarchical levels based on head/tail breaks—a classification and visualization tool for data with a heavy-tailed distribution (Jiang 2013a, 2015a). The locations of the cities at the different hierarchical levels are respectively used to create Thiessen polygons (Fig. 2b). Based on the polygon–polygon relationships, we set up a complex network, in which small polygons point to large ones at a same level, and contained polygons point to containing polygons across two consecutive levels (Fig. 2c). This complex network can help compute the degrees of wholeness for individual cities (Jiang 2015b, 2016), which are represented by the dot sizes (Fig. 2c). Unlike the city sizes, the degrees of wholeness are well adapted to their surroundings. It is clear that the cities statistically demonstrate the scaling property of far more smalls than larges, because of Zipf's law (1949). However, they are not correctly arranged geometrically or according to central place theory (Christaller 1933, 1966), since the cities are given some random locations. Nevertheless, given this spatial configuration and according to the theory of centers (Alexander 2002–2005), individual cities obtain strength or wholeness from others. The computed degrees of wholeness or beauty are still put at the three hierarchical levels (Fig. 2c).

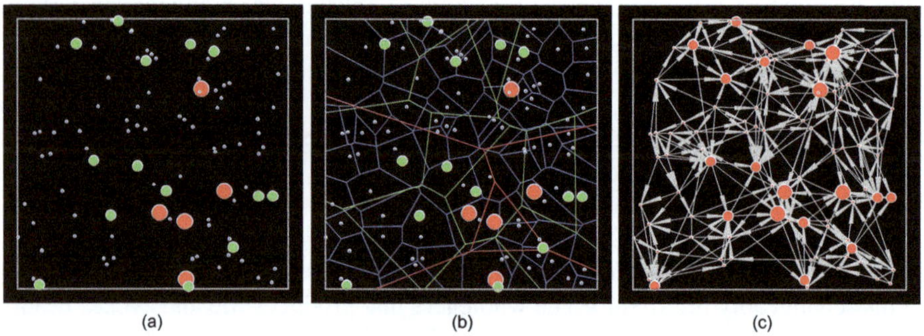

Fig. 2 The topological representation viewing cities as a whole (*Note* One hundred fictional cities with the sizes of 1, 1/2, 1/3, …, and 1/100 in three hierarchical levels, indicated by red, green and blue, are given some random locations (**Panel a**), the Thiessen polygons of the cities with the cities on the top (**Panel b**), and the complex network showing scaling hierarchy with dot sizes representing the degree of wholeness (**Panel c**). The cities in Panels a and b are differentiated from each other but lack local adaptation. The representation of cities in Panel c shows not only cities differentiation but also their adaptation to each other and to the whole network. In this connection, the point pattern in Panel c is more whole than that of those cities in Panels a and b.)

There are essentially two types of beauty or harmony shown in the pattern of Fig. 2c: one type among nearby things that are well adapted to each other with more or less similar sizes, and the other type across all scales that are differentiated from each other with far more small sizes than large ones. Comparing the initial city sizes and degrees of wholeness, we notice that some cities move up or down in the hierarchical levels, while some remain unchanged. Observing carefully, those cities that remain unchanged or move up have good support from others, while those that move down lack or have less support. From this observation, we can understand that the degrees of wholeness are essentially exogenous, rather than endogenous. In other words, cities move up or down along the hierarchal levels because of the other cities in the interconnected whole, rather than just themselves. In fact, these cities moving up or down in the hierarchical levels can be directly linked to cities' expanding and shrinking, which often occurs in reality. For example, the uppermost green city was growing, whereas the lowermost red city was shrinking. Cities' expansion and shrinkage, or urban structure and dynamics in general, can be better understood from the point of view of wholeness. Cities are not isolated, but support (and are supported by) others within a whole. In this regard, the degrees of wholeness can be thought of as ideal "city sizes" that ought to be, under the spatial configuration. These city sizes are harmonized with surroundings, with differentiation across all scales and adaptation among those nearby. In the following case studies, we examine the city sizes and the degrees of wholeness in three different ways: (1) Power law detection (with three parameters α, Xmin, p) and ht-index (differentiation), (2) calculation and visualization of ht-index (differentiation and adaptation), and (3) correlation between sizes and wholeness (adaptation).

The topological representation is developed primarily for showing cities' structure from the perspective of topology and wholeness. However, it can be applied to carpets, city plans, building facades, and virtually any other designs. They can be quantitatively examined if they are a living structure and further be computed for the degrees of beauty for individual coherent centers and the whole. When it is applied to designs, the key issue is to identify all latent centers, and their support relationships or topology, to set up a complex network representing the wholeness. Degrees of wholeness can then be computed relying on the mathematical model of wholeness (Jiang 2015b, 2016). The computed degrees of wholeness possess both properties of differentiation and adaptation, reflecting the underlying spatial configuration or arrangement of the centers. Alexander (2002–2005) referred to spatial configuration as geometry or living geometry. This should be more clearly called *topology*, hence the topological representation. The representation and topological analysis enable us to see the two transformation processes of differentiation and adaptation; see more details in the next section. The city sizes, or their equivalent occupied space of the Thiessen polygons, demonstrate differentiation, while the cities' growth and shrinkage show how cities adapt to their surroundings in the whole. We will further discuss these two transformation processes in Sect. 5. Before that, we apply the representation to some case studies.

4 Living Structures of China and UK Natural Cities

We applied the topological representation to China (mainland) and UK (main island) natural cities to demonstrate that they are living structures. Natural cities in the context of this paper refer to naturally, objectively derived high-density patches using large amounts of streets nodes or tweeted locations, and based on head/tail breaks (Jiang 2013a, 2015a). Natural cities, different from conventional concepts of cities defined or imposed by statistical or census authorities, provide a new instrument for better understanding urban structure and dynamics. This study examines the cities' structure and, specifically, how cities of each country are differentiated from, and adapt to, each other to form a coherent whole from both statistical and topological perspectives. These perspectives enable us to see cities as a geometrically coherent whole, rather than an arbitrary assembly. Some visualization methods are developed to illustrate the underlying scaling hierarchy of the cities' structure. In this paper, we adopted a recursive definition of geographic space, in which all cities of a country constitute a living structure, as do all hotspots of a city (Fig. 3). We took China and UK each as a whole, as well as Beijing and London natural cities each as a whole, for detailed investigations.

4.1 Data and Data Processing

The case studies are based on two datasets: 2,728,143 streets nodes in China (Long et al. 2016), and 3,281,935 tweeted locations in the UK during June 1–8, 2014 (previously studied in Jiang et al. 2016) (Table 2). The 3 million locations were used to build up a huge triangulated irregular network (TIN) for each country. The TIN consists of far more short edges than long ones. The short and long edges respectively indicate high and low densities. All the short edges constitute individual patches called natural cities (see Jiang and Miao 2015 for a tutorial). After having extracted all the natural cities, we then took Beijing and London natural cities, respectively,

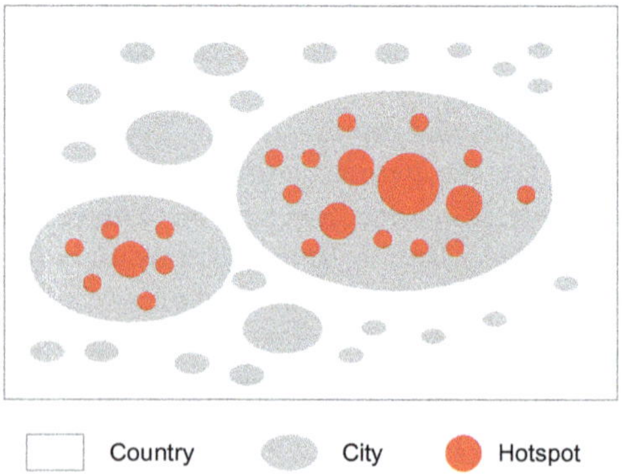

Fig. 3 A recursive definition of geographic space (*Note* Geographic space is not closed, which implies that a country is part of a continent or the globe.)

Table 2 Numbers of locations and derived natural cities and hotspots

Case studies	Locations	Cities/hotspots
China	2,728,143	4322
Beijing	157,755	6054
UK	3,281,935	4664
London	203,986	8630

containing 157,755 and 203,986 locations, and each of the natural cities repeated the above processes to get individual hotspots in the cities. The hotspots to a city are what cities are to a country, so countries, cities, and hotspots constitute nested relationships as illustrated in Fig. 3.

Following the procedure in Fig. 2, these natural cities and hotspots were then topologically represented. The data processing was carried out mainly through ArcGIS and other related scripts for computing the degrees of wholeness. To make the paper self-contained, we briefly describe the data processing here. Use the head/tail breaks and base on city sizes to classify the centroids of the natural cities into different hierarchical levels. Create Thiessen polygons from the centroids, as shown in Fig. 2. For convenience of ArcGIS users, this step can be done by: (1) ArcToolbox > Proximity > Create Thiessen Polygons (using a country's bounding box); and (2) Geoprocessing > Clip (using a country boundary). Write a simple script to create a complex network for all the Thiessen polygons; small cities point to adjacent large cities for those at a same level, and contained polygons point to containing polygons across two consecutive levels. Based on the complex network, compute the degrees of wholeness for individual Thiessen polygons or equivalently individual cities or hotspots.

After the above data processing, we derived large amounts of natural cities: 67,700 for China and 124,608 for the UK. In order for the amounts of natural cities and hotspots to be of the same magnitude, we respectively chose the large cities by following the head/tail breaks for detailed investigation (Table 2)—the first head 4,322 large cities in China and the second head 4,664 large cities in the UK. These cities or hotspots are used to further examine the properties of differentiation and adaptation.

4.2 *Visualization of Differentiation and Adaptation*

We found that cities are differentiated from each other in terms of the sizes and their degrees of wholeness. The differentiation can be characterized by a power law distribution and ht-index (Jiang and Yin 2014) for respectively measuring the hierarchy and the hierarchical levels. Figure 4 illustrates the natural cities in China, with the background and foreground respectively showing the six hierarchical levels of Thiessen polygons and the largest cities on the top three levels. We can clearly see that the pattern of wholeness is just slightly different from that of sizes, indicating that

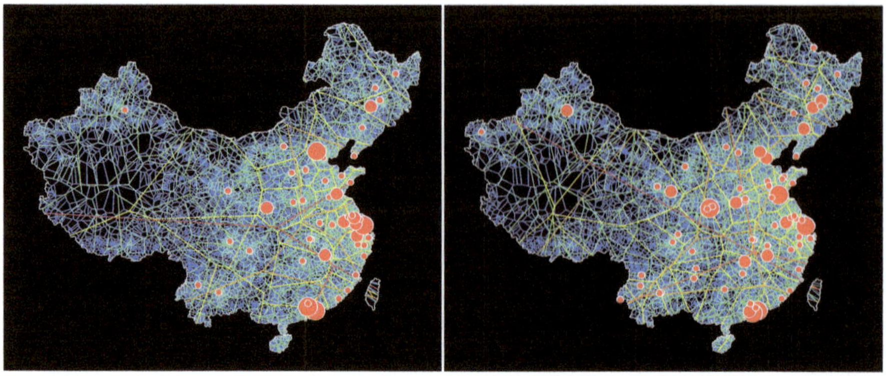
(a) (b)

Fig. 4 China natural cities (4,322 in total) as a living structure from both cities' sizes and their degrees of wholeness (*Note* There are six hierarchical levels in terms of (**a**) the cities' sizes, with the 65 largest cities on the top four levels; and (**b**) the degrees of wholeness, with the 70 most beautiful cities on the top four levels.)

sizes are well adapted to their surroundings. There are a few expanded cities, such as Shanghai and Shenyang, for they are surrounded or supported by many others. The pattern for the degrees of wholeness (Fig. 4b) is considered the most harmonized and adapted. On the other hand, the pattern for city sizes (Fig. 4a) resembles that of the degrees of wholeness. In fact, the city sizes and degrees of wholeness are well correlated with an R square of 0.82.

We visualized the complex network of cities with the foreground and background respectively representing links across scales or levels (blue lines) and within a same scale or level (gray lines) to further illustrate the underlying scaling hierarchy or the fact that cities are differentiated from and adapted to each other. The six hierarchical levels of China's natural cities are shown in Fig. 5. Compared to the hierarchy based on the cities' sizes (Fig. 5a), more cities are forced to the highest hierarchical levels in terms of the degrees of wholeness (Fig. 5b). Therefore, there are three cities at the two highest levels in terms of the cities' sizes, whereas there are six at the two highest levels in terms of the degrees of wholeness.

We further examined the power law distributions (with three parameters) and ht-indices (Table 3). While examining p value in some cases in which the largest city or the hottest spot is too big (sometimes more than three times larger than the second largest city), we excluded it. This is to follow the concept of *primate city*, which was empirically well observed and mathematically proven (Jefferson 1939; Chen 2012). Although there are two cases in which the p value is zero, their ht-indices are still very high around 5 and 6, indicating some heavy-tailed distributions. It should be noted that the power law of cities' sizes is purely statistical, while the degrees of wholeness is not only statistical but also geometrical, indicating the property of adaptation. All these analytical results point to the fact that cities in a country constitute a living structure.

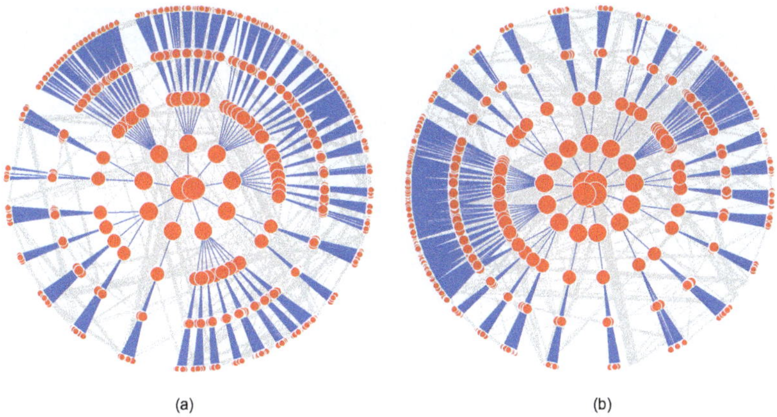

Fig. 5 Scaling hierarchy of China cities' sizes and their degrees of wholeness (*Note* Both the cities' sizes and their degrees of wholeness possess six hierarchical levels **a** for cities' sizes, and **b** for degrees of wholeness, indicating that the properties of differentiation and adaptation.)

Table 3 Power law detection results and ht-indices for cities sizes and their degrees of wholeness (*Note* The power law detection is based on the robust method introduced in Clauset et al. (2009), while the ht-index is based on Jiang and Yin (2014))

Case studies	Sizes/wholeness	α	Xmin	p	ht
China	Sizes	1.86	37	0	6
	Wholeness	2.14	0.0005103	0.18	6
UK	Sizes	2.66	106	0.05	5
	Wholeness	2.28	0.0000487	0.07	6
Beijing	Sizes	2.34	16	0.71	4
	Wholeness	2.44	0.0000430	0.05	6
London	Sizes	2.20	23	0.04	6
	Wholeness	2.26	0.0000244	0	5

The living structure appears not only at the country level but also at the city level. To illustrate, the London natural city consists of 8,630 hotspots, the smallest of which are a patch with three locations. Both the sizes and the degrees of wholeness follow very well a power law distribution, with α being 2.20 and an ht-index greater than 5 (Table 3). The spatial distribution of some of the largest hotspots is illustrated in Fig. 6a, with five at the two highest levels, and the most beautiful hotspots (Fig. 6b) with eight at the highest level. Similar to Fig. 5 at the country level, Fig. 7 illustrates the complex network with the hierarchical tree in the foreground. It should be noted in Fig. 6 that some small hotspots have the highest degrees of wholeness, thus becoming the most beautiful. This is understandable, since they get many supports from their surroundings, so become well adapted to wholeness. The Beijing natural city has a similar structure, although the ht-index is only 4. Despite the fact that the two

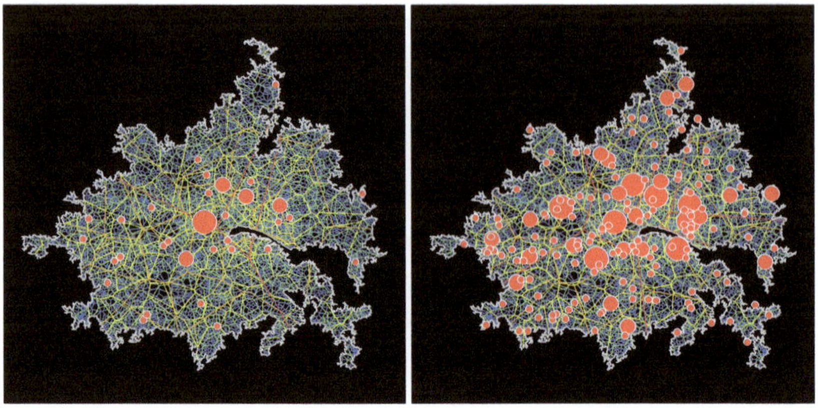

(a) (b)

Fig. 6 London natural city consisting of 8,630 hotspots as a living structure from both cities' sizes and their degrees of wholeness (*Note* There are six hierarchical levels in terms of **a** the sizes of hotspots with the 32 largest on the top three levels, and five hierarchical levels according to **b** the degrees of wholeness with the 165 most alive spots on the top three levels.)

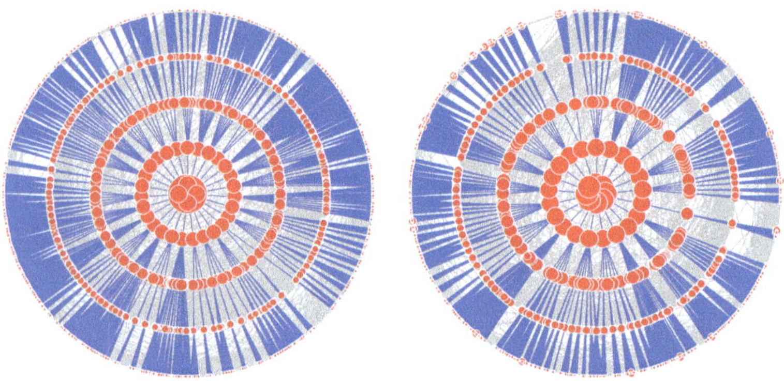

(a) (b)

Fig. 7 Scaling hierarchy of London hotspot sizes and their degrees of wholeness (*Note* The hotspot sizes and their degrees of wholeness respectively possess **a** six and **b** five hierarchical levels, indicating the nature of differentiation and adaptation. Note that there are the top two levels at the center in (**a**))

patterns in Fig. 6 look dramatically different, the hotspot sizes correlate well with the degrees of wholeness. Except for the largest spot, the R square value is 0.87. To some extent, the R square indicates how cities' sizes adapted to each other; the higher the R square, the more adapted the cities.

Through the case studies, we found that cities in a big country as a whole have a higher degree of wholeness than a city by itself. This is understandable, just as the human body is more complex than the human brain because the latter is part of the former. The results of the case studies provide further evidence to support the wholeness-oriented design, in particular, the differentiation and adaptation processes. Through the topological representation and analysis, we illustrated and demonstrated that cities are a living structure, and so is a city by itself. The living structure illustrated at the city level is closely related to the concepts of vitality, imageability, and legibility, which underlie the theory of the image of the city (Lynch 1960). These concepts, together with beauty, life, order, and wholeness, all sound subjective but are actually measurable and computable through the degrees of wholeness. For example, the image of the city can be formed simply because of the underlying living structure (Jiang 2013b). There is little doubt that those largest or most beautiful nodes (representing cities or hotspots) in the inner circles of Figs. 5 and 7 constitute the image of the country and city.

We can add a few remarks on the living structure to this point. First, the point pattern in Fig. 2c is more whole than that in Fig. 2a or b. This is because the degrees of wholeness under the influence of spatial configuration are more adapted to each other than to the cities' sizes. Second, with respect to Figs. 5 and 7, the foregrounds are hierarchal trees, while the foregrounds and backgrounds together constitute complex networks (Jiang 2015c). In other words, cities are near decomposable in terms of Simon (1962), or cities are not trees but semi-lattices (Alexander 1965). Third, the results from the case studies point to the two properties of differentiation and adaptation at both country and city levels. According to Alexander (2002–2005), differentiation and adaptation, or the living structure in general, appears at all scales ranging from the Planck length (10^{-43} m) to the scale of the universe itself (10^{27} m), virtually from the infinitely small to the infinitely large. Based on these analytical insights, we further discuss and elaborate the implications of the topological representation and analysis.

5 Discussions on the Topological Representation and Analysis

The topological representation is rooted in the living structure (Alexander 2002–2005), in the central place theory (Christaller 1933, 1966), and in fractal geometry (Mandelbrot 1982), as well as in the new definition of fractal: *a set or pattern is fractal if the scaling of far more small things than large ones recurs multiple times or with ht-index being at least three* (Jiang and Yin 2014). All these theories have one common thing in which they differ from Euclidean geometry, which essentially deals with regular or simple shapes. The geometry dealing with the living structure is referred to as *living geometry*, which "*follows the rules, constraints, and contingent conditions that are, inevitably, encountered in the real world*" (Alexander et al. 2012, p. 395). Beyond fractal geometry, living geometry aims not only for understanding things but also for making things. Unfortunately, modernist urban planning and design is very much (mis-)guided by Euclidean geometric thinking (Mehaffy and Salingaros 2006). As a devastated result, modern architecture and city planning (Corbusier 1989) inevitably lead to dead or lifeless buildings and cities. Their design or structure lacks differentiation and adaptation and is therefore not living structure. To create sustainable built environments, we therefore must abandon the Euclidean geometric thinking. What we want to abandon is the Euclidean geometric thinking rather than Euclidean geometry. Euclidean geometry is essential for fractal geometry, because one must first measure things with Euclidean geometry in order to see far more small things than large ones. In this connection, Jiang and Brandt (2016) provide detailed arguments as to why the Euclidean geometric thinking is limited in understanding complex geographic forms and processes.

The analytical evidence and insights about the living structure developed in the present paper add further implications on the wholeness-oriented design. Any design or planning should respect the wholeness, and retain it and further extend it. This is the design philosophy Alexander (Alexander 2002–2005) advocated through his theory of centers, in order to effectively create living structures in buildings, cities, and artifacts. This wholeness-oriented design philosophy applies not only to cities and buildings but also to any artifacts in pursuing the living structure or beauty. Any design or structure must include many substructures across many scales that are differentiated from each other to form a scaling hierarchy. On the one hand, there is adaptation across all scales by a constant scaling ratio of approximately 2 or 3, also called scaling coherence (Salingaros 2005). On the other hand, the substructures must adapt to each other at same or similar scales. This design philosophy is in line with complexity theory, which states that cities emerged either autonomously or in some self-organized manners (e.g., Portugali 2000). The wholeness-oriented design reflects what humans, across different countries and cultures, did in the past centuries before the 1920s in creating buildings, cities, and artifacts. The two properties of differentiation and adaptation help us not only correctly conceive of but also beautifully create architecture. If we differentiate a space in a way that treats it as a coherent whole, it would lead to a living structure for the space and even beyond towards a larger space. Practically speaking, wholeness-oriented design must be guided by the 15 properties discussed in Sect. 2. In this regard, many practical ways of creating living structure that are guided by scientific principles, rather than artistic standards, have been thoroughly discussed (e.g., Salingaros 2013; Mehaffy and Salingaros 2015).

The topological representation or the analysis in general helps us clearly understand some misperceptions in modern urban planning and design. First, the order comes only from regular shapes or grids, governed by geometrical fundamentalism (Mehaffy and Salingaros 2006). In fact, what looks chaotic or disorderly on the surface could hold hidden or deep order. This hidden order is characterized by the scaling law of far more small things than large ones. This order recurs across all scales, ranging from the smallest to the largest. Second, the order comes from a tree structure. This is a deadly misperception (Alexander 1965). As demonstrated in the above case studies and elsewhere (Jiang 2015c), a city is not a tree, but a complex network. In other words, a complex network is no less orderly than a tree. Different from a tree, parts in a complex network often overlap, nest, and interact each other.

However, both complex networks and trees share the same scaling hierarchy of far more small things than large ones. The third misperception is that symmetry is always global. In fact, many living structures are full of local symmetries, yet lack global symmetry. For example, neither the Alhambra plan nor the Nolli map is globally symmetric. The global shape must adapt to both natural and built environments, and numerous local symmetries can enhance degrees of wholeness or order (Alexander 2002–2005). Fourth, cities are less orderly than the Sierpinski carpet; similarly, coastlines are less orderly than Koch curves. In fact, both cities and the Sierpinski carpet, or coastlines and Koch curves, share the same scaling or fractal or living or beautiful order: the former being statistical (Mandelbrot 1982), and the latter being rigid. To summarize and to paraphrase Alexander (2002–2005), the order in nature is essentially the same as that in what we build or make: buildings, cities or artifacts.

The two properties of differentiation and adaptation are governed by two fundamental laws: Scaling law and Tobler's law. Scaling law refers to some heavy-tailed distributions for many natural and societal phenomena. It can be simply phrased as far more small things than large ones; or numerous smallest things, a few largest things, and some in between the smallest and the largest. The existing geometry-based representations cannot well address this scaling property of geographic features. Instead, the topological representation developed in this paper provides a useful instrument for seeing the scaling property. Tobler's law—the first law of geography—refers to the effect of spatial autocorrelation or dependence, widely studied in the geography literature. Tobler's law states that "*everything is related to everything else, but near things are more related than distant things*" (Tobler 1970, p. 236). Current spatial analysis and spatial statistics in particular concentrates too much on the effect of spatial dependence, yet little on scaling law. Scaling law and Tobler's law complement each other for characterizing the Earth's surface: the former across all scales being global, universal, while the latter on a single scale being local; the former on heterogeneity, while the latter on homogeneity; the former characterized by power law statistics and fractal geometry, as well as living geometry, while the latter by Gaussian statistics and Euclidean geometry. These two laws are fundamental not only to geographical phenomena but also to any other living structure that recurs between the Planck length and the size of the universe itself.

6 Conclusion

Cities are not isolated but are coherent entities within an interconnected whole. This situation is the same for a single city that consists of many coherent hotspots. This paper developed a topological representation that takes cities as a whole or hotspots as a whole to uncover their underlying scaling hierarchy, and to further understand the city as a problem in organized complexity (Jacobs 1961). We adopted a recursive definition of geographic space, under which a country acts as a coherent whole, as does any individual city. This paper provided analytical insights to advocate for wholeness-oriented design, particularly the differentiation and adaptation principles. To a great extent, this study provided scientific evidence for the claim that the order in nature is essentially the same as that in what we build or make (Alexander 2002–2005). The kind of topological representation and analysis can also be applied to modern architecture and urban design to objectively judge whether or not certain buildings and cities are stiff and lifeless in terms of the underlying structure. This is in line with what Alexander (2002–2005) claimed that the goodness of built environments is not a matter of opinion, but a matter of fact.

This paper clarified some misperceptions in urban design. The order or structure illustrated through the case studies cannot be characterized by Euclidean geometry or Gaussian statistics. For example, many living structures (except for carpets because of their rectangular shapes) lack global symmetry but full of local symmetries. A complex network is no less orderly than a tree. To put it in another way, a city is not a tree, but a complex network (Alexander 1965; Jiang 2015c). Living structures are governed by two laws: Tobler's law at the same scale and scaling law across all scales, or equivalently the two spatial properties of dependence and heterogeneity. We further pointed out two types of coherence or adaptation: across all scales for those being far more small things than large ones, and at the same scale for those being more or less similar. The kind of topological representation can be applied to any works of art. The key issue is to identify the coherent entities or centers, and importantly their nested and overlapping relationships, so that a complex network can be set up for further computing their degrees of beauty or life. Ultimately, this paper provides analytical evidence and insights to support the wholeness-oriented design, particularly the two fundamental design processes of differentiation and adaptation. In future, we will seek applications of the topological representation for analysis and prediction of geographic phenomena.

Acknowledgements This chapter is a reprint of the journal article (Jiang 2018) with the permission kindly granted by the publisher Wiley. I would like to thank Zheng Ren for his research assistance and Ding Ma for helping with some of the figures. The three anonymous referees and the editor deserve my special thanks for their constructive comments. However, any errors and inadequacies of the paper remain solely my responsibility.

References

Alexander C. (1965), A city is not a tree, *Architectural Forum*, 122(1 + 2), 58–62.

Alexander C. (1993), *A Foreshadowing of 21st Century Art: The color and geometry of very early Turkish carpets*, Oxford University Press: New York.

Alexander C. (2002–2005), *The Nature of Order: An essay on the art of building and the nature of the universe*, Center for Environmental Structure: Berkeley, CA.

Alexander C., Neis H. and Alexander M. M. (2012), *The Battle for the Life and Beauty of the Earth*, Oxford University Press: Oxford.

Alexander C., Neis H., Anninou A. and King I. (1987), *A New Theory of Urban Design*, Oxford University Press: London.

Batty M. and Longley P. (1994), *Fractal Cities: A geometry of form and function*, Academic Press: London.

Chen Y. (2012), The mathematical relationship between Zipf's law and the hierarchical scaling law, *Physica A*, 391(11), 3285–3299.
Christaller W. (1933, 1966), *Central Places in Southern Germany*, Prentice Hall: Englewood Cliffs, N. J.
Clauset A., Shalizi C. R. and Newman M. E. J. (2009), Power-law distributions in empirical data, *SIAM Review*, 51, 661–703.
Corbett J. P. (1979), *Topological Principles in Cartography*, Technical paper 48, U.S. Dept. of Commerce, Bureau of the Census: Washington D. C.
Corbusier L. (1989), *Towards a New Architecture*, Butterworth Architecture: London.
Egenhofer M. J. and Herring J. R. (1990), A mathematical framework for the definition of topological relationships, *Proceedings of the Fourth International Symposium on Spatial Data Handling*, International Geographical Union, Zurich 1990, 803–813.
Goodchild M. F. and Mark D. M. (1987), The fractal nature of geographic phenomena, *Annals of the Association of American Geographers*, 77(2), 265–278.
Goodchild M. F., Yuan M., and Cova T. J. (2007), Towards a general theory of geographic representation in GIS, *International Journal of Geographical Information Science*, 21(3), 239–260.
Jacobs J. (1961), *The Death and Life of Great American Cities*, Random House: New York.
Jefferson M. (1939), The law of the primate city, *Geographical Review*, 29(2), 226–232.
Jiang B. (2013a), Head/tail breaks: A new classification scheme for data with a heavy-tailed distribution, *The Professional Geographer*, 65 (3), 482–494.
Jiang B. (2013b), The image of the city out of the underlying scaling of city artifacts or locations, *Annals of the Association of American Geographers*, 103(6), 1552–1566.
Jiang B. (2015a), Head/tail breaks for visualization of city structure and dynamics, *Cities*, 43, 69–77. Reprinted in Capineri C., Haklay M., Huang H., Antoniou V., Kettunen J., Ostermann F., and Purves R. (editors, 2016), *European Handbook of Crowdsourced Geographic Information*, Ubiquity Press: London.
Jiang B. (2015b), Wholeness as a hierarchical graph to capture the nature of space, *International Journal of Geographical Information Science*, 29(9), 1632–1648.
Jiang B. (2015c), A city is a complex network, in: M. W. Mehaffy (editor, 2015), *Christopher Alexander A City is Not a Tree: 50th Anniversary Edition*, Sustasis Press: Portland, OR, 89–100.
Jiang B. (2016), A complex-network perspective on Alexander's wholeness, *Physica A*, 463, 475–484.
Jiang B. (2018), A topological representation for taking cities as a coherent whole, *Geographical Analysis*, 50(3), 298–313.
Jiang B. and Brandt A. (2016), A fractal perspective on scale in geography, *ISPRS International Journal of Geo-Information*, 5(6), 95, https://doi.org/10.3390/ijgi5060095.
Jiang B. and Claramunt C. (2004), Topological analysis of urban street networks, *Environment and Planning B: Planning and Design*, 31(1), 151–162.
Jiang B. and Miao Y. (2015), The evolution of natural cities from the perspective of location-based social media, *The Professional Geographer*, 67(2), 295–306. Reprinted in Plaut P. and Shach-Pinsly D. (editors, 2019), *ICT Social Networks and Travel Behaviour in Urban Environments*, Routledge.
Jiang B. and Yin J. (2014), Ht-index for quantifying the fractal or scaling structure of geographic features, *Annals of the Association of American Geographers*, 104(3), 530–541.
Jiang B., Ma D., Yin J. and Sandberg M. (2016), Spatial distribution of city tweets and their densities, *Geographical Analysis*, 48, 337–351.
Jiang B., Zhao S. and Yin J. (2008), Self-organized natural roads for predicting traffic flow: a sensitivity study, *Journal of Statistical Mechanics: Theory and Experiment*, July, P07008.
Köhler W. (1947), *Gestalt Psychology: An Introduction to New Concepts in Modern Psychology*, LIVERIGHT: New York.
Long Y., Shen Y. and Jin X. (2016), Mapping block-level urban areas for all Chinese cities, *Annals of the American Association of Geographers*, 106(1), 96–113.
Lynch K. (1960), *The Image of the City*, The MIT Press: Cambridge, Massachusetts.

Mandelbrot B. B. (1982), *The Fractal Geometry of Nature*, W. H. Freeman and Co.: New York.
Mehaffy M. W. and Salingaros N. A. (2006), Geometrical fundamentalism, In: Salingaros N. A. (2006), *A Theory of Architecture*, Umbau-Verlag: Solingen, 172–194.
Mehaffy M. W. and Salingaros N. A. (2015), *Design for a Living Planet: Settlement, science, and the human future*, Sustasis Press: Portland, Oregon.
Portugali J. (2000), *Self-Organization and the City*, Springer: Berlin.
Salingaros N. A. (2005), *Principles of Urban Structure*, Techne: Delft.
Salingaros N. A. (2013), *Algorithmic Sustainable Design: Twelve lectures on architecture*, Sustasis Press: Portland, Oregon.
Simon H. A. (1962), The architecture of complexity, *Proceedings of the American Philosophical Society*, 106, 468–482.
Theobald D. M. (2001), Topology revisited: representing spatial relations, *International Journal of Geographical Information Science*, 15(8), 689–705.
Tobler W. (1970), A computer movie simulating urban growth in the Detroit region, *Economic geography*, 46(2), 234–240.
Yuan M., Mark D. M., Egenhofer M. J. and Peuquet D. J. (2005), Extensions to Geographic Representations, in: McMaster R. B. and Usery E. L. (editors, 2005), *A Research Agenda for Geographic Information Science*, CRC Press: New York, 129–156.
Zipf G. K. (1949), *Human Behaviour and the Principles of Least Effort*, Addison Wesley: Cambridge, MA.

Part V
Other Forms of Quantification

A Multiscale Clustering of the Urban Morphology for Use in Quantitative Models

Patrick M. Schirmer and Kay W. Axhausen

Abstract Geo-spatial data available to researchers and practitioners has increased substantially over the last decades, offering new opportunities for analyzing and characterizing locations and their spatial environment in an objective way. Various studies in the field of urban planning and design have given recommendations on "good urban form", suggesting that especially characteristics of the urban morphology inform the quality of locations and additionally influence spatial behavior. While we find a growing number of quantitative spatial models such as hedonic price models, location choice models, or applied machine learning algorithms, characteristics describing the urban morphology are still rarely reported in these. One reason can be found in the limited knowledge on how to characterize the urban morphology best. In this chapter, we address this deficit and define attributes that are suitable to characterize the urban morphology in quantitative means at different scales. We show how these can be processed from a data model that is simple enough to allow for reproducibility in most study areas and process them for the case study of Switzerland. Finally, we use the attributes to define urban typologies through clustering methods. These are compared on their outcome, their consistency, and interpretation.

1 Introduction and State of the Art

While the methodologies to model spatial effects in quantitative terms have substantially increased throughout the last decades—reaching from discrete location choice to machine learning algorithms, researchers failed to focus on how to best describe spatial characteristics through objective measures in such models (Marshall 2012).

The representation of the urban environment has often been simplified to socioeconomic distributions available in census data (Schirmer et al. 2014), whereas urban

P. M. Schirmer (✉)
UrbanDataLab Schirmer, ETH Zürich, Zürich, Switzerland
e-mail: ps@urbandatalab.ch

K. W. Axhausen
Institute for Transport Planning and Systems (IVT), ETH Zürich, Zürich, Switzerland

L. D'Acci (ed.), *The Mathematics of Urban Morphology*,
Modeling and Simulation in Science, Engineering and Technology,
https://doi.org/10.1007/978-3-030-12381-9_16

morphology has long be ignored. This is incomprehensible as the built environment forms a long-lasting element of urban landscapes allowing for analyses at different scales. This setting is not only important for the character of an area, but in consequence defines the outline of its possible usage and the options of activities, i.e., it has substantial impact on spatial decisions and their economic effects.

Alexander et al. (1977) give an early description of patterns for cities, neighborhoods, buildings as answers to design problem, but miss to describe algorithms on how to evaluate these. Later publications focus on analyzing the configuration of a network and definition of street typologies (Hillier and Hanson 1984; Marshall 2004) which can be linked to wayfinding and location choice of retail (Hillier 1996; Bafna 2003; Sevtsuk 2010). In recent years, we also find attempts to classify buildings and blocks using morphological attributes (Meinel et al. 2008).

However, as Gil et al. (2012) state, "the analytical process is laborious and not entirely objective" and "there is a need to integrate different morphological approaches to obtain a more complete and complex set of urban environmental attributes." One issue is the missing representation of different scales. Within this chapter, we want to address this limitation and propose set of attributes that represent the built environment at different scales and can be derived through geoprocessing routines.

The structure of urban areas is often described and compared through aggregated characteristics (top-down). Lampugnani et al. (2007) compare the structure of different cities using a square-kilometer of their city center. The *bottom-up view* is looking at the location instead. Each location is evaluated based on its individual characteristics. Neighboring locations do not necessarily have the same characteristics, e.g., the distance to the block border can vary and the building types may be very different. The bottom-up view respects these differences and groups locations that are comparable into urban typologies, independent of their spatial distribution. Following Gil et al. (2012) one can make use of cluster analysis to define such urban typologies. We propose to define urban typologies for different scales using the morphologic attributes. We report on this approach using three clustering algorithms on a case study covering a large part of the canton of Zurich.

2 Data Model

2.1 *Local and Global Data*

The amount of available geolocated data has exponentially risen throughout the last decade. Most of these form *local datasets* that are specific to a region; they vary in structure and content for different regions while *global datasets* use the identical data model. Google Earth, Google Maps, Google Inc. (2013a, b), and OpenStreetMap (OpenStreetMap contributors 2017) form examples of global data. Studies of different regions depend on such data or have to extract comparable information from local data. As the access to global data is still limited to researchers we propose a

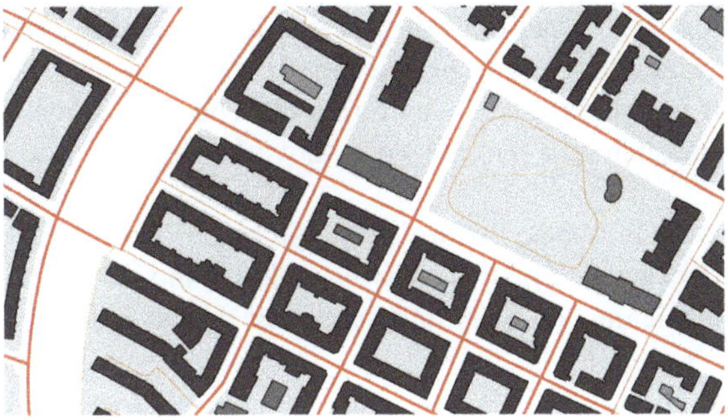

Feature	Columns	Type	Examples
buildings	building_id	primary key	
	stories	integer	1,2,10,35
	height	f oat	24.5
	postcode	integer	271, 8048
	the_geom	geometry	
blocks	block_id	primary key	
	postcode	integer	271, 8048
	the_geom	geometry	
streets	street_id	primary key	
	type	string	primary, secondary
	bridge	boolean	TRUE/FALSE
	tunnel	boolean	TRUE/FALSE
	maxspeed	integer	50
	hierarchy	integer	1,2,3,4,5
	the_geom	geometry	

Fig. 1 Initial objects represented in the proposed data model: buildings, blocks, and streets. *Source* Cadastral survey of Canton Zurich and ©2011 Swisstopo (JA100120)

data model that is flexible enough to allow for integration of datasets from different origins.

2.2 *An Object Driven Data Model*

Our data model relies on buildings, blocks, and streets, which form components of the built environment that are available as geodata in most countries. As the transformation of these objects in the real world requires construction activity, they form long-lasting structural components of the built environment. The mayor information is derived from the geometries in the data—the use of further attributes is avoided to the greatest possible extent, to keep the process as generic as possible. An overview of the data model uses and attributes is given in Fig. 1.

Buildings are represented through the building composite, which groups touching buildings into one volumetric object. They have a polygon as geometry and the information on their height (2.5D). Streets are represented through the connecting network of centerlines. Their configuration and capacity follow the definition as given in OpenStreetMap, i.e., streets get classified into motorway, primary (arterials), secondary (collectors or distributors), and tertiary roads (local streets). The differentiation of bridges and tunnels is recommended for simplifying the cleaning of the network topology. Blocks are defined as the polygon which is formed through enclosing streets and include only the attributes given by their geometry.

2.3 *Methods of Geoprocessing*

While geodata allow to link user-defined attributes to the geometry, also the geometry as such contains information. The kind of information one can derive depends on the geometry's feature type. A point will only define a location, but forms the basis for Euclidean distance calculation and kernel densities. A line has a starting point and end point which defines an orientation and a specific length, a polygon consists of lines forming a closed line-string and covering a certain area. The connection of several lines forms a closed shape such as a polygon or a network which allows to apply graph analysis.

Through means of geoprocessing methods, feature types can be transformed and additional characteristics can be extracted, e.g., it is possible to define the orientation of a polygon through partitioning it into its lines and summing up the individual lengths per orientation (also see Sect. 3.1).

3 Scales and Attributes

In most quantitative spatial models—such as location choice models—the focus lies on the characteristics of individual locations and not on arbitrary zonal definitions. This should also be reflected at the higher scales, so that we propose the use of scales of perception that refer to the actual location. Six scales of perception are proposed for the description of the objects in the data model: building, composition, neighborhood, district, municipality, and region (Fig. 2).

This classification is believed to reflect main characteristics of the planning guidelines and designs in the fields of architecture, urban design, and urban planning, although the implied data model cannot cover all characteristics and details of these disciplines.

3.1 Building

The geometry and volume of buildings, i.e., their morphological characteristics, also define the architectural type of a building. These attributes can be grouped according to the initial feature type:

Polygon Characteristics Buildings and blocks are represented as volumes in 2.5D, i.e., through polygons and an attribute on its height. The geometry of a polygon includes information on the area and the perimeter as well as its compactness. This

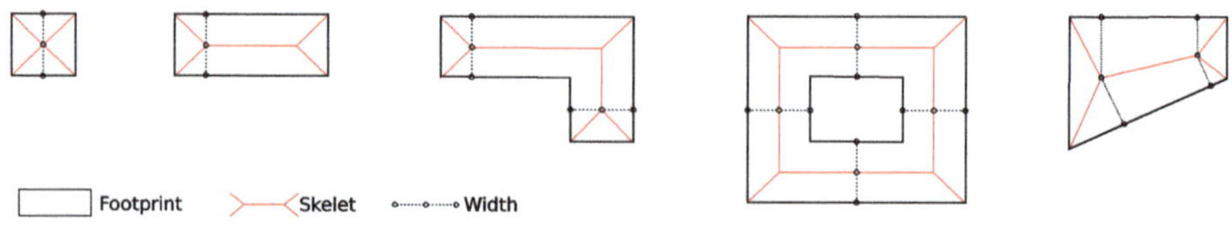

Fig. 2 The shape and morphologic skeleton of example buildings

allows to distinguish linear housing from closed forms or compact, detached housing. Besides we find the number and size of inner polygons, i.e., courtyards in a building.

Following Meinel et al. (2008), we extract the bounding box and enclosing circle of buildings and derive their area, perimeter, and compactness. Furthermore, the characteristics of the enclosing objects and the initial objects are evaluated against each other in form of ratios.

Volumetric Information The volume is used to build the ratio of volume to area of facade, and height allows to estimate floor space, assuming a general floor height per land use. If given, the use of the number of stories is of advantage.

Topological Skeleton A skeleton is a line-string that represents a thinned version of shapes boundary, but keeps the topological information (Lee 1982; Ogniewicz and Kbler 1995), i.e., where each point of the line-string lies within the polygon and has the maximum distance to the closest boundary (equidistant to boundary). The skeleton of a building allows to extract typological information as the length, width, and orientation of a building but also to characterize its shape through evaluating the number of unconnected lines ("dead-ends") and the number of connected lines.

3.2 Composition

By relating blocks, networks, and buildings to each other, we characterize the immediate environment of a location, i.e., its "composition". While the building form does mainly represent architectural characteristics, the composition does reflect the spatial configuration of buildings and their perception, i.e., elements of urban design. One can distinguish the configuration of a building to other buildings, blocks, and streets.

Influence Zone The distance to the next building allows to differentiate loosely spread buildings from grouped buildings. We propose a topological skeleton of the unbuilt space to determine the "morphological influence zone" of a building (see Fig. 3). This influence zone forms a polygon that allows to measure the same characteristics as those in Sect. 3.1, including the ratio to the area covered by the building and the volume of the building.

Courtyards A courtyard forms a semipublic space which can be of substantial benefit for a location. Depending on buildings' footprints and heights the space between buildings can form a courtyard, even when they do not touch each other, as long as its access is perceived as "closed". This perception strongly depends on the width of the access in relation to the height of the nearby buildings.

Street and Block The positioning of a building in relation to the streets defines its perception from the public space. One can differentiate the distance from the block border, the number of enclosing streets, the percentage of a facade that is visible

Fig. 3 Influence zone of buildings, permeability of block (mid) and degree of closure (right). *Source* Based on Cadastral survey of Canton Zurich

from the streets and the orientation toward the streets. They can be derived from the building skeleton and the outer polygon shape.

Similar to courtyards, also the character of a street depends on its section: therefore not only on the width of the street is relevant, but the ratio to the height of the associated buildings.

The Voronoi on buildings also reflects the visual permeability of a block and allows to measure the number of axial looks through the block (see Fig. 3).

The degree of closure is defined as the length of the building facades along the street in ratio to the actual perimeter of a block. Urban blocks are expected to have more construction along the block border, than rural blocks, i.e., a higher degree of closure.

Figure 3 visualizes the degree of closure, permeability, and influence zone as proposed here.

3.3 Neighborhood and District

The neighborhood represents the walkable surrounding while the district is defined as the accessible space when using various transport modes in a given time. Both include primarily densities, measured as Euclidean distance (neighborhood) or network distance (district). The Euclidean distance of 300 m is used as approximation for a walkable surrounding and the network distance of 1 min freeflow travel time as extent for the district (Fig. 4).

Densities Density of building footprints or of floor space is a common measure in community design. It is normally measured in reference to other polygonal objects, e.g., the parcel or a zone. With the growing size of these reference objects, this measure can be inconsistent with individual location characteristics, especially with a high variance of construction across the reference polygon. To avoid these, a predefined Euclidean distance can be used instead, i.e., within a walkable neighborhood of 300 m.

Fig. 4 Neighborhood characteristics capture the densities of the built environment and the network structure in the close surrounding of a location. *Source* Generated data based on Cadastral survey of Canton Zurich and ©2011 Swisstopo (JA100120)

The Euclidean density is appropriate for the direct neighborhood of buildings, but ignores barriers to access within the neighborhood. A network distance based density represents a cumulative opportunities measure which respects barriers, i.e., the number of objects accessible in a certain time (Bhat et al. 2000; Vickerman 1974; Wachs and Kumagai 1973). Our definition of a district reflects these considerations: densities get measured as the number of opportunities in the area that is accessible in one minute driving.

Network Structure Various networks analyses on graphs have been developed and discussed (Newman 2010). Marshall (2004) gives a good overview of these with a focus on street networks. One can differentiate attributes characterizing the network structure and those relating the network to destinations along it as discussed in Sect. 3.4.

Dill (2004) reviews the literature and distinguishes connectivity measures based on intersection density, street density, percent of four-way intersections, connected intersection ratio, the percentage of grids, and pedestrian route directness. The network attributes we propose are based on these previous studies and evaluate the number of intersections, links, length, and number of dead-ends in the district of a location, i.e., within one minute freeflow distance.

Local Access Clifton et al. (2008) highlight that the accessibility measures on the level of a neighborhood are diverse and include among others the number of commercial and retail locations within walking distance, pedestrian connectivity or connectivity accessibility (Holtzclaw 1994; Song and Knaap 2003; Randall and Baetz 2001; Dill 2004).

Sevtsuk (2010) introduces further network-based characteristics as extensions of the gravity-based model as remoteness, the cumulative distance, cumulative number of turns or cumulative number of intersections required to reach specific points of interest in a given network radius.

These characteristics refer to information not given in our morphologic data model, so that we can only sum up the densities of buildings and their floor space.

3.4 Municipality and Region

The position of a location in a municipality or region defines the uniqueness and role of the location. On a municipal level, the topological information serves to characterize the urban landscape, network structure, accessibility, and proximity to urban center around locations. The weight and prominence of these urban centers across multiple municipalities are part of the analyses in a regional scale (Christaller 1933). Both scales are mainly operationalized by urban and regional planning disciplines.

Settlement Area Administrative borders of municipalities do not have necessarily represent morphological characteristics. They form political regions, but the current extent of the settled area often differs substantially from these. Through merging the buffer around buildings (dissolved buffer) we can define the morphologic settlement area; small areas need to be filtered to represent only conglomerations of buildings.

The shape and extent of this one can then be described through means of the attributes already mentioned in Sect. 3.1. The position of a location inside the settlement area is described through the distance to the boundary.

Centrality and Accessibility While object specific network measures have already been discussed in relation to the district (Sect. 3.3), the topological information can be used to quantify structural characteristics of cities and regions. Centrality measures, namely, closeness, betweenness and degree Newman 2010 are used to study the relevance of a node within a network.

Different weights or costs can be applied to the road segments in the calculation of centrality measures. Three cost functions are proposed: a constant of one per link, the length of a link, and the freeflow travel time per link. The latter can be estimated through assuming a certain speed per street type.

Accessibility calculations on floor space and buildings are used to represent local and regional differences within the urban landscape (Hansen 1959). The accessibility

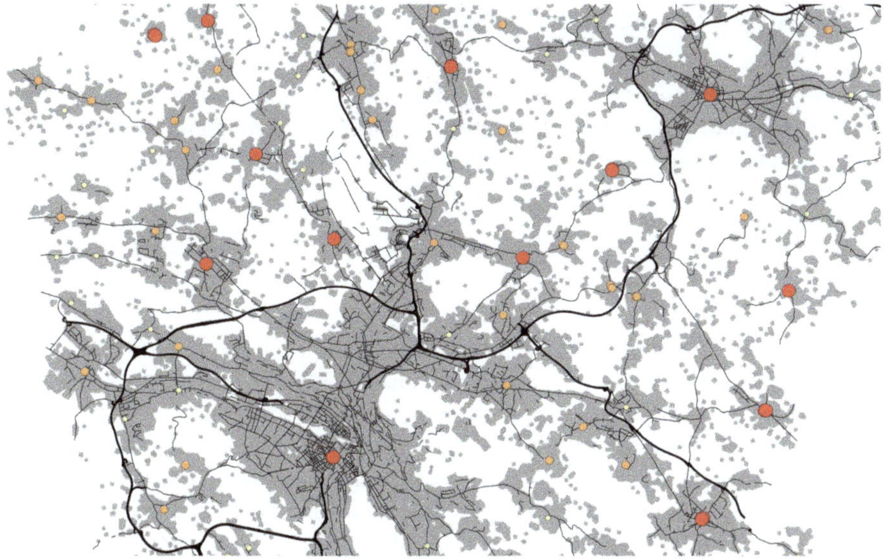

Fig. 5 Municipalities can be characterized by their settlement area (gray), networks (black), and global or local centers (red and orange). *Source* Generated data based on Cadastral survey of Canton Zurich and ©2011 Swisstopo (JA100120)

as distance weighted sum of opportunities (or gravity type measure) uses the freeflow travel time in the distance decay function to sum the opportunities of a location.

Urban Centers Assuming a higher density of construction for urban centers, one can analyze the built structure and define the location of urban centers in an algorithmic definition. The spatial distribution of built densities can be quantified through means of the interpolated accessibility measures of buildings. This represents a smoothed surface, which can be analyzed and described using methods of terrain analysis (Wilson and Gallant 2000).

The topographic or orometric prominence defines the height of a peak in the geomorphological classification of mountains. It is measured in relation to the lowest contour line encircling it and no higher summit (see Fig. 5). Applying this method on the accessibility measures allows to distinguish the location of a center, i.e., its peak, but also its local weight (peak height), global weight (height above sea level), and its catchment area (area of lowest contour line). Each location is evaluated by the distance to its closest center and the abovementioned values of the center. On a regional scale, we can evaluate the catchment area of the cities largest center and the distance to the next center with a higher global value (Fig. 5).

4 Examples of Location Types

After having defined a set of attributes to characterize the morphology in the different scales, we want to explore their suitability to define urban typologies in these scales. First, we want to summarize what kind of typologies should be the expected outcome.

4.1 Building

In their study of Stuttgart in Germany, Herrmann and Humpert (2001, p. 235ff) differentiate buildings based on morphologic characteristics into point-like buildings, linear buildings, buildings clusters, open and closed construction in a row, space defining developments on large scales and courthouses. They define a total of 22 subgroups for these, which differentiate modern and traditional housing. Similarly, Hecht et al. (2013) differentiate residential from nonresidential and define 10 types of buildings based on the main use and the shape of a building (Fig. 6). An earlier version of their paper differentiates multifamily houses into detached and closed forms and classifies "industrial multifamily housing" according to their height, which results in a total of 16 types (Meinel et al. 2008). None of these classifications make a finer differentiation of morphological characteristics, e.g., through reflecting the length of a building or the orientation of a building. On the other side, they all refer to a mix of building-specific characteristics and composition-specific characteristics, which are modeled as distinct scales in this chapter. We thus expect to find between 15 and 20 types for the building morphology. Through grouping the attributes of composition and buildings the outcome of the combined clustering is also evaluated. Up to 30 different types are expected for this typology.

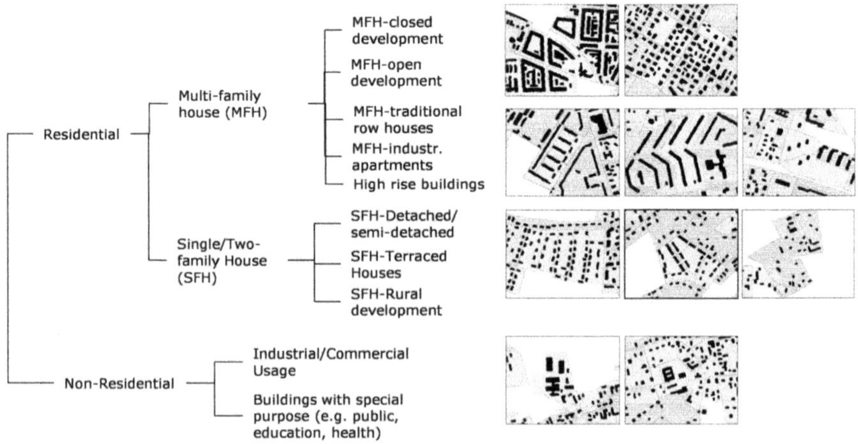

Fig. 6 Building types as described by Hecht et al. (2013). *Source* Hecht et al. (2013)

4.2 Composition

The composition describes the spatial interaction of buildings, blocks, and the streets. To the authors' knowledge, these kinds of attributes have not been reported previously to define urban types. However, some of the proposed characteristics are found in combination with building attributes to define types of buildings, e.g., when defining that detached buildings form a block (see above). Hecht et al. (2009) use several of the proposed attributes on buildings and blocks to define building types and later block types, but ignore the network structure.

Early tests showed that the degree of closure and permeability of a block, as well as the orientation of the building to the street are the dominating effects that can be differentiated in the clusters at that scale. It is unclear how many types of composition exist, but based on these findings one would expect between 10 and 15 types to be a suitable assumption.

4.3 Neighborhood

Neighborhood attributes evaluate built densities and network structure. One would expect to find core areas, disconnected areas and the border of settlements as typological definitions of these, but to the authors' knowledge, such descriptions with the proposed attributes have not been reported in the literature yet. The examples in Fig. 7 illustrate different schemes of networks, parcels and buildings, which give an idea of the variation to expect. It should at least be possible to differentiate these types, which vary in connectivity measures, network densities, the number of intersections or

Fig. 7 Examples of urban structures in Zurich. *Source* ©2011 Swisstopo (JA100120)

dead-ends and the density measures of buildings. However, one might further expect to find local differences related to these measures and thus a substantially higher number of neighborhood types than the five presented by Herrmann and Humpert (2001, p. 210).

4.4 *Municipality*

The scale of municipalities refers to the location, weight and catchment area of centers, as well as the accessibility and connectivity of locations. The building floor space and the network of streets are used as basis for this evaluation, as the points of interests or information on the use in a building are not given. Herrmann and Humpert (2001, p. 102) distinguish five types of urban centers based on their catchment area, which can be compared to the Central Place Theory proposed by Christaller (1933): cities (100,000 inhabitants), districts (50,000–100,000 inhabitants), quarters (25,000–50,000 inhabitants), major neighborhood center (15,000–25,000 inhabitants), minor neighborhood center (5000–15,000 inhabitants).

The Federal Office of Statistics of Switzerland differentiates nine types of municipalities based on density, size, and accessibility and 25 categories using socioeconomic criteria (Swiss Federal Statistical Office (BFS) 2012).

The attributes at the scale of municipality presented in Sect. 3.4 combine macroscopic characteristics, e.g., the size of the settlement or the catchment area of an urban center with local differences which can differ for neighboring locations, such as the connectivity measures. We thus expect to find between 10 and 20 types in the case study.

5 Clustering

5.1 *Classification Versus Clustering*

Hecht et al. (2009) introduce a rule-based classification based on the morphological characteristics of buildings, blocks, and their interaction. They predefine the types of buildings through weighting the impact and range of their morphological attributes. This method involves the training of the data on a reference sample. The advantage of supervised classification is the comparability of typologies across study areas: As their definition is static, we can observe the variation of behavior, price, and urban structure according to these types. Common classification techniques that allow to extract rules of a training set are decision trees, random forest trees, and regression trees (Hecht et al. 2013). The drawback of this method is the lack of usability in morphologically substantially different areas, as the training set needs to refer to all

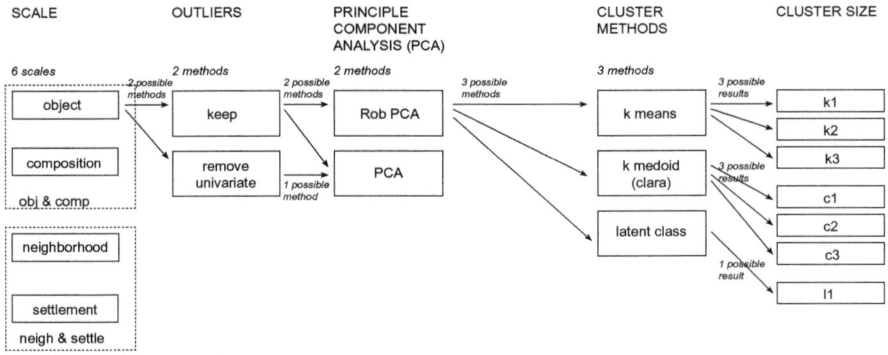

Fig. 8 Processing work frame as reported in this case study

existing types. If the predefined typologies do not cover all observations, we fail in assigning locations to individual types or misclassify objects.

Gil et al. (2012) take an alternative approach in form of unsupervised learning techniques, where the classification is not known a priori. They use a k-means cluster analysis to distinguish what they call *urban typologies*, i.e., each cluster represents one typology. In unsupervised classification, the number and content of classes are not predefined but data-driven. In consequence, the number and characteristics of attributes used to define a cluster are less rigid. On the other hand, the resulting clusters can vary across studies and reduce their comparability: The comparison can only be achieved through running the clustering on a dataset which groups all the observations. While this is a substantial drawback we do not know what kind of types to expect in our case study. As the types are unknown, cluster analysis is used, but the number of expected clusters will be guided by our expectations. Figure 8 gives an overview of the applied processing which will be explained.

5.2 *Methodology*

Various techniques of cluster analysis have been presented (Xu et al. 2005). Typical clustering methods include hierarchical clustering (connectivity based), centroid based clustering (k-means and k-medoids), and density-based clustering.

Hierarchical clustering proceeds in stages, beginning by defining every observation as an individual cluster. These get incrementally grouped based on their distance to other clusters until all observations are grouped into one cluster. The user gets a hierarchical tree of clusters and can define the appropriate clustering. A drawback of this method is its sensibility to noise. Furthermore, the required memory is proportional to the square of the number of groups in the initial partition, which strongly reduces the practical applicability to large datasets.

Centroid-based clustering uses the number of clusters defined a priori by the user to group the observations. These algorithms are also referred to as relocation methods due to the assignment of observations to cluster centers, which get iteratively optimized so that the squared distance from the observations to these get minimized. We differentiate k-means from k-medoid (Hartigan and Wong 1979; Kaufmann and Rousseeuw 1990). The earlier uses centroids and their Voronoi diagrams to find the optimal clustering while the later one uses data points of the observations as cluster centers instead. According to its implementation k-medoid is more robust to noise and outliers as compared to k-means because it uses the "mean (equivalent to the sum) of the dissimilarities of the observations to their closest medoid" instead of squared Euclidean distances (R Core Team 2014). The drawback of these techniques is that the algorithms prefer clusters of similar size: The values of the observations need to be scaled and to be of the same metric scale.

Density-based spatial clustering defines clusters as areas of higher density at the observations point (Ester et al. 1996). This handles noise and outliers and makes the clustering less sensitive to the irregular spatial distribution of data points: While k-means has a tendency to cluster in form of ellipsoids, density-based clustering can handle irregular cluster forms. The drawback is that the user has to make definitions for the size of neighborhood and the minimum number of points, which serve as density criterion. These require adaption to every dataset, which reduces the generic implementation across case studies, so that we did not implement this approach.

Latent class clustering as a probabilistic, model-based clustering has become popular throughout the last decade (Vermunt and Magidson 2002). These statistical models assume that the observed data "are generated by a mixture of underlying probability distributions in which each component represents a different group or cluster" (Fraley and Raftery 1998). They are sometimes referred to as mixture models, Bayesian classification, latent class cluster analysis or distribution-based clustering. These distributions are optimized to fit the data using an expectation–maximization algorithm and can be compared using statistical criteria. Characteristics of the distributions, i.e., the shape, volume, and orientation, are estimated and can be constrained or varied across the clusters, resulting in different model outputs (Fraley and Raftery 1998). A criterion such as the Bayesian information criteria is used to define the best model. In consequence, the algorithm allows overlap of clusters and additionally gives a measure of uncertainty that represents the quality of the clustering for each observation. Magidson and Vermunt (2002) compare it to k-means and find it not only to perform better than k-means, but to have substantial benefits due to its probabilistic classification.

5.3 *Principal Component Analysis*

Principal component analysis (PCA) is a linear transformation of variables into a lower dimensional space, i.e., it reduces the number of variables while retaining a maximum amount of information. Such PCA on the standardized variables is used to remove the correlations.

Following the Kaiser criteria, only those principal components (PC) are included that have an eigenvalue higher than 1. In our case study, this works well, except for the scale of composition, where many attributes have a low correlation, leading to an eigenvalue smaller than 1. To keep the applicability of the method as generic as possible the variation of algorithms across scales is avoided. Instead of the Kaiser criteria, all PC are used that explain a minimum of 95% of the variation. Hubert et al. (2005) introduce a robust principal component analysis, which is less sensitive to outliers and is also applied in our case study.

5.4 *Number of Clusters and Runtime*

The reprojected data into the principal components, i.e., the scores, form uncorrelated data that are used for the cluster analysis. A key to k means cluster analyses is selecting the number of clusters. Different methods to define these have been applied:

The Elbow Method Running the cluster analysis with the k-means algorithm on different number of clusters allows to observe the change in explained variance. This explained variance is then plotted against the number of clusters. These plots show the expected convergence with a growing number of clusters and in theory allow to define the ideal number of clusters (Fig. 9a, b). One should choose the number of clusters where adding another cluster does not give much better modeling results. This method is sometimes called the *elbow method*, because of the expected angle in the graph at the location of the *elbow criterion*. Our tests with samples of 10,000 observations showed variation, but we do not see any elbow criteria when running

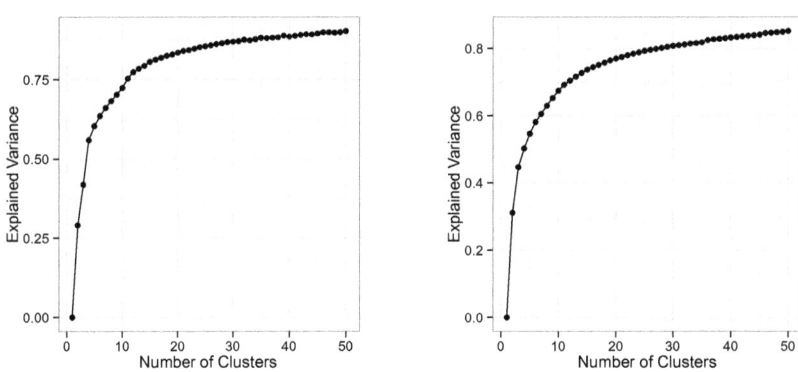

(a) 1000 observations (b) 10000 observations

Fig. 9 Explained variance per cluster for scale object

it on all observations. The variety in the data results in a smooth curve that does not allow for definition of elbow criteria and also highlights the ambiguous definition of number of clusters to take for k-means. A further drawback of this approach is the manual definition of ideal number of clusters, which does reduce its generic application on different scales and study areas.

The Indexes of NbClust The library *NbClust* of *R* groups various algorithms to find the ideal number of clusters in a data-driven approach. Referring to the initial expectation for the scale of objects, we found the indexes of Hartigan, the cubic clustering criterion (CCC), and Calinski and Harabasz (CH) to be the most appropriate. However, the NbClust package did not converge when run on the large dataset, so that we had to run it on a subset of 1000 observations. Due to the variation in small datasets observed for the elbow method, we repeated this run in 50 iterations and recorded the results of each run. Table 1 shows the results that were reported most often and its frequency.

The Silhouette Following Rousseeuw (1987), the silhouette of a k-means clustering can be used to control the quality of a clustering result. In the case study, this tends to result in inappropriate small number of clusters as compared to the earlier

Table 1 Estimated number of clusters per scale

Method	Building	Composition	Neighborhood	Municipality
K-means centroid				
Silhouette (best)	2	2	2	2
Silhouette (second)	3	3	3	3
Silhouette (third)	4	4	4	4
NbClust (hartigan)	3 (14/50)	3 (20/50)	3 (14/50)	3 (14/50)
NbClust (CCC)	2 (37/50)	2 (44/50)	2 (14/50)	2 (35/50)
NbClust (CH)	18 (4/50)	34 (4/50)	49 (7/50)	47 (5/50)
K-means medoid				
Silhouette (best)	2	2	3	2
Silhouette (second)	20	3	2	9
Silhouette (third)	22	10	4	3
Latent class				
BIC ($kmax = 50$)	50	50	40	50
Predefined				
based on expectation	25	15	15	20

expectation.

The summary of the proposed number of clusters is given in Table 1. These are not uniform across the different methods tested, i.e., one might question if there is a clear definition of the clusters at all. While our early tests showed that, the clustering results follow the expectations rather well, further studies may concentrate on transformation o limitation of attributes to facilitate the clustering. On the easy to evaluate scale of buildings, this is not required—they distinguish the expected types when using a high number of cluster. To guide the clustering the ideal number of clusters is predefined in different versions per scale and their outcome is evaluated: object (15, 20, 25), composition (5, 10, 15), object and composition (20, 25, 30), neighborhood (5, 10, 15), municipality (10, 15, 20), neighborhood and municipality (15, 20, 25).

The runtime for latent class clustering is tremendously higher than for k-means or k-medoid. Whether the results are any better as compared to these faster methods will be evaluated later.

As reported above the k-means and k-medoid clustering was run for a different number of clusters, to evaluate the effect of the cluster sizes. Due to the long calculation time, the latent class clustering was only run for the number of clusters proposed by the Bayesian information criteria (BIC). The user only defines the maximum number of clusters *kmax* to test, which was set to 50 in our case study. This showed a tendency to define *kmax* as the ideal number of clusters, except for the scale of neighborhood and the combined scales (object and composition, neighborhood, and municipality). Some of the individual clusters contain only very few observations, and additionally, we find cluster with high correlation. To avoid such effects, a lower *kmax* was defined based on the definitions already used for k-means and k-medoid.

6 Clusters Results

In the following the cluster outcomes of k-means, k-medoid and the latent class clustering will be discussed. For simplification, the models get abbreviated to k3 (k-means clustering), c3 (k-medoid clustering), and l1 (latent class clustering), while k1, k2, c1, c3 stand for the alternative cluster numbers with k-means and k-medoid.

Three approaches are taken to evaluate the cluster outcomes: 1. To view the spatial distribution and spatial interpretation of the clusters we print them in maps, showing one cluster per map. 2. The boxplot per cluster allows to get a deeper insight into the distribution and variance per PC. 3. Starplots (spider diagrams) give an overview of the median for each principal component and allow a comparison of clusters on multiple dimension (Figs. 10 and 11).

Object Size and dispersity as well as linearity and orientation are the dominant components to distinguish the building types across the different algorithms. K-means defines the clusters primarily on size and dispersity (PC1) and respects the extremes in the other dimensions—only linearity and orientation (PC4) gets less respected in

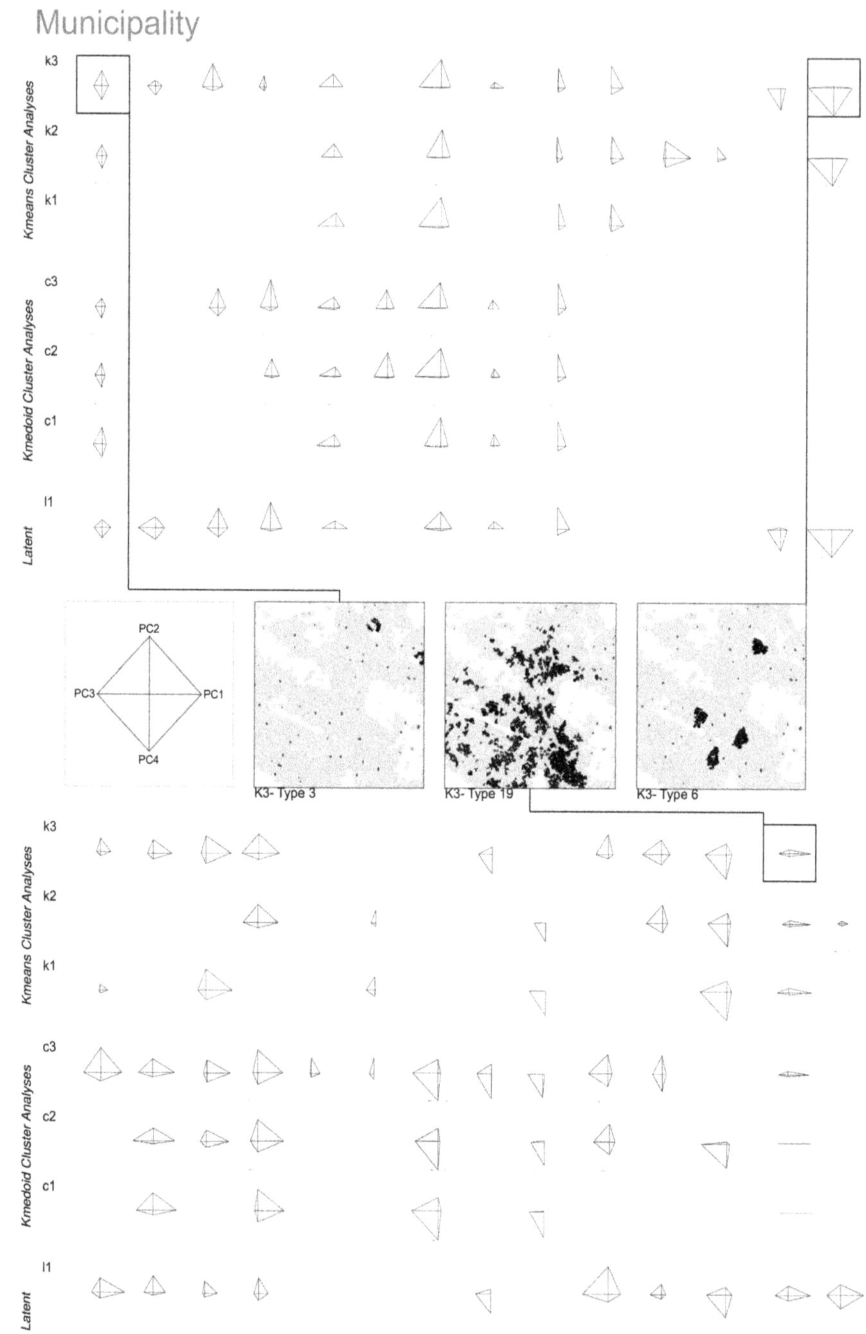

Fig. 10 Using starplots for comparing clusters. Example on the scale municipality

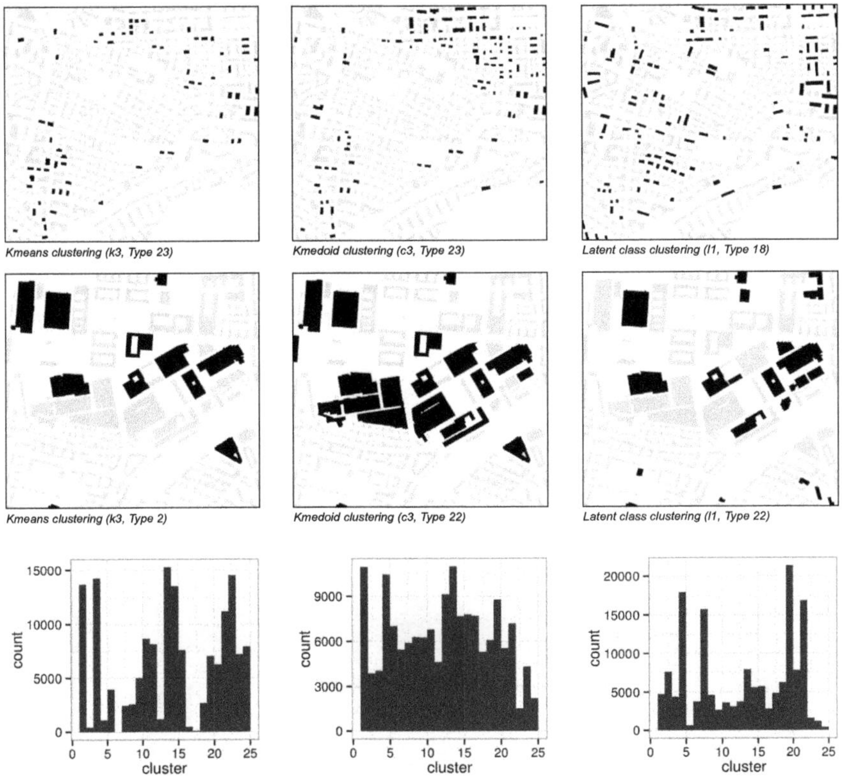

Fig. 11 Examples of clustering results and number of observations on the scale of buildings. *Source* Cadastral survey of Canton Zurich

this approach. These clusters allow for easy visual interpretation in maps and follow intuitive expectations on building types best. Meanwhile, k-medoid tends to separate using linearity and orientation (PC4) in the clusters. The results partly follow the ones of k-means, but we find a mix of different object sizes and building forms within each cluster. Furthermore, c3 will show many clusters with similar characteristics, indicating that a low number of clusters is more appropriate (c2, c1) with this approach. The simulation-based approach of the latent class clustering leads to higher variation in the distribution per dimension as compared to k-medoid and k-means. These clusters show a good recognition of urban blocks and open block structures, but mix building sizes and building forms even more than k-medoid (Figs. 12 and 13).

Composition The dimensions proposed in the PCA indicate that the clusters should distinguish the clusters based on the permeability of the block structure, the isolation, and density of construction. Similar to the scale of object, the descriptive analyses of the clusters show the best separation of the PC and also the highest consistency on different numbers of clusters for the k-means clusterings. All dimensions get respected in this clustering.

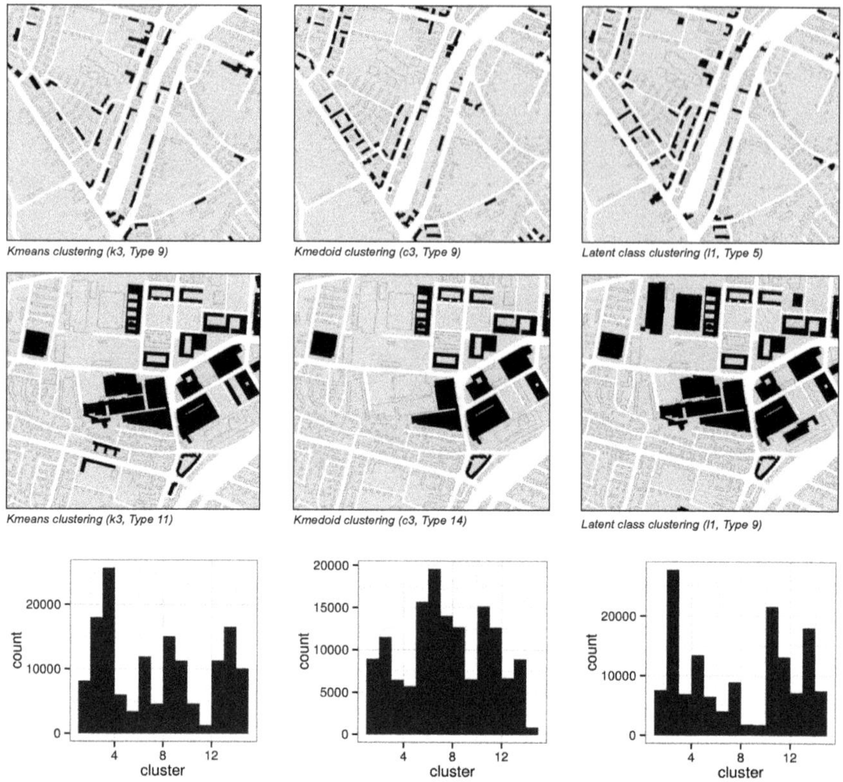

Fig. 12 Examples of clustering results and number of observations on the scale of composition. *Source* Cadastral survey of Canton Zurich

The high variance across neighboring locations reduced the interpretation in maps. However, we often find them comparable to building types, especially urban housing (urban villas), historic centers and urban blocks (irregular morphology of buildings, but clear block structure and densities) get well defined. Furthermore, buildings in second row, i.e., in the core of blocks without direct street access, get well identified. As many attributes explain characteristics that are specific to buildings types, we review also the combined the attributes of building features and composition features.

Object and Composition The results for the different algorithms suggest that clusters are primarily separated through their building morphology, and only then further separated based on their spatial relations. As in previous observations on scale building, k-means shows a more coherent differentiation of building types than k-medoid and the latent class clustering. Both latter form clusters of mixed morphologies but comparable composition features.

The combination of object and composition attributes thus reduces the clear separation of the building typologies. However, the combined use of both set of features is useful for a subset of building types, which have already been identified in scale

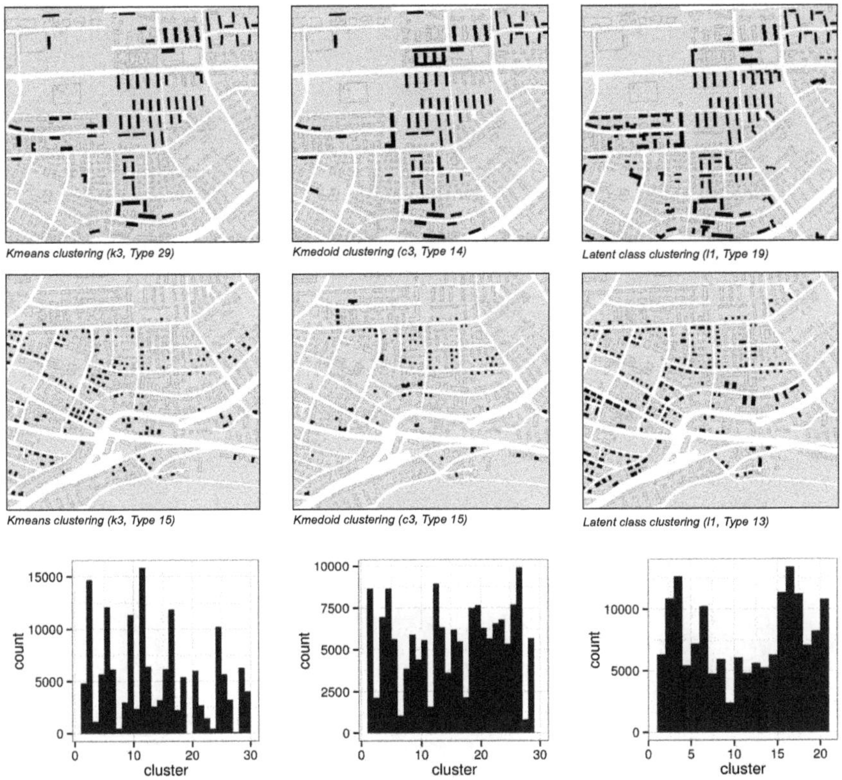

Fig. 13 Examples of clustering results and number of observations on the scale of object and composition. *Source* Cadastral survey of Canton Zurich

composition.

Neighborhood The neighborhood attributes combine densities and network characteristics in different extends for each location, i.e., smoothed (continuous) spatial distributions with almost discrete spatial distributions. Characterizing both types of features proves to create different results per clustering approach: The k-means clustering represents the variation in larger extends, i.e., of densities very well, but we find local differences of the network having low impact. Such local differences of the network (PC3) get well identified by k-medoid, which also will characterize the isolation (PC4) of a location in the cluster outcomes. The latent class clusters form a mix of both, respecting all dimensions, but forming more overlaps in the distributions per cluster.

Viewing the representation as maps we observe the k-means and the latent class approach lead to spatial distribution of clusters that form areas of relatively homogeneous boundary and reflect the distance to settlement border, whereas this is less

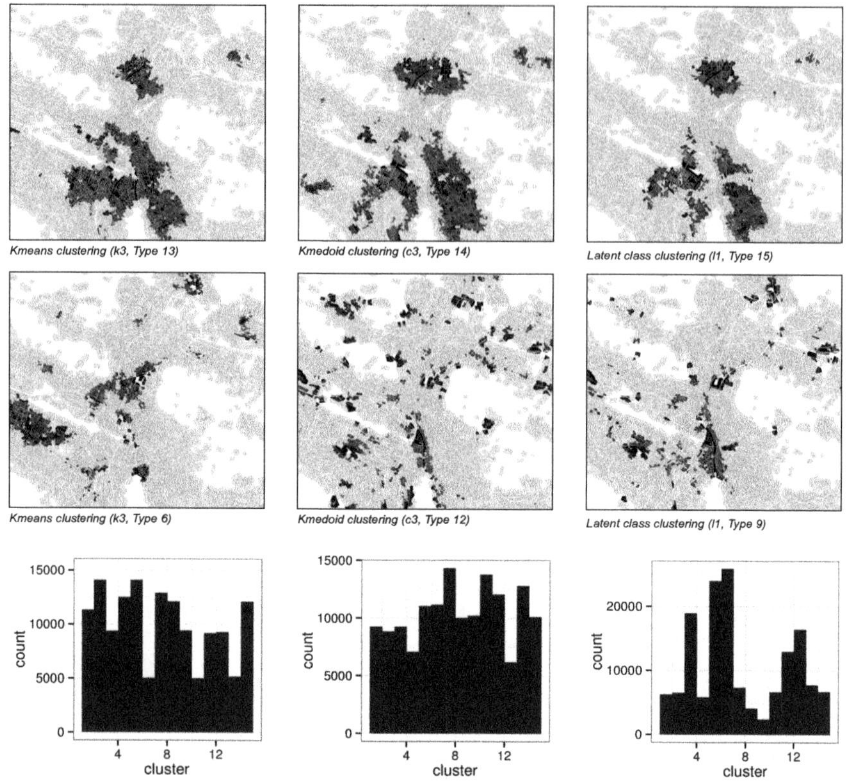

Fig. 14 Examples of clustering results and number of observations on the scale of neighborhood. *Source* Cadastral survey of Canton Zurich

found for k-medoid.

Municipality The features in scale municipality combine descriptions that vary locally (microscopic) with descriptions that vary only on wider scale (macroscopic).

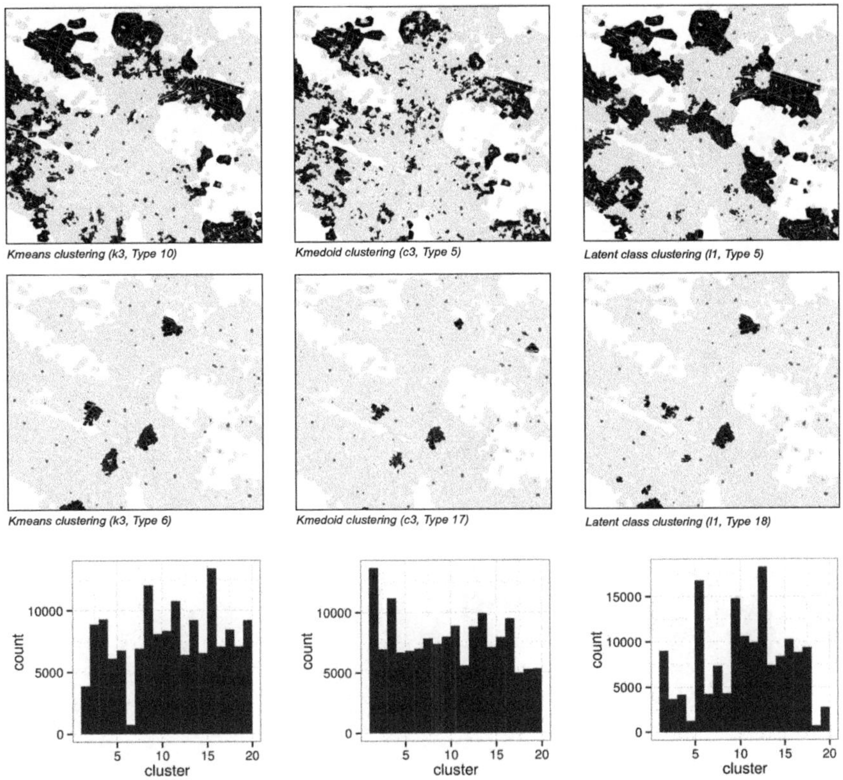

Fig. 15 Examples of clustering results and number of observations on the scale of municipality. *Source* Cadastral survey of Canton Zurich

Example for the earlier is the network betweenness, example for the latter the accessibility of buildings. Similar to the scale neighborhood we observe that the cluster algorithms weight these dimensions differently. K-means primarily represent the macroscopic attributes while k-medoid respects microscopic differences. The clusters of the latent class approach respect all dimensions, which leads to a less clear separation per cluster (Figs. 14, 15 and 16).

Neighborhood and Municipality The combined scales of neighborhood and municipality show that k-means and k-medoid clustering distinguish the same PCs as in the individual scales while the latent class respects again all PCs, but having a wider variation per cluster. In the maps, the latent class clusters are the easiest to read and represent the catchment areas of urban centers.

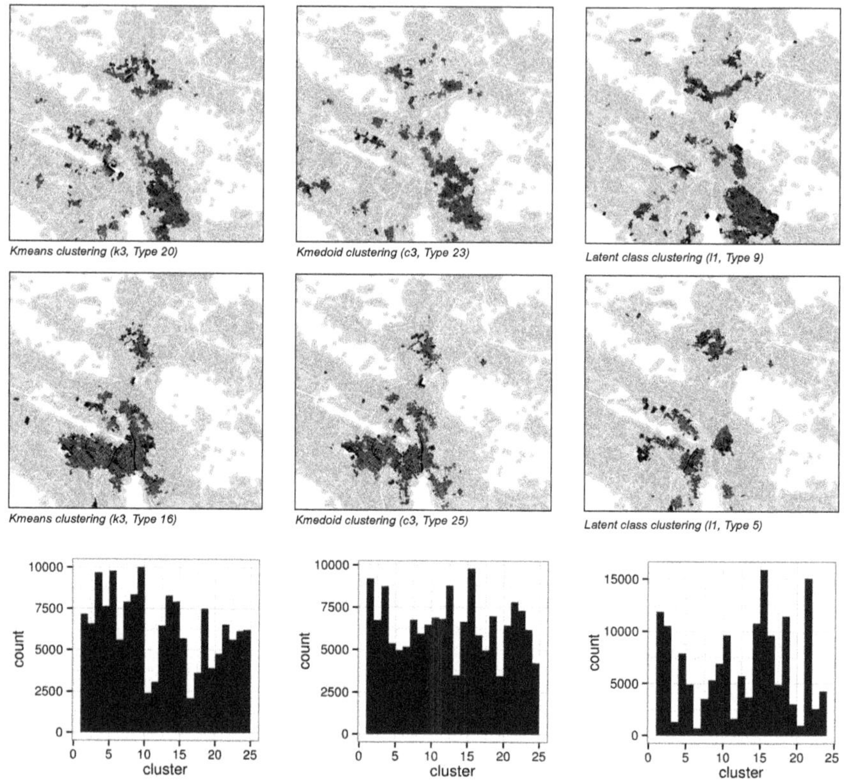

Fig. 16 Examples of clustering results and number of observations on the scale of neighborhood and municipality. *Source* Cadastral survey of Canton Zurich

7 Summary and Discussion

In this chapter, we show that urban morphology can be used to characterize the spatial configuration of urban settlement at different scales, through conducting a case study of the canton of Zurich. The quantitative attributes proposed allow for use in most quantitative models. Moreover, the processing is coded and thus objective and reproducible. It runs independent of the scale of the study area and allows to visualize and quantify key feature of the built environment. In that sense, it is a step forward to allow for systematic application in quantitative models and comparison of study areas.

Besides the direct use of the attributes, principal component analyses and cluster analysis have been applied. The robust principal component analyses have proven to be suited to handle outliers and reduce correlation among attributes. The resulting principal components create an additional insight to read the spatial structure through more generic descriptions such as connectivity, urbanity, or isolation. For the current case study, we used the full set of attributes for each principal component—a possible extension would be a selection of loadings to use per principal component.

Through means of clustering analyses, the principal components were used to define urban typologies. Three clustering algorithms have been applied and compared: k-means clustering, k-medoid clustering, and latent class clustering, all of which allow for application in large datasets. While showing differences for the methods and cluster sizes test, all of these results show clear trends for the resulting urban typologies. Buildings get well separated into architectural types. The composition supports the definition of specific building types, such as the historic centers, urban villas, and isolated locations, but is not of use for definition of typologies. Neighborhood typologies are less clear to define and either form areas of similar densities or distinct locations with common network characteristics—representing both would require to define a large number of typologies. Last the typologies on scale of municipality form districts through integrating centers, catchment areas, and accessibilities.

The proposed data model is kept simple to allow for application in any study area. Obviously, this limits the spatial dimensions that are represented, e.g., points of interests, public transport networks, or sociodemographic characteristics are not represented. In our opinion, most important extensions to this model would be the integration of the topography and public transport networks as well as a systematic definition of points of interests and land use information. However, the low availability of these information on a homogeneous level currently reduces their application in different study areas, and the proposed approach proves to be a good alternative. Furthermore, our recent studies on location choice and real estate prices have shown that the urban morphology explains a large part of these—while further attributes can help to fine-tune the models, it is recommended to use the morphologic description as a key element.

This chapter shows that the proposed methods allow to describe locations and distinguish similar locations at the different scales. Assuming that a household has a preference for a certain location type, e.g., a building type, this allows to capture such preferences. Furthermore, it allows to "map" the spatial structure through transforming multiple continuous attributes into discrete types. Finally, the urban types remove the correlations across the individual variables. All these reasons make it an interesting and valid approach to analyze a study area, even though they do not allow for comparison across different regions.

The urban types of different scales proposed here show correlations which represent planning constraints and developers behavior. For example, single family housing (scale object) is rarely found in the city center (scale municipality) but often located near open fields (scale composition). These correlations can only be seen through the definition of urban types as proposed here. They allow to better understand the urban structure and to evaluate the results of planning constraints, e.g., as undertaken in zoning maps.

However, the proposed definitions of urban types also were found to be problematic for multiple reasons. 1. The interpretation of the urban types is not obvious. While building types form discrete observations known to us, the urban types of the other scales require detailed analyses to become readable. This reduces their applicability and the comparability to other study areas. A predefinition of standard types for the different scales could be of benefit—urban types then would have to be classified in a rule-based classification. 2. The processing of the clusters is not objective if the number of clusters is user-defined. While such results can guide the user to define standard types that represent different dimensions of space, a standardized preprocessing of the attributes might help to better distinguish clusters automatically. 3. The value of urban types in the scale composition needs further evaluation. Most variables appear to be uncorrelated so that the PCA is not needed. Furthermore, the resulting types are hard to understand, partly correlated with building types. However, some of the variables proposed in this scale can help to better identify specific building types in future work.

The spatial attributes proposed allow for objective comparison and modeling of spatial behavior, but require a high effort to calculate these. Their calculation demand a large set of computationally intensive geoprocessing operations and prior knowledge on these. This shows the need to move forward through making such spatial characteristics easily accessible for research and practice in a standardized way. With the creation of UrbanDataLab (UrbanDataLab.net), we want to offer this service in making data available to research and practicals, so that a standardized set of spatial descriptions can be applied in a wide range of studies. Visit us at www.urbandatalab.net, for additional information or if interested in applications.

Foreword and Acknowledgments This chapter is the first attempt toward a systematic comparison of the morphology of urban areas, which has been developed as part of a Ph.D.-Thesis at the IVT ETH Zürich (Schirmer 2015).

The morphologic attributes described in this chapter were initially published in extended version in the Journal of Transport and Landuse—JTLU (Schirmer and Axhausen 2016). We would like to thank JTLU and especially David Levinson for the permission to include our earlier work in this chapter. Our research on the urban morphology and the spatial behavior has been ongoing since then. Throughout the last year, the geoprocessing routines have been extended with the start-up company UrbanDataLab Schirmer. Our focus lies on a generic use in other study areas and the analyses of implication to location choice and real estate prices.

References

Alexander, C., Ishikawa, S., Silverstein, M., Jacobson, M., Fiksdahl-King, I., & Angel, S. (1977). *A pattern language*. Oxford: Oxford University Press.

Bafna, S. (2003). Space syntax a brief introduction to its logic and analytical techniques. *Environment and Behavior*, *35*(1), 17–29.

Bhat, C. R., Kockelman, K., Chen, Q., Handy, S., Hani, M., & Weston, L. (2000). Urban accessibility index: Literature review. Research Report TX_01/7-4938-1, University of Texas, Austin.

Christaller, W. (1933). *Die zentralen Orte in Sueddeutschland: eine oekonomisch-geographische Untersuchung ueber die Gesetzmaessigkeit der Verbreitung und Entwicklung der Siedlungen mit staedtischen Funktionen*. Jena: Fischer Verlag.

Clifton, K., Ewing, R., Knaap, G., & Song, Y. (2008). Quantitative analysis of urban form: a multidisciplinary review. *Journal of Urbanism: International Research on Placemaking and Urban Sustainability*, *1*(1), 17–45.

Dill, J. (2004). Measuring network connectivity for bicycling and walking. In *83th Annual Meeting of the Transportation Research Board*. Washington, D.C.

Ester, M., Kriegel, H.-P., Sander, J., & Xu, X. (1996). A density-based algorithm for discovering clusters in large spatial databases with noise. In E. Simoudis, J. Han, & U. Fayyad (Eds.) *Proceedings of 2nd International Conference on Knowledge Discovery and Data Mining (KDD-96)*, (pp. 226–231). Portland: AAAI Press.

Fraley, C., & Raftery, A. E. (1998). How many clusters? Which clustering method? Answers via model-based cluster analysis. *The Computer Journal*, *41*(8), 578–588.

Gil, J., Beiro, J. N., Montenegro, N., & Duarte, J. P. (2012). On the discovery of urban typologies: data mining the many dimensions of urban form. *Urban Morphology*, *16*(1), 27–40.

Google Inc. (2013a). Google earth. http://www.google.de/intl/de/earth/.

Google Inc. (2013b). Google maps. https://www.google.com/maps.

Hansen, W. (1959). How Accessibility Shapes Land Use. *Journal of the American Institute of Planners*, *25*(2), 73–76.

Hartigan, J. A., & Wong, M. A. (1979). A k-means clustering algorithm. *Journal of the Royal Statistical Society. Series C (Applied Statistics)*, *28*(1), 100–108.

Hecht, R., Herold, H., & Meinel, G. (2009). Automated Analysis of Settlement Development using Topographic Maps. In *GI2009 - 9. Schsisches GIS-Forum*. Dresden.

Hecht, R., Herold, H., & Meinel, G. (2013). Automatic derivation of urban structure types from topographic maps by means of image analysis and machine learning. In M. Buchroithner (Ed.) *26th International Cartographic Conference—Proceedings: International Cartographic Association*. Dresden: International Cartographic Association.

Herrmann, T., & Humpert, K. (2001). Typologie der Stadtbausteine. In J. Jessen, U. Grammel, & A. Hussermann (Eds.) *Lehrbausteine Stdtebau: Basiswissen fr Entwurf und Planung*. Universitt Stuttgart, Stdtebau-Institut.

Hillier, B. (1996). *Space is the machine: a configurational theory of architecture*. Cambridge: Press Syndicate of the University of Cambridge.

Hillier, B., & Hanson, J. (1984). *The social logic of space*. Cambridge: Cambridge University Press.

Holtzclaw, J. (1994). *Using residential patterns and transit to decrease auto dependence and costs*. San Francisco: Natural Resources Defense Council.

Hubert, M., Rousseeuw, P. J., & Vanden Branden, K. (2005). ROBPCA: A new approach to robust principal component analysis. *Technometrics*, *47*(1), 64–79. https://doi.org/10.1198/004017004000000563.

Kaufmann, L., & Rousseeuw, P. J. (1990). *Finding groups in data*. New York.: J. Wiley & Sons.

Lampugnani, M. L., Keller, T. K., & Buser, B. (2007). *STADTISCHE DICHTE*. Zurich: NZZ Libro.

Lee, D. T. (1982). Medial axis transformation of a planar shape. *IEEE Transactions on Pattern Analysis and Machine Intelligence*, *4*(4), 363–369.

Magidson, J., & Vermunt, J. (2002). Latent class models for clustering: A comparison with k-means. *Canadian Journal of Marketing Research*, *20*(1), 36–43.

Marshall, S. (2004). *Streets and patterns*. London: Routledge.

Marshall, S. (2012). Science, pseudo-science and urban design. *Urban Design International*, *17*(4), 257–271.

Meinel, G., Hecht, R., Herold, H., & Schiller, G. (2008). Automatische Ableitung von stadtstrukturellen Grundlagendaten und Integration in einem Geographischen Informationssystem. Tech. Rep. Heft 134, Bundesamt fr Bauwesen und Raumordnung, Bonn.

Newman, M. E. J. (2010). *Networks: an Introduction*. Oxford: Oxford University Press.

Ogniewicz, R. L., & Kbler, O. (1995). Hierarchic voronoi skeletons. *Pattern Recognition*, *28*(3), 343–359.

OpenStreetMap contributors (2017). Planet dump retrieved from https://planet.osm.org, https://www.openstreetmap.org.

R Core Team (2014). R: Clustering large applications. http://stat.ethz.ch/R-manual/R-patched/library/cluster/html/clara.html.

Randall, T. A., & Baetz, B. W. (2001). Evaluating pedestrian connectivity for suburban sustainability. *Journal of Urban Planning and Development*, *127*(1), 1–15.

Rousseeuw, P. J. (1987). Silhouettes: a graphical aid to the interpretation and validation of cluster analysis. *Journal of computational and applied mathematics*, *20*, 53–65.

Schirmer, P., & Axhausen, K. W. (2016). A multiscale classification of the urban morphology. *Journal of Transport and Land Use*.

Schirmer, P. M. (2015). *Classification of the Urban Morphology for Use in Residential Location Choice Models*. Ph.D. thesis, ETH Zürich, Zürich.

Schirmer, P. M., Eggermond, M. A. B. v., & Axhausen, K. W. (2014). The role of location in residential location choice models: a review of literature. *Journal of Transport and Land Use*, *7*(2), 3–21.

Sevtsuk, A. (2010). *Path and Place: A Study of Urban Geometry and Retail Activity in Cambridge and Somerville*. Ph.D. thesis, Massachusetts Institute of Technology, Cambridge.

Song, Y., & Knaap, G.-J. (2003). New urbanism and housing values: a disaggregate assessment. *Journal of Urban Economics*, *54*(2), 218–238.

Swiss Federal Statistical Office (BFS) (2012). *Gemeindetypologie und Stadt/Land-Typologie*. Swiss Federal Statistical Office (BFS), Neuchatel. https://www.bfs.admin.ch/bfs/de/home/statistiken/raum-umwelt/nomenklaturen/gemtyp.html.

Vermunt, J. K., & Magidson, J. (2002). Latent class cluster analysis. *Applied latent class analysis*, (pp. 89–106).

Vickerman, R. W. (1974). Accessibility, attraction, and potential: a review of some concepts and their use in determining mobility. *Environment and Planning A*, *6*(6), 675–691.

Wachs, M., & Kumagai, T. (1973). Physical accessibility as a social indicator. *Socio-Economic Planning Sciences*, *7*(5), 437–456.

Wilson, J. P., & Gallant, J. C. (2000). *Terrain Analysis: Principles and Applications*. New York: John Wiley & Sons.

Xu, R., Wunsch, D., and others (2005). Survey of clustering algorithms. *IEEE Transactions on Neural Networks*, *16*(3), 645–678.

An Urban Morphogenesis Model Capturing Interactions Between Networks and Territories

Juste Raimbault

Abstract Urban systems are composed of complex couplings of several components, and more particularly between the built environment and transportation networks. Their interaction is involved in the emergence of the urban form. We propose in this chapter to introduce an approach to urban morphology grasping both aspects and their interaction. We first define complementary measures and study their empirical values and their spatial correlations on European territorial systems. The behavior of indicators and correlations suggest underlying non-stationary and multi-scalar processes. We then introduce a generative model of urban growth at a mesoscopic scale. Given a fixed exogenous growth rate, population is distributed following a preferential attachment depending on a potential controlled by the local urban form (density, distance to network) and network measures (centralities and generalized accessibilities), and then diffused in space to capture urban sprawl. Network growth is included through a multi-modeling paradigm: implemented heuristics include biological network generation and gravity potential breakdown. The model is calibrated both at the first (measures) and second (correlations) order, the later capturing indirectly relations between networks and territories.

Keywords Urban morphology · Road network topology · Spatial correlations · Urban morphogenesis · Reaction–diffusion · Coevolution

J. Raimbault (✉)
UPS CNRS 3611 ISC-PIF, Paris, France
e-mail: juste.raimbault@polytechnique.edu

J. Raimbault
UMR CNRS 8504 Géographie-cités, Paris, France

L. D'Acci (ed.), *The Mathematics of Urban Morphology*,
Modeling and Simulation in Science, Engineering and Technology,
https://doi.org/10.1007/978-3-030-12381-9_17

1 Introduction

1.1 Urban Morphology

The structure of urban systems determine their functional properties at several scales, and the *urban morphology* both participates in the emergence of urban functions and determines how they evolve and how they are used by the agents (Batty and Longley 1994). For example, relationships between urban form, mobility practices, and sustainability of metropolitan areas can be established (Le Néchet 2010).

Several approaches can be taken to study urban morphology. At the microscopic scale, in operational urbanism, for example, urban morphology is defined as "the characteristics of the material form of cities and fabrics" (Paquot 2010). A similar positioning is taken by architecture, considering forms at the scale of buildings (Moudon 1997). At the mesoscopic scales, Tsai (2005) introduces indicators to quantify urban form, considered as the spatial distribution of population. Le Néchet (2009) recalls the necessity of a multidimensional measure. It is however possible to obtain a robust description of urban form with a small number of independent indicators by a reduction of the dimension (Schwarz 2010). It is also possible to use indexes from fractal analysis, such as systematically applied by Chen (2016) to classify urban forms. Other more original indexes can be proposed, such as by Lee et al. (2017) which use the variations of trajectories for routes going through a city to establish a classification and show that it is strongly correlated with socioeconomic variables. An other entry is through the topology of urban networks, and more particularly road networks, for which (Lagesse 2015) gives a broad overview. Finally, geographers also consider at the macroscopic scale the spatial distribution of cities and their populations as the form of the urban system (Pumain 2011).

The link between urban morphology and the topology of the underlying relational network has been suggested in a theoretical approach by Badariotti et al. (2007). D'Acci (2015) models cities as isobenefit lines for agents and suggests in opening the link between the distribution of amenities and networks and flows, highlighting the open question of coevolutive processes between the urban form and spatial networks, in the sense of circular causalities.

There is to the best of our knowledge no contribution in the literature considering the interaction between the built environment and networks as an intrinsic process of the emergence of the urban form. This chapter thus aims at sketching an exploration of this concept at the mesoscopic scale, through empirical data analysis and modeling. More precisely, our contribution is twofold: (i) we introduce indicators quantifying simultaneously the morphology of population spatial distribution and the topology of road networks, and study empirically their values and spatial correlations, including thus the interactions between both in the quantification of the form; (ii) we describe and explore a model of urban morphogenesis, aiming at capturing these interaction processes in the emergence of the form.

The rest of this chapter is organized as follows. We first detail the rationale of our approach to urban form, in particular the scale and objects considered. The second section develops the methodology used to quantify urban form, from the point of view of morphological indicators, network topology indicators, and their correlations. Empirical results for Europe are then described, to then be used to calibrate a model of urban morphogenesis that we detail and explore. We finally discuss potential developments and applications of this work.

1.2 Urban Morphology and Interactions Between Networks and Territories

Through relocation processes, sometimes induced by networks, we can expect the latest to influence the distribution of populations in space (Wegener and Fürst 2004). Reciprocally, network characteristics can be influenced by this distribution. We propose therefore here to study a coupled approach to the urban form, given by synthetic indicators for these two subsystems, and by correlations between these indicators. At the scale of the system of cities, the spatial nature of the urban system is captured by cities position, associated with aggregated city variables. We propose to work here at the mesoscopic scale, at which the precise spatial distribution of activities is necessary to understand the spatial structure of the territorial system. We will therefore use the term of morphological characteristics for population density and the road network.

The choice of "relevant" boundaries for the territory or the city is a relatively open problem which will often depend on the question we are trying to answer (Páez and Scott 2005). In this way, Guérois and Paulus (2002) show that the entities obtained are different if we consider an entry by the continuity of the built environment (morphological), by urban functions (employment area, for example) or by administrative boundaries. Boeing (2017a) furthermore shows that statistics of network measures significantly change when the scale of study switches from neighborhood to cities and metropolitan areas. Similarly, Cottineau et al. (2017) show that scaling exponents also strongly depends on boundaries and thresholds chosen to define the urban area.

To tackle this ontological issue of the object under study, we (i) stay at a high resolution; and (ii) compute field values with a small enough offset to obtain continuous values. At the chosen scale, we can assume that territorial characteristics, for population and network, are locally defined and vary in an approximately continuous way in space. We will compute therefore the indicators on spatial windows of fixed size, taken of the order of magnitude of 100 km (in practice 50 km, following Raimbault (2018a) which furthermore shows a low sensitivity of indicators when comparing with windows of 30 and 100 km), but with a small enough offset. Thus, the construction of fields of morphological indicators will allow us to endogenously reconstruct territorial entities through the emergent spatial structure of indicators at larger scales.

2 Measuring Morphology: Method

We detail in this section the indicators used to quantify morphology, both for population distribution (urban morphology) and for network topology.

2.1 *Urban Morphology*

Our quantification of the urban form in itself, in the sense of properties of the spatial distribution of populations, is taken from a previous work developed in Raimbault (2018a). We recall here the formal definition of morphological indicators. We work with gridded population data $(P_i)_{1\leq i\leq N^2}$, where $M = N^2$, d_{ij} is the Euclidean distance between cells i, j, $P = \sum_{i=1}^{M} P_i$ the total population and $\bar{P} = P/M$ the average population per cell. We use the following indicators: (i) rank-size slope γ, expressing the degree of hierarchy in the distribution; (ii) entropy of the distribution (Le Néchet 2015), given by

$$\mathcal{E} = \sum_{i=1}^{M} \frac{P_i}{P} \cdot \ln \frac{P_i}{P} \tag{1}$$

(iii) spatial autocorrelation given by Moran index (Tsai 2005) defined as

$$I = M \cdot \frac{\sum_{i\neq j} w_{ij} \left(P_i - \bar{P}\right) \cdot \left(P_j - \bar{P}\right)}{\sum_{i\neq j} w_{ij} \sum_i \left(P_i - \bar{P}\right)^2} \tag{2}$$

where the spatial weights are taken as $w_{ij} = 1/d_{ij}$; and (iv) average distance between individuals (Le Néchet 2009) given by

$$\bar{d} = \frac{1}{d_M} \cdot \sum_{i<j} \frac{P_i P_j}{P^2} \cdot d_{ij} \tag{3}$$

where d_M is a normalization constant taken as the diagonal of the area on which the indicator is computed in our case.

The first two indexes are not spatial and are completed by the last two that take space into account. Following Schwarz (2010), the effective dimension of the urban form justifies the use of all.

2.2 *Network Measures*

We consider network aggregated indicators as a way to characterize transportation network properties on a given territory, the same way morphological indicators yielded information on urban structure. They are assumed thus to capture another dimension of urban form. We propose to compute some simple indicators on same spatial extents as for population density, to be able to explore relations between these static measures.

Static network analysis has been extensively documented in the literature, such as through the typology of urban street networks obtained by Louf and Barthelemy (2014) on a cross-sectional study of cities. Similarly, Moosavi (2017) uses techniques from deep learning to establish a typology of urban road networks for a large number of cities across the world. The questions behind such approaches are multiple: they can aim at finding typologies or at characterizing spatial networks, at understanding underlying dynamical processes in order to model morphogenesis, or even at being applied in urban planning such as *Space Syntax* approaches (Hillier and Hanson 1989). We aim here at characterizing urban morphological properties.

We introduce indicators to have a broad idea of the form of the network, using a certain number of indicators to capture the maximum of dimensions of properties of networks, more or less linked to their use. These indicators summarize the mesoscopic structure of the network and are computed on topological networks obtained through simplification steps that will be detailed later. If we denote the network with $N = (V, E)$, nodes have spatial positions $\vec{x}(V)$ and populations $p(v)$ obtained through an aggregation of population in the corresponding Voronoï polygon, and edges E have *effective distances* $l(E)$ taking into account impedances and real distances (to include the primary network hierarchy). We then use:

- Characteristics of the graph, obtained from graph theory, as defined by Haggett and Chorley (1970): number of nodes $|V|$, number of links $|E|$, density d, average length of links $\bar{d_l}$, average clustering coefficient $\bar{c}$, number of components c_0.
- Measures directly related to distances within the network: diameter r, Euclidean performance v_0 (defined by Banos and Genre-Grandpierre (2012)), average length of shortest paths $\bar{l}$.
- Centrality measures: these are aggregated at the level of the network by taking their average and their level of hierarchy, computed by an ordinary least squares of a rank-size law, for the following centrality measures:
  - Betweenness centrality (Crucitti et al. 2006), average $\bar{bw}$ and hierarchy α_{bw}: given the distribution of centrality on all nodes, we take the slope of a rank-size adjustment and the average of the distribution.
  - Closeness centrality (Crucitti et al. 2006), average $\bar{cl}$ and hierarchy α_{cl}.
  - Accessibility (Hansen 1959), which is in our case computed as a closeness centrality weighted by populations: average $\bar{a}$ and hierarchy α_a.

Network performance is close to the rectilinearity measure (*straightness*) proposed by Josselin et al. (2016), which show that it efficiently differentiates rectilinear networks and radio-concentric networks that are both recurring urban networks. Our indicators are conceived to capture network topology but not the use of the network: developments with suited data could extend these analyses to the functional aspect of networks, such as performance measures computed by Trépanier et al. (2009) using massive data for a public transportation network.

2.3 Correlations

Local spatial correlations are computed on spatial windows gathering a certain number of observations, and thus of windows on which indicators have been computed. We denote by l_0 the resolution of the distribution of indicators. The estimation of correlations is then done on squares of size $\delta \cdot l_0$ (with δ which can vary typically from 4 to 100). Correlations are estimated using a standard Pearson estimator. Our approach is equivalent to computing geographically weighted correlations (Brunsdon et al. 2002) with a Heaviside window.

δ gives simultaneously the number of observations used for the local estimation of correlation, and the spatial range of the corresponding window. Its value thus directly influences the confidence of the estimation. We can indeed derive the behavior of the correlation estimator as a function of the size of the sample. Under the assumption of a normal distribution of two random variables X, Y, then the Fisher transform of the Pearson estimator $\hat{\rho}$ such that $\hat{\rho} = \tanh(\hat{z})$ has a normal distribution. If z is the transform of the real correlation ρ, then a confidence interval for ρ is of size

$$\rho_+ - \rho_- = \tanh(z + k/\sqrt{N}) - \tanh(z - k/\sqrt{N}) \tag{4}$$

where k is a constant. As $\tanh z = \frac{\exp(2z)-1}{\exp(2z)+1}$, we can develop this expression and reduce it. We obtain

$$\rho_+ - \rho_- = 2 \cdot \frac{\sinh(2k/\sqrt{N})}{\cosh(2z) + \cosh(2k/\sqrt{N})} \tag{5}$$

Using the fact that $\cosh u \sim_0 1 + u^2/2$ and that $\sinh u \sim_0 u$, we indeed obtain that

$$\rho_+ - \rho_- \sim_{N \gg 0} k'/\sqrt{N} \tag{6}$$

This expected asymptotic confidence interval will be of use when studying the behavior of correlations as a function of δ.

3 Empirical Application

We can now give implementation details and results for the application of this method to local territorial systems of Europe. All source code and results for this section are available on the open repository of the project at https://github.com/JusteRaimbault/CityNetwork/tree/master/Models/StaticCorrelations.

3.1 Urban Morphology

As this work extends (Raimbault 2018a), the empirical values of morphological indicators for population distribution are taken from it. We recall that the implementation of indicators must be done carefully, since computational complexities can reach $O(N^4)$ for the Moran index, for example: we use convolution through fast Fourier transform, which is a technique allowing the computation of the Moran index with a complexity in $O(\log^2 N \cdot N^2)$.

Indicators are computed on 50 km width square windows with grids of resolution 500 m, computed from the Eurostat population dataset (EUROSTAT 2014). According to Batista e Silva et al. (2013) which details its construction, our aggregation should allow us to avoid biases at high resolutions. The spatial distribution of indicators unveils typical local and regional regimes, which (Raimbault 2018a) synthesizes using unsupervised learning. They typically contain a metropolitan regime, a medium-sized city regime, a rural regime (split in two between North and South), and a mountainous regime.

3.2 Network Topology

3.2.1 Data Preprocessing

The implementation of network indicators requires a preprocessing from raw data that we detail first. We assume to work only with the road network. Indeed, data for the current road network is openly available through the OpenStreetMap (OSM) project (2012). Its quality was investigated for different countries such as England (Haklay 2010) and France (Girres and Touya 2010). It was found to be of a quality equivalent to official surveys for the primary road network. We will however simplify the network at a sufficient level of aggregation to ensure the robustness of results.

The network constituted by primary road segments is aggregated at the fixed granularity of the density grid to create a graph. It is then simplified to keep only the topological structure of the network, normalized indicators being relatively robust to this operation. This step is necessary for a simple computation of indicators and a thematic consistence with the density layer.

Recent tools such as the one proposed by Boeing (2017b) provide algorithms to operate an extraction of network topology. The algorithm we use is very similar but is necessary as our approach necessitates special tuning for the following points: (i) aggregation of data at the level of raster cells, which resolution can be variable; and (ii) construction of networks on significant areas (continental scale), made possible in our case by a parallelization of the computation.

We keep only the nodes with a degree strictly greater or smaller than two, and corresponding links, by taking care to aggregate the real geographical distance when constructing the corresponding topological link. Given the order of magnitude of data size (for Europe, the initial database has $\simeq 44.7 \cdot 10^6$ links, and the final simplified database $\simeq 20.4 \cdot 10^6$), a specific parallel algorithm is used, with a *split-merge* structure. It separates the space into areas that can be independently processed and then merged. We detail it in the following.

3.2.2 Network Simplification Algorithm

We detail here the road network simplification algorithm from OpenStreetMap data. The general workflow is the following: (i) data import by selection and spatial aggregation at the raster resolution; (ii) simplification to keep only the topological network, processed in parallel through *split/merge*.

OSM data are imported into a `pgsql` database (`Postgis` extension for the management of geometries and to have spatial indexes). The import is done using the software `osmosis` (2016), from an image in compressed `pbf` format of the OpenStreetMap database (the dump was retrieved from http://download.geofabrik.de, in July 2016). We filter at this stage the links (`ways`) which possess the tag `highway`, and keep the corresponding nodes.

The network is first aggregated at a 100 m granularity in order to be consistently used with population grids. It furthermore allows to be robust to local coding imperfections or to very local missing data. For this step, roads are filtered on a relevant subset of tags that we take within `motorway`, `trunk`, `primary`, `secondary`, `tertiary`, `unclassified`, and `residential`. For the set of segments of corresponding lines, a link is created between the origin and the destination cell, with a real length computed between the center of cells and a speed taken as the speed of the line if it is available.

The simplification is then operated the following way:

1. The whole geographical coverage is cut into areas on which computations will be partly done through parallel computation (*split* paradigm). Areas have a fixed size in number of cells of the base raster (200 cells).
2. On each subarea, a simplification algorithm is applied the following way: as long as there still are vertices of degree 2, successive sequences of such vertices are determined, and corresponding links are replaced by a unique link with real length and speed computed by cumulation on the deleted links.
3. As the simplification algorithm keeps the links having an intersection with the border of areas, a fusion followed by a simplification of resulting graphs is necessary. To keep a reasonable computational cost, the size of merged areas has to stay low: we take merge areas composed of two contiguous areas. A paving by four sequences of independent merging allows then to cover the full set of joints between areas, these sequences being executed sequentially.

We have then at our disposition a topological graph given by the links between cells of the base raster, having distance and speed attributes corresponding to the underlying real links. This topological graph for Europe has been made available as an open database on the dataverse repository at https://doi.org/10.7910/DVN/RKDZMV.

3.2.3 Empirical Values of Network Indicators

Network indicators have been computed on the same areas as urban form indicators, in order to put them in direct correspondence and later compute the correlations. We show in Fig. 1 a sample for France.

The spatial behavior of indicators unveils local regimes as for the urban form (urban, rural, metropolitan), but also strong regional regimes. They can be due to the different agricultural practices depending on the region for the rural, for example,

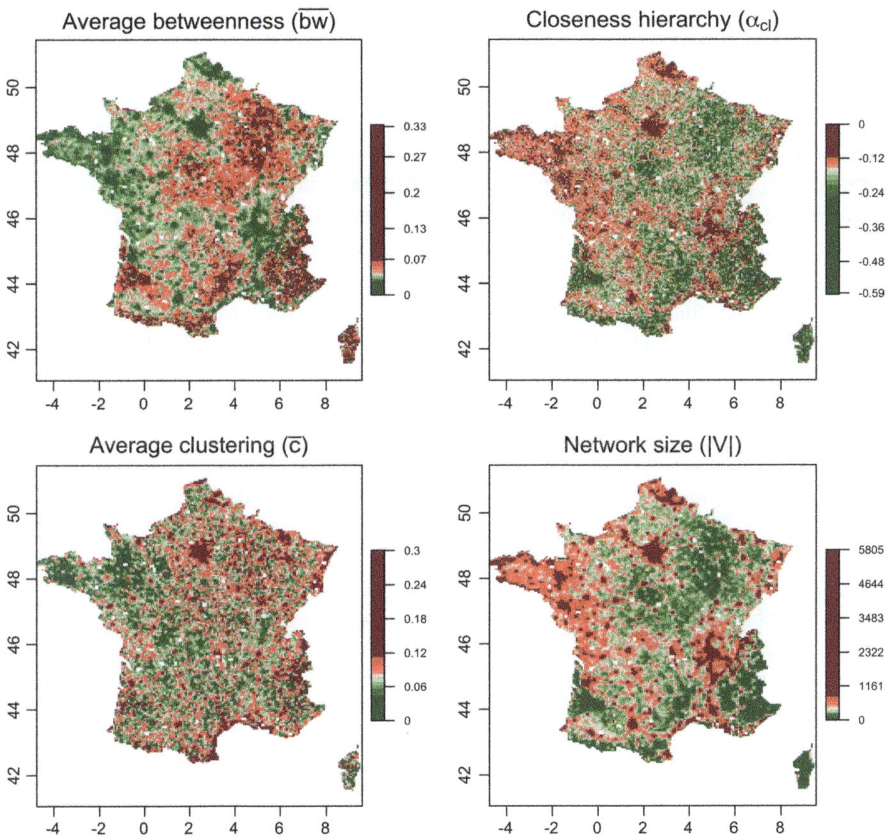

Fig. 1 Spatial distribution of network indicators. We show indicators for France, in correspondence with morphological indicators described previously. We give here the average betweenness centrality $\bar{bw}$, the hierarchy of closeness centrality $\alpha_c l$, the average clustering coefficient $\bar{c}$, and the number of nodes $|V|$

implying a different partition of parcels and also a particular organization of their serving. For network size, Brittany is a clear outlier and rejoins urban regions, witnessing very fragmented parcels (and a fortiori also of a land property fragmentation in the simplifying assumption of corresponding parcels and properties). This is partly correlated to a low hierarchy of accessibility. The South and the East of the extended *Bassin Parisien* are distinguishable by a strong average betweenness centrality, in accordance with a strong hierarchy of the network.

The same way as for urban form, this spatial variability suggests the search of variables regimes of interactions between indicators, as we will do for later through their correlations.

3.3 Effective Static Correlations and Non-stationarity

3.3.1 Spatial Correlations

We show in Fig. 2 examples of correlations estimated with $\delta = 12$ in the case of France. With 20 indicators, the correlation matrix is significantly large in size, but the effective dimension (the number of components required to reach the majority of variance) is reduced: principal components analysis shows that 10 components already capture 62% of the variance, and the first component already captures 17%, what is considerable in a space where the dimension is 190.

Figure 3 gives the spatial distribution for all Europe, for a sample of correlations between indicators: $\rho[\alpha_{cl}, I]$, $\rho\left[\gamma, \alpha\right]$, $\rho\left[\bar{bw}, \gamma\right]$, $\rho[\alpha_{bw}, \alpha_{cl}]$, $\rho\left[\left|V\right|, \bar{l}\right]$, $\rho\left[\gamma, r_\gamma\right]$ (with r_γ adjustment coefficient for γ). We see interesting structures emerging, such as the hierarchy and its adjustment which exhibit an area of strong correlation in

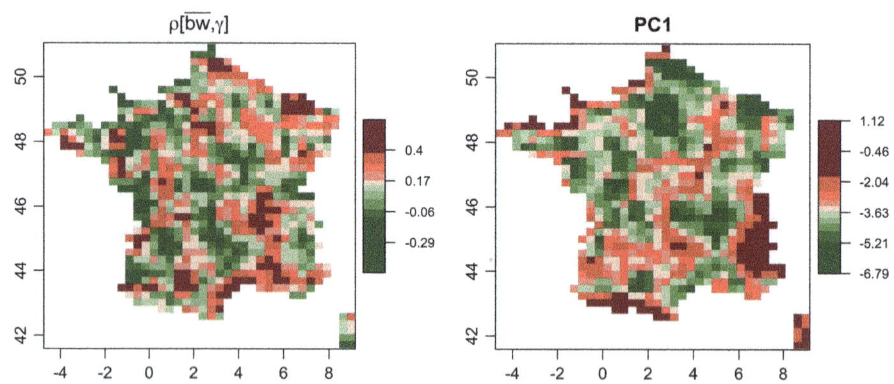

Fig. 2 Examples of spatial correlations. For France, the maps give $\rho\left[\bar{bw}, \gamma\right]$, correlation between the average betweenness centrality and the hierarchy of population (*left*) and the first component of the reduced matrix (*right*)

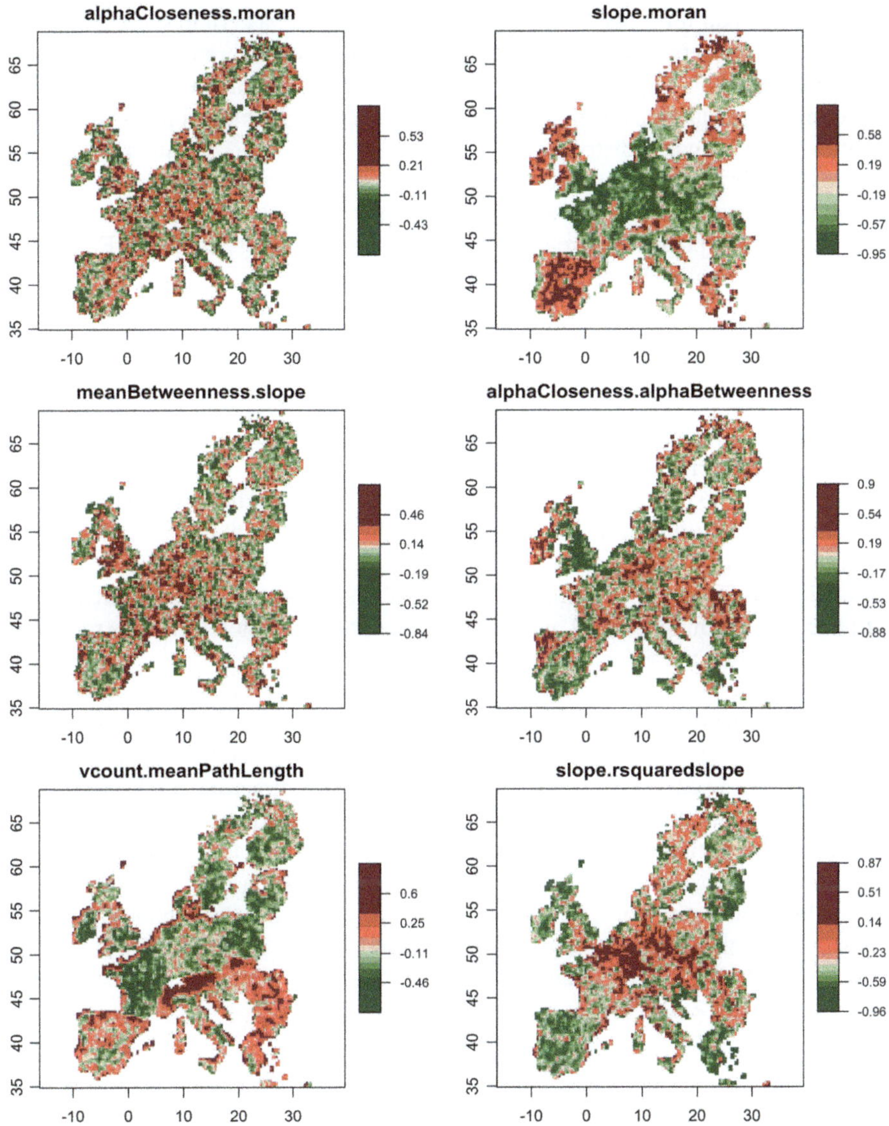

Fig. 3 **Spatial correlations for Europe**. The estimation is done here with $\delta = 12$

the center of Europe and negative correlation areas, or the number of nodes and the path length which correlate in mountains and along the coasts (what is expected since roads do several detours in such topographies) and have a negative correlation otherwise.

It is possible to examine within the correlation matrix the bloc for urban form, for the network, or for crossed correlations, which directly express a link between properties of the urban form and of the network. For example, a certain correspondence between average betweenness centrality and morphological hierarchy that we obtain allows to understand the process corresponding to the correspondence of hierarchies: a hierarchical population can induce a hierarchical network or the opposite direction, but it can also induce a distributed network or such a network create a population hierarchy—this must be well understood in terms of correspondence and not causality, but this correspondence informs on different urban regimes. Metropolitan areas seem to exhibit a positive correlation for these two indicators, as shows Fig. 2, and rural spaces a negative correlation.

In order to give a picture of global relations between indicators, we can refer to the matrix obtained for $\delta = \infty$: for example, a strong population hierarchy is linked to a high and hierarchical betweenness centrality, but is negatively correlated to the number of edges (a diffuse population requires a more spread network to serve all the population). However, it is not possible this way to systematically link indicators, since they especially strongly vary in space.

Furthermore, to give an idea of the robustness of the estimation, we investigate for the correlations estimated on the full dataset (corresponding to $\delta = \infty$) the relative size of confidence intervals at the 95% level (Fisher method) given by $\frac{|\rho_+ - \rho_-|}{|\rho|}$, for correlations such that $|\rho| > 0.05$. The median of this rate is at 0.04, the ninth decile at 0.12 and the maximum at 0.19, which means that the estimation is always relatively good compared to the value of correlations.

This suggests a very high variety of interaction regimes. The spatial variation of the first component of the reduced matrix confirms it, what clearly reveals the spatial non-stationarity of interaction processes between forms, since the first and second moments vary in space. The statistical significance of stationarity can be verified in different ways, and there does not exist to the best of our knowledge a generic test for spatial non-stationarity. Zhang and Zhou (2014) develop, for example, a test for rectangular regions of any dimension, but in the specific case of *point processes*. We use here the method of Leung et al. (2000) which consists in estimating through bootstrap the robustness of geographically weighted regression models. These will be developed below, but we obtain for all tested models a significant non-stationarity without doubt ($p < 10^{-3}$).

3.3.2 Variations of the Estimated Correlations

We show in Fig. 4 the variation of the estimation of correlation as a function of window size. More precisely, we observe a strong variation of correlations as a function of δ, what is reflected in the average value of the matrix given here (which extends, for example, from $\rho(4) = 0.22$ to $\rho(80) = 0.12$ for average absolute cross-correlations). An increase of δ leads for all measures a shift toward positive values, but also a narrowing of the distribution, these two effects resulting in a decrease of average absolute correlations, which approximatively stabilize for large values of δ.

Such a variation could be a clue of a multi-scalar behavior: a change in window size should not influence the estimation if a single process would be implied; it should

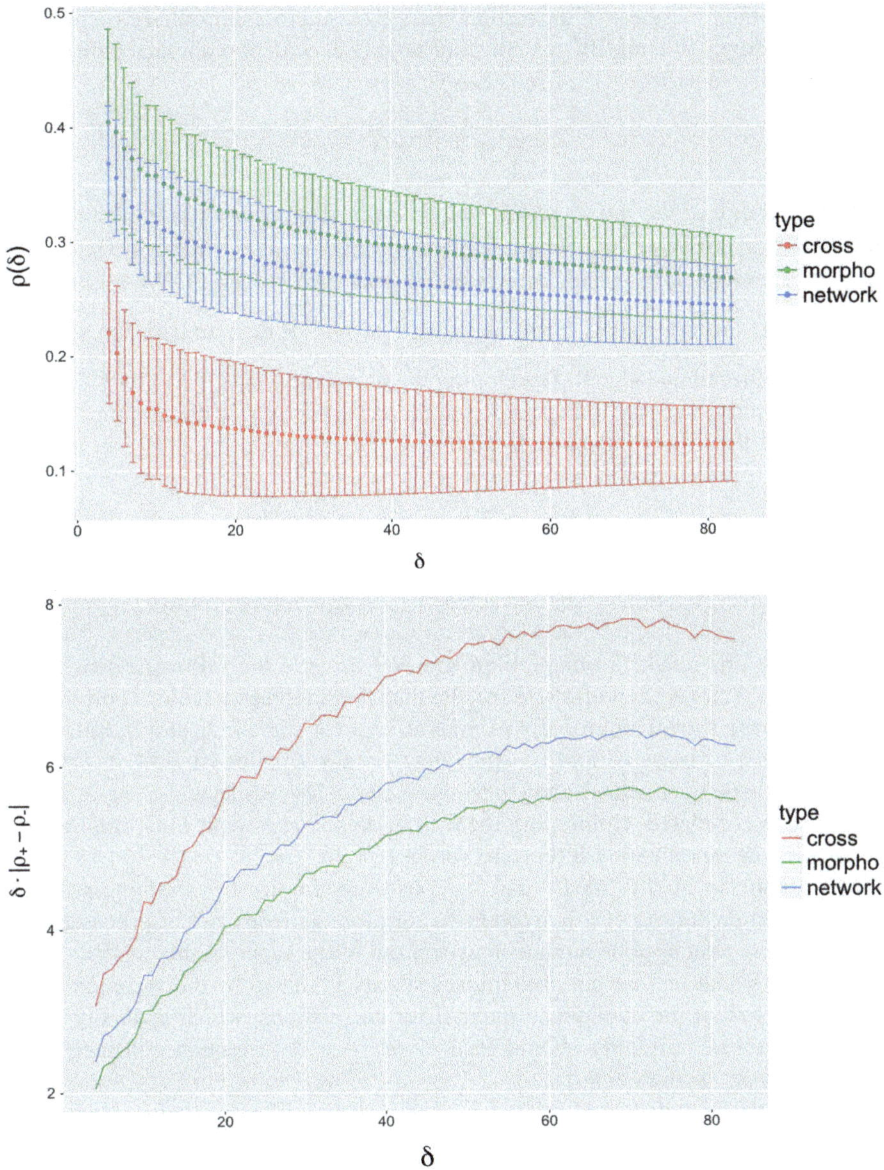

Fig. 4 Variation of correlations with scale, for correlations computed on Europe. (*top*) Average absolute correlations and their standard deviations, for the different blocs, as a function of δ; (*bottom*) normalized size of the confidence interval $\delta \cdot |\rho_+ - \rho_-|$ (confidence interval $\left[\rho_-, \rho_+\right]$ estimated by the Fisher method) as a function of δ

only change the robustness of the estimation. Let sketch in a simplified case how this link can be done. To simplify, we consider processes with two characteristic scales which linearly superpose, i.e., which can be written as

$$X_i = X_i^{(0)} + \tilde{X}_i \tag{7}$$

with $X_i^{(0)}$ trend at the small scales with a characteristic evolution distance d_0, and $\tilde{X}_i$ signal evolving at a characteristic distance $d \ll d_0$. We can then compute the decomposition of the correlation between two processes. We assume that $\text{Cov}\left[X_i^{(0)}, \tilde{X}_j\right] = 0$ for all i, j, and denoting $\varepsilon_i = \frac{\sigma\left[X_i^{(0)}\right]}{\sigma\left[\tilde{X}_i\right]}$ the rate between standard deviations of trend and signal. Developing the correlation $\rho\left[X_1, X_2\right]$ by bilinearity and developing the resulting expression at the first order under the assumption that $\varepsilon_i \ll 1$ gives the following approximation:

$$\rho\left[X_1, X_2\right] = \left(\varepsilon_1 \varepsilon_2 \rho\left[X_1^{(0)}, X_2^{(0)}\right] + \rho\left[\tilde{X}_1, \tilde{X}_2\right]\right) \cdot \left(1 - \frac{1}{2}(\varepsilon_1^2 + \varepsilon_2^2)\right) \tag{8}$$

The overall correlation is thus corrected by both an attenuation and an interference factor.

Let apply this result to our problematic. We observe the following reasonable assumptions: (i) for areas with a size smaller than the stationarity scale of correlations (which we have shown empirically to exist at least for some indicators), estimating the covariance of noise should be equivalent on any smaller window in terms of estimator values; (ii) trends, if they exist, have a very low variance, i.e., $\varepsilon_i \ll 1$; (iii) trends are uncorrelated. Combining these three hypotheses with Eq. 8 implies that $\rho(\delta)$ should decrease with δ if they are verified.

This confirms in this ideal case the observed empirical variation in Fig. 4. A formal demonstration of this hypothesis for more general type of processes and less restricting assumptions remains however out of the scope of this work.

An other signature of a possible multi-scalarity is shown by the variation of the normalized size of the confidence interval for correlations, which in theory under an assumption of normality should lead $\delta \cdot |\rho_+ - \rho_-|$ to remain constant, since boundaries vary asymptotically as $1/\sqrt{N} \sim 1/\sqrt{\delta^2}$ as obtained in Eq. 6, follows the direction of this hypothesis of processes superposed at different scales as proposed previously.

Thus, processes are non-stationary, and this stylized insight suggests that they are the product of the superposition of processes at different scales. We however recall that the notion of multi-scalar process is otherwise very broad, and can manifest itself in scaling laws, for example, West (2017). An approach closer to the one we took is given by Chodrow (2017) which measures intrinsic scales to segregation phenomena by using measures from information theory.

3.3.3 Typical Scales

We also propose to explore the possible property of multi-scalar processes by the extraction of endogenous scales in the data. A geographically weighted principal component analysis (GWRPCA) (Harris et al. 2011) as exploratory analysis suggests weights and importances that vary in space, which is in consistence with the non-stationarity of correlation structures obtained above. There is no reason a priori that the scales of variation of the different indicators are strictly the same. We propose thus to extract typical scales for crossed relations between the urban form and network topology.

We implement therefore the following method: we consider a typical sample of indicators (four for each aspect, see the list in Table 1), and for each indicator we formulate all the possible linear models as a function of opposite indicators (network for a morphological indicator, morphological for a network indicator), aiming at directly capturing the interaction without controlling on the type of form or of network. These models are then adjusted by a geographically weighted regression (GWR) with an optimal range determined by a corrected information criteria (AICc), using the R package GWModel (Gollini et al. 2013). For each indicator, we keep the model with the best value of the information criteria. We adjust the models on data for France, with a *bisquare* kernel and an adaptative bandwidth in a number of neighbors.

Results are presented in Table 1. It is first interesting to note that all models have only one variable, suggesting relatively direct correspondences between topology and morphology. All morphological indicators are explained by network performance, i.e., the quantity of detours it includes. On the contrary, the network topology is explained by Moran index for centralities, and by entropy for performance and the number of vertices. There is thus a dissymmetry in relations, the network being conditioned in a more complex way to the morphology than the morphology to the network. The adjustments are rather good ($R^2 > 0.5$) for most indicators, and

Table 1 Interrelations between network indicators and morphological indicators. Each relation is adjusted by a geographically weighted regression, for the optimal range adjusted by AICc

Indicator	Model	Range (km)	Adjustment (R^2)
Average distance $\bar{d}$	$\bar{d} \sim v_0$	11.6	0.31
Entropy $\mathcal{E}$	$\mathcal{E} \sim v_0$	8.8	0.75
Moran I	$I \sim v_0$	8.8	0.49
Hierarchy γ	$\gamma \sim v_0$	8.8	0.68
Average betweenness $\bar{bw}$	$\bar{bw} \sim I$	12.3	0.58
Average closeness $\bar{cl}$	$\bar{cl} \sim I$	13.9	0.26
Performance v_0	$v_0 \sim \mathcal{E}$	8.6	0.86
Number of nodes $\|V\|$	$\|V\| \sim \mathcal{E}$	8.6	0.88

p-values obtained for all models (for the constant and the coefficient) are lower than 10^{-3}. Concerning the scales corresponding to the optimal model, they are very localized, of the order of magnitude of ten kilometers, i.e., a larger variation than the one obtained the correlations. This analysis confirms thus statistically, on the one hand, the non-stationarity and, on the other hand, gives a complementary point of view on the question of endogenous scales.

4 Urban Morphogenesis Model

After having characterized empirically urban morphology from the point of view of interactions between the built environment and networks, we propose to gain indirect knowledge on processes involved in the emergence of the form through modeling, by introducing a model of urban morphogenesis which aims at capturing empirical results obtained above.

4.1 Model Rationale

Urban settlements and transportation networks have been shown to be coevolving, in the different thematic, empirical, and modeling studies of territorial systems developed up to here. As we saw, modeling approaches of such dynamical interactions between networks and territories are poorly developed. We propose in this section to realize the first entry at an intermediate scale, focusing on morphological and functional properties of the territorial system in a stylized way. We introduce a stochastic dynamical model of urban morphogenesis which couples the evolution of population density within grid cells with a growing road network.

The general principles of the model are the following. With an overall fixed growth rate, new population aggregate preferentially to a local potential, for which parameters control the dependence to various explicative variables. These are in particular local density, distance to the network, centrality measures within the network, and generalized accessibility. Rui and Ban (2014) show in the case of Stockholm the very strong correlation between centrality measures in the network and the type of land-use, what confirms the importance to consider centralities as explicative variables for the model at this scale. We generalize thus the morphogenesis model studied in Raimbault (2018a), with aggregation mechanisms similar to the ones used by Raimbault et al. (2014). A continuous diffusion of population completes the aggregation to translate repulsion processes generally due to congestion. Because of the different time scales of evolution for the urban environment and for networks, the network grows at fixed time steps: a first fixed rule ensures connectivity of newly populated patches to the existing network. Different network generation heuristics are then included in the model, which are expected to be complementary. Figure 5 summarizes the general structure of the morphogenesis model.

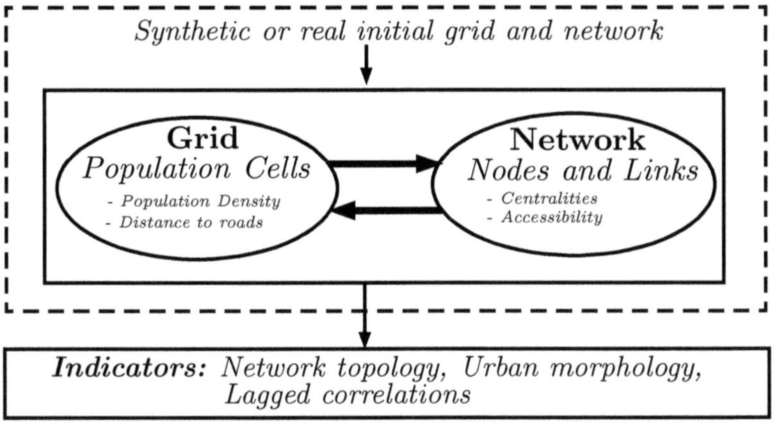

Fig. 5 Structure of the coevolution model at the mesoscopic scale

4.2 Model Description

The model is based on a squared population grid of size N, whose cells are defined by populations (P_i). A road network is included as a vectorial layer on top of the grid, similarly to Raimbault et al. (2014). We assume at the initial state a given population distribution and a network.

The evolution of densities is based on a utility function, influenced by local characteristics of the urban form and function, that we call *explicative variables*. Let $x_k(i)$ be a local explicative variable for cell i, which will be among the following variables: population P_i; proximity to roads, taken as $\exp(-d/d_n)$ where d is the distance by projection on the closest road, and $d_n = 10$ is fixed; betweenness centrality; closeness centrality; accessibility.

For the last three variables, they are defined as previously for network nodes, and then associated to cells by taking the value of the closest node, weighted by a decreasing function of the distance to it, i.e., of the form $x_k = x_k^{(n)}(\text{argmin}_j\, d(i, j)) \cdot \exp\left(-\min_j d(i, j)/d_0\right)$, with $x_k^{(n)}$ the corresponding variable for nodes, the index j being taken on all nodes, and the decay parameter d_0 is in our case fixed at $d_0 = 1$ to keep the property that network variables are essentially significant at close distances from the network. We consider then normalized explicative variables defined by $\tilde{x}_k(i) = x_k(i) - \min_j x_k(j)/(\max_j x_k(j) - \min_j x_k(j))$.

The utility of a cell is then given by the following linear aggregation:

$$U_i = \sum_k w_k \cdot \tilde{x}_k(i) \tag{9}$$

where $\tilde{x}_k$ are the normalized local explicative variables, and w_k are weight parameters, which allow to weight between the different influences. Alternatives to this simple utility function include, for example, Cobb–Douglas functions, what are equivalent to a linear aggregation on the logarithms of variables.

A time step of model evolution includes then the following stages.

1. Evolution of the population following rules similar to Raimbault (2018a). Given an exogenous growth rate N_G, individuals are added independently following an aggregation done with a probability $U_i^\alpha / \sum_k U_k^\alpha$, followed by a diffusion of strength β to neighbor cells, done n_d times.
2. Network growth is done through multi-modeling as proposed by Raimbault (2017b) and explored by Raimbault (2018c), knowing that this takes place is the time step is a multiple of a parameter t_N, which allows to integrate a differential between temporal scales for population growth and for network growth. The different heuristic used for adding links are: (i) nothing (baseline); (ii) random links; (iii) deterministic potential breakdown; (iv) random potential breakdown (Schmitt 2014); (v) cost-benefits (Louf et al. 2013); and (vi) biological network growth (Tero et al. 2010).

The fact that the aggregation of population follows a power law of the utility provides flexibility in the underlying optimization problem by agents, since as Josselin and Ciligot-Travain (2013) recall, the use of different norms in spatial optimal location problems corresponds to different logics of optimization.

The parameters of the model that we will make vary are then: (i) aggregation-diffusion parameters α, β, N_g, n_d; (ii) the four weight parameters w_k for the explicative variables, which vary in [0; 1]; network growth parameters for the different heuristics (see Raimbault (2018c)). Output model indicators are the urban morphology indicators, topological network indicators, and lagged correlations between the different explicative variables.

4.3 Simulation Results

The model is implemented in NetLogo, given the heterogeneity of aspects that have to be taken into account, and this language is particularly suitable to couple a grid of cells with a network. Urban morphology indicators are computed thanks to a NetLogo extension specifically developed. Source code and results for the modeling part are available at https://github.com/JusteRaimbault/CityNetwork/tree/master/Models/MesoCoevol.

We propose to focus on the ability of the model to capture relations between networks and territories, and more particularly the coevolution. Therefore, we will try to establish if (i) the model is able to reproduce, beyond the form indicators, the static correlation matrices computed previously; and (ii) the model produces a variety of dynamical relations in the sense of causality regimes developed by Raimbault (2017a).

The model is initialized on fully synthetic configurations, with a grid of size 50. Configurations are generated through an exponential mixture in a way similar to Anas et al. (1998): $N_c = 8$ centers are randomly located, to which a population is attributed following a scaling law $P_i = P_0 \cdot (i+1)^{-\alpha_S}$ with $\alpha_S = 0.8$ and $P_0 = 200$. The population of each center is distributed to all cells with an exponential kernel of shape $d(r) = P_{max} \exp(-r/r_0)$ where the parameter r_0 is determined to fix the population at P_i, with $P_{max} = 20$ (density at the center). We have indeed $P_i = \iint d(r) = \int_{\theta=0}^{2\pi} \int_{r=0}^{\infty} d(r) r dr d\theta = 2\pi P_{max} \int_r r \cdot \exp(-r/r_0) = 2\pi P_{max} r_0^2$, and therefore $r_0 = \sqrt{\frac{P_i}{2\pi P_{max}}}$. An initial network skeleton is generated by sequential closest components connection.

We explore a Latin Hypercube Sampling of the parameter space, with 10 repetitions for around 7000 parameter points, corresponding to a total of around 70,000 model repetitions, realized on a computation grid by using the OpenMole model exploration software (Reuillon et al. 2013). The simulation results are available at https://doi.org/10.7910/DVN/OBQ4CS.

4.3.1 Static and Dynamical Calibration

The model is calibrated at the first order, on indicators for the urban form and network measures, and at the second order on correlations between these. We introduce an ad hoc calibration procedure in order to take into account the first two moments that we detail below. More elaborate procedures are used, for example, in economics, such as Watson (1993) which uses the noise of the difference between two variables to obtain the same covariance structure for the two corresponding models, or in finance, such as Frey et al. (2001) which define a notion on equivalence between latent variables models which incorporates the equality of the interdependence structure between variables. We avoid here to add supplementary models and consider simply a distance on correlation matrices. The procedure is the following.

- Simulated points are the ones obtained through the sampling, with average values on repetitions.
- In order to be able to estimate correlation matrices between indicators for simulated data, we make the assumption that second moments are continuous as a function of model parameters, and split for each heuristic the parameter space into areas to group parameter points (each parameter being binned into $15/k$ equal segments, where k is the number of parameters: we empirically observed that this allowed to always have a minimal number of points in each area), what allows to estimate for each group indicators and the correlation matrix.
- For each estimation done this way, that we write $\bar{S}$ (indicators) and $\rho[S]$ (correlations), we can then compute the distance to real points on indicators $d_I(R_j) = d(\bar{S}, R_j)$ and on correlation matrices $d_\rho(R_j) = d(\rho[S], \rho[R_j])$ where R_j are the real points with their corresponding correlations, and d an Euclidean distance normalized by the number of components.
- We consider then the aggregated distance defined as $d_A^2(R_j) = d_I^2(R_j) + d_\rho^2(R_j)$. Indeed, the shape of Pareto fronts for the two distances considered suggests the

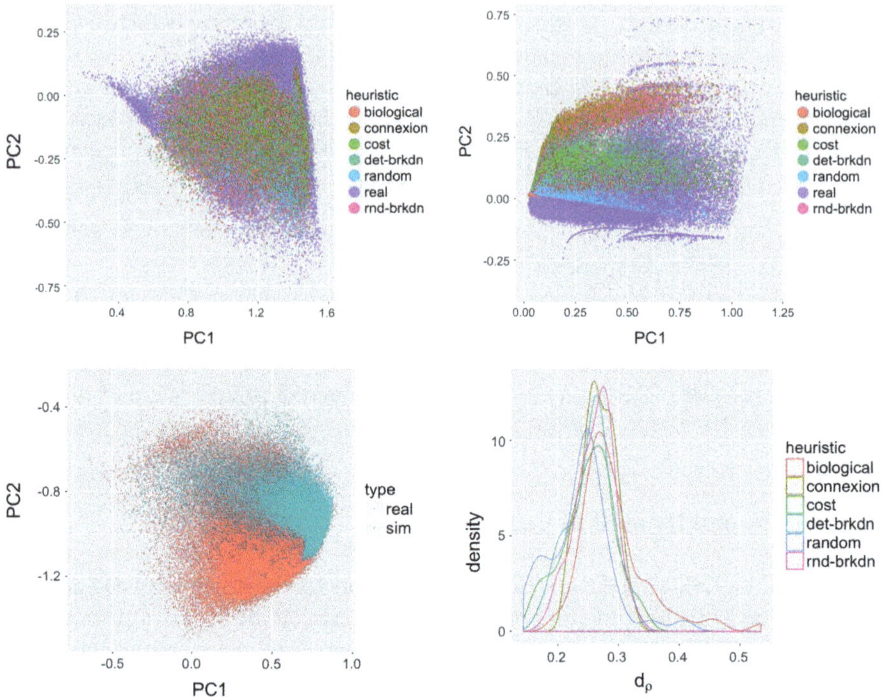

Fig. 6 Calibration of the morphogenesis model at the first and second order. (*top left*) Simulated and observed point clouds in a principal plan for urban morphology indicators. (*top right*) Simulated and observed could points in a principal plan for network indicators. (*bottom left*) Simulated and observed point clouds in a principal plan for all indicators. (*bottom right*) Distributions of distances on correlations d_ρ, for the different heuristics

relevance of this aggregation. The real point closest to a simulated point is then the one in the sense of this distance.

Figure 6 summarizes calibration results. Morphological indicators are easier to approach than network indicators, for which a part of the simulated clouds does not superpose with observed points. We find again a certain complementarity between network heuristics. When considering the full set of indicators, few simulated points are situated far from the observed points, but a significant proportion of these is beyond the reach of simulation. Thus, the simultaneous capture of morphology and topology is obtained at the price of less precision.

We however obtain a good reproduction of correlation matrices as shown in Fig. 6 (histogram for d_ρ, bottom right). The worse heuristic for correlations is the biological one in terms of maximum, whereas the random produces rather good results: this could be due, for example, to the reproduction of very low correlations, which accompany a structure effect due to the initial addition of nodes which imposes already a certain correlation. On the contrary, the biological heuristic introduces

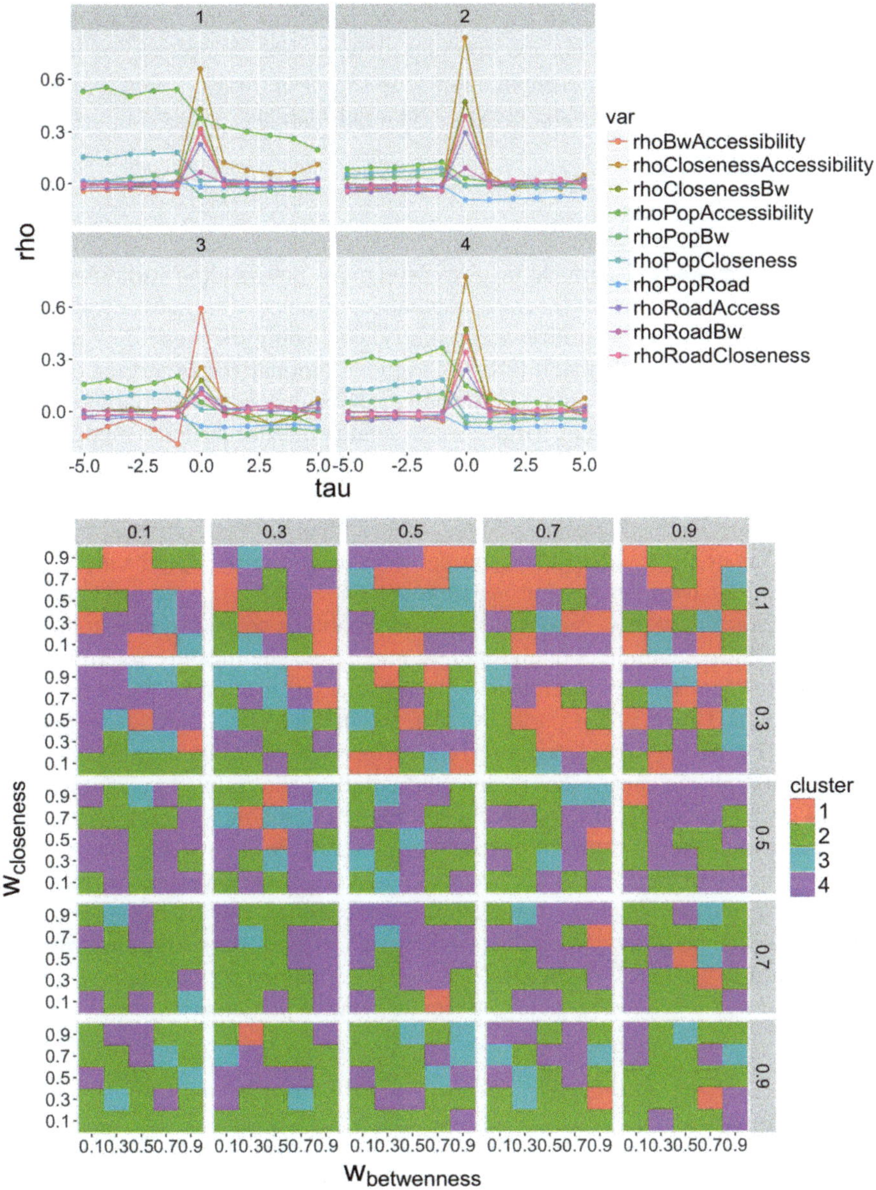

Fig. 7 Causality regimes for the coevolution model. (*top*) Trajectories of classes centers in terms of $\rho[\tau]$ between the different explicative variables. (*bottom*) Phase diagram of regimes in the parameter space for w_k, represented here as the variation of diagrams for (w_{bw}, w_{cl}), along the variations of w_{road} (in rows) and of w_{pop} (in columns)

supplementary processes which can possibly be beneficial to the network in terms of independence (or following the opposed viewpoint be detrimental in terms of correlations). In any case, this application shows that our model is able to resemble real configurations both for indicators and their correlations.

4.3.2 Causality Regimes

We furthermore study dynamical lagged correlations between the variations of the different explicative variables for cells (population, distance to the network, closeness centrality, betweenness centrality, accessibility). We apply the method of causality regimes introduced in Raimbault (2017a). Figure 7 summarizes the results obtained with the application of this method on simulation results of the coevolution model. The number of classes inducing a transition is smaller than for the RDB model, translating a smaller degree of freedom, and we fix in that case $k = 4$. Centroid profiles allow to understand the ability of the model to more or less capture a coevolution.

The regimes obtained appear to be less diverse than the ones obtained by Raimbault (2017a) or for a macroscopic model of coevolution by Raimbault (2018b). Some variables have naturally a strong simultaneous correlation, spurious from their definitions, such as closeness centrality and accessibility, or the distance to the road and the closeness centrality. For all regimes, population significantly determines the accessibility. The regime 1 corresponds to a full determination of the network by the population. The second is partly circular, through the effect of roads on populations. The regime 3 is more interesting, since closeness centrality negatively causes the accessibility: this means that in this configuration, the coupled evolution of the network and the population follow the direction of a diminution of congestion. Furthermore, as population causes the closeness centrality, there is also circularity and thus coevolution in that case. When we locate it in the phase diagram, this regime is rather sparse and rare, contrary, for example, to the regime 1 which occupies a large portion of space for a low importance of the road ($w_{road} \leq 0.3$). This confirms that the coevolution produced by the model is localized and not a characteristic always verified, but that it is however able to generate some in particular regimes.

5 Discussion

5.1 Quantifying Urban Form

We have first shown empirically the non-stationarity of interactions between the morphology of the distribution of populations and the topology of the road network. Various developments of this analysis are possible.

Population density grids exist for all regions of the world, such as the ones provided by Stevens et al. (2015). The analysis may be repeated on other regions of the world, to compare the correlation regimes and test if urban system properties stay the same, keeping in mind the difficulties linked to the differences in data quality.

The research of local scales, i.e., with an adaptive estimation window in terms of size and shape for correlations, would allow to better understand the way processes locally influence their neighborhood. The validation criteria for window size would still be to determine: it can be as above an optimal range for explicative models that are locally adjusted.

The question of ergodicity, crucial to urban systems (Pumain 2012), should also be explored from a dynamical point of view, by comparing time and spatial scales of the evolution of processes, or more precisely the correlations between variations in time and variations in space, but the issue of the existence of databases precise enough in time appears to be problematic. The study of a link between the derivative of the correlation as a function of window size and of the derivatives of the processes is also a direction to obtain indirect informations on dynamics from static data.

Finally, the search of classes of processes on which it is possible to directly establish the relation between spatial correlations and temporal correlations is a possible research direction. It stays out of the scope of this present work, but would open relevant perspectives on coevolution, since it implies evolution in time and an isolation in space, and therefore a complex relation between spatial and temporal covariances.

5.2 *Modeling Urban Morphogenesis*

We have then proposed a coevolution model at the mesoscopic scale, based on a multi-modeling paradigm for the evolution of the network. The model is able to reproduce a certain number of observed situations at the first and second order, capturing thus a static representation of interactions between networks and territories. We are therefore able to produce the emergence of the urban form in a coupled way, suggesting the relevance of this approach.

It also yields different dynamical causality regimes, being however less diverse than for the simple model studied by Raimbault (2017a): therefore, a more elaborate structure in terms of processes must be paid in flexibility of interaction between these. This suggests a tension between a "static performance" and a "dynamical performance" of models.

An open question is to what extent a pure network model with preferential attachment for nodes would reproduce results close to what we obtained. The complex coupling between aggregation and diffusion (shown by Raimbault (2018a)) could not be easily included, and the model could in any case not answer to questions on the coupling of the dynamics.

5.3 *Implications for Policies*

Although remaining theoretical and stylized in the case of the model, our results could give useful insights for policies. First, in terms of empirical computation of correlations, a direct interpretation of their values can inform on the underlying dynamics of the territorial system and suggest issues on which planners must be alerted: for example, a high positive correlation between population hierarchy and network betweenness hierarchy will imply structural congestion effects. The coupling with sustainability indicators such as Le Néchet (2010) does with energy consumption, could also provide useful insights for sustainable territorial planning. Second, a more refined calibration of the model should allow understanding better territorial dynamics in precise cases: for example, focusing on the best network heuristic would provide the most probable processes ruling network evolution, and make experiments on these (e.g., implementing them differently by changing their parameters, or replacing with other processes), given performance indicators. This last point is of particular interest, as it could provide insights into the combination of top-down (planned) and bottom-up network evolution processes, as both are included in the model.

Conclusion

We have introduced a novel methodology to quantify urban form by coupling the built environment (captured by the spatial distribution of population) with road networks, which uses corresponding indicators and their spatial correlations. We studied their empirical values for Europe and the behavior of correlations as a function of scale. This coupled approach was complemented by a morphogenesis model based on a coevolution paradigm which precisely considers strong relations between both aspects. The calibration of the model on empirical data suggests that such coupled processes indeed occur in the emergence of the urban form and function. The two complementary axes of this work therefore pave the way toward more integrative approaches to urban morphology.

Acknowledgements Results obtained in this paper were computed on the vo.complex-system.eu virtual organization of the European Grid Infrastructure (http://www.egi.eu). We thank the European Grid Infrastructure and its supporting National Grid Initiatives (France-Grilles in particular) for providing the technical support and infrastructure.

References

Anas, A., Arnott, R., and Small, K. A. (1998). Urban spatial structure. *Journal of Economic Literature*, 36(3): pp. 1426–1464.

Badariotti, D., Banos, A., and Moreno, D. (2007). Conception d'un automate cellulaire non stationnaire à base de graphe pour modéliser la structure spatiale urbaine: le modèle remus. *Cybergeo: European Journal of Geography*.

Banos, A. and Genre-Grandpierre, C. (2012). Towards new metrics for urban road networks: Some preliminary evidence from agent-based simulations. In *Agent-based models of geographical systems*, pages 627–641. Springer.

Batista e Silva, F., Gallego, J., and Lavalle, C. (2013). A high-resolution population grid map for europe. *Journal of Maps*, 9(1):16–28.

Batty, M. and Longley, P. A. (1994). *Fractal cities: a geometry of form and function*. Academic Press.

Boeing, G. (2017a). A multi-scale analysis of 27,000 urban street networks. arXiv preprint arXiv:1705.02198.

Boeing, G. (2017b). Osmnx: New methods for acquiring, constructing, analyzing, and visualizing complex street networks. *Computers, Environment and Urban Systems*, 65:126–139.

Brunsdon, C., Fotheringham, A., and Charlton, M. (2002). Geographically weighted summary statistics–a framework for localised exploratory data analysis. *Computers, Environment and Urban Systems*, 26(6):501–524.

Chen, Y. (2016). Normalizing and Classifying Shape Indexes of Cities by Ideas from Fractals. arXiv preprint arXiv:1608.08839.

Chodrow, P. S. (2017). Structure and information in spatial segregation. *Proceedings of the National Academy of Sciences*, 114(44):11591–11596.

Cottineau, C., Hatna, E., Arcaute, E., and Batty, M. (2017). Diverse cities or the systematic paradox of urban scaling laws. *Computers, environment and urban systems*, 63:80–94.

Crucitti, P., Latora, V., and Porta, S. (2006). Centrality measures in spatial networks of urban streets. *Physical Review E*, 73(3):036125.

D'Acci, L. (2015). Mathematize urbes by humanizing them. cities as isobenefit landscapes: psycho-economical distances and personal isobenefit lines. *Landscape and Urban Planning*, 139:63–81.

EUROSTAT (2014). Eurostat geographical data. http://ec.europa.eu/eurostat/web/gisco.

Frey, R., McNeil, A. J., and Nyfeler, M. (2001). Copulas and credit models. *Risk*, 10(111114.10).

Girres, J.-F. and Touya, G. (2010). Quality assessment of the french openstreetmap dataset. *Transactions in GIS*, 14(4):435–459.

Gollini, I., Lu, B., Charlton, M., Brunsdon, C., and Harris, P. (2013). Gwmodel: an r package for exploring spatial heterogeneity using geographically weighted models. arXiv preprint arXiv:1306.0413.

Guérois, M. and Paulus, F. (2002). Commune centre, agglomération, aire urbaine: quelle pertinence pour l'étude des villes? *Cybergeo: European Journal of Geography*.

Haggett, P. and Chorley, R. J. (1970). *Network analysis in geography*. St. Martin's Press.

Haklay, M. (2010). How good is volunteered geographical information? a comparative study of openstreetmap and ordnance survey datasets. *Environment and planning B: Planning and design*, 37(4):682–703.

Hansen, W. G. (1959). How accessibility shapes land use. *Journal of the American Institute of planners*, 25(2):73–76.

Harris, P., Brunsdon, C., and Charlton, M. (2011). Geographically weighted principal components analysis. *International Journal of Geographical Information Science*, 25(10):1717–1736.

Hillier, B. and Hanson, J. (1989). *The social logic of space*. Cambridge university press.

Josselin, D. and Ciligot-Travain, M. (2013). Revisiting the optimal center location. a spatial thinking based on robustness, sensitivity, and influence analysis. *Environment and Planning B: Planning and Design*, 40(5):923–941.

Josselin, D., Labatut, V., and Mitsche, D. (2016). Straightness of rectilinear vs. radio-concentric networks: modeling simulation and comparison. arXiv preprint arXiv:1609.05719.

Lagesse, C. (2015). Read Cities through their Lines. Methodology to characterize spatial graphs. arXiv preprint arXiv:1512.01268.

Le Néchet, F. (2009). Quantifier l'éloignement au modèle de bussière: monocentrisme contre "acentrisme". In *Neuvièmes rencontres de Théo Quant*, pages 19–p.

Le Néchet, F. (2010). *Approche multiscalaire des liens entre mobilité quotidienne, morphologie et soutenabilité des métropoles européennes: cas de Paris et de la région Rhin-Ruhr*. PhD thesis, Université Paris-Est.

Le Néchet, F. (2015). De la forme urbaine à la structure métropolitaine: une typologie de la configuration interne des densités pour les principales métropoles européennes de l'audit urbain. *Cybergeo: European Journal of Geography*.

Lee, M., Barbosa, H., Youn, H., Holme, P., and Ghoshal, G. (2017). Morphology of travel routes and the organization of cities. *Nature communications*, 8(1):2229.

Leung, Y., Mei, C.-L., and Zhang, W.-X. (2000). Statistical tests for spatial nonstationarity based on the geographically weighted regression model. *Environment and Planning A*, 32(1):9–32.

Louf, R. and Barthelemy, M. (2014). A typology of street patterns. *Journal of The Royal Society Interface*, 11(101):20140924.

Louf, R., Jensen, P., and Barthelemy, M. (2013). Emergence of hierarchy in cost-driven growth of spatial networks. *Proceedings of the National Academy of Sciences*, 110(22):8824–8829.

Moosavi, V. (2017). Urban morphology meets deep learning: Exploring urban forms in one million cities, town and villages across the planet. arXiv preprint arXiv:1709.02939.

Moudon, A. V. (1997). Urban morphology as an emerging interdisciplinary field. *Urban morphology*, 1(1):3–10.

OpenStreetMap (2012). Openstreetmap. http://www.openstreetmap.org.

Osmosis (2016). Osmosis. http://wiki.openstreetmap.org/wiki/Osmosis.

Páez, A. and Scott, D. M. (2005). Spatial statistics for urban analysis: a review of techniques with examples. *GeoJournal*, 61(1):53–67.

Paquot, T. (2010). *L'abc de l'urbanisme*. IAU - UPEC.

Pumain, D. (2011). Systems of cities and levels of organisation. In *Morphogenesis*, pages 225–249. Springer.

Pumain, D. (2012). Urban systems dynamics, urban growth and scaling laws: The question of ergodicity. In *Complexity Theories of Cities Have Come of Age*, pages 91–103. Springer.

Raimbault, J. (2017a). Identification de causalités dans des données spatio-temporelles. In *Spatial Analysis and GEOmatics 2017*.

Raimbault, J. (2017b). Modeling the co-evolution of urban form and transportation networks. In *Conference on Complex Systems 2017*.

Raimbault, J. (2018a). Calibration of a Density-based Model of Urban Morphogenesis. *forthcoming in PlOS ONE*. arXiv:1708.06743.

Raimbault, J. (2018b). Modeling the co-evolution of cities and networks. arXiv preprint arXiv:1804.09430.

Raimbault, J. (2018c). Multi-modeling the morphogenesis of transportation networks. In *Artificial Life Conference Proceedings*, pages 382–383. MIT Press.

Raimbault, J., Banos, A., and Doursat, R. (2014). A hybrid network/grid model of urban morphogenesis and optimization. In *4th International Conference on Complex Systems and Applications*, pages 51–60.

Reuillon, R., Leclaire, M., and Rey-Coyrehourcq, S. (2013). Openmole, a workflow engine specifically tailored for the distributed exploration of simulation models. *Future Generation Computer Systems*, 29(8):1981–1990.

Rui, Y. and Ban, Y. (2014). Exploring the relationship between street centrality and land use in stockholm. *International Journal of Geographical Information Science*, 28(7):1425–1438. https://doi.org/10.1080/13658816.2014.893347

Schmitt, C. (2014). *Modélisation de la dynamique des systèmes de peuplement: de SimpopLocal à SimpopNet.* PhD thesis, Paris 1.

Schwarz, N. (2010). Urban form revisited—selecting indicators for characterising european cities. *Landscape and Urban Planning*, 96(1):29 – 47.

Stevens, F. R., Gaughan, A. E., Linard, C., and Tatem, A. J. (2015). Disaggregating census data for population mapping using random forests with remotely-sensed and ancillary data. *PLoS ONE*, 10(2):1–22.

Tero, A., Takagi, S., Saigusa, T., Ito, K., Bebber, D. P., Fricker, M. D., Yumiki, K., Kobayashi, R., and Nakagaki, T. (2010). Rules for biologically inspired adaptive network design. *Science*, 327(5964):439–442.

Trépanier, M., Morency, C., and Agard, B. (2009). Calculation of transit performance measures using smartcard data. *Journal of Public Transportation*, 12(1):5.

Tsai, Y.-H. (2005). Quantifying urban form: compactness versus' sprawl'. *Urban studies*, 42(1):141–161.

Watson, M. W. (1993). Measures of fit for calibrated models. *Journal of Political Economy*, 101(6):1011–1041.

Wegener, M. and Fürst, F. (2004). Land-use transport interaction: state of the art. *Available at SSRN 1434678.*

West, G. (2017). *Scale: The Universal Laws of Growth, Innovation, Sustainability, and the Pace of Lifein Organisms, Cities, Economies, and Companies*. Penguin.

Zhang, T. and Zhou, B. (2014). Test for the first-order stationarity for spatial point processes in arbitrary regions. *Journal of agricultural, biological, and environmental statistics*, 19(4):387–404.

Continuum Percolation and Spatial Point Pattern in Application to Urban Morphology

Hoai Nguyen Huynh

Abstract One of the important problems in studying urban morphology is to be able to quantify urban objects using mathematical descriptions or measures. Doing so allows one to make a comparison amongst different types of geometry/morphology in urban systems, and effectively classify them. A number of techniques have indeed been developed in urban/geographical analysis. Here in this work, we propose a method to quantify spatial patterns of a set of points based on the idea of percolation, that is rooted in mathematical physics and has been applied to various fields. Employing the idea of percolation enables us to characterise the relative spatial arrangement of points in a set. In particular, the method could be applied to analyse the distribution of different point datasets in urban context and illustrate how it could characterise and classify their spatial patterns. We show that different spatial distributions of points can generally be classified into four groups with distinctive features: clustered, dispersed or regularly distributed at single or multiple length-scales. In applying to urban morphology, the results could enable quantitative discussion on the existence of two different forms of urban system: well-planned and organically grown.

Keywords Urban morphology · Spatial point pattern · Buffer radius · Cluster · Percolation · Geometry

1 Introduction

For a long time, the pattern of human settlement has always been a subject of much interest across different fields of Science, from Geography, Archaeology to Statistics (Small et al. 2005; Helbing et al. 1997; Anschuetz et al. 2001; Mellars 2006; Schumacher 1976; Zucchetto 1983). Over time, urban systems have become ever

H. N. Huynh (✉)
Institute of High Performance Computing, Agency for Science Technology and Research, Singapore, Singapore
e-mail: huynhhn@ihpc.a-star.edu.sg

Department of Mathematics, Imperial College London, London, UK

L. D'Acci (ed.), *The Mathematics of Urban Morphology*,
Modeling and Simulation in Science, Engineering and Technology,
https://doi.org/10.1007/978-3-030-12381-9_18

more important as a form of human settlement on the globe. As such, the study of urban morphology, the study of the spatial structure of people's physical living space in which their daily activities take place, forms a major component of settlement or urban research as a whole. A good understanding of the morphology of an urban system provides us with the comprehension of its current status of development or even the living condition of people inside it. On the other hand, the knowledge of spatial organisation of an urban system could unveil the dynamical processes behind its growth and development, and allows us to gain insights into the way the system operates and evolves.

Traditionally, as mentioned, urban study has been performed by geographers, archaeologists or statisticians. However, urban systems, or 'cities' in modern terms, are typical examples of highly complex systems in which overwhelmingly many agents interact in non-trivial and nonlinear manners over a wide spectrum of spatial and temporal scales (West 2017; Bettencourt et al. 2007; Batty 2017). The results of such tangled interactions are the emergence of unexpected global patterns that cannot be derived from solely knowing the local behaviours of individual agents. It is this emergence property that has attracted researchers in theoretical fields like Mathematics or Physics to apply their tools to provide different and unconventional perspectives on urban systems.

One of the toolboxes which proves very useful in tackling complexity problems is that from Statistical Physics. The framework of Statistical Physics is particularly suitable for dealing with complex systems which have multiple agents, interacting locally with each other in simple but nonlinear manners (Viswanathan et al. 2014; Bertin 2016). The result of such interactions is a complex pattern at the level of the entire system, usually with long-range correlation that normally coexists with a critical state, signified by a phase transition and characterised by the notion of universality class. One such tool is the well-studied percolation phenomena with long-established literature in both Applied Mathematics and Theoretical Physics. As will be seen in this chapter, we will apply the techniques in percolation to analysing the cluster structure of sets of points, which could represent either transport points, be it road junctions or transport stations, or various sites in the urban context.

Spatial point pattern analysis is a well-studied field with various applications in biology (e.g. spatial distribution of cells obtained from bio-imaging (Burguet and Andrey 2014)), ecology (e.g. location and distribution of species (Knegt et al. 2010)) or geography (e.g. spatial patterns of people's location or movement and physical entities, both natural like terrain and artificial like infrastructure (Fotheringham et al. 2000)). There have been many different methods developed either specific to each problem or aimed to be a general approach applicable to a wide range of problems (Illian et al. 2008). Most of them have been mainly concerned with identifying whether or not a point pattern distribution is complete spatial randomness (CSR), which implies that the mechanism underlying the process of generating the points is likely to be a homogeneous Poisson with little inter-event correlation. It appears that most of the analysis methods utilise the nearest-neighbour information based on some metric. Well-known methods like Ripley's K-function (Ripley 1976, 1977) or pair correlation function (Stoyan and Penttinen 2000) also require proper identification of observation window which can become very complicated in the case of locations of urban entities because the domain of such is normally not a simple square or rectangle due to the geography and topography of the system. Furthermore, these methods are also known to be sensitive with respect to the boundary conditions for the domain (Li and Zhang 2007). Here we propose a method that is simply based on the relative position among points in a domain without requiring the knowledge of its boundary. The method is inspired by the ideas in ordinary critical phenomena (e.g. Uzunov 2010), percolation in particular (Stauffer et al. 1994). In percolation or ordinary critical phenomena in general, there is a control parameter that can be tuned to drive the system of interest to different desired states possessing distinct properties. The transition from one state to another takes place at some special value of the control parameter which is called a critical value. A percolation system is a collection of many sites each of which can be either percolating or non-percolating with a certain probability. This probability of a site being percolating (or non-percolating) is the control parameter in the percolation system. When the percolating probability is low, most of the sites are non-percolating, blocking the path to go from one end of the system to the other. When the percolating probability is high, most of the sites are percolating, allowing such path to exist. The length of the shortest such path is typically the (linear) size of the largest contiguous cluster connecting the neighbouring percolating sites. It has been shown that there exists a critical value of the percolating probability—the critical point—at which the system transits from being not percolating to being so. The behaviours of the system approaching the critical point can then be used to classify the system, i.e. identifying its universality class (Stanley 1999). These behaviours include the critical exponents (and scaling function (Hinrichsen 2000)) characterising the scaling of the size of the percolating cluster, among many other quantities.

Bringing this idea of percolation to studying the spatial point pattern, we introduce a control parameter that changes the state of a spatial point. To serve this purpose, the state of a point should reflect its connectivity to the surrounding points. A cluster should then be a group of neighbouring sites that share the same state. As we vary the control parameter and, hence, change the state of the points, the clusters would change. Analysing the properties of these clusters would allow us to classify the system of points of interest.

In what follows, we first present the method we employ to study the pattern of a spatial point distribution, including the quantities we define to analyse its structure. The analysis yields the characteristic distances of a distribution of points and how they can be used to classify different distributions. Finally, we discuss several technical aspects of the proposed method including the comparison with other known methods in the literature, and the applicability and interpretation when applying to study urban morphology.

2 Percolation in Mathematical Physics

2.1 Description

The idea of percolation is indeed very simple. On a lattice (in discretised space), which can be thought of as a porous media, every site[1] can be either occupied, meaning a liquid is passing through it, with a certain probability p or otherwise empty. At a small value of p, it is very difficult for the liquid to pass through the sites and only small clusters of (adjacent) occupied sites exist (see Fig. 1a–c). But at large value of p, a large cluster is formed spanning one edge to the opposite (Fig. 1d–f). That means the liquid can now easily percolate through the media. The graph of percolation probability against the occupation probability would show a sharp pick up near p equal to 0.6, which is known as the percolation threshold ($p_c \approx 0.5927$) (Newman and Ziff 2000) on a two-dimensional square lattice, when there 'suddenly' exists a giant cluster that spans one side to the other of the lattice.

Simply put, at small value of p, most sites are empty and no large clusters exist. At the other end of large value of p, most sites are occupied and a so-called giant cluster exists. So, what happens in between? By the argument of continuity, we know the transition from small clusters to a giant cluster must take place somewhere. It has been studied and found that there exists a critical value p_c of the occupation probability p. When p approaches p_c both from above and below, all sorts of scaling behaviours come in to play, for example the divergence of correlation length in the system that is characterised by a power law (Stauffer et al. 1994).

2.2 Phase Transition

It is clear that the higher the percolating probability is, the easier the sites are connected to one another other, and *vice versa*. The percolating probability is then viewed as a control parameter in the system. The transition occurs when the control parameter is adjusted to a critical value (called the critical point) at which the system transits from one state (or phase) to another with distinct properties, namely the non-percolating and percolating phases respectively at low and high percolating

[1]The description is for site percolation, the similar idea applies to bond percolation. In bond percolation, a link between a pair of sites is traversable, meaning a liquid can flow through it, with probability p or otherwise blocked.

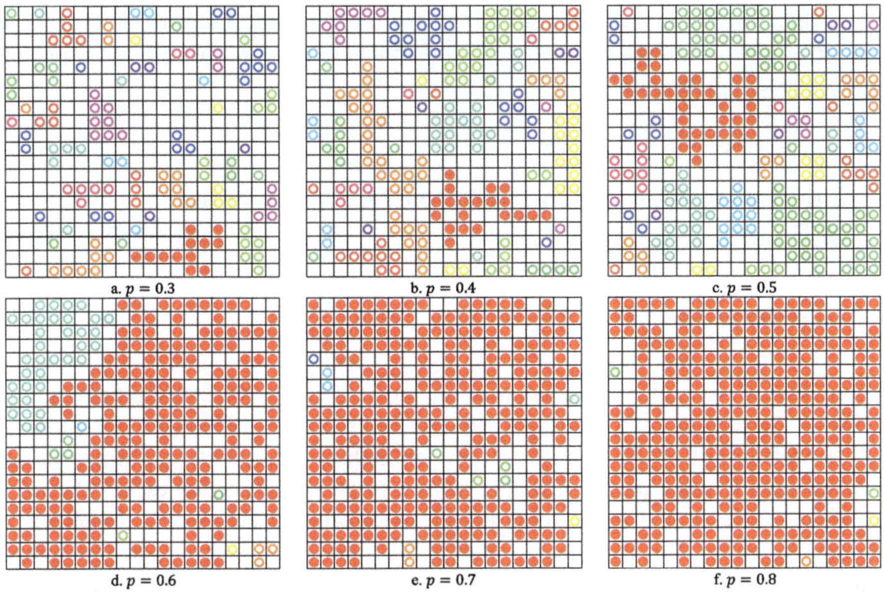

Fig. 1 Realisations of sites' configuration on a square lattice with different occupation probabilities p. On average, for uniform site-percolation model, the density (or fraction) of occupied sites equals the probability p of every site being occupied

probabilities. The behaviour of the system approaching the critical point can then be used to classify the system, i.e. identifying its universality class (Stanley 1999).

When the system is at the critical point, many physical quantities diverge in a manner that can be described by a power law

$$X(p) \propto |p - p_c|^{-\beta_X}, \quad |p - p_c| \ll 1, \tag{1}$$

whose behaviour is solely characterised by the exponent $\beta_X > 0$. X could quantify some relevant properties of the system, for example, correlation length (distance beyond which the states of two points, on average, become uncorrelated) or average cluster size, etc. A small value of the exponent β_X indicates a slow divergence of X, whilst a larger one means that X diverges fast. In physics literature, the set of exponents β_X together with other mathematical descriptions are used to characterise the universality class of a physical system (Hinrichsen 2000).

Applying that idea of transition in standard percolation, we find the clusters of points in our data by defining a parameter distance ρ that sets the maximum nearest neighbour distance between a pair of points in the same cluster. Once a cluster is identified, its size is defined as the number of points in the cluster, and area as the union area of circles of radius ρ centred at all the points. Intuitively, we know that for small parameter distance ρ, most points belong to distinct clusters while at large parameter distance, most points belong to a single cluster. A transition is expected to take place in between.

3 Continuum Percolation and Spatial Arrangement of Points

3.1 General Ideas

We propose a procedure to characterise the spatial pattern of a set of points, that is based on the idea of continuum percolation (Meester et al. 1996) (see also random disk models or Gilbert disk (Boolean) model (Balister et al. 2008)). The procedure involves identifying clusters of points, whose pairwise distance does not exceed a parameter value ρ, and quantifying the growth of the clusters as the parameter ρ changes its value. This distance parameter (also known as buffer radius in spatial analysis) ρ sets out the neighbourhood of a point, which could be related to its connectivity. A larger value of ρ means an extensive vicinity for the point, allowing access or connection to more other points to form a common cluster.

Mathematically, we consider a domain $\mathfrak{D}$ in $\mathbb{R}^2$, in which there are N points, each of which is located at coordinates (x_i, y_i). We then construct the clusters based on the buffer radius ρ. Any point j, whose Euclidean distance

$$d_{ij} = \sqrt{\left(x_i - x_j\right)^2 + \left(y_i - y_j\right)^2} \tag{2}$$

from point i is less than or equal to ρ, belongs to the same cluster as i (Fig. 2).[2] We denote $\eta(\rho)$ as the number of clusters given the buffer radius ρ. For each cluster, we define its size $\xi(\rho)$ as the number of points in the cluster whereas its area $A(\rho)$ the union of area of circles of uniform radius ρ centred at the points in the cluster. To make different domains comparable, we normalise the clusters' size $\xi(\rho)$ by the number of points N in the domain (i.e. as fraction of points), while the clusters' area $A(\rho)$ by the union of area $\Xi(\rho)$ of circles of radius ρ centred at *all* N points in the domain $\mathfrak{D}$. In this analysis, we are interested in the cluster with either largest size $\xi_{max}(\rho)$ or largest area $A_{max}(\rho)$, which may not be necessarily the same one. For simplicity of discussions, and unless stated otherwise explicitly, the descriptions of cluster size ξ below also hold for cluster area A.

In applying the idea of the percolation cluster to analyse the spatial pattern of a set of points, the buffer radius ρ plays the role of the control parameter. For different values of ρ, the system could be argued to be in different phase, namely, a *segregate*

[2]This step is similar to the procedure in DBSCAN algorithm in computer science literature (Ester et al. 1996) when no noise is considered, i.e. smallest cluster size is 1 or a point is allowed to form a cluster of its own.

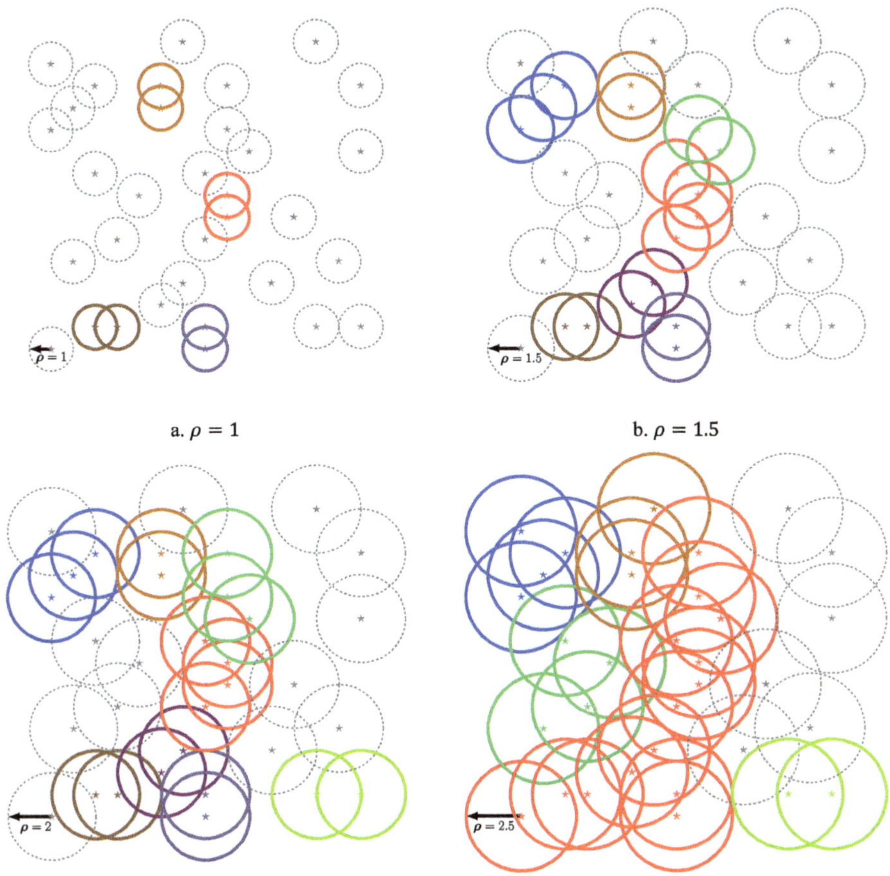

a. $\rho = 1$ b. $\rho = 1.5$

c. $\rho = 2$ d. $\rho = 2.5$

Fig. 2 Illustration of the cluster identification based on buffer radius ρ. Points (indicated as $\star$, with circle of radius ρ centred at it) belonging to the same cluster are marked with same colour. The cluster with largest size is coloured red, and it grows as ρ increases. A cluster of size 1 is coloured grey with dashed circle around it. (Tip to read clusters: a point ($\star$) must be of the same colour as the circle ($\odot$) enclosing it.)

phase for small values of ρ and an *aggregate* phase for large values of ρ. The transition from one phase to the other takes place in the intermediate regime of ρ. When ρ is small, there are many small clusters as most points are not connected due to limited vicinity, resulting in a large number of clusters $\eta(\rho)$. The size of largest cluster $\xi_{max}(\rho)$, therefore, stays small and slowly increases due to this disjoint structure. As ρ enters an intermediate regime, $\xi_{max}(\rho)$ increases faster than it does in the small-ρ regime. This is when the larger clusters merge together making a significant expansion in size of the largest cluster. At the same time, $\eta(\rho)$ starts to decrease when small clusters merge to form larger ones. By this time, the largest cluster would have already encompassed most of the points in the domain, leading to very mild change in its size $\xi_{max}(\rho)$ as ρ increases further into the regime of large ρ.[3] The picture above illustrates that as ρ is tuned, the system undergoes a change in its state, from a segregate to an aggregate phase, via a region of 'phase transition' corresponding to the intermediate regime of ρ, which has been well studied in physics (Domb et al. 2001), with a prime example of percolation (Stauffer et al. 1994; Meester et al. 1996).

The idea of percolation, in fact, has been employed in a number of urban-related studies. These include modelling the growth of cities (Makse et al. 1998) as a variant of percolation (i.e. correlated percolation), analysis of road network in Britain using percolation cluster to determine the hierarchical and organisational structure of cities and regions (Arcaute et al. 2016), analysis of the global urban land cover pattern to show a transition from separated clusters to a gigantic component at the country scale (Fluschnik et al. 2016), or application to modelling of urban retail locations (Piovani et al. 2017). However, while the cluster size appears in most of the studies, the cluster area has not previously been considered. In this chapter, we shall see that the combination of analysis of profiles of both the largest system size $\xi_{max}(\rho)$ and area $A_{max}(\rho)$ transit through the intermediate region would allow us to characterise the spatial pattern of a set of points.

3.2 *Characterisation of Spatial Distribution of Points*

The growth pattern of the clusters of points as the buffer radius parameter ρ varies could be used to quantify the spatial pattern of the set of points. When ρ increases, more points at farther distance can now join to belong to the same cluster as previously disjoint points, making the clusters merge to increase their size. Since the largest cluster size $\xi_{max}(\rho)$ increases monotonically with ρ, there exists a representative *characteristic distance* $\rho_\xi^\star$ at which $\xi_{max}(\rho)$ effectively exhibits the most significant increase. As ρ approaches $\rho_\xi^\star$, the profile $\xi_{max}(\rho)$ could exhibit different behaviours, showing either a sharp narrow or a gradual wide-range increase. To account for that, it is also meaningful to introduce a quantity σ_ξ, called *spread of transition*, to measure the overall width of the increases in the profile of ξ_{max}.

3.2.1 Peak Analysis

In order to calculate the characteristic distance $\rho_\xi^\star$ and its spread σ_ξ, we make use of the profile of the first derivative $\xi'_{max}(\rho) = \frac{\mathrm{d}\xi_{max}(\rho)}{\mathrm{d}\rho}$ of the largest cluster size (similarly, we use $A'_{max}(\rho) = \frac{\mathrm{d}A_{max}(\rho)}{\mathrm{d}\rho}$ for the cluster area), which measures its growth. If the profile of $\xi_{max}(\rho)$ exhibits a sharp increase, its first derivative produces a single

[3] Strictly speaking, the size and area of a cluster are step functions of the buffer radius ρ, but we assume in this study that the profiles $\xi_{max}(\rho)$ and $A_{max}(\rho)$ can be approximated by smooth functions so that their derivatives exist at all points (see Sects. 3.2.2 and 3.2.3).

dominant peak. On the other hand, if $\xi_{max}(\rho)$ increases gradually over a wide range of ρ, multiple scattering peaks are present in the profile of $\xi^{'}_{max}(\rho)$. To identify these peaks, we consider the profile of $\xi^{'}_{max}(\rho)$ (and $A^{'}_{max}(\rho)$) at every value of the buffer radius ρ ranging from $\rho_{min} = \rho_1$ to $\rho_{max} = \rho_M$ in the step of $\delta\rho = \rho_{i+1} - \rho_i, \forall i$. Since the values of the buffer radius are discrete, a point $\rho_i, \xi^{'}_{,max}(\rho_i))$ is a peak if and only if

$$\begin{cases} \xi^{'}_{max}(\rho_i) > \xi^{'}_{max}(\rho_{i-1}) \\ \xi^{'}_{max}(\rho_i) > \xi^{'}_{max}(\rho_{i+1}) \end{cases} \tag{3}$$

This discrete nature also produces a lot of small noisy peaks. For practical purpose, these noisy peaks could be filtered out by offsetting the entire profile of $\xi^{'}_{max}(\rho)$ by a sufficiently small amount (usually less than 0.5% of the maximum possible peak, which is 1, after normalisation) and considering only the positive remaining peaks.

3.2.2 Characteristic Distances Among Points

Once the peaks have been properly identified, the characteristic distance $\rho^{\star}_{\xi}$ could be calculated to be the mean of ρ of all peaks, as the representative value of ρ the transition in largest cluster size takes place, since all peaks in the derivative $\xi^{'}_{max}(\rho)$ contribute to the growth of the cluster $\xi_{max}(\rho)$. However, since not every peak is contributing equally to the growth, a measure of the characteristic distance $\rho^{\star}_{\xi}$ must take into account the effects of their strength. It follows that a high peak indicates a more significant increase in cluster size (a major merger) than those indicated by a lower one. Hence, the average of all values of $\rho^{\dagger}_{\xi,i}$ at which a peak *i* occurs, weighted by the height $\xi^{'}_{max}\left(\rho^{\dagger}_{\xi,i}\right) = \left.\frac{d\xi_{max}(\rho)}{d\rho}\right|_{\rho=\rho^{\dagger}_{\xi,i}}$ of the peaks, is an appropriate measure of this characteristic distance, i.e.

$$\rho^{\star}_{\xi} = \frac{\sum_i \xi^{'}_{max}\left(\rho^{\dagger}_{\xi,i}\right)\rho^{\dagger}_{\xi,i}}{\sum_i \xi^{'}_{max}\left(\rho^{\dagger}_{\xi,i}\right)} \tag{4}$$

Similarly, we have for the cluster area

$$\rho^{\star}_{A} = \frac{\sum_i A^{'}_{max}\left(\rho^{\dagger}_{A,i}\right)\rho^{\dagger}_{A,i}}{\sum_i A^{'}_{max}\left(\rho^{\dagger}_{A,i}\right)} \tag{5}$$

It should be noted that every peak in $\xi^{'}_{max}(\rho)$ signifies the existence of one or a few clusters of points located at a farther distance than those in the current largest cluster. This thus provides us with information on the length scales of distribution of points within the set. If there are many peaks, the points are distributed in clusters separated by different distances. On the other hand, the existence of a few peaks implies an almost uniform distribution of points that are (approximately) equidistant from one another. In either case, it is without a doubt that there exists a characteristic distance in the spatial distribution of points. This characteristic distance should tell us the length scale above which the points are (largely) connected and below which they are disconnected.

It is noteworthy that $\rho_\xi^\star$ and $\rho_A^\star$ are different from the average of pairwise distance among all points in the set because they encode the connectivity of the points in terms of spatial distribution. In other words, the two characteristic distances are the measure of typical distance between points in the set in the perspective of global connectivity of all points. In the context of points in an urban system, they translate to the distance one has to traverse to get from one point to another in order to explore the entire system. It then follows that a large value of characteristic distance implies a sparsely distributed set of points, i.e. a low density of points per unit area (Huynh et al. 2018).

The analyses of peaks in size profile $\xi'_{max}(\rho)$ and area profile $A'_{max}(\rho)$ provide different perspectives on the spatial distribution of points. The size quantifies the number of points with respect to the distance while the area further takes into account the relative position of the points. There is no redundancy in the consideration of the two measures, but rather, one is complementary to the other. This comes to light in the next Sect. 3.2.3 when the combination of the two allows us to classify distinct types of distribution of points.

3.2.3 Quantification and Classification of the Spatial Distributions of Points

The characteristic distances introduced above inform us on the points of transition of cluster size and area but they do not tell us how the size and area of the cluster transit from small to large value, i.e. how the cluster grows. This, however, can be easily characterised by further exploiting the analysis of peaks in $\xi'_{max}(\rho)$ (and $A'_{max}(\rho)$) discussed in Sec. 3.2.1. As discussed above, if the cluster grows rapidly through the transition, there are very few peaks in $\xi'_{max}(\rho)$, and all of which are sharp and localised. On the other hand, the peaks are scattered over a wide range of ρ should the cluster grow gradually. The standard deviation of the location $\rho^\dagger_{\xi,i}$ of the peaks, or the spread of transition σ_ξ for cluster size, is a good measure of such scattering. Again, however, a low peak that is distant from a group of localised high peaks should not significantly enlarge the spread. Therefore, the standard deviation of $\rho^\dagger_{\xi,i}$ needs to be weighted by the height $\xi'_{max}\left(\rho^\dagger_{\xi,i}\right)$ of the peaks, i.e.

$$\sigma_\xi = \sqrt{\frac{\sum_i \xi'_{max}\left(\rho^\dagger_{\xi,i}\right)\left(\rho^\dagger_{\xi,i} - \rho^\star_\xi\right)^2}{\sum_i \xi'_{max}\left(\rho^\dagger_{\xi,i}\right)}} \tag{6}$$

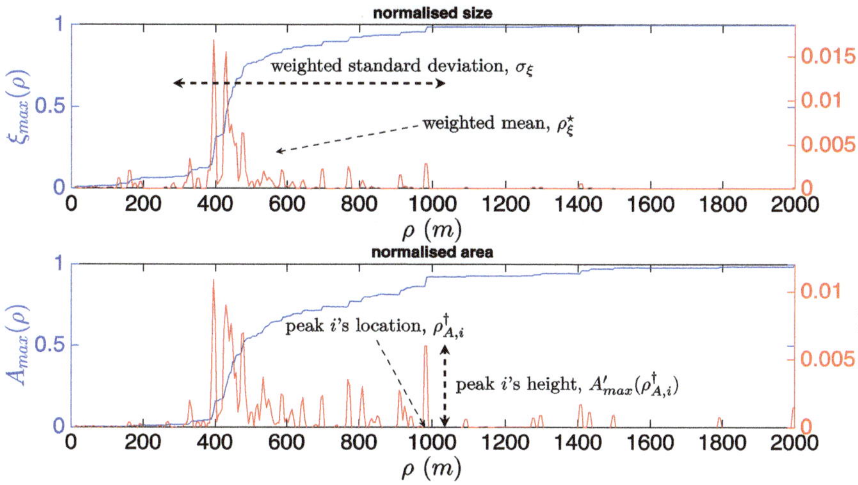

Fig. 3 Illustration of the quantities yielded from analysis of peaks in the size and area profile of the largest cluster

Similarly, we have the spread of transition for cluster area

$$\sigma_A = \sqrt{\frac{\sum_i A'_{max}\left(\rho^\dagger_{A,i}\right)\left(\rho^\dagger_{A,i} - \rho^\star_A\right)^2}{\sum_i A'_{max}\left(\rho^\dagger_{A,i}\right)}} \tag{7}$$

An illustration of the characteristic distance $\rho_\xi^\star$ (and $\rho_A^\star$) and the spread σ_ξ (and ρ_A) is depicted in Fig. 3.

Once we have quantified the measures of the largest cluster size and largest cluster area, we shall argue that the combination of these quantities allows us to characterise the pattern of relative position of points in a set. In particular, the combination of the two spreads of transition defined in Eqs. (6) and (7) enables us to construct the (σ_ξ, σ_A) diagram (see Fig. 4) to discuss different regions corresponding to different types of spatial point distribution. In general, there are four types of distribution that can be identified from the spreads of transition in size σ_ξ and area σ_A. The first one is the region of small $\sigma_\xi \approx \sigma_A$ in which both $\xi_{max}(\rho)$ and $A_{max}(\rho)$ exhibit a sharp rise. This is the case of a system of regularly distributed points in which the points are (approximately) equally distant from each other, e.g. grid points. Outside this region, going along the diagonal line of $\sigma_\xi \approx \sigma_A$, the second type could be identified in which both $\xi_{max}(\rho)$ and $A_{max}(\rho)$ exhibit a gradual increase and almost every peak in $\xi'_{max}(\rho)$ has a respective peak in $A'_{max}(\rho)$. When the σ_ξ and σ_A differ significantly, we have two other types of pattern, one below and one above the diagonal line. To discuss these two types of pattern, we recall that the size of a cluster is the number of points in that cluster, whereas its area the union area of circles of radius ρ centred

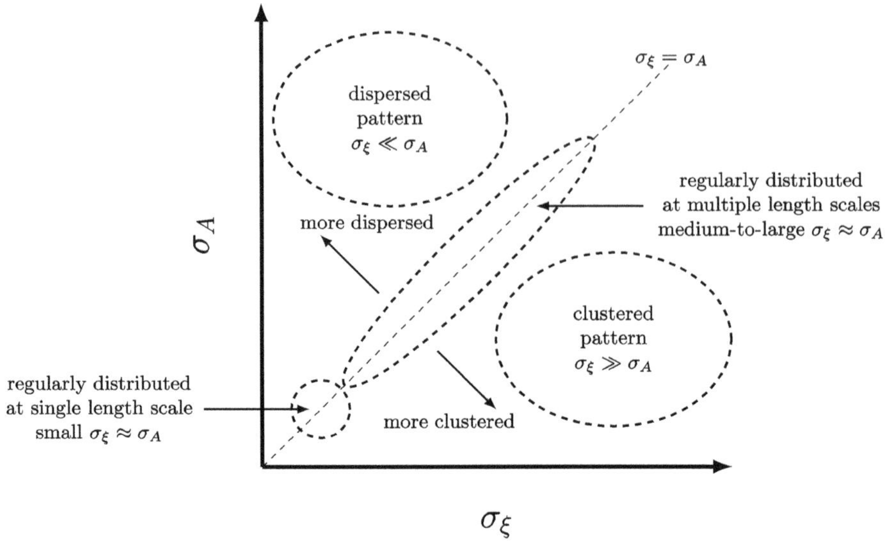

Fig. 4 Interpretation of different patterns of spatial point distributions given different values of the pair (σ_ξ, σ_A)

at the points in the cluster. Above the diagonal, in the region of $\sigma_\xi \gg \sigma_A$, the peaks in $\xi'_{max}(\rho)$ tend to spread over a wider range of ρ than those in $A'_{max}(\rho)$, implying clustered points and compact coverage area. In such distributions, there are jumps in the size of the largest cluster size that do not give rise to a jump in its area. This happens when the points of an acquired cluster are compact, contributing very little increase in the area of the largest cluster. If the acquired cluster is not compact, i.e. its points span a larger area, there might be a significant increase in the area of the largest cluster. On the other side, below the diagonal, in the region of $\sigma_\xi \ll \sigma_A$, the peaks in $\xi'_{max}(\rho)$ tend to be more localised than those in $A'_{max}(\rho)$, implying dispersed points and broad coverage area. In such distributions, there are jumps in the area of the largest cluster area that do not give rise to a jump in its size. This happens when the points of an acquired cluster are stretched apart (but still within the buffer radius so that they belong to the same cluster). This way, the increase in the area of the largest cluster is more significant than that in its size.

3.3 Comparison with Other Methods

There are a few well-established methods in the literature to analyse the spatial pattern of a set of points. Among them, the most popular method in the geographical analysis is probably the Ripley's K-function (Ripley 1976, 1977). The method of pair correlation function (Stoyan and Penttinen 2000) is also widely used in the field of forestry to understand the pattern of trees' location in an area. These methods are mainly concerned with the null hypothesis whether the pattern of the points in the set is complete spatial randomness (CSR). If not CSR, the pattern can further be determined to be either regular or clustered.

The Ripley's K-function method centralises about the function K, which is defined as the expected number of neighbours within distance d from a randomly chosen point rescaled by the density λ of points in the domain of interest. The function $K(d)$ of the set of points is then compared against that of a homogeneous Poisson process[4] (or CSR pattern), which is known to be $K_{Poisson}(d) = \pi d^2$. When the value $K(d)$ is larger than that of $K_{Poisson}(d)$, the points are said to be clustered at distance d because more neighbours are expected within distance d of a point than in a Poisson process. Conversely, when the value $K(d)$ is smaller than that of $K_{Poisson}(d)$, the points are said to be regular at distance d. On the other hand, the pair correlation function method utilises the pair correlation function g between two points at distance u (Penttinen et al. 1992) which is related to the Ripley's K-function via

$$K(d) = \int_0^d g(u) 2\pi d \mathrm{d}u$$

In both cases, the function K (or g), however, involves the density λ of points in the domain and, hence, requires a proper identification of the area of the domain. An overestimate or underestimate of the density can significantly alter the behaviour of the function K. Visually, an immense domain seeing the points clustering in a small area would yield very small λ and, hence, significantly boost up $K(d)$, making it incorrectly larger than $K_{Poisson}(d)$. Contrarily, a tight domain may artificially make $K(d)$ smaller than $K_{Poisson}(d)$. In the study of forestry, the domain is usually taken to be the range of the coordinates of the trees' location, which is a rectangular bounding box, due to the nature of the observation. However, in the case of urban entities like road junctions, the bounding box might not be a proper identification of the domain of the points because of their irregular distribution pattern due to unknown underlying urban growth processes. These methods could be difficult to apply in analysing the spatial patterns of points in an urban context like in the distribution of road junctions.

4 Interpretation of the Patterns in Urban Morphology

In the coming section, we discuss how the analysis method described in Sect. 3.2 could be applied to quantify different patterns in urban morphology. For this discussion, we refer to the points in an urban system as spatial locations of important entities that could be used as a representative set of points in an urban system. Such

[4]In this process, each point is randomly and independently located in the domain. The process has it name from the fact that the number of points in a sub-region of a certain size follows a Poisson distribution.

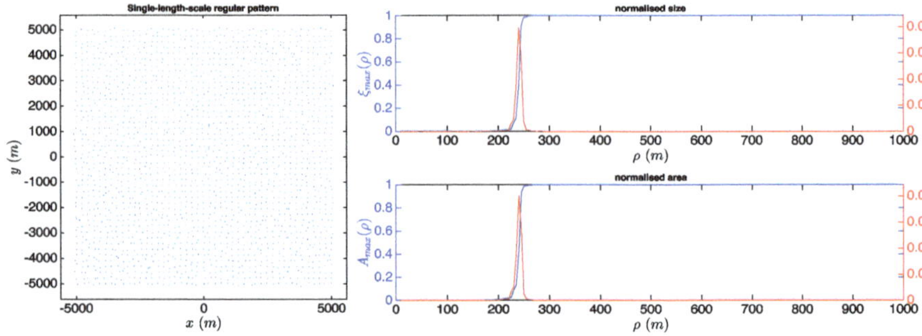

Fig. 5 An example of single-scale regular point pattern. The pattern is generated by adding small random displacement to points of a regular lattice, which comprises 2025 (45 × 45) points spanning an area of 10 km × 10 km. The values of the characteristic distances are $\rho^{\star}_{\xi} = 240$ m and $\rho^{\star}_{\mathrm{A}} = 240$ m. The measures of spread of transition are $\sigma_{\xi} = 0$ m and $\sigma_{\mathrm{A}} = 0$ m (there is only a single peak, resulting in vanishing standard deviation as obtained from Eqs. 6 and 7; in general, σ_{ξ} and σ_{A} should stay small)

a set could nominally be, for example the locations of road junctions. In a recently published study, the method has been applied to a case study of spatial distribution of locations of public transport points in different cities in the U.S. (Huynh et al. 2018).

4.1 Single-scale Regular Pattern

The bottom-left corner of the (σ_{ξ}, σ_A) plot (Fig. 4) represents the set of points whose values of the spreads σ_{ξ} and σ_A are very small. That means the profiles of $\xi_{max}(\rho)$ and $A_{max}(\rho)$ exhibit sharp transition with localised peaks in both $\xi^{'}_{max}(\rho)$ and $A^{'}_{max}(\rho)$. This signifies a characteristic length scale at which most of the points are (approximately) equally spaced from each other, e.g. grid points (see Fig. 5). There is a sharp transition at the value of the characteristic distances $\rho^{\star}_{\xi} \approx \rho^{\star}_{A}$: the largest cluster transits from occupying a modest fraction to almost the entire of the points in the set. One could take the boroughs of Bronx, Brooklyn and Manhattan of New York city as typical examples of such kind of distribution. The pattern of spatial points in these cities appears very regular. In fact, by inspecting their street patterns, one can easily tell the pattern of parallel roads in one direction cutting those in the other, dividing the land into well-organised polygons with almost perfect square and rectangular shapes. Apparently, this feature was a result of top-down planning and design that happened before the infrastructure was being built in the city (Barthelemy et al. 2013).

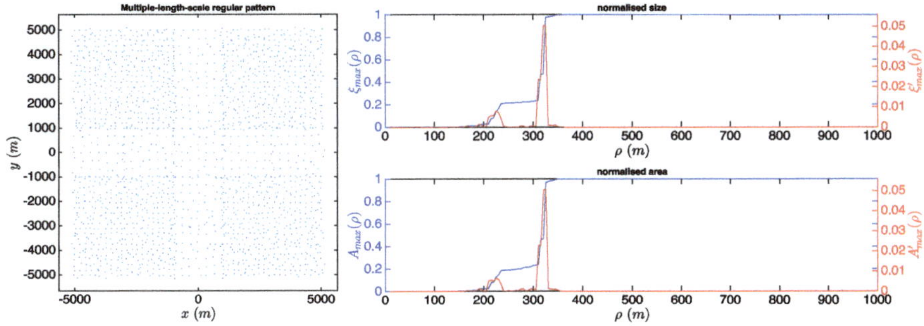

Fig. 6 An example of multiple-scale regular point pattern. The pattern is generated by superimposing different sets of single-scale pattern (of different densities) like in Sect. 4.1. There are in total 2041 points spanning an area of 10 km × 10 km. The values of the characteristic distances are $\rho^\star_\xi = 315.080$ m and $\rho^\star_{\text{A}} = 314.357$ m. The measures of spread of transition are $\sigma_\xi = 23.185$ m and a similar value for $\sigma_{\text{A}} = 25.172$ m

4.2 *Multi-scale Regular Pattern*

The set of points in a city can also be distributed in a regular manner but at different length scales (Chen and Wang 2013; Nie et al. 2015). For example, the entire set of points can be divided into several subsets and within each subset, the points are (quasi-)equally distant from each other (see Fig. 6). At larger length scale, i.e. ρ increases further, these subsets of points are again (quasi-)equally distant from each other, i.e. hierarchical structure. The buffer radius ρ can thus be thought to play the role of a zooming parameter. In this multi-scale regular pattern, the profile of the largest cluster size $\xi_{max}(\rho)$ and area $A_{max}(\rho)$ experience a significant jump every time ρ changes its zooming level. At the lowest level are individual transport point. When ρ zooms out to the second level, the points that are closest to each other start to form their respective clusters. Moving to the next level, the nearby clusters start joining to form larger cluster but there will be many of these 'larger clusters', i.e. the largest cluster is of comparable size or area to several other clusters. The most important feature of this spatial pattern is that the jumps in the profile of $\xi_{max}(\rho)$ correspond well to those in $A_{max}(\rho)$, even though the locations of the jumps are spread apart. That leads to the (approximate) equality of the spread of transitions σ_ξ and σ_A despite their not being small.

4.3 *Clustered Pattern*

There are cases in which the jumps in the profile of largest cluster size $\xi_{max}(\rho)$ do not correspond to those in the area $A_{max}(\rho)$ and vice versa. In such cases, the spatial distribution of the transport points deviates from regular patterns. We first consider the scenarios in which $\sigma_\xi \gg \sigma_A$. For such distributions, the points are clustered and

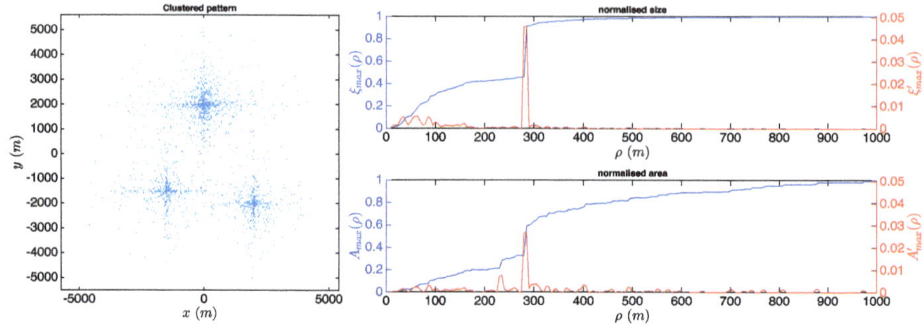

Fig. 7 An example of clustered point pattern. The pattern is generated by first defining a fixed number of centres (3 in this case) and then randomly adding points surrounding them with density decaying away from the centres. There are in total 2000 points spanning an area of 10 km × 10 km. The values of the characteristic distances are $\rho_\xi^\star = 258.747$ m and $\rho_\mathrm{A}^\star = 277.752$ m. The measures of spread of transition are $\sigma_\xi = 65.474$ m and a smaller value for $\sigma_\mathrm{A} = 17.603$ m

tend to minimise the coverage area (see Fig. 7). When $\sigma_\xi \gg \sigma_A$, there are jumps in the size of the largest cluster size that do not give rise to a jump in its area. This happens when the points of an acquired cluster are compact, contributing very little increase in the area of the largest cluster. If the acquired cluster is not compact, i.e. its points span a larger area, there might be significant increase in the area of the largest cluster and, hence, a peak would be reflected by its contribution to σ_A. However, the size measure is not affected as it only tells the number of points that are included in the cluster but not their relative location with respect to each other.

4.4 Dispersed Pattern

At the other end, we have the scenarios of $\sigma_\xi \ll \sigma_A$, in which the points are dispersed and tend to maximise the coverage area. When $\sigma_\xi \ll \sigma_A$, there are jumps in the area of the largest cluster that do not give rise to a jump in its size. This happens when the points of an acquired cluster are dispersed (but still within the buffer radius so that they belong to the same cluster, see Fig. 8). This way, the increase in the area of the largest cluster is more significant than that in its size, resulting $\sigma_\xi \ll \sigma_A$.

If the feature of single-scale regular spatial pattern (when both σ_ξ and σ_A are small) is a result of well-designed and top-down planning in an urban system, the other spatial patterns (either σ_ξ or σ_A is not small) can be interpreted as a consequence of urban system development under local constraints. In the former case, the urban system appears to be of organised type while in the latter, it can be said to be of organic type because its spatial features develop in an ad hoc manner as the city grows. Thus, our analysis on the spatial patterns in urban systems provides more nuances of their morphology and insights into different possible growth processes as the systems progress through their development.

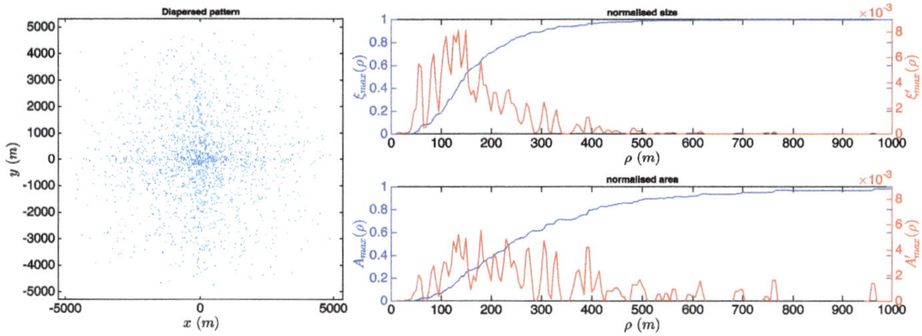

Fig. 8 An example of dispersed point pattern. The pattern is generated by scattering points with uneven density. There are in total 2000 points spanning an area of 10 km × 10 km. The values of the characteristic distances are $\rho_{\xi}^{\star} = 124.322$ m and $\rho_{\mathrm{A}}^{\star} = 191.146$ m. The measures of spread of transition are $\sigma_{\xi} = 30.875$ m and a larger value for $\sigma_{\mathrm{A}} = 67.361$ m

The proposed method in this chapter provides quantitative measures of spatial pattern of a set of points, that can be applied to, for example, the road junctions in the context of urban systems. The public road junctions play an important role in forming the backbone of any modern urban system. For example, street network serves the essential role of enabling flow or exchange of various processes in the city, whose overall patterns could be captured by the intersections, which in turn define the pattern of land parcels.

The results from the analysis in this work unveil different types spatial patterns in urban systems. The patterns are shown to be either of organised type, in which the entities are well spaced as if they were built top-down, or of organic type, in which the entities are spaced with multiple length scales as if they grew spontaneously (Cheng et al. 2003; Makse et al. 1998; Longley et al. 1992). It should be noted that while these two types of patterns have been discussed in other literature like architecture or urban geography, the emphasis of this study is the quantification of the perceived spatial patterns found in different urban systems, via σ_{ξ} and σ_{A}, that allows comparison among them.

Acknowledgements HNH acknowledges the support of A*STAR International Fellowship (Ref: AGA/SE-17/FS-023/SE/AIF/14/003).

References

K. F. Anschuetz, R. H. Wilshusen and C. L. Scheick, "An Archaeology of Landscapes: Perspectives and Directions," *Journal of Archaeological Research,* vol. 9, p. 157, 2001.

E. Arcaute, C. Molinero, E. Hatna, R. Murcio, C. Vargas-Ruiz, A. P. Masucci and M. Batty, "Cities and regions in Britain through hierarchical percolation," *Royal Society Open Science,* vol. 3, p. 150691, 2016.

P. Balister, A. Sarkar and B. Bollobás, "Percolation, Connectivity, Coverage and Colouring of Random Geometric Graphs," in *Handbook of Large-Scale Random Networks*, Berlin, Springer, 2008, pp. 117–142.
M. Barthelemy, P. Bordin, H. Berestycki and M. Gribaudi, "Self-organization versus top-down planning in the evolution of a city," *Scientific Reports,* vol. 3, p. 2153, 2013.
M. Batty, The New Science of Cities, MIT Press, 2017.
E. Bertin, Statistical Physics of Complex Systems, Springer, 2016.
L. M. A. Bettencourt, J. Lobo, D. Helbing, C. Kühnert and G. B. West, "Growth, innovation, scaling, and the pace of life in cities," *Proceedings of the National Academy of Sciences,* vol. 104, p. 7301, 2007.
J. Burguet and P. Andrey, "Statistical Comparison of Spatial Point Patterns in Biological Imaging," *PLoS ONE,* vol. 2, p. e87759, 2014.
Y. Chen and J. Wang, "Multifractal characterization of urban form and growth: the case of Beijing," *Environment and Planning B: Planning and Design,* vol. 40, p. 884, 2013.
J. Cheng and I. Masser, "Modelling urban growth patterns: a multiscale perspective," *Environment and Planning A,* vol. 35, p. 679, 2003.
C. Domb, M. S. Green and J. L. Lebowitz, Phase Transitions and Critical Phenomena, London: Academic Press, 2001.
M. Ester, H.-P. Kriegel, J. Sander and X. Xu, "A density-based algorithm for discovering clusters in large spatial databases with noise," in *Proceedings of the Second International Conference on Knowledge Discovery and Data Mining*, 1996.
T. Fluschnik, S. Kriewald, A. García Cantú Ros, B. Zhou, D. E. Reusser, J. P. Kropp and D. Rybski, "The Size Distribution, Scaling Properties and Spatial Organization of Urban Clusters: A Global and Regional Percolation Perspective," *ISPRS International Journal of Geo-Information,* vol. 5, p. 110, 2016.
A. S. Fotheringham, C. Brunsdon and M. Charlton, Quantitative Geography: Perspectives on Spatial Data Analysis, Sage, 2000.
D. Helbing, J. Keltsch and P. Molnár, "Modelling the evolution of human trail systems," *Nature,* vol. 388, p. 47, 1997.
H. Hinrichsen, "Non-equilibrium critical phenomena and phase transitions into absorbing states," *Advance in Physics,* vol. 49, pp. 815–958, 2000.
H. N. Huynh, E. Makarov, E. F. Legara, C. Monterola and L. Y. Chew, "Characterisation and comparison of spatial patterns in urban systems: A case study of US cities," *Journal of Computational Science,* vol. 24, pp. 34–43, 2018.
J. Illian, A. Penttinen, H. Stoyan and D. Stoyan, Statistical Analysis and Modelling of Spatial Point Patterns, Wiley, 2008.
H. J. d. Knegt, F. v. Langevelde, M. B. Coughenour, A. K. Skidmore, W. F. d. Boer, I. M. A. Heitkönig, N. M. Knox, R. Slotow, C. v. d. Waal and H. H. T. Prins, "Spatial autocorrelation and the scaling of species-environment relationships," *Ecology,* vol. 91, pp. 2455–2465, 2010.
F. Li and L. Zhang, "Comparison of point pattern analysis methods for classifying the spatial distributions of spruce-fir stands in the north-east USA," *Forestry,* vol. 80, pp. 337–349, 2007.
P. Longley, M. Batty, J. Shepherd and G. Sadler, "Do Green Belts Change the Shape of Urban Areas? A Preliminary Analysis of the Settlement Geography of South East England," *Regional Studies,* vol. 26, p. 437, 1992.
H. A. Makse, J. S. Andrade, M. Batty, S. Havlin and H. E. Stanley, "Modeling Urban Growth Patterns with Correlated Percolation," *Physical Review E,* vol. 58, pp. 7054–7062, 1998.
H. A. Makse, J. S. J. Andrade, M. Batty, S. Havlin and H. E. Stanley, "Modeling urban growth patterns with correlated percolation," *Physical Review E,* vol. 58, p. 7054, 1998.
R. Meester and R. Roy, Continuum Percolation, New York: Cambridge University Press, 1996.
P. Mellars, "Going east: New genetic and archaeological perspectives on the modern human colonization of Eurasia," *Science,* vol. 313, p. 796, 2006.
M. E. J. Newman and R. M. Ziff, "Efficient Monte Carlo Algorithm and High-Precision Results for Percolation," *Physical Review Letters,* vol. 85, pp. 4104–4107, 2000.

Q. Nie, X. Jianhua and L. Zuhan, "Fractal and multifractal characteristic of spatial pattern of urban impervious surfaces," *Earth Science Informatics,* vol. 8, no. 2, p. 381, 2015.

A. Penttinen, D. Stoyan and H. M. Henttonen, "Marked Point Processes in Forest Statistics," *Forest Science,* vol. 38, pp. 806–824, 1992.

D. Piovani, C. Molinero and A. Wilson, "Urban retail location: Insights from percolation theory and spatial interaction modeling," *PLoS ONE,* vol. 12, p. e0185787, 2017.

B. D. Ripley, "Modelling Spatial Patterns," *Journal of the Royal Statistical Society: Series B,* vol. 39, pp. 172–212, 1977.

B. D. Ripley, "The Second-Order Analysis of Stationary Point Processes," *Journal of Applied Probability,* vol. 13, pp. 255–266, 1976.

E. F. Schumacher, "Patterns of Human Settlement," *Ambio,* vol. 5, p. 91, 1976.

C. Small, F. Pozzi and C. Elvidge, "Spatial analysis of global urban extent from DMSP-OLS night lights," *Remote Sensing of Environment,* vol. 96, p. 277, 2005.

H. E. Stanley, "Scaling, universality, and renormalization: Three pillars of modern critical phenomena," *Review of Modern Physics,* vol. 71, pp. S358–S366, 1999.

D. Stauffer and A. Aharony, Introduction to percolation theory, London: Taylor & Francis, 1994.

D. Stoyan and A. Penttinen, "Recent applications of point process methods in forestry statistics," *Statistical Science,* vol. 15, pp. 61–78, 2000.

D. I. Uzunov, Introduction to the Theory of Critical Phenomena: Mean Field, Fluctuations and Renormalization, 2nd ed., Singapore: World Scientific, 2010.

G. M. Viswanathan, E. P. Raposo and M. G. E. da Luz, Perspectives and Challenges in Statistical Physics and Complex Systems for the Next Decade, World Scientific, 2014.

G. West, Scale: The Universal Laws of Life and Death in Organisms, Cities and Companies, Weidenfeld & Nicolson, 2017.

J. Zucchetto, "Energy and the future of human settlement patterns: Theory, models and empirical considerations," *Ecological Modelling,* vol. 20, p. 85, 1983.

Urban Compactness: New Geometric Interpretations and Indicators

Stephen Marshall, Yi Gong and Nick Green

Abstract The 'compact city' concept is prominent in contemporary planning policy debates about ideal urban forms. However, the property of compactness itself is not well defined, and is sometimes confused or conflated with density. This chapter develops a new geometric interpretation of compactness with specific indicators—relating to diameter and perimeter—that can capture this property in the urban context. The chapter demonstrates these compactness indicators first by application to theoretical geometric shapes and then a range of English urban areas. The chapter reflects on the interpretation of the core concept of compactness, and suggests additional indicators such as 'built compactness' and 'population compactness'.

1 Introduction

The idea of a compact city as an ideal city has a long tradition. Idealised cities of the Renaissance were often depicted as bounded polygons, with walled enclosure separating the urban from the rural surrounds. Historically, the compact city has not only been a Utopian ideal but a pragmatic one grounded in geometric reality: walled cities of compact shape had the advantage of minimising the length of perimeter wall to enclose the maximum area of city (Barrow 2008). More recently, the late twentieth century rise of environmentalism and concern for 'sustainable' development led to renewed favour for compact cities whose dense form could in principle minimise land take and reduce travel distances, minimise energy and so on (for example, in

S. Marshall (✉)
Bartlett School of Planning, University College London, London WC1H 0NN, UK
e-mail: s.marshall@ucl.ac.uk

Y. Gong
Sustainable Places Research Institute, Cardiff University, Cardiff CF10 3AT, UK
e-mail: GongY2@cardiff.ac.uk

N. Green
Manchester, UK
e-mail: c_nick_green@mac.com

L. D'Acci (ed.), *The Mathematics of Urban Morphology*,
Modeling and Simulation in Science, Engineering and Technology,
https://doi.org/10.1007/978-3-030-12381-9_19

the UK, Urban Task Force 1999; ODPM 2003; DCLG 2007; Dempsey 2010; Westerink et al. 2013; Wolsink 2016). As mainstream advocacy for compact cities grew, researchers and commentators questioned the supposed benefits of compactness (see, for example, Breheny 1992a; Frey 1999; Echenique et al. 2009; Ferreira and Batey 2011). There is now an ongoing, outstanding debate on the merits or otherwise of the compact city, and the idea is sufficiently mature to have encouraged empirical studies of its ongoing validity and effectiveness (for example, Chettri and Corcoran 2013; Westerink et al. 2013; Bunker 2014; Wolsink 2016; Tappert and Drilling 2018).

However, this debate is hindered by the lack of a common consistent basis for agreeing what a compact city is, or specifying particular attributes such as 'compactness'. Without agreement on what compactness means, the articulation of policy options and the evaluation of the performance of compact cities are likely to remain confused and contested.

In fact, we can identify two distinct uses of the term compact. The first broad sense is to use compact as a label for a typical kind of city, where a 'compact city' has a package of attributes, usually featuring a combination of density, containment and proximity of services, often associated with mixed land uses and transport accessibility; this general definition can reasonably be dubbed a 'standard' definition (Jenks et al. 1996; Burton 2002; Dempsey 2010). The second narrower sense is of compactness being a specific geometric property of urban form. The two senses are not mutually exclusive: a compact city may include geometric compactness as one of its attributes. While the compact city is well aired in the urban literature, compactness as geometric property is less well articulated, and it is this second topic that this chapter focuses on. This focus is not to claim that the term compact should only be used to refer to a specific geometric property; but we suggest that when compactness is used as a property, it should be clearly defined, conceptually distinct from density. Better articulation of the narrower geometric conception of compactness should, in turn, support the better articulation of arguments in the broader compact city debate.

This chapter explores the concept of compactness and how this property might be captured and applied to settlements. The aim is to provide a clearer articulation of the geometric property of compactness through new interpretations and indicators, which may then be used for onward application, testing and debate. Following this introduction, Sect. 2 reviews existing concepts of urban compactness within the 'compact city' policy debate, and then Sect. 3 reviews compactness as a geometric property. Then, we develop new interpretations and indicators of compactness, which are illustrated with reference to some theoretical geometric shapes (Sect. 4), and subsequently applied to selected urban settlements (Sect. 5). In Sect. 6, we then reflect on the core meaning of compactness and suggest further interpretations of different kinds of compactness. Finally, we draw conclusions on the potential use of such compactness indicators, in future urban analysis and policy debates.

2 Compact Cities

Whatever the merits of compact cities, we need to be clear about what is meant by a compact city is in the first place, and interpretations of urban compactness in contradistinction to density. In this section, we shall first look briefly at the context of the compact city policy debate, before focusing in on existing conceptions of urban compactness, and then finally discuss briefly, what we can learn from the corresponding case of density.

2.1 The Compact City Policy Debate

The compact city is generally promoted as a form of 'sustainable' city and an antidote to 'sprawl'. In the UK, the compact ideal attempts to address the perceived problems brought about by the long-held English preference for a suburban lifestyle (Rasmussen 1982; Breheny 1997; Breheny and Hall 1999; Champion 1989, 2001; Power 2001). The compact city has been associated positively with 'energy, opportunity, diversity and excitement' (Rogers and Power 2000; Power 2001; Rogers 2011), fashionable notions of 'loft living' (Zukin 1982), healthy social networking (Breheny et al. 1993), walking, cycling and the use of public transport, and (hence) 'green urbanism' and 'sustainability' (Jenks et al. 1996; Edwards 2005).

And yet, the compact city remains a contested notion. As Neumann points out, it is a concept that seems slightly unsure of itself: there is a preponderance of question marks in the titles of articles dealing with the compact city (Neuman 2005), perhaps because many of the merits of urban intensification and the compact city have been based on assertion and theory rather than empirical evidence (Breheny 1992a, b; Jenks et al. 1996; Williams 2000; Williams et al. 2000; Vallance et al. 2005). Several chapters have questioned whether the compact city is a 'sustainable' urban form from a variety of points of view (Burton 2002; Bramley et al. 2006; Gaigné 2009; Dempsey 2010). Even if a compact city is in principle desirable, there are question marks over how to achieve 'compaction' through infill and intensification, not only in terms of physically accommodating new buildings and access roads within the existing urban fabric, but also taking account of consumer demand for living at high density in specific locations (Arup 2005; Echenique et al. 2009). Judgement on these matters relies on an understanding of what is meant by 'compact' in the first place.

2.2 Urban Compactness

In a chapter called 'Measuring Urban Compactness in UK Towns and Cities', Burton (2002) notes that there are various ways of defining both urban compactness and urban density; but none of them are completely satisfactory. For example, compactness can be defined in terms of how well connected it is; in other words, it could be physically diffuse, but 'compact' in the sense that travelling around it is quick and easy (Burton 2002). Burton goes on to point out an even looser definition of compact (self-contained) before noting that in the UK context, compact carries a less abstract and more literal interpretation: the compact city is a moderately sized 'free-standing, [self-]contained urban settlement' that is nonetheless 'large enough to support the whole range of services and facilities' (Burton 2002, p. 220).

Burton argues that urban compactness could be analysed using a wide variety of indicators, some of which are to do with density of different uses rather than density of urban form per se (Burton 2002). She suggests that besides high population density, a compact city might be expected to have a high density of built form (which equates to net density); a pattern of decentralised concentration; and high-density residential forms such as apartment blocks, or terraced houses (Burton 2002). To capture compactness, Burton combines indicators, arguing in effect that compactness is a somewhat holistic concept that provides a conceptual summary of the wide variety of different facets that most settlements have (Burton 2002). In this schema, urban form is considered in terms of the density of built form or of population; or of variety and proximity of facilities. Crucially, the actual physical shape of a place is not mentioned.

This inconclusive state of affairs is reflected in more recent attempts to pin down urban compactness (for example, Tsai 2005; Angel et al. 2005)—or its antithesis, sprawl (Galster et al. 2001)—using various coefficients and indices. These tend to be devised on an ad hoc basis, for particular places and purposes, using particular datasets; but basically, there seems to be no single satisfactory measure of compactness, a symptom that there is not even agreement in principle as to what compactness really means. When it comes to measuring compactness using specific quantitative indicators, it is interesting that so often density is used as a component of or proxy for compactness (see, for example Burton 2000, 2002; Abdullahi et al. 2015; Mojaheri et al. 2016). It is our contention that density is an unsatisfactory proxy for compactness, but the pursuit of specific indicators of compactness could yet learn from how density is defined and applied.

2.3 *Learning from Density*

Density itself is a contested issue, with a plethora of different kinds of measures of density available, which have significance for different parts of the debate. At the broadest level, we can distinguish between physical (e.g. buildings per hectare) or demographic (e.g. inhabitants per hectare) properties, between interpretations of net and gross densities and between physical and perceived density (see, for example Berghauser Pont and Haupt 2007; Boyko and Cooper 2011; Dovey and Pafka 2014). Even a specific seeming category such as net residential density may be measured in various ways, such as dwellings per hectare, or bedrooms per hectare, or people per hectare TCPA (2003).

Without going into all the details of the density debate, we can nevertheless form three useful lessons. First, we have to be clear how a property such as density (or compactness) is defined, with appropriate indicators, in any situation, to allow meaningful interpretation and comparison. Second, there are many indicators possible; there is no single correct indicator but different indicators suit different purposes. Third, despite the plethora of density indicators, there is yet a clear sense of what density means that all density indicators have in common: they all refer to 'something' enclosed within a given area. Ideally, we should be able to do the same for compactness. The challenge is to develop a concept of compactness that is conceptually explicit (and *inter alia* distinct from density), and quantifiable, and is also meaningful for application in the context of urban debates.

3 Geometric Interpretations of Compactness

Compactness is considered as probably 'the single most important aspect of geographic shapes' (MacEachren 1985, p. 65). Its geographical application includes not only urban physical form, but also river catchment basins (Bárdossy and Schmidt 2002), spatial decision support systems (Vanegas et al. 2010) electoral ward boundaries (Flaherty and Crumplin 1992) and linguistic regions (Angel et al. 2010). Compactness has mathematical interpretations (Sundström 2010) both in topology with respect to compact space (Arkhangel'skii 2001) and in geometry with respect to isoperimetry (Pólya and Szegő 1950). The concept also has application in such diverse contexts as electronic components (Lefkaditis and Tsirigotis 2009), galaxies, lava flows, buildings and organisms (Angel et al. 2010).

Due to the multiplicity of contexts involved, and the previous attention given to reviewing indicators of compactness, we do not attempt to review all measures here, but provide a summary overview referring to existing sources, to which the reader is referred for fuller discussion. Here, we give most attention to those indicators that are most useful in giving insights of concern to *urban* compactness.[1]

Table 1 lists 31 compactness measures. This is not intended as a complete indication of the exact meaning of every interpretation of area, diameter and so on, nor the assumptions behind the calculations (which also relate to some of the constants), and for these the reader is directed to the original sources. Rather, the table is intended as a summary overview, to give a rapid impression of the diversity of different combinations of the same kinds of indicators, almost all of which contain area, and perimeter or diameter or radius, in some combination.

[1]Ironically, that most abstract of disciplines, mathematics uses the case of *urban* compactness to illustrate the 'isoperimetric problem', concerning optimisation of area while minimising perimeter (Ashbaugh and Benguria no date), also referred to as the 'Queen Dido' problem, from Virgil's account in the Aeneid concerning the challenge to build the city of Carthage within the bounds laid out by a strip of oxhide (for example, as discussed by Lord Kelvin 1894).

Table 1 Catalogue of 31 selected compactness indicators or proxies

Compactness indicator or proxy		Alternative name	Source
Area			
1	A_I/A_C		(M); (F)
2	A_I/A		(M); (F)
3	A/A_C		(F)
4	$A_\cap/A_U$		(M); (F)
5	A_O/A	Exchange	(A)
Diameter or radius			
6	D_I/D_C		(M); (F)
7	D_{min}/D_{max}		(F)
8	r_A/r_C		(F)
9	$1 - [\Sigma\|100(r_i/\Sigma\ r_i) - (100/n)\|/200]$		(M); (F)
Perimeter			
10	P_A/P		(F)
Area with diameter or radius			
11	$A/\pi(0.5D)^2$		(M)
12	$\sqrt{(A/\pi)}/(0.5D) = \sqrt{(4A/\pi)}/D$		(M); (L)
13	$4A/\pi D^2$	Roundness	(T); (L)
14	$(2\sqrt{(\pi A)})/r$	Perimeter[a]	(A)
15	$2\sqrt{(A/\pi)}/D_C$	Range	(A)
16	$r_C/\sqrt{(A/\pi)}$	Girth	(A)
Area with perimeter			
17	$0.282P/\sqrt{(A)}$		(M)
18	$A/(0.282P)^2$		(M); (F)
19	P^2/A	Circularity	(T)
20	$4\pi A/P^2$	Thinness; Circularity	(T); (L)
21	$P/2\sqrt{(\pi A)}$		(B)
22	$P^{1/F}/\sqrt{(4\pi A)}$		(B)
23	$(2\sqrt{(\pi A)}/P_h$	Detour	(A)
24	$(\Sigma i P_i/p_i)/n$		(S)

(continued)

Table 1 (continued)

Compactness indicator or proxy		Alternative name	Source
Area with distance increments			
25	$A/\pi ds$	Cohesion	(A)
26	$(2\sqrt{(A/\pi)})/3d_{cs}$	Proximity	(A)
27	$3ds/\sqrt{(A/\pi)}$	Depth	(A)
28	$1 - d_r/r$	Dispersion	(A)
29	$1.2732\sqrt{(A/\pi ds)}$	Traversal	(A)
Area with X, Y coordinate variance			
30	$A/2\pi(\sigma_X^2 + \sigma_Y^2)$		(M); (F)
31	$\sqrt{(A/2\pi(\sigma_X^2 + \sigma_Y^2))}$		(M); (F)

Symbols A area, I inscribed circle, C circumscribed circle, A_O overlap area, $A_\cap$ Area of intersection of shape and circle of same area, A_U Area of union of shape and circle of same area, D Diameter or longest axis or Feret diameter, ds distance increment, d_{cs} distance to the 'proximate centre' of a shape, d_r average radial distance, F fractal dimension, *min* minimum, *max* maximum, n number of radial lines or number of patches used in calculation, P Perimeter, P_A circumference of equal area circle, P_h perimeter of convex hull, P_i circle perimeter, p_i patch perimeter, r radius, r_i radius i of n, r_A radius of circle having same area as shape, σ_X^2 and σ_Y^2 are the variance of X and Y coordinates of each elemental area dA. *Note* Symbols may be changed from original for consistency within this table. In some cases, exact or rough equivalents are used (e.g. circumference = 'perimeter' of circle)

[a]This is equivalent to measure 10 above (Angel et al. 2010, p. 459)

Sources (A) Angel et al. (2010); (B) Bárdossy and Schmidt (2002); (F) Flaherty and Crumplin (1992); (L) Lefkaditis and Tsirigotis (2009); (M) MacEachren (1985); (S) Silva et al. (2014); (T) Teknomo et al. (2004)

All indicators in Table 1 are dimensionless.[2] The simplest indicators of compactness are those involving a simple ratio of two properties. More typically, compactness is expressed as some kind of ratio typically involving area (A), perimeter (P) or diameter (D), and some constants. In a subset of cases, the measure of compactness compares some property of a shape—be it area, perimeter or diameter—to some reference shape, such as a circle of the same area (or perimeter) as the given shape, or relative to an inscribed or circumscribed circle. Curiously, none of the indicators involves a combination of area, diameter *and* perimeter.

MacEachren (1985) evaluated the utility of the different kinds of indicators for measuring compactness. He found that those indicators capturing the dispersion of elements of a shape's area were more accurate than those reflecting the overall properties of the whole shape (such as area, perimeter or diameter). However, the latter are more easily calculated, and MacEachren acknowledges that another measure of

[2]This is not to say all possible indicators are dimensionless: Teknomo et al. (2004) equate compactness with the simple ratio of perimeter to area (P/A); and Flaherty and Crumplin (1992) exhibit an indicator $A/0.282P$.

compactness $(\sqrt{(A/\pi)}/0.5D$ originally suggested by Schumm (1963) to be reasonably useful in terms of trading off accuracy for ease of calculation.

Of course, the usefulness all depends on which aspect of compactness one is interested in. Some of the indices, while able to disaggregate the positions of different parts of a geographical shape, may be overly concerned with their distance to an abstract centroid, but the idea of a centroid is practically unheard of in the real-world compact city debate. Meanwhile, Schumm's index although simple omits perimeter, which may well be of less interest for some purposes (e.g. political boundaries) but nonetheless of interest for urban area boundaries. Indeed, MacEachren concedes (1985, p. 66) that there may be cases where irregularity or length of a border may be of significance.

The measurement of perimeter is sensitive to the scale of resolution at which the measurement is made—this relates to Mandelbrot's classic problem 'How long is the coast of Britain?' (1967). That said, as Bárdossy and Schmidt (2002, p. 938) point out, this is not a problem as long as the same scale of resolution is used for measurement is the same when comparing two or more cases. The possibility of using perimeter should therefore not be discarded.

Angel et al. (2010, p. 444) emphasise that there is—and can be—no single best indicator of compactness. The search for a single metric or a composite metric that fully characterises shape compactness in a manner that always corresponds to out intuiting of what makes shapes compact is doomed to fail. Accordingly, these authors do not claim that their indicators of compactness are the only possible 10 or 11, and acknowledge that there may be more. Usefully, they provide five criteria which any measure of compactness should be able to fulfil, to be considered (in effect) useful and meaningful. These are very reasonable criteria to apply to compactness measures, and shall be borne in mind when developing our own:

1. The index must correspond to a recognisable property of the shape that is associated with a recognisable function or set of forces.
2. There must be real-world examples that illustrate this property… at both the low end and the high end of the index.
3. The index must apply to all two-dimensional geometric shapes, including those made up of several non-contiguous patches.
4. The index must be dimensionless (independent of the size of the shape) as well as directionless (independent of its orientation).
5. The index must vary between 0 and 1, with the value of 1 assigned to the circle as the shape with maximum compactness (ibid.).

There are some clear lessons emerging from this review. First, compactness seems to be typically regarded as a dimensionless measure, something to do with area, diameter and/or perimeter, in which the circle is the most compact shape. Second, however, there is no single measure of geometric compactness—it is not a single strictly defined mathematical entity (in the way that, say, a radius or centroid is). Rather there is a 'proliferation' of possible measures (MacEachren 1985, p. 65). Finally, compactness can have a wide variety of uses in different contexts, and the utility and value of the measures chosen will depend on the purpose to which it is put. In this chapter, the purpose to which we now return to is interpreting urban compactness.

4 A New Interpretation and Indicator of Compactness

Here, we move forward to develop new interpretations and indicators of compactness for application to urban areas. The aim is to develop indicators that are precisely definable and quantifiable, and yet whose interpretation is simple and intuitive for use by urban planners. First, a qualitative interpretation of compactness is offered; this is followed by a quantitative formulation, which is then illustrated with reference to some theoretical geometric shapes. Throughout, we interpret the compactness with respect to the shape of the built-up area.

4.1 *Interpreting Compactness*

If we accept that the most compact theoretical shape is the circle, then we need to consider ways in which an urban area boundary is *not* compact. Here, we suggest three instrumental ways of not being compact, due to (i) elongation, (ii) convolution and (iii) dispersal.[3] These cases are illustrated in Figs. 1, 2 and 3, respectively. Each involves an increase in diameter and/or perimeter relative to area.

The foregoing suggests developing an indicator of compactness that includes some relation of the area enclosed to *both* the perimeter and the diameter—something that none of the indicators from the literature do (Table 1). The perimeter is required to ensure that more convoluted shapes (and not merely those with the most distant extremities) are considered less compact, while diameter is required to ensure that those urban shapes with greatest elongation or dispersal are accounted for, and not

Fig. 1 Elongation. The diameter increases relative to area, from the most compact (**a**) to least compact shape (**c**)

[3]This is not to say that we are concerned with all of the ways in which an urban shape does or does not reflect the properties of a circle—for example, a circle has translational and rotational symmetry, and is consistently curved and perfectly convex; but these are properties that no urban area (boundary) has.

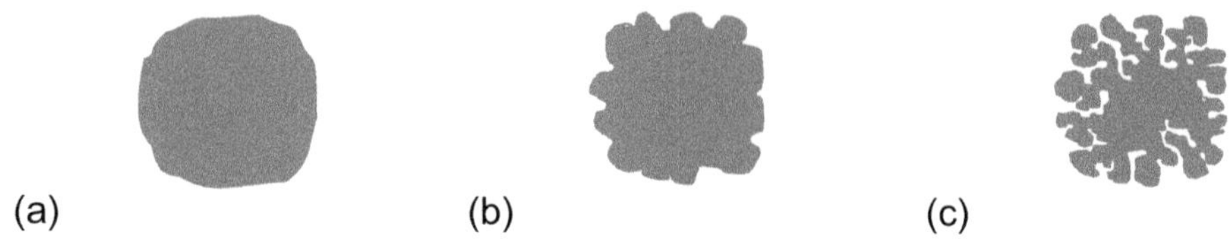

Fig. 2 Convolution. The perimeter increases relative to area, from the most compact (**a**) to least compact shape (**c**)

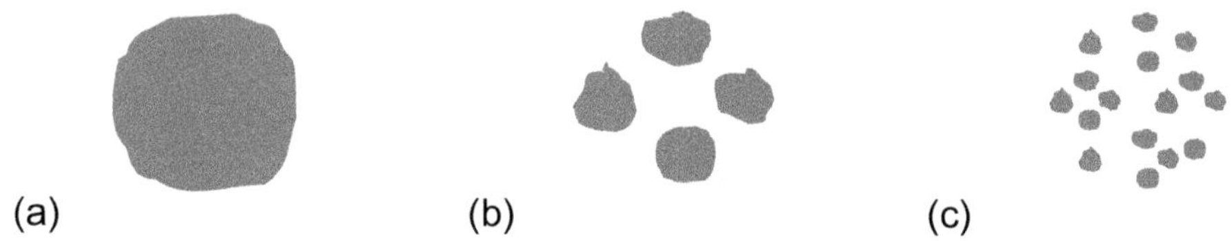

Fig. 3 Dispersal. The perimeter (and possibly overall diameter) increases relative to area, from the most compact (**a**) to least compact shape (**c**)

just those whose boundaries are most convoluted. A quantitative formulation is now proposed.

4.2 *Quantification of Compactness*

Implicit in the foregoing demonstrations are the following observations: (i) compactness seems to be related to minimising diameter relative to area, (ii) compactness seems to be related to minimising perimeter relative to area and (iii) the circle is the most compact form. Hence, a potential quantitative indicator for compactness is to define compactness as follows:

$$C = \frac{4A}{DP} \tag{1}$$

where

C compactness,
A area,
D diameter, and
P perimeter.

By this formulation, incorporating the constant 4, a circle has a maximum compactness equalling unity:

$$C = \frac{4A}{DP} = \frac{4\pi r^2}{2r \cdot 2\pi r} = 1.0 \tag{2}$$

Table 2 Examples of compactness values for different geometric shapes

	Shape	Area (A)	Diameter (D)	Perimeter (P)	Compactness (C) $= 4A/DP$
(a)	 Radius r	πr^2	$2r$	$2\pi r$	1.00
(b)	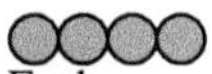 Each radius ½r	πr^2	$4r$	$4\pi r$	0.25
(c)	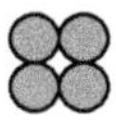	πr^2	$2.41r$	$4\pi r$	0.415
(d)	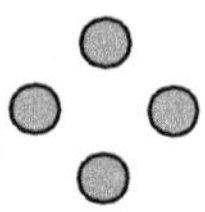	πr^2	$4r$	$4\pi r$	0.25
(e)	 Side length x	x^2	$1.41x$	$4x$	0.71
(f)	 Length $5x$	$5x^2$	$5.1x$	$12x$	0.327
(g)	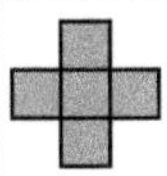	$5x^2$	$3.16x$	$12x$	0.527
(h)		$5x^2$	$4.24x$	$20x$	0.236

The compactness values of a range of geometric shapes are given in Table 2. The compactness indicator is successful insofar as the compactness values have a good fit with an intuitive interpretation of the compactness of these shapes. The first four cases in Table 2 are based on circles, where each configuration has the same area. The first (a) is a single circle whose compactness is 1.0. The subsequent three configurations each is less compact, due to (b) elongation with convolution, (c) convolution and (d) dispersal. The second four cases are based on squares, where the last three have the same area. Compared with the single square (e), the cross shape (g) is relatively compact, although it has some convolution, case (f) is less compact due to elongation and case (g) due to convolution verging on non-contiguity and dispersal. Insofar as the cases in Table 2 are purely geometric shapes, they could apply to any geometric objects, including solid objects or building floor plans. However, we now need to see how the indicator would apply to urban settlements.

5 Application to Urban Areas

Our new compactness indicator is now applied to some selected settlements. First, an initial illustration and calculation is given for the case of Cambridge. Then, we show application to 20 English settlements. Finally, we contrast the compactness values with density values. In each case, we use UK 2001 Census data, based on the definition of an urban area as 'an area of urban land use of 20 hectares or more with 1,500 or more residents'.[4] In other words, 'urban areas' here are actually based on a criterion relating to density, rather than administrative boundaries.

5.1 *Initial Illustration*

Figure 4b displays the Cambridge Urban Area (in white) on top of the Google Map (a). The boundary of the Cambridge urban area can be seen to lie snugly with its built-up area, especially in Fig. 4c.

From the database, the area (A) of Cambridge urban area is 36.04 km^2. The diameter (D) is 12.62 km and perimeter (P) is 133.94 km. Using Eq. 1 returns a compactness value of $4A/DP = 0.085$.

5.2 *Compactness for a Range of Urban Areas*

We now demonstrate a larger set of results for 20 English settlements (Table 3). These include ten of the larger cities in England (Fig. 5a), plus ten others that represent a variety of different kinds of boundary shapes which yield a range of compactness values (Fig. 5b). It is convenient to express urban compactness as a percentage in this context where, as we can see, most values lie under 0.1 (10%). This general finding of itself suggests that settlements turn out to be not very compact, relative to any possible theoretical shape of object.

Beverley has a high compactness, it is suggested, because it is small and so its boundary will be less convoluted than larger settlements. The urban area of Loughborough is compact, and its approximation to a solid circle is apparent. Kingston upon Hull has a relatively high compactness partly because it has a long stretch of relatively straight waterfront boundary. Brighton also has this, but the overall effect is a linear settlement, with a relatively low value. Longton is less compact both by being elongated and convoluted, and by being dispersed into separate urban area 'blocs' (polygons). West Yorkshire—a conurbation—is straggling with linear elements and punctured with many voids, and is the least compact of all.

---

[4]English Urban Areas, 2001 from Edina UKBorders (https://www.census.ac.uk/search/Full_display.aspx?id=1081) This work is based on data provided with the support of the EPSRC and JISC and uses boundary material which is copyright of the Crown and the ED-LINE Consortium.

Fig. 4 Cambridge Urban Area (Google Maps)

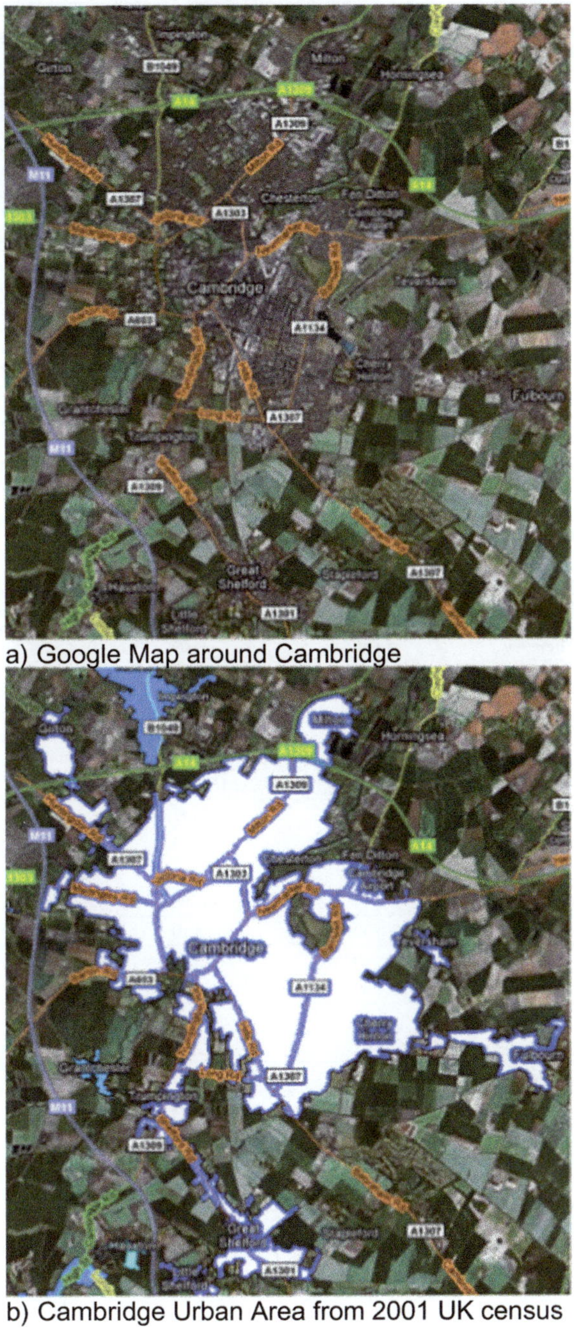

Fig. 4 (continued)

c) Close up to Cambridge Airport

Table 3 Density and compactness values for selected English settlements

Urban area	Residential density (persons per ha)	Area A (km^2)	Diameter D (km)	Perimeter P (km)	Compactness C (%)
Beverley	41	7.09	4.52	20.11	31.2
Brighton/Worthing/ Littlehampton	49	94.09	37.99	241.98	4.1
Bristol	39	139.76	20.14	316.78	8.8
Cambridge	36	36.04	12.62	133.94	8.5
Greater London	51	1623.13	78.00	1940.68	4.3
Greater Manchester	40	556.47	43.44	1310.91	3.9
Kingston upon Hull	37	80.39	17.21	144.83	12.9
Leicester	43	101.64	20.95	215.66	9.0
Liverpool	44	186.06	28.24	305.26	8.6
Longton	30	4.10	8.06	63.52	3.2
Loughborough	40	13.78	5.63	39.86	24.5
Nottingham	39	158.43	27.24	436.62	5.3
Sheffield	42	162.19	24.10	419.37	6.4
Southampton	42	72.76	14.35	182.67	11.1
Teesside	32	114.30	21.91	268.30	7.8
The Potteries	38	96.64	21.59	320.02	5.6
Tyneside	42	210.72	33.27	607.31	4.2
West Midlands	38	599.66	41.38	666.33	8.7
West Yorkshire	41	369.97	40.30	1285.92	2.9
Weymouth	41	13.56	8.99	87.62	6.9

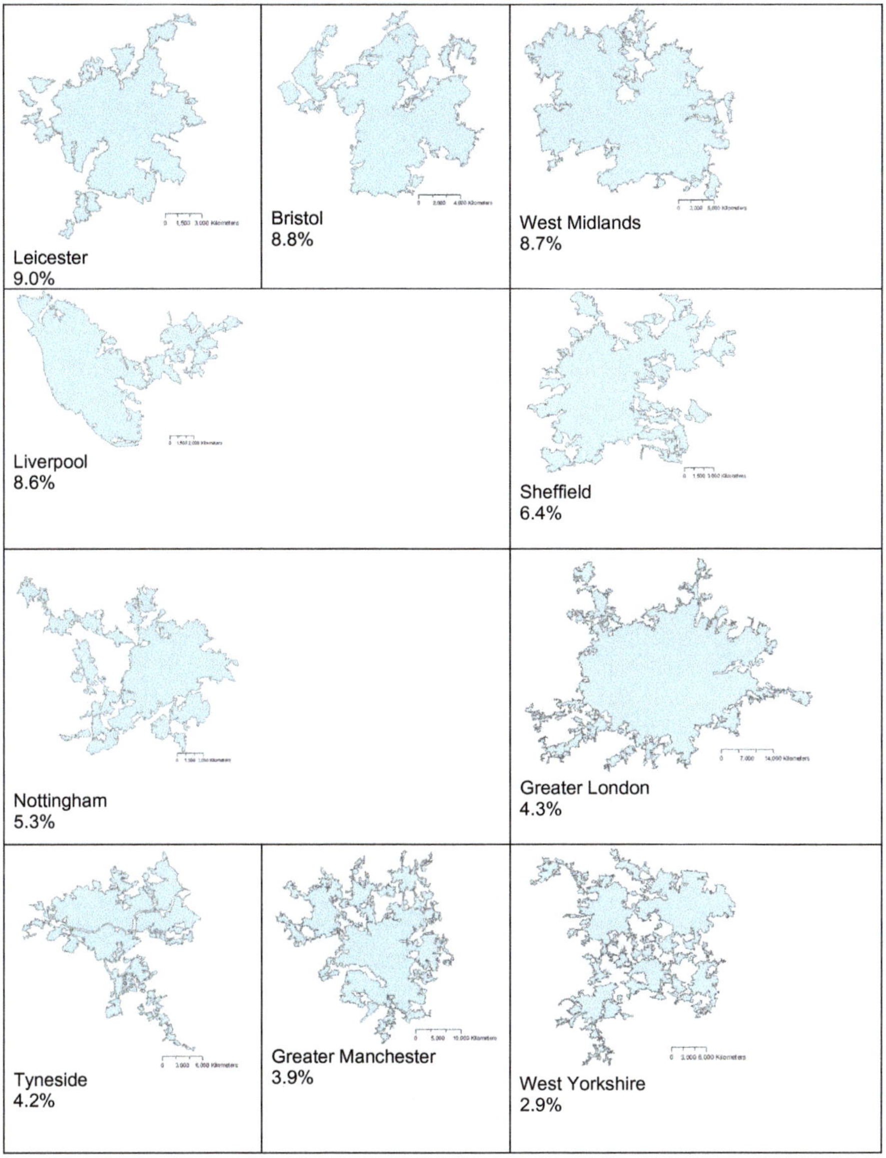

Fig. 5 **a** Compactness values for ten major cities or conurbations of England. **b** Compactness values for selected urban areas

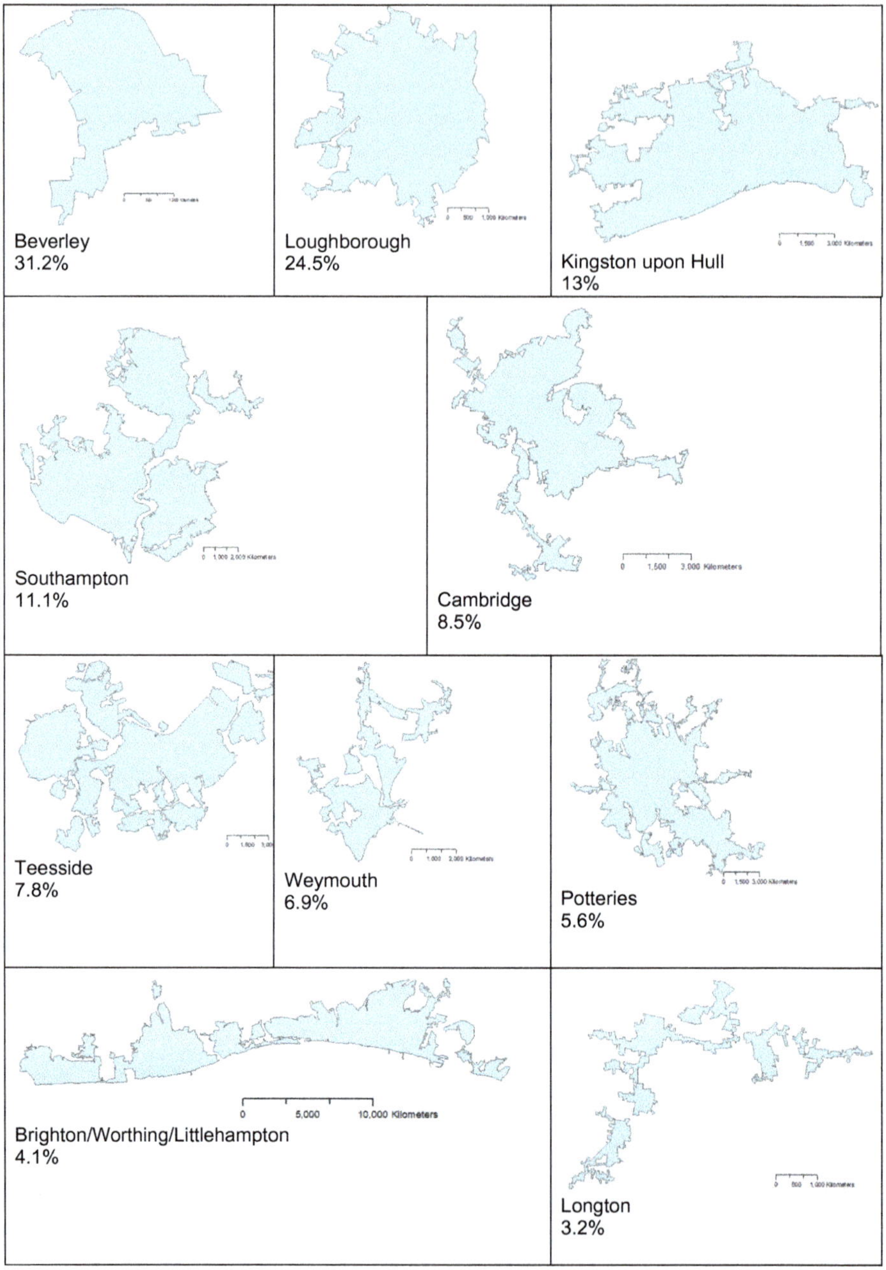

Fig. 5 (continued)

In general, larger urban areas tend to be less compact compared with smaller urban areas. This may be explained by the fact that urban areas tend to be built of relatively small units (such as buildings and plot boundaries) that do not significantly or systematically increase in size with settlement size. Whereas an individual building footprint or urban plot may be rectangular or square (or even circular), and hence be relatively compact, an aggregation of buildings, as more buildings are added, is increasingly less likely to retain its compactness, as its boundary perimeter is likely to become increased, due to being more and more convoluted.

As one can infer from Fig. 5, some of these values of compactness are somewhat sensitive to the way the urban area boundaries are drawn as polygons. However, this arguably goes for any data relating to any 'artificial' cartographic features such as boundaries; the question becomes whether the boundary, and hence the compactness indicator is capturing something useful or not. In some cases, settlements will be less compact than they would otherwise be because they include polygons outside the core urban area. But arguably this is as it should be: the existence of such outliers, and their contribution to the effect of dispersal, is after all one of the things that compactness is trying to capture. Conversely, 'subtracting' from an urban area can reduce compactness. Southampton, Tyneside and Teesside are all less compact than they might have otherwise been due to their being divided by rivers where the boundaries along both sides of the river increase the perimeter. Yet this reflects reality: these cities really are less compact than they would have been had they not been divided by rivers.

Overall, the demonstration shows in principle the intuitive sense in which the property of compactness is captured; in Fig. 5, the boundary shapes of Beverley and Loughborough are clearly intuitively more compact than Brighton or Longton, however those shapes were arrived at or whatever they are representing in detail.

5.3 Compactness Distinct from Density

We have argued that compactness is conceptually distinct from density and we have developed an indicator of compactness that logically should give different results from density since it uses different parameters. The independence of the two properties is apparent when we plot compactness against residential density for our 20 settlements (Fig. 6). This is confirmed by a Spearman correlation which suggests that the relationship is statistically insignificant ($r\ (20) = -0.053$, $p = 0.826$).

6 Compactness Revisited

Here, we revisit the interpretation of compactness. First, we examine how compactness can be interpreted with respect to parts of a settlement. Then, we scrutinise the

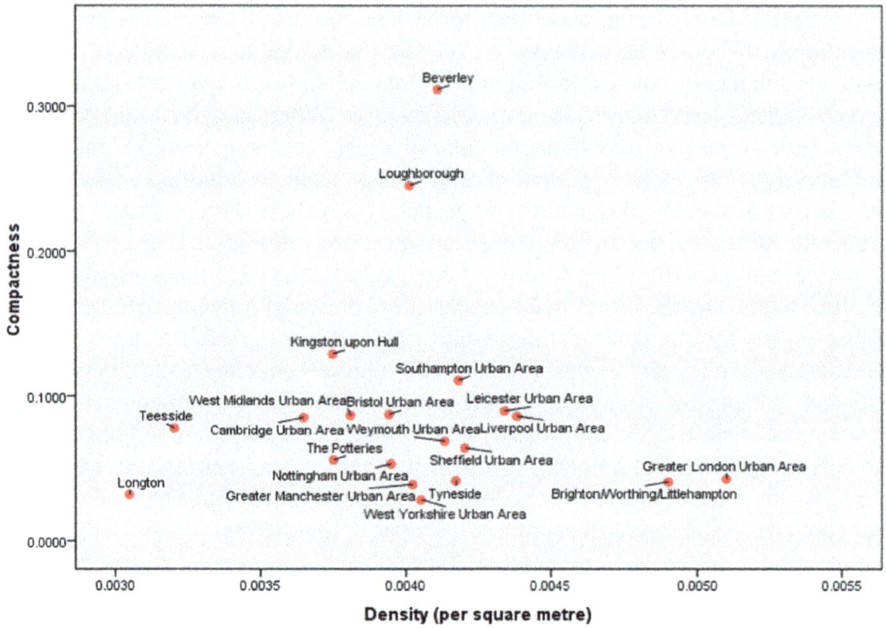

Fig. 6 Compactness versus residential population density *Source* 2001 Census, Office for National Statistics

core meaning of compactness in relation to area, diameter and perimeter. And finally, we suggest further indicators of compactness.

6.1 *Wholes and Parts*

How does the compactness of a whole settlement relate to the compactness of the parts? Let us find out by looking in more detail at the Cambridge case. The Cambridge urban area has six individual polygons as shown in Fig. 7, of which compactness varies from 0.07 to 0.23, with an average compactness of 0.15. However, the compactness of Cambridge urban area as a whole is 0.085, which is substantially different from the average compactness of the components (Table 4).

This situation is different from area and hence density. Areas are simply additive, so that the whole is the sum of the parts. Accordingly, the weighted average of the densities of two or more separate areas will (unlike compactness) equal the density of the sum of the areas. This will mean the decision to include 'outliers' (non-contiguous areas) which will affect values of compactness or density in different ways. For example, if we take 'Cambridge' to mean only the contiguous core area (polygon 2), then we reduce the area by just under a quarter, but the compactness

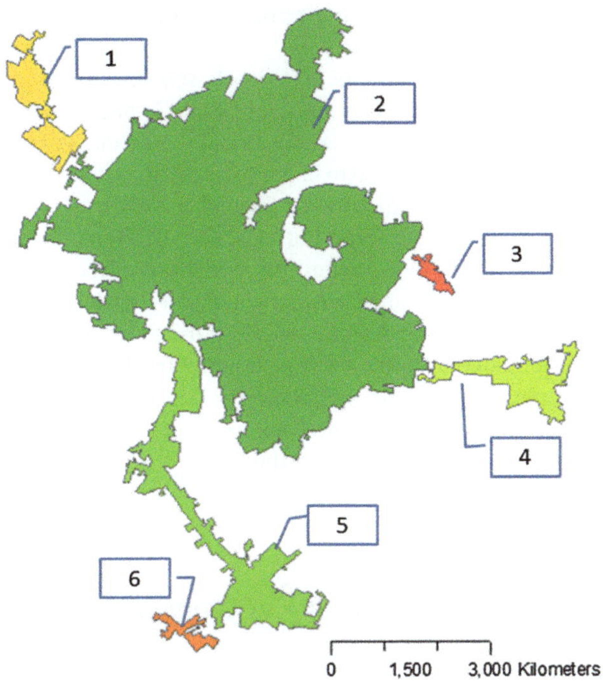

Fig. 7 Cambridge urban area showing six discrete bounded areas

doubles. This implies that compactness is more 'elastic' than area and measures based more simply on area, such as density.

Compactness can therefore be interpreted as a property of the whole that affects how we analyse sample parts of urban areas. For example, we can take a sample area 'x' from city X, where the overall city X is not compact, but the sample area x could be perfectly compact. Conversely, we can take a noncompact sample area 'y' from a compact city Y (Fig. 8). This situation is different from density, in the sense that

Table 4 Breakdown of compactness values by polygon

	Area A (km^2)	Diameter D (km)	Perimeter P (km)	Compactness C (%)
Polygon 1	0.3	1.5	11.0	15 ·
Polygon 2	29.0	11.5	69.8	14
Polygon 3	0.2	1.1	3.3	23
Polygon 4	1.39	3.4	12.7	13
Polygon 5	3.8	6.9	31.2	7
Polygon 6	0.3	1.5	5.9	15
Average	5.8	4.3	22.3	15
Total	36.0	12.6	133.9	8.5

if area X is dense overall we could reasonably expect a sample 'x' to be dense too, and similarly for Y and y.

In this case, compactness is clearly sensitive to the choice of boundary of the sample, but as argued before, this sensitivity is arguably not a weakness, but reinforces the idea that compactness is all about boundedness of discrete urban wholes—or discrete parts—as opposed to being (like density) conceivable as a continuously varying 'material property' distributed through an urban area. The apparent compactness values of shapes in Fig. 8 are not 'artifacts' or 'figments' of the way the boundaries are selected; rather, they are faithful reflections of the boundary shapes. Shapes x and Y really *are* rather compact, whereas X and y really *aren't*. In each case, compactness is a property of the discrete whole or the discrete part.

The boundary issue can be summed up this way. Many attributes of a city—for example, physical size, population or density—will be sensitive to how the boundary is drawn. In such cases, one does not want idiosyncrasies of the boundary to 'intrude' in influencing unduly the attribute under scrutiny. But in the case of compactness, the boundary itself is positively what is under scrutiny.

6.2 The Core Meaning of Compactness

Compactness, then, can be seen as a property of a whole bounded shape (or whole set of shapes), in a way that reflects the character of its boundary. Indeed, with the suggested definition of compactness ($4A/DP$), compactness is actually *directly proportional to area*, which at first sight may seem counter-intuitive or paradoxical since intuitively we might expect something compact to occupy less area, and hence expect that compactness would be inversely related to area, as density is.

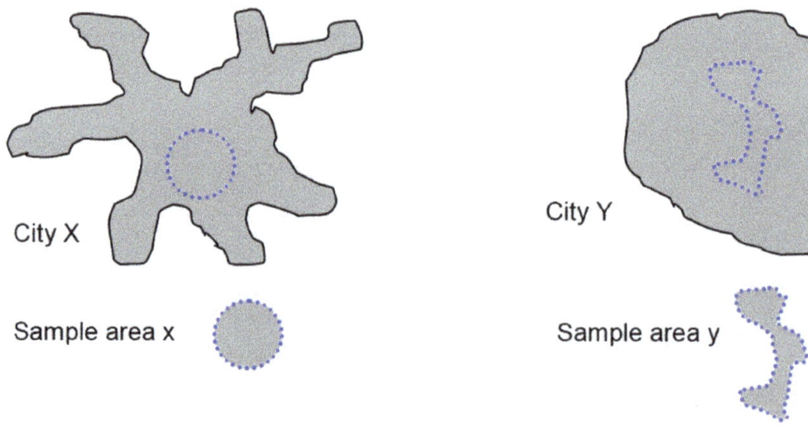

Fig. 8 Sample areas x, y cut out from overall urban areas X, Y

The answer suggested is that the essential sense of compactness is captured in the denominator: compactness represents 'something' divided by the product of diameter and perimeter. This is equivalent to the way that the fundamental sense of density is to do with 'something' divided by area—for example, population density or employment density.[5] In these conventional manifestations of density, we can have any parameter on the numerator, and as long as area is on the denominator, then the whole indicates some kind of density. By extension, we suggest that we can have any property divided by the product of diameter and perimeter, and call that property some kind of compactness.

What the foregoing definition of compactness (Eq. 1) does is to make *area* itself the quantum that is fitted into a given perimeter and diameter. And this does seem to correspond with intuitive concepts of compactness. A compact shape here means one that encloses a high amount of area relative to its perimeter and diameter. By such means, compactness may be considered rightfully to be directly proportional to area; where area is in effect a proxy for some urban quantity to be accommodated within a given footprint. After all, the positive sense of urban compactness is not just about internal distance minimisation, but also about minimising the impact of the urban footprint, in terms of minimising perimeter and diameter. This interpretation of compactness also fits with the intuitive sense of something being virtuously 'small on the outside, big on the inside'. In fact, we can see that compactness need not involve area at all, but that the 'active ingredient' or essence of compactness is captured by the reciprocal of the product of diameter and perimeter. As such, the numerator could be any property of interest, just as with density.

6.3 Further Kinds of Compactness

In our original general definition, compactness relates to how much area can be fitted within a boundary of a given diameter and perimeter. We can relabel this original property $4A/DP$ as 'area compactness' or 'A-compactness', denoted by C_A, where circumstances demand a more specific term. Then, just as there is not just one kind of density—we could have been building density or population density and so on—we can have *different kinds of compactness*—building compactness and population compactness and so on. Some examples are now suggested (Eqs. 3–6).

Built compactness

$$C_B = \frac{A_B}{A}C = \frac{A_B}{A} \bullet \frac{4A}{DP} = \frac{4A_B}{DP} \tag{3}$$

where

[5] This is referring to the urban and geographical contexts of properties such as population density, employment density and so on. In physics, of course, density is associated with *volume*.

C_B built compactness or B-compactness,
A_B built-up area (area of building footprints).

Floorspace compactness

$$C_F = \frac{4A_F}{DP} \tag{4}$$

where

C_F floorspace compactness,
A_F total floorspace (area) within given perimeter P.

Population compactness

$$C_P = \frac{4I}{DP} \tag{5}$$

where

C_P population compactness,
I number of inhabitants within given perimeter P.

Employment compactness

$$C_E = \frac{4E}{DP} \tag{6}$$

where

C_E employment compactness,
E number of employees within given perimeter P.

Note that both built compactness (C_B) and floorspace compactness (C_F) will give a more three-dimensional feeling of compactness, than the original area-based version of compactness (C_A) because these take account not only of properties associated with the two-dimensional boundary, but also the third dimension through what is built 'up' on it. Built compactness implies the presence of a third dimension, while floorspace compactness can reflect the influence of multi-storey buildings.

The foregoing suggests that, in principle, anything that could be expressed as a kind of density (population, employment, etc.) could have a corresponding value of compactness, where in each case the numerator is the same but the denominator is changed from area to the product of diameter and perimeter. Hence, we see that compactness is not only distinct from density, and not just another one of a plethora of variants of density proxies, but is a property in its own right, almost like a whole new property, with its own corresponding plethora of variants.

7 Conclusions

This chapter has suggested a new way of interpreting compactness geometrically, relating to the reciprocal of the product of the diameter and perimeter of a given area. We believe that the core indicator of compactness ($4A/DP$) is simple enough that it is intuitively easy to understand and relatively easy to compute. Being based on three variables—area, diameter and perimeter—it is able to capture something useful about *urban* compactness. This relates to dispersal, elongation and convolution, which, in turn, relate to policy concerns about settlements straggling, sprawling, coalescing with each other, which are not routinely captured quantitatively in 'compact city' debates.

We believe that the $4A/DP$ indicator, as well as being 'new' (and as far as we know unique in combining A, D and P), meets all five of the attributes of a useful indicator suggested by Angel et al. (2010), and can capture a certain sense of compactness that is under-reported in the literature: to do with the maximising content with minimum containment—relating some kind of functional 'capacity' against physical 'cost'. Our interpretation of compactness leads not just to one extra indicator, but opens up a new 'dimension' populated by a number of compactness indicators, including built compactness (B-compactness) (C_B) and floorspace compactness (C_F), which can reflect urban compactness in the third dimension.

The measure of compactness developed here ($4A/DP$) is focused on the shape of the urban area. This is considered appropriate since we believe this is ultimately what planners are intuitively trying to get at when talking in the narrow sense about compactness. This core concept of geometric compactness can be used in conjunction with other characteristics to help articulate wider aspects of 'compact city' debates.

Because of the way compactness is defined, this property is clearly conceptually distinct from density, and its independence in practice has been confirmed through empirical results for urban settlements (Fig. 6). This means planners and others can no longer afford to assume that compact means dense or vice versa. The separate identity of density and compactness makes a useful 'linguistic division of labour' as it allows us to use these terms consistently to refer to different underlying properties. It implies that policymakers should evaluate more than one kind of 'compact city'; we could have urban areas that are compact but not dense, or dense but not compact, as well as cities that are both dense and compact.

We believe that the articulation of compactness herein constitutes a concrete contribution to the articulation of urban pattern properties and it is not an end point, but a stepping stone towards further applications. Future research could include the following:

- The expansion of empirical investigations, including comparison of compactness against other urban form variables such as density or polycentricity, and also sensitivity testing, to understand, for example, the sensitivity of compactness to settlement size and scale of measurement.
- The extension of compactness interpreted in the third dimension of the urban fabric, including potential application to the compactness of buildings.

- The consideration of the performance of urban areas of different degrees of compactness, such as relating compactness to travel distance, energy use and other measures relating to 'sustainability'.

As it stands, the conceptual focus on the meaning of compactness can on its own help to clarify and even redefine certain aspects of the compact city debate. The analysis herein suggests that urban settlements—other than the smallest, simplest entities—are not particularly compact, relative to theoretical geometric shapes. This suggests that planners focus attention on the extent to which compactness is an optimisable planning goal. To what extent are idealised 'compact cities' really about density or compactness? And what different kinds of compactness might be desirable? Compaction could imply containment (maintaining discretely bounded settlements) but it could also imply infill, consolidation and creeping coalescence—outcomes pointing in different directions. Should planners prioritise accessibility within a compact boundary, or between discrete compact settlements? An elongated or convoluted urban–rural boundary could actually be welcomed for its exposure of residents to open land locally, but not so good in terms of urban footprint impacting on the countryside. What are the relative merits of roughly round settlements (which could be low density) versus linear settlements (which could be high density) suitable for serving by public transport? Resolving these questions of compactness—leading beyond contemporary preoccupations with density, or ancient ones about city walls—could help answer to what extent compact cities really could be ideal cities.

Acknowledgements This chapter reports on research originally carried out as part of the SOLUTIONS project (Sustainability Of Land Use and Transport In Outer NeighbourhoodS), funded by the UK Engineering and Physical Sciences Research Council (EPRSC).

References

Abdullahi S, Pradhan B, Jebur M N, 2015 "GIS-based sustainable city compactness assessment using integration of MCDM, Bayes theorem and RADAR technology" *Geocarto International* 30 (4) 365–387

Angel S, Parent J, Civco, D L, 2010, "Ten compactness properties of circles: measuring shape in geography" *The Canadian Geographer* **54** 441–461

Angel S, Sheppard S C, Civco D L, 2005 *The Dynamics of Global Urban Expansion* (The World Bank, Washington, D.C.)

Arkhangel'skii A V, 2001, "Compact space", in *Encyclopedia of Mathematics* Ed M Hazewinkel (Kluwer, Dordrecht)

Arup, 2005 *Regional Futures* (Ove Arup and Partners Ltd, London)

Ashbaugh M, Benguria R, no date, "The problem of Queen Dido. Overview of the subject of Isoperimetry" http://www.fis.puc.cl/~rbenguri/117eng.pdf (accessed 12.2.19)

Bárdossy A, Schmidt F, 2002, "GIS approach to scale issues of perimeter-based shape indices for drainage basins" *Hydrological Sciences*, **47** 931–942

Barrow J D, 2008, "The importance of being peripheral", lecture at Gresham College, November 2008. Transcript at: http://www.gresham.ac.uk/lectures-and-events/the-importance-of-being-peripheral (accessed 12.2.19)

Berghauser Pont M, Haupt P, 2007 "The Relation between Urban Form and Density" *Urban Morphology* 11 (1): 62–66

Boyko C T, Cooper R, 2011, "Clarifying and Re-Conceptualising Density" *Progress in Planning,* **76** (1): 1–61.

Bramley G, Dempsey N, Power S, Brown C, 2006, "What is 'social sustainability' and how do our existing urban forms perform in nurturing it", in 'Sustainable Communities and Green Futures' conference proceedings, Bartlett School of Planning, University College London, London

Breheny M, 1992a, "The Contradictions of the Compact City: A Review" in *Sustainable Development and Urban Form,* Ed. M Breheny (Pion, London) pp 138–159

Breheny M, 1992b *Sustainable Development and Urban Form* (Pion, London)

Breheny M, 1997 *The People: Where Will They Work?* (Town and Country Planning Association, London)

Breheny M, Gent T, Lock D, 1993 *Alternative Development Patterns: New Settlements* (HMSO, London)

Breheny M, Hall P, 1999 *The People: Where Will They Live?* (Town and Country Planning Association, London)

Bunker R, 2014, "How Is the Compact City Faring in Australia?" *Planning Practice & Research* **29**(5) 449–460

Burton E, 2000, "The compact city: just or just compact? a preliminary analysis" *Urban Studies* **37** 1969–2000

Burton E, 2002, "Measuring urban compactness in UK towns and cities" *Environment and Planning B: Planning and Design* **29** 219–250

Champion A G, 1989, "Counterurbanization in Britain" *Geographical Journal* **155** 52–9

Champion A G, 2001, "A changing demographic regime and evolving polycentric urban regions: consequences for the size, composition and distribution of city populations" *Urban Studies* **38** 657–677

Chhetri P and Corcoran J, 2013, "Mapping urban residential density patterns: Compact city model in Melbourne, Australia" *City, Culture and Society* **4**(2) 77–85

DCLG, 2007 *Eco-Towns Prospectus* (Department of Communities and Local Government, London)

Dempsey N, 2010, "Revisiting the compact city?" *Built Environment* **36** 5–8

Dovey K, Pafka E, 2014 "The Urban Density Assemblage: Modelling Multiple Measures" *Urban Design International* **19** (1): 66–76

Echenique M, Hargreaves T, Mitchell G, 2009 "Spatial planning, sustainability and long run trends" *Town and Country Planning* **78** 380–385

Edwards B, 2005 *Rough Guide to Sustainability* (RIBA Enterprises Ltd, London)

Ferreira A, Batey P, 2011, "On why planning should not reinforce self-reinforcing trends: a cautionary analysis of the compact-city proposal applied to large cities" *Environment and Planning B: Planning and Design* **38** 231–247

Gaigné C, Riou S, Thisse J-F, 2009 *Are Compact Cities Environmentally Friendly*? (INRA, Rennes)

Flaherty M S, Crumplin W W, 1992, "Compactness and electoral boundary adjustment: an assessment of alternative measures" *The Canadian Geographer* **36** 159–171

Frey H, 1999 *Designing the City. Towards a More Sustainable Urban Form* (Routledge, London)

Galster G, Hanson R, Ratcliffe M R, Wolham H, Coleman S, Freihage J, 2001, "Wrestling sprawl to the ground: defining and measuring an elusive concept" *Housing Policy Debate* **12** 681–717

Jenks M, Burton E, Williams K, 1996 *The Compact City: A Sustainable Urban Form?* (E & F N Spon, London)

Kelvin, Lord, 1894, "Isoperimetrical problems, Being a Friday evening lecture delivered to the Royal Institution, 12th May 1893", in Popular lectures and addresses by Sir William Thomson (Baron Kelvin) in Three Volumes, Vol. II, Geology and General Physics, Nature Series (Macmillan & Co., London) http://math.arizona.edu/~dido/LordKelvin1894.pdf. (accessed 1.8.11)

Lefkaditis D, Tsirigotis G, 2009, "Morphological feature selection and neural classification for electronic components" *Journal of Engineering Science and Technology Review* **2** 151–156

MacEachren A M, 1985, "Compactness of geographic shape: comparison and evaluation of measures" *Geografiska Annaler* **67** 53–67
Mandelbrot B B, 1967, "How long is the coast of Britain? Statistical self-similarity and fractional dimension" *Science* **156** 636–638
Mohajeri N, Upadhyay G, Gudmundsson A, Assouline D, Kämpf J, Scartezzini J-L (2016) "Effects of urban compactness on solar energy potential" *Renewable Energy* **93**, 469–482.
Neuman M, 2005, "The compact city fallacy" *Journal of Planning Education and Research* **25** 11–26
ODPM, 2003 *Sustainable Communities: Building for the Future* (Office of the Deputy Prime Minister, London)
Pólya G, Szegő, G, 1950 *Isoperimetric Inequalities in Mathematical Physics* (Princeton University Press, Princeton)
Power A, 2001, "Social exclusion and urban sprawl: is the rescue of cities possible?" *Regional Studies* **35** 731–742
Rasmussen S E, 1982 *London, The Unique City* (The MIT Press, Cambridge, MA)
Rogers R, 2011, "Learning from continuity and change" *Proceedings of the Institution of Civil Engineers: Urban Design and Planning* **164** (DP1) 3–5
Rogers R, Power A, 2000 *Cities for a Small Country* (Faber and Faber, London)
Schumm S N, 1963, "Sinuosity of alluvial rivers on the Great Plains" *Bulletin of the Geological Society of America* **74** 1089–1100
Silva L T, Oliveira M, Silva J F, 2014 "Urban form indicators as proxy on the noise exposure of buildings" *Applied Acoustics* 76, 366–37.
Sundström M R, 2010, "A pedagogical history of compactness" arXiv:1006.4131v1 [math.HO] (accessed 1.8.11)
Tappert S and Drilling M, 2018, "Contested urban green spaces in the compact city: The (re)-negotiation of urban gardening in Swiss cities" *Landscape and Urban Planning* **170** 69–78
TCPA (2003) *Residential Densities – Policy Statement* (Town and Country Planning Association, London)
Teknomo K, Gerilla G P, Hokao K, 2004 "Cellular urban descriptors of lowland urban model" in *Proceedings of the International Symposium of Lowland Technology*, Bangkok, pp 297–302
Tsai Y H, 2005, "Quantifying urban form: compactness versus 'sprawl'" *Urban Studies* **42** 141–161
Urban Task Force (1999) Towards an Urban Renaissance (DETR and Spon, London)
Vallance S, Perkins H C, Moore K, 2005, "The results of making a city more compact: neighbours' interpretation of urban infill" *Environment and Planning B: Planning and Design* **32** 715–733
Vanegas P, Cattrysse D, van Orshoven J, 2010, "Compactness in spatial decision support: a literature review", in *Proceeding of ICCSA* Part 1 (Springer-Verlag, Berlin, Heidelberg) pp 414–429
Westerink J Haase D Bauer A Ravetz J Jarrige F and Aalbers CBM, 2013, "Dealing with Sustainability Trade-Offs of the Compact City in Peri-Urban Planning Across European City Regions" *European Planning Studies* **21**(4) 473–497
Williams K, 2000, "Does Intensifying Cities Make them More Sustainable?" in *Achieving Sustainable Urban Form* Eds K Williams, E Burton, M Jenks (E & FN Spon, London) pp 30–45
Williams K, Burton E, Jenks M, 2000 *Achieving Sustainable Urban Form?* (E & FN Spon, London)
Wolsink M, 2016, "Sustainable City' requires 'recognition'—The example of environmental education under pressure from the compact city" *Land Use Policy* **52** 174–180
Zukin S, 1982 *Loft Living: culture and capital in urban change* (Johns Hopkins University Press, Baltimore)

Using Google Street View for Street-Level Urban Form Analysis, a Case Study in Cambridge, Massachusetts

Xiaojiang Li and Carlo Ratti

Abstract City streets are a focal point of human activities in urban areas. As an important element of urban form, streets are also a major interface of the social interaction between urban dwellers and urban built environment. Quantifying the urban built environment is thus important for us to understand the potential impact of the urban built environment on urban dwellers. The publicly accessible Google Street View (GSV), which captures the streetscape appearances of cities around the world, provides a very good tool for urban studies at a fine level. In this study, we illustrated using GSV for describing and mapping urban form at street-level in terms of the enclosure of street canyons in Cambridge, Massachusetts. We further mapped and analyzed the influence of street enclosure on solar radiation reaching the street canyons by estimating the sunlight duration in street canyons. Some other potential applications of GSV data were also introduced in this paper. The results of this study would shed new light on future urban studies using the publicly accessible and globally available GSV data. Other researchers may find the method illustrated in this study is directly deployable for different studies related to urban form analysis.

1 Introduction

City streets are a focal point of human activities in urban areas (Li et al. 2017). As an important element of urban form, streets are a major interface of the social interaction between urban dwellers and urban built environment. The openness of street canyons influences human perception of the environment, and enclosed street canyons may give a feeling of oppressiveness to pedestrians (Asgarzadeh et al. 2012, 2014). The urban form would also influence the energy balance in street canyons, which would further affect human thermal comfort and exposure to sunlight (Carrasco-Hernandez

X. Li (✉) · C. Ratti
MIT Senseable City Lab, Department of Urban Studies and Planning,
Massachusetts Institute of Technology, Room 9-250, 77 Massachusetts Avenue,
Cambridge, MA 02139, USA
e-mail: xiaojian@mit.edu

L. D'Acci (ed.), *The Mathematics of Urban Morphology*,
Modeling and Simulation in Science, Engineering and Technology,
https://doi.org/10.1007/978-3-030-12381-9_20

et al. 2015; Li et al. 2017). This effect on human thermal comfort is more obvious during hot summer.

Quantifying the urban form is thus important for us to understand the potential impacts of the urban built environment on urban dwellers. Based on digital city models, various metrics can be calculated to describe and quantify different aspects of urban form. With the availability of high-resolution digital city models, it would be also possible to simulate the transmission of solar radiation within street canyons (Gal et al. 2009; Ratti and Richens 2004). However, digital city models cannot fully represent the streetscapes because most digital city models oversimplify the complex geometries of street canyons (Carrasco-Hernandez et al. 2015). In addition, the urban vegetation, which is a very important part of the urban natural system, is usually not included in those digital city models (Li et al. 2018). What is more, the high-resolution building city models are not always available in many cities.

The publicly accessible Google Street View (GSV), which captures the streetscape appearances of cities around the world, provides a very good tool for urban studies at a fine level. GSV was first introduced in 3D city modeling (Torii et al. 2009; Lee 2009; Micusik and Kosecka 2009) because the panoramic sequences in GSV can be used to reconstruct the 3D model of streetscapes. Since GSV represents the physical appearance of streetscapes and may have more direct connection with human perception of environment, GSV was further used to map human perception of environment using crowdsourcing method (Salesses et al. 2013; Naik et al. 2014). Based on time-series GSV images, Naik et al. (2017) measured the changes in the physical appearances of neighborhoods in five U.S. cities. The strong associations between the social characteristics and the streetscape appearance changes show that the GSV-based method can help to predict neighborhood improvement.

In this study, we illustrated using GSV for describing and modeling the enclosure of street canyons in Cambridge, Massachusetts. We estimated and mapped the sky view factor (SVF), which is a very important parameter of urban form, at street-level using GSV. We further mapped and analyzed the spatial distributions of sunlight duration in urban street canyons in Cambridge, MA during leaf-on seasons based on the generated hemispherical images from GSV panoramas.

2 Data Preparation

2.1 Google Street View (GSV) Panorama Collection

GSV panoramas can be collected from Google Server using Google Maps Application Programming Interfaces (APIs). In this study, in order to collect GSV panoramas to represent the urban form, we first created sample sites every 100 m along the streets. Figure 1a shows the generated sample sites along streets in Cambridge, Massachusetts. Based on the coordinates of these sample sites, we further collected

(a)

panoid	lon	lat	year	month	yaw
lnaMj49WfB4t8BAde57eBw	-71.138718	42.399655	2007	08	265.268372
SRCvTTCdtOcHwy8klBwopw	-71.138658	42.39967	2011	12	266.387085
P-L4zZGxcRxw_XsmWW7Frw	-71.138656	42.399672	2012	01	266.983398
zJ7VBv6pFixr7nY8U2fJbQ	-71.138694	42.399666	2014	09	264.010559
y3Wr7AW4td7f_I7Fa3FV0A	-71.136078	42.399155	2007	08	358.268829
NRy5UDjBUh_lwQQQeRcT2Q	-71.136095	42.399143	2011	08	358.977631
P6sm7no8OOpPQRzwMHBp	-71.13609	42.399156	2012	01	359.18927
3ceXROggUgCQxe6lerol2Q	-71.136099	42.399143	2013	09	359.33606
_M6XnP_D69coMmGalsnRB	-71.136095	42.399144	2014	09	359.063629
X44SlLBzk8ob7LLAu3r32A	-71.136103	42.399154	2016	03	359.875671
g0NaLDLD0tSiylCJTn1Ulw	-71.135882	42.397507	2007	08	274.225098
731ebcCtYEP28kdnAo-QyA	-71.135884	42.397511	2011	08	275.571533
8KAVlhtKtXfdzuQ9jUxQPA	-71.135897	42.397512	2013	09	274.839783
dNb9UPQQKXEa8s-HtfBnZQ	-71.135886	42.397511	2014	09	274.432129
LKPWGb7-bkAqXgiXjN5zDw	-71.127241	42.396629	2007	08	140.692108
M6qNBuFHvPLhpMyz4lLmog	-71.127221	42.396618	2011	08	140.652771
Nr-peaQG5l3-k1iolDilyg	-71.127232	42.396625	2013	09	141.104202
llYOBdCHiKbJXGVULFBFXQ	-71.127235	42.396628	2014	09	140.779892
022VshHlLHWJB2MvFsye9A	-71.127228	42.396623	2016	03	321.321777
VJStgPPGAjW8axV8HldFXg	-71.128885	42.39623	2007	07	322.318146

(b)

Google Maps API

https://geo0.ggpht.com/cbk?cb_client=maps_sv.tactile&authuser=0&hl=en&gl=us&**panoid**=bmyVsWWBXtgYVdnWsFgkqg&output=tile&**x**=0&**y**=0&zoom=1&nbt&fover=2

https://geo0.ggpht.com/cbk?cb_client=maps_sv.tactile&authuser=0&hl=en&gl=us&**panoid**=bmyVsWWBXtgYVdnWsFgkqg&output=tile&**x**=1&**y**=0&zoom=1&nbt&fover=2

***x*=0 *y*=0**

***x*=1 *y*=0**

(c) Google Street View panorama

Fig. 1 The workflow for GSV panorama collection, **a** the created sample sites in Cambridge, Massachusetts, **b** the metadata of GSV panoramas, **c** a GSV panorama of one sample site

the metadata of GSV panoramas in the study area. Here is an example of collecting the metadata of a GSV panorama located at (42.359048, −71.093574),

URL: *http://maps.google.com/cbk?output=xml&ll=42.359048,-71.093574*
Metadata of the a GSV panorama

{
"copyright" : "© 2017 Google",
"date" : "2017-9",
"location" : {
"lat" : 42.358964,
"lng" : -71.093537
},
"pano_id" : " 4G5km0yE7QsmzxE7YBPRYw*",*
"pano_yaw_deg": " 341.80734"
}

Based on the panorama IDs in the metadata, GSV panoramas can also be downloaded. Figure 1 shows the workflow of collecting GSV panorama metadata and the final GSV panoramas. In this study, we developed a Python script (**Appendix A**)

to download tiles of GSV panoramas and mosaic them to a panorama for each site using the panorama ID as input.

2.2 Geometric Transform of Google Street View (GSV) Panoramas

The collected GSV panoramas are in the form of equidistant cylindrical projection as shown in Fig. 1c. For urban form studies, the cylindrical projection GSV panoramas need to be transformed to equidistant azimuthal projection. Figure 2 shows the geometric model of transforming cylindrical projection to azimuthal projection. A GSV panorama with width of W_c and height of H_c can be re-projected to an azimuthal hemispherical image with the width and height of W_c/π. For any pixel (x_f, y_f) in the generated hemispherical image, the corresponding pixel in the cylindrical panorama should be (x_c, y_c),

$$\begin{aligned} x_c &= \frac{\theta}{2\pi} W_c \\ y_c &= \frac{r}{r_0} H_c \end{aligned} \tag{1}$$

where r and θ are the distance of the pixel (x_f, y_f) to the center of the hemispherical image and the zenith angle, respectively (Fig. 2).

Considering the fact that the central column in the cylindrical image represents the driving direction of the GSV vehicle rather than the true north direction. Therefore, the generated hemispherical images need to be further rotated by the *yaw* angle to make sure the generated hemispherical images represent the north, east, south, and the west direction correctly. The *yaw* angle can be accessed from the metadata of GSV panorama in Sect. 2.1. The pixel (x_f, y_f) in the synthetic hemispherical images should be further converted into $(x_f{}', y_f{}')$ in the rotated hemispherical images as,

$$\begin{aligned} x_f' &= x_f \cos\varphi - y_f \sin\varphi \\ y_f' &= x_f \sin\varphi + y_f \cos\varphi \\ \varphi &= 360 - yaw \end{aligned} \tag{2}$$

where *yaw* is the yaw angle from the metadata of GSV panorama.

2.3 Image Classification

The sky extraction is a requisite step to derive urban form information from hemispherical images. In this study, we applied the object-based image classification

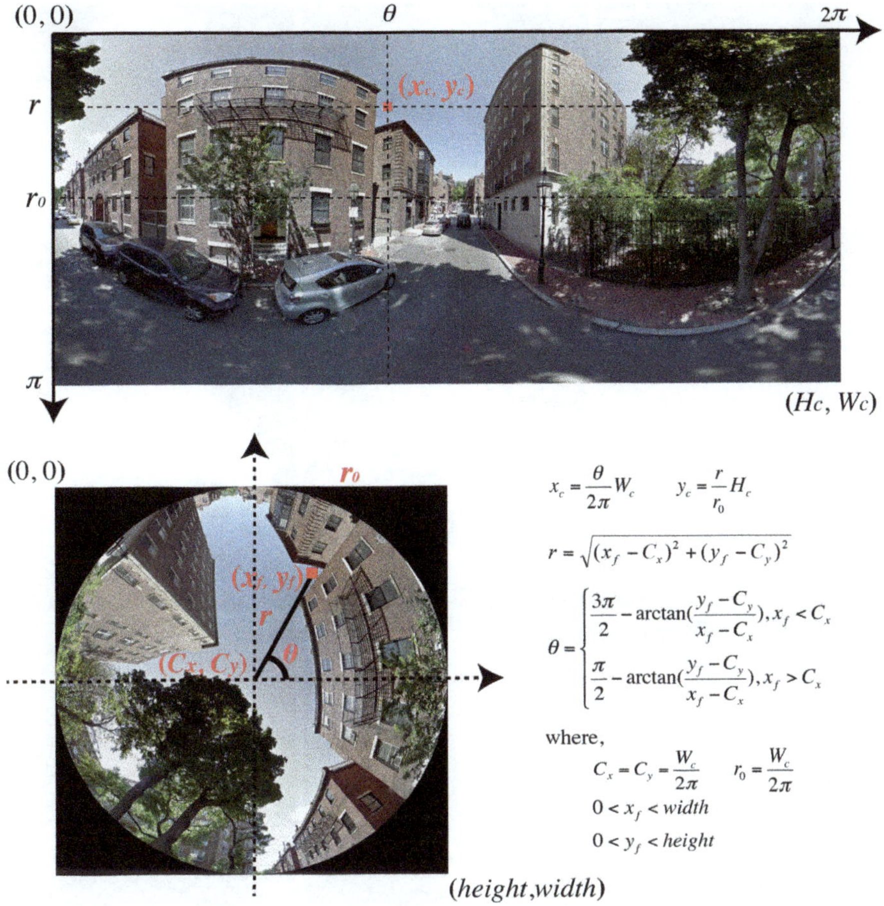

Fig. 2 Geometrical transform of equidistant cylindrical projection to equidistant azimuthal projection (hemispherical image)

method to classify the hemispherical into sky pixels and non-sky pixels the sky (Li et al. 2018). Hemispherical images were first segmented into homogeneous and physically meaningful objects based on the mean-shift algorithm (Comaniciu and Meer 2002; Li et al. 2018). Figure 3b shows segmentation results on hemispherical images. Compared with the original hemispherical images (Fig. 3a), the segmented images have enhanced difference between sky pixels and non-sky pixels.

Since sky pixels are usually brighter than non-sky pixels and non-sky greenery pixels usually have higher values in the *ExG* ($2 \times$ green − blue − red) image, we used the brightness and *ExG* to extract the sky pixels from the segmented hemispherical images . The Otsu's method (Otsu 1979) was then used to find the optimum thresholds

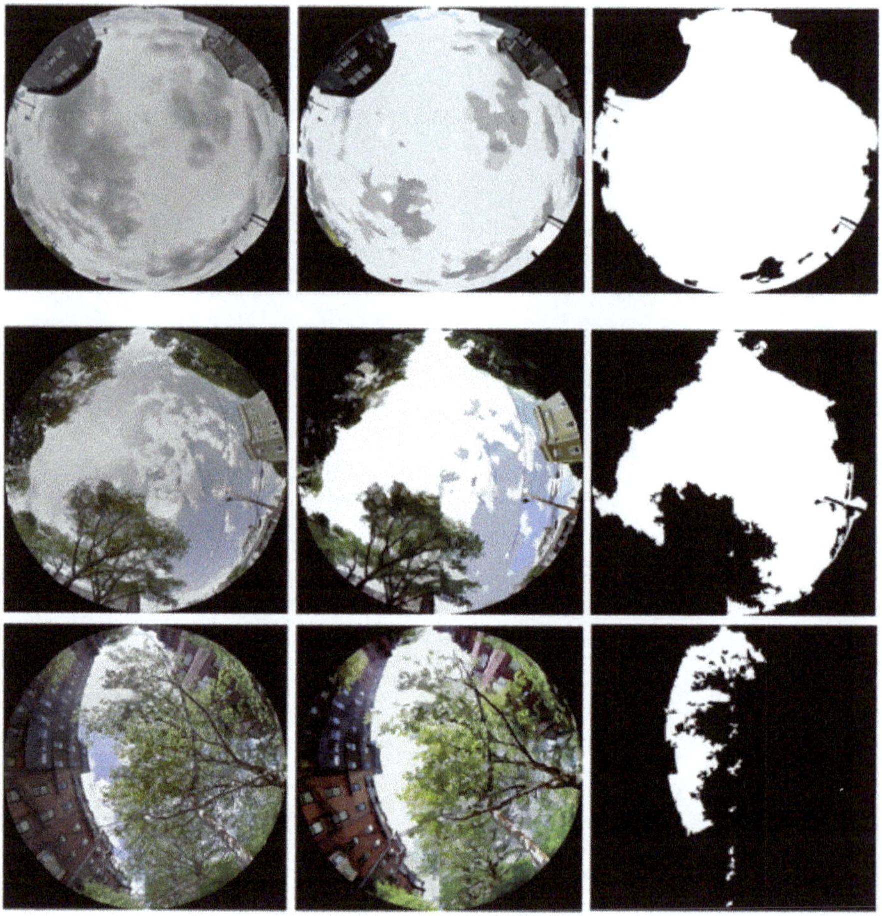

(a) (b) (c)

Fig. 3 The classification of sky pixels in three generated hemispherical images, **a** the original hemispherical images generated from GSV panoramas, **b** the segmented images using mean-shift algorithm, **c** the sky classification results

to separate sky pixels and non-sky pixels. Those pixels that have higher *Brightness* and lower *ExG* values than optimum thresholds are sky pixels. Figure 3c shows the classification results of sky pixels on three generated hemispherical images based on the above spectral and geometrical rules.

3 Estimating and Mapping the Sky View Factor in Different Seasons

The sky view factor (SVF) is an important parameter of urban form. The SVF was proposed by urban climatologists to describe the amount of solar radiation reaching the street canyons. The SVF is defined as (Steyn 1980),

$$SVF = \frac{1}{\pi r_0^2} \int_{S^p} dS^p \tag{3}$$

where r_0 is the radius of the hemispheric radiating environment, S^p is the area of the circle sky area projected on the ground. When the sky is totally obstructed the SVF is zero, and the SVF is one when there is no obstruction for a site. In summer, building blocks and tree canopies act as major obstructions of sky in street canyons, while building block is the main obstruction in winter since trees are leafless. Therefore, in this study, we used the GSV-based photographic method and the building height model-based simulation method to estimate and SVF in summer and winter, respectively.

The photographic method is one of the standard methods for SVF estimation. In this study, we used the hemispherical images generated from GSV panoramas taken in summer to estimate the SVF in street canyons of Cambridge during summer. The photographic method (Steyn 1980; Johnson and Watson 1984) first divides the fisheye image into n concentric annular rings of equal width, and then sums up all annular sections representing the visible sky. The SVF is then calculated as

$$SVF = \frac{1}{2\pi} \sin\left(\frac{\pi}{2n}\right) \sum_{i=1}^{n} \sin\left(\frac{\pi(2i-1)}{2n}\right) \alpha_i \tag{4}$$

where n is the total number of rings, i is the ring index, and α_i is the angular width in ith ring. Based on previous studies (Chen et al. 2012), in this study, we set the n to 37. Since GSV panoramas used in this study were captured during leaf-on seasons, therefore, building blocks and street trees both act as obstructions of solar radiation in street canyons. The estimated SVF using GSV-based photographic method represents the openness of street canyons in summer.

In winter, trees are leafless and building blocks act as the main obstruction. Therefore, we further estimated the SVF in winter using the simulation method based on building height model and ray-tracing algorithm. In the simulation method, the SVF can be estimated based on the simulation of light radiation in the building height model with consideration of the obstruction of building blocks only. The SVF can be calculated as (Gal et al. 2009),

$$SVF = 1 - \frac{1}{360} \sum_{\alpha=0}^{359} sin^2 \beta_\alpha \tag{5}$$

where β_α is the obstruction angle of the obstruction building at horizontal direction of α, and can be calculated as the following formula considering the fact that GSV panoramas were captured at a height of 2.5 m,

$$\beta_\alpha = \max(\beta_{\alpha,i}) = \max\left(\arctan \frac{H_{\alpha,i} - 2.5}{D_{\alpha,i}}\right) \tag{6}$$

where $\beta_{\alpha,i}$ is the obstruction angle of building i along the horizontal direction α, $H_{\alpha,i}$ is the height of building i, and $D_{\alpha,i}$ is the distance between the building i and the site of the GSV panorama.

Figure 4a shows the spatial distribution of the SVF values in winter with the consideration of the obstruction of building blocks only. The central and southern parts of the study area, which around Harvard Square and Kendall Square, have significantly lower SVF values than other regions. This is because of the large numbers of high-rise buildings in there. In summer, the street tree canopies would also act as obstruction and should be considered in calculating the real SVF. Figure 4b shows the spatial distribution of the SVF map in summer using the GSV-based photographic method, which considers the obstruction effects of both the building blocks and the street tree canopies. There is no obvious pattern in the SVF distribution after considering the obstruction effects of both buildings and street trees in the study area.

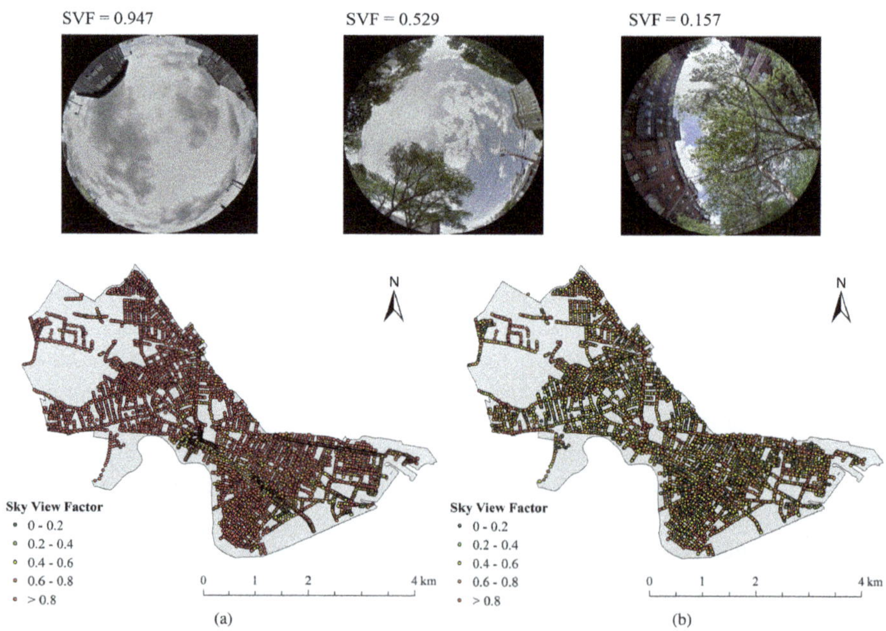

Fig. 4 The spatial distributions of SVF in Cambridge, MA and hemispherical images with different SVF values, **a** the SVF map in winter when building blocks are the major obstruction, **b** the GSV-based photographic SVF map in summer when tree canopies and buildings both act as obstruction

4 Estimating Direct Sunlight Duration in Street Canyons

The sunlight duration, which measures the duration of sunlight in a given period, is an important climatological parameter influencing the human thermal comfort within street canyons. It would also influence pedestrian activities in street canyons during hot summer. The sunlight duration in street canyons is influenced by the spatial configuration of urban features and the orientation of streets. Based on the hemispherical images generated from GSV panoramas, it is possible to estimate the spatio-temporal distribution of the sunlight duration at street level.

The sun positions can be calculated with high accuracy, as it varies with the location of those sites and time. By overlaying the hemispherical image and sun path for a given day at one site, it is possible to measure the duration of sunshine throughout a day for the site. Figure 5 shows the overlays of sun positions on hemispherical images from 5:00 a.m. to 7:00 p.m. on September 1st at three sites.

By assuming that the sunlight would be blocked if the sun were not located in the open sky areas on the hemispherical images, we calculated the sunlight duration for one site as the duration of sunlight not blocked by obstructions for this site within the street canyon. The calculation was based on the assumption that the weather is sunny and cloudless. Although the result may not represent the real sunlight duration in cloudy or raining day, this simulation would give an estimate of the potential sunlight duration theoretically, which would provide a reference for urban planning.

Figure 6 shows the spatial distributions of direct sunlight duration at site level and block group levels on August 1st. Different parts of the study area have very different spatial distributions of sunlight duration. Generally, the northwestern and eastern parts of the study area have longer sunlight duration. This study also shows that it is possible to estimate the spatio-temporal distribution of sunlight duration within street canyons at large scales with a fine temporal resolution. Considering the abundance of GSV data in cities around the world, the proposed automatic method would provide a huge impetus to all studies relating street canyons level solar radiation.

Fig. 5 The overlay of the sun path of September 1st, 2014 on three hemispheircal images taken in September

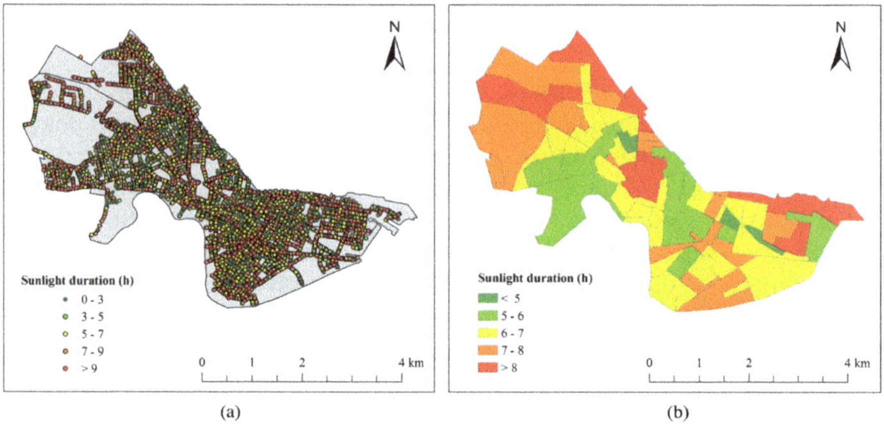

Fig. 6 The spatial distribution of direct sunlight duration in Cambridge, MA

5 Other Potential Applications

The globally available and publicly accessible street-level images would benefit urban environmental studies at street-level dramatically for cities around the world. For example, based on the proposed green view index calculated from GSV images (Li et al. 2015), the *Treepedia* (http://senseable.mit.edu/treepedia) project was launched to map and compare the spatial distributions of street greenery for cities around the world. Currently, the *Treepedia* project has mapped street greenery for more than 30 cities around the world. Figure 7 shows the spatial distribution of the green view index in Boston. Environmental metrics derived from GSV images would help urban planners, public health researchers, and social scientists to investigate the interaction of physical environment and human beings from a new perspective at fine level.

The progress in artificial intelligence research would make deriving more information possible from street-level images. Figure 8 shows the image classification results using the PSPnet (pyramid scene parsing network) deep learning algorithm (Zhao et al. 2016) on GVS images. Different kinds of urban features along the streets can be recognized accurately, which would be useful for urban planning.

Google has collected street-level images for most countries of the world and Google keeps updating the street-level images database periodically. In 2014, Google launched the time-machine tool in Google Street View service, and users can access the historical GSV images. It would make studying the temporal changes of the streetscapes possible using the street-level images. Figure 9 shows the changes of streetscapes at two sites of Cambridge, Massachusetts in GSV panoramas.

Other than Google Street View, there are more and more street-level images providers, such as Tencent Street View, Baidu Street View, Bing Maps Streetside, and Mapillary, etc. The coming autonomous vehicle technologies would also provide

Fig. 7 The green View index map in Boston from *Treepedia*

abundant street-level images for urban studies in future. Future urban form studies will benefits significantly from more abundant street-level images and more advanced image processing algorithms.

6 Discussion and Conclusion

This study presented a case study in Cambridge, MA using the publicly accessible Google Street View (GSV) panoramas and building height model to derive quantitative information of urban form at street-level. Based on the quantitative information derived from GSV panoramas and the building height model, it is possible to estimate the spatio-temporal distributions of sky view factor (SVF) and the sunlight duration at the street canyon level. Different from the building height model, the GSV panoramas represent the actual appearance of streetscape, therefore, the GSV images would be a perfect data source for mapping the urban form with consideration of the urban natural ecosystem. With combination of the building height model, it is possible to map the spatio-temporal distribution of SVF and sunlight duration in different seasons. The GSV would also be a surrogate for those study areas with high-resolution digital city models not available.

This study showed that that GSV is a very promising data source for urban studies considering its public accessibility and global availability. The developed workflow

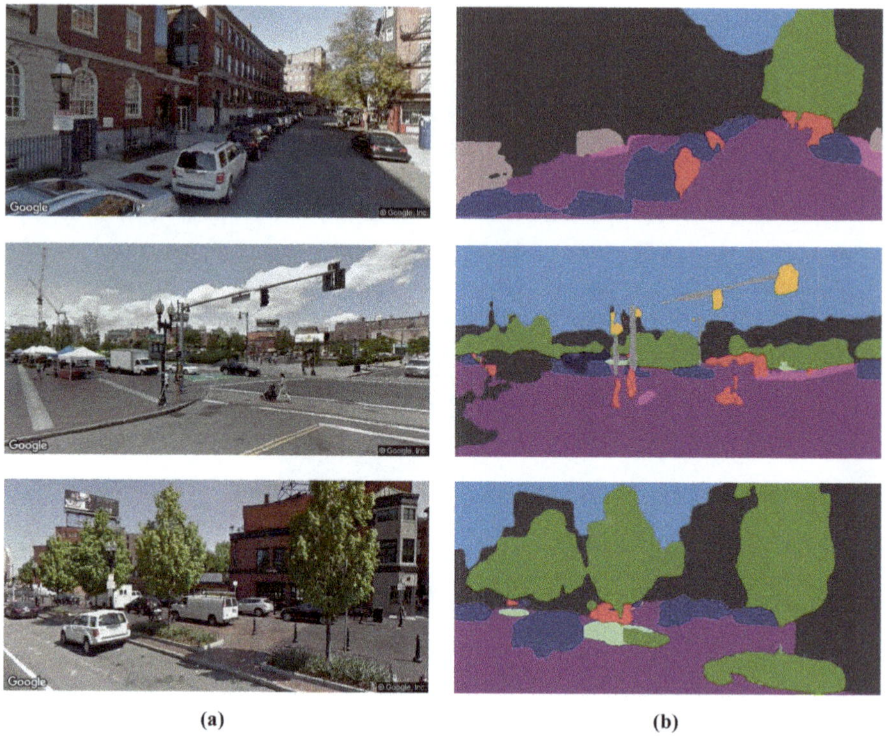

Fig. 8 Image classification results on GSV images using pyramid scene parsing convolutional neural network

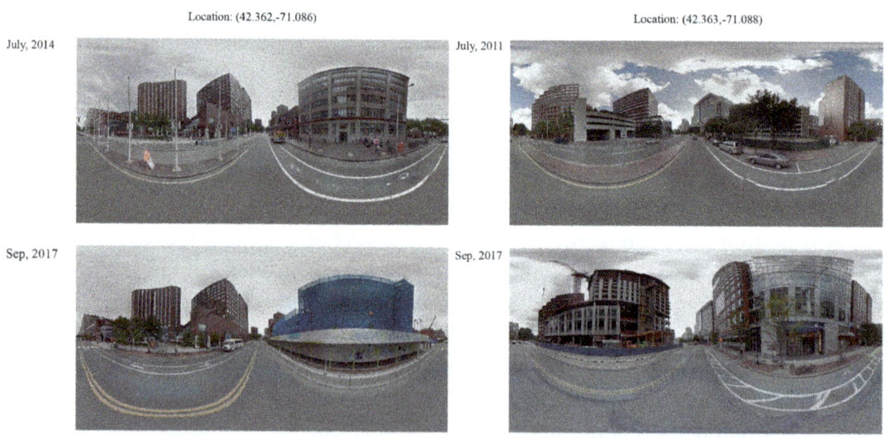

Fig. 9 The changes of streetscape at two sites of Cambridge, MA in GSV panoramas

for urban form analysis in this study is totally automatic and without any human intervention. Therefore, it is possible to simply and rapidly estimate the SVF and sunlight duration at street-level for any city with GSV service available. The GSV-based method is suitable for large-scale SVF estimation can help researchers, urban planners and managers better understand the influence of urban form on the urban microclimate, urban air pollution migration, and human perception of urban environment.

While this study has showed using GSV to estimate and map the urban form information quantitatively, there are still some limitations that should be addressed in future applications. First, GSV images were captured within street canyons, therefore, GSV is only suitable for estimating the urban form information within street canyons. Combining with different data sources would help us to make a better understanding of the urban form.

In addition, although the SVF and direct sunlight duration are important parameters for understanding the urban form and its performance in cities, the performances may vary in different cities. In the future study, more parameters need to be considered to better represent the human experience in street canyons.

Although GSV images make it possible accurately derive fine-level urban form information at a large scale, future studies should also consider giving more focus on humans. With the advancement in deep learning and abundance of human GPS trajectories, it is possible to derive more information from street-level images and better investigate the interaction between humans and urban built environment at a fine level.

Acknowledgements We would like to thank Bill. Y. Cai for providing the technical support in processing Google Street View images based on convolutional neural networks.

Appendix A

The code for collecting Google Street View panoramas from Google's server based on the Google Maps API, https://github.com/xiaojianggis/skyview.

References

Asgarzadeh, M., Lusk, A., Koga, T., & Hirate, K. (2012). Measuring oppressiveness of streetscapes. *Landscape and Urban Planning*, 107(1), 1–11.

Asgarzadeh, M., Koga, T., Hirate, K., Farvid, M., & Lusk, A. (2014). Investigating oppressiveness and spaciousness in relation to building, trees, sky and ground surface: A study in Tokyo. *Landscape and Urban planning*, 131, 36–41.

Carrasco-Hernandez, R., Smedley, A. R., & Webb, A. R. (2015). Using urban canyon geometries obtained from Google Street View for atmospheric studies: Potential applications in the calculation of street level total shortwave irradiances. *Energy and Buildings*, 86, 340–348.

Chen, L., Ng, E., An, X., Ren, C., Lee, M., Wang, U., & He, Z. (2012). Sky view factor analysis of street canyons and its implications for daytime intra urban air temperature differentials in high rise, high density urban areas of Hong Kong: a GIS based simulation approach. *International Journal of Climatology*, 32(1), 121–136.

Comaniciu, D., & Meer, P. (2002). Mean shift: A robust approach toward feature space analysis. *IEEE Transactions on Pattern Analysis and Machine Intelligence*, 24(5), 603–619.

Gál, T., Lindberg, F., & Unger, J. (2009). Computing continuous sky view factors using 3D urban raster and vector databases: comparison and application to urban climate. *Theoretical and applied climatology*, 95(1–2), 111-123.

Johnson, G. T., & Watson, I. D. (1984). The determination of view-factors in urban canyons. *Journal of Climate and Applied Meteorology*, 23(2), 329–335.

Lee, T. (2009, September). Robust 3D street-view reconstruction using sky motion estimation. In Computer Vision Workshops (ICCV Workshops), 2009 IEEE 12th International Conference on (pp. 1840–1847). IEEE.

Li, X., Ratti, C., & Seiferling, I. (2017, July). Mapping Urban Landscapes Along Streets Using Google Street View. In International Cartographic Conference (pp. 341–356). Springer, Cham.

Li, X., Ratti, C., & Seiferling, I. (2018). Quantifying the shade provision of street trees in urban landscape: A case study in Boston, USA, using Google Street View. *Landscape and Urban Planning*, 169, 81–91.

Li, X., Zhang, C., Li, W., Ricard, R., Meng, Q., & Zhang, W. (2015). Assessing street-level urban greenery using Google Street View and a modified green view index. Urban Forestry & Urban Greening, 14(3), 675–685.

Micusik, B., & Kosecka, J. (2009, June). Piecewise planar city 3D modeling from street view panoramic sequences. In Computer Vision and Pattern Recognition, 2009. CVPR 2009. IEEE Conference on (pp. 2906–2912). IEEE.

Naik, N., Philipoom, J., Raskar, R., & Hidalgo, C. (2014). Streetscore predicting the perceived safety of one million streetscapes. In Proceedings of the IEEE Conference on Computer Vision and Pattern Recognition Workshops (pp. 779–785).

Naik, N., Kominers, S. D., Raskar, R., Glaeser, E. L., & Hidalgo, C. A. (2017). Computer vision uncovers predictors of physical urban change. *Proceedings of the National Academy of Sciences*, 114(29), 7571–7576.

Otsu, N. (1979). A threshold selection method from gray-level histograms. *IEEE transactions on Systems, Man, and Cybernetics*, 9(1), 62–66.

Ratti, C., & Richens, P. (2004). Raster analysis of urban form. *Environment and Planning B: Planning and Design*, 31(2), 297–309.

Salesses, P., Schechtner, K., & Hidalgo, C. A. (2013). The collaborative image of the city: mapping the inequality of urban perception. *PloS* ONE, 8(7), e68400.

Steyn, D. G. (1980). The calculation of view factors from fisheye lens photographs. Atmosphere-Ocean, 18, 254–258.

Torii, A., Havlena, M., Pajdla, T. (2009). From Google Street View to 3D city models. In: IEEE 12th International Conference on Computer Vision Workshops, Kyoto, Japan, pp. 2188–2195.

Zhao, H., Shi, J., Qi, X., Wang, X., & Jia, J. (2016). Pyramid scene parsing network. arXiv:1612. 01105.

Examining Spatial Structure Using Gravity Models

Martijn Burger, Frank van Oort and Evert Meijers

Abstract In this chapter, we discuss the use of gravity models in the study of spatial structure. Using the recent discussion on functional polycentricity as a background, we argue that the gravity model approach has one obvious advantage when examining spatial structure: it can simultaneously assess functional polycentricity and spatial interdependencies within one modelling framework. The chapter concludes with a discussion of methods that can be applied to estimate the gravity model.

1 Introduction

Fifty years ago, the French geographer Jean Gottmann (1957, 1961) envisioned the rise of a super-metropolitan region along the northeastern seaboard of the United States, stretching from just north of Boston all the way to Washington DC. Gottmann named this new urban form after the Peloponnesian city Megalopolis, founded by Epaminondas of Thebes as the seat of the Arcadian league in an attempt to form a political counterweight to Sparta. According to Gottmann (1961, p. 4), '*the name applied to* [this area] *should … be new as a place name but old as a symbol of the long tradition of human aspirations and endeavour*'. Indeed, the Greek Megalopolis was planned on an enormous scale; the city was populated through the enforced transfer of inhabitants from 40 local villages and was encompassed by 9 km in circumference of strong walls (Baigant 2004). Although Epaminondas' Megalopolis did not succeed as hoped and gradually faded into oblivion, Gottmann was optimistic about the future of the new Boston–Washington corridor Megalopolis. He felt the

M. Burger (✉) · F. van Oort
Erasmus University Rotterdam, Rotterdam, The Netherlands
e-mail: mburger@ese.eur.nl

F. van Oort
e-mail: vanoort@ese.eur.nl

E. Meijers
Delft University of Technology, Delft, The Netherlands
e-mail: E.J.Meijers@tudelft.nl

L. D'Acci (ed.), *The Mathematics of Urban Morphology*,
Modeling and Simulation in Science, Engineering and Technology,
https://doi.org/10.1007/978-3-030-12381-9_21

region could function effectively as an interregional polycentric urbanised system that still had many characteristics of a single city. Gottmann (1961, p. 5) argued that:

> We must abandon the idea of the city as a tightly settled and organized unit in which people, activities, and riches are crowded into a very small area clearly separated from its nonurban surroundings. Every city in this region spreads out far and wide around its original nucleus; it grows amidst an irregularly colloidal mixture of rural and suburban landscapes; it melts on broad fronts with other mixtures, of somewhat similar though different texture, belonging to the suburban neighborhoods of other cities.

Gottmann considered the Megalopolis to be the emergent form of spatial organisation, characterised by high average population densities and the flow of high volumes of people, goods, capital and information. Functional relationships between the different parts of the Megalopolis would be of the utmost importance for its ability to function as a single city. The Megalopolis reflected the enlarged scale of urban life and the shift from a single metropolis with a principal centre to an *urban network with multiple centres*. Gottmann also emphasised the importance in the Megalopolis of inter-state cooperation and governance organised on larger geographical scales than the local scale. He claimed local governments would inadequately fulfil the needs of these large cities and their ever-expanding suburbs and sub-centres. The Megalopolis would be characterised by a marriage of urban and rural modes of life, leading to maximum freedom of movement and a perfection of the modern urban lifestyle, one that would remedy the problems of the congested city and the backward village. Visions for the city similar to Gottmann's were also expressed in (earlier) planning concepts such as the Garden City (Howard 1902), Broadacre City (Wright 1935), and the Regional City (Stein 1964).

2 Functional Spatial Structure

Gottmann's vision of the Boston–Washington corridor as a polycentric urban network was radical in the 1950s and broke with the conventional conceptualisations of cities as local hierarchical urban systems.[1] This was stressed again by Short (2007), updating much of Gottmann's analysis and referring to the area as a 'liquid city' to stress the fluidity of the city and region at large. Today, super-regions like the Northeastern Seaboard in the United States, the Greater Southeast in the United Kingdom, the Flemish Diamond in Belgium, the Randstad in the Netherlands, and the Rhein-Ruhr and Rhein-Main areas in Germany have gained considerable attention in the academic literature (see e.g. Hoyler et al. 2008; Florida et al. 2008; Burger et al. 2014c). Although it should be acknowledged that Gottmann's original concept was predominantly morphological in nature, he later (in response to critiques) stressed the functional aspects of the various centres in the Megalopolis (Hall 1997). On the one hand, the Megalopolis is a polycentric super-region of cities in close proximity

[1]For an exception at the city-level, see Harris and Ullman (1945) and Alonso (1956).

to each other. On the other hand, without the functional and complementary relationships between places like Boston, New York, Philadelphia and Washington DC it is hard to argue that Megalopolis could be regarded as a truly integrated polycentric region, since it does not function as such. Accordingly, it can be argued that for a super-region to function as a coherent polycentric networked urban entity:

(1) There should not only be a balance in the city size distribution, but also a certain balance in the distribution of functional linkages between places.
(2) There should be a certain extent to which the places within the Megalopolis are functionally linked.

The first condition has been referred to as 'functional polycentricity' (Green, 2007). The second addresses the requirement that there is a significant degree of spatial integration or spatial interdependencies. Hence, (a) spatial structure should not only be addressed by looking at the mere existence of multiple centres within one area, but also by looking at the functional linkages between places within an area and (b) functional polycentricity and spatial interdependencies can be regarded as two defining elements of a polycentric urban super-region (Burger and Meijers 2012).

In analytical work on spatial structure, it is important not to conflate the degree of spatial interdependence with the degree of functional polycentricity: they are different theoretical constructs (Burger and Meijers 2012; Vasanen 2013). There are spatial systems that are strongly networked as well as monocentric, and there are spatial systems that are not networked at all but are polycentric. In fact, previous empirical research has shown that there is no correlation between the degree of functional polycentricity and degree of spatial interdependence, indicating that they should be treated as two distinct aspects of the spatial organisation of regions (Burger et al. 2011; Burger and Meijers 2012).

2.1 Functional Polycentricity

Over the past decades, the urban systems literature has seen a surge in papers that attempt to measure spatial structure along the monocentricity–polycentricity dimension. Accordingly, measuring the degree of a balanced distribution with respect to the importance of centres within a given territory is the focus in these papers. In this literature, it has been debated whether monocentricity–polycentricity refers to only morphological aspects of the urban system or whether it should also incorporate relational aspects between the centres that constitute the urban system in question (Green 2007; Meijers 2008; Burger and Meijers 2012). In other words, should one measure the importance of centres on locational (internal) characteristics or on the basis of flows? This distinction is also represented in Fig. 1. Morphological measures of monocentricity–polycentricity capture the size distribution of the urban centres within a territory, where a more balanced distribution of centre size (usually expressed in terms of population sizes) equates with a polycentric spatial structure

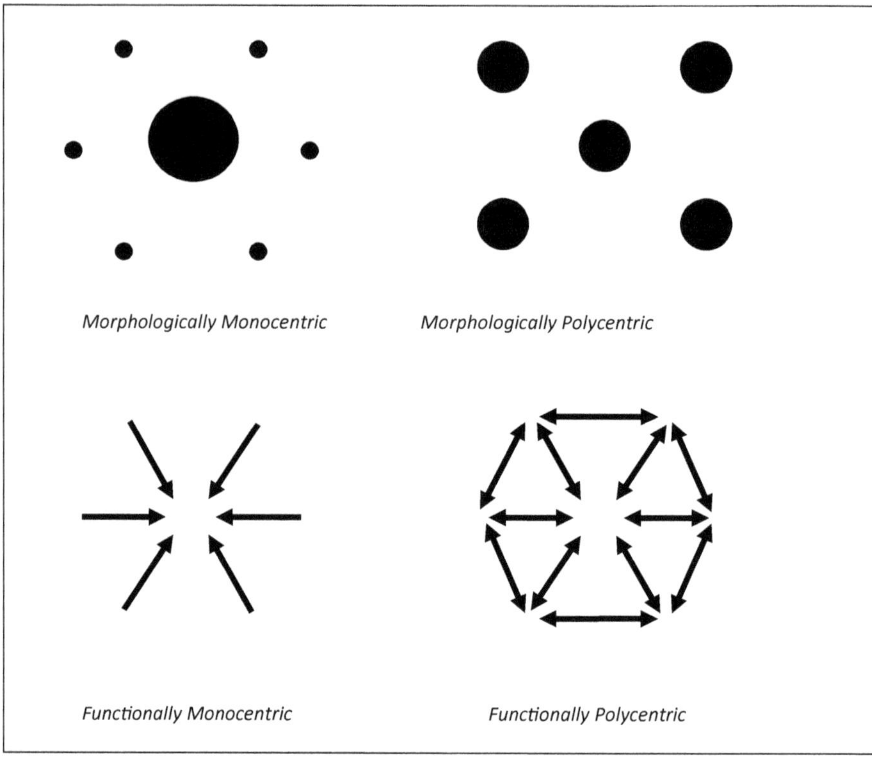

Fig. 1 Morphological versus functional spatial structures *Source* Burger and Meijers (2012)

(see e.g. Kloosterman and Lambregts 2001; Parr 2004; Meijers and Burger 2010). Functional monocentricity–polycentricity measures, on the contrary, take the functional connections (e.g. commuting, shopping, knowledge collaboration and trade flows) between the settlements into account, and consider a more balanced, multi-directional set of relations to be more polycentric (Green 2007; Burger and Meijers 2012). Within a functional polycentric system, there is a more equal balance in the distribution of inflows, meaning that functional relationships are not only directed at one centre (like in a monocentric urban system), but two-sided (reciprocal) and crisscross (also existing between smaller centres) (Van der Laan 1998; De Goei et al. 2010; Burger et al. 2011; Li and Phelps 2018).

It is important to note that studies that adhere to the functional dimension of monocentricity–polycentricity do not dismiss the morphological approach, but extend it to include also the pattern of functional interaction between the urban centres. The approach generally taken has many similarities with the morphological approach, using—for example—also urban primacy measures and rank-size distributions (based on network data) to assess spatial structure (see, e.g. Van der Laan 1998;

Meijers 2008; Meijers and Burger 2010; Burger et al. 2011; Veneri and Burgalassi 2012).

2.2 Spatial Interdependencies

Functional polycentricity does not so much address the existence or strength of functional relationships between centres within a given territory, but rather the balance in the distribution of these functional relationships. However, another pre-condition for a region like the Megalopolis to exist would be a certain degree of spatial integration or spatial interdependencies. Without functional relationships between the historically and geographically separate parts of a region, it is hard to argue that the places within a territory function as a region. Methods to measure the degree of spatial interdependencies include network density (Green 2007), connectivity fields (Vasanen 2013), geographical scope of functional relations (Burger et al. 2013) and the gravity model (De Goei et al. 2010; Van Oort et al. 2010; Hanssens et al. 2014; Coombes and Champion 2016).

3 Gravity Models and Spatial Structure

The gravity model approach has one obvious advantage when examining spatial structure: it can simultaneously assess functional polycentricity and spatial interdependencies within one modelling framework. Using the gravity modelling framework, one can employ Newton's law of universal gravitation to gauge the interaction between spatial units. These interactions can be any kind of functional relationships between places within a region, ranging from commuting and shopping flows to business trade and investment flows. The model holds that the gravitational force between two spatial units is directly proportional to the product of the mass of the interacting spatial units and inversely proportional to the physical distance between them. Traditionally, the gravity model can be expressed by

$$I_{ij} = K \frac{M_i^{\beta_1} M_j^{\beta_2}}{d_{ij}^{\beta_3}},$$

where I_{ij} is the interaction intensity, e.g. the number of people travelling between places i and j, K a proportionality constant, M_i the size of place i, M_j the size of place j, d_{ij} the physical distance between the two places, β1 the potential to generate flows, β2 the potential to attract flows and β3 an impedance factor reflecting the rate of increase of the friction of distance. If an area is functionally polycentric and the places are spatially interdependent, then network structures of commuting, trade, shopping, and other types of functional relations within this area should be solely determined by the size of the places and the distance between them. In other words, once the size of places and distance between places are controlled for, in an equilibrium situation there should be no additional flows or interactions.

With regard to spatial interdependencies, one would expect that the interdependencies between places are two-sided (or exchange) and crisscross (periphery–periphery; Burger et al. 2011) in character: places should be both sender and receiver of relationships and interdependencies between places at different levels of the original urban hierarchy (e.g. core–periphery relationships) should not be stronger than the interdependencies between places at the same level of the hierarchy (e.g. core–core relationships or periphery–periphery relationships). The degree of spatial interdependencies within an area can be determined by assessing to what extent a region functions as one place. If this would be the case, we would see that controlling for size and distance, interdependencies within places (or between places within one part of the region) should not be stronger than interdependencies between places (or places across different parts of the region).

4 Estimation of the Gravity Model

By taking logarithms of both sides of gravity equation and including a disturbance term, the multiplicative form can be transformed into a linear stochastic form. It results in an equation that is testable using ordinary least squares, in which the disturbance term ε_{ij} is assumed to be identical and independently distributed (i.i.d):

$$\ln I_{ij} = \ln K + \beta_1 M_i + \beta_2 M_j - \beta_3 d_{ij} + \varepsilon_{ij}$$

The model above can be extended to a panel data framework, so that it becomes possible to study the development of spatial interactions over time (see e.g. De Goei et al. 2010). In addition, the empirical gravity model can be easily augmented to include other factors than size and distance. Most notably, dummy variables that reflect the type of relationship between places (e.g. core-periphery vs crisscross; or within a subregion vs. between subregions; see Fig. 2), can be included in the model to test the degree of functional polycentricy and spatial interdependencies in a region. Also barriers between places, like lack of physical accessibility, or language and cultural differences between places, can be introduced to test whether these hamper or stimulate interaction.

The models can be estimated using OLS, but the application of a linear regression model often results in inefficient, inconsistent, and biased estimates (Flowerdew and Aitkin 1982) since the underlying assumptions of normal distribution and homoskedasticity are often not satisfied. For this reason, the use of alternative regression techniques such as count data model is then judged more appropriate. Applications of count data model in assessing spatial structure can be found in the work of De Goei et al. (2010), Van Oort et al. (2010) and Hanssens et al. (2014). A

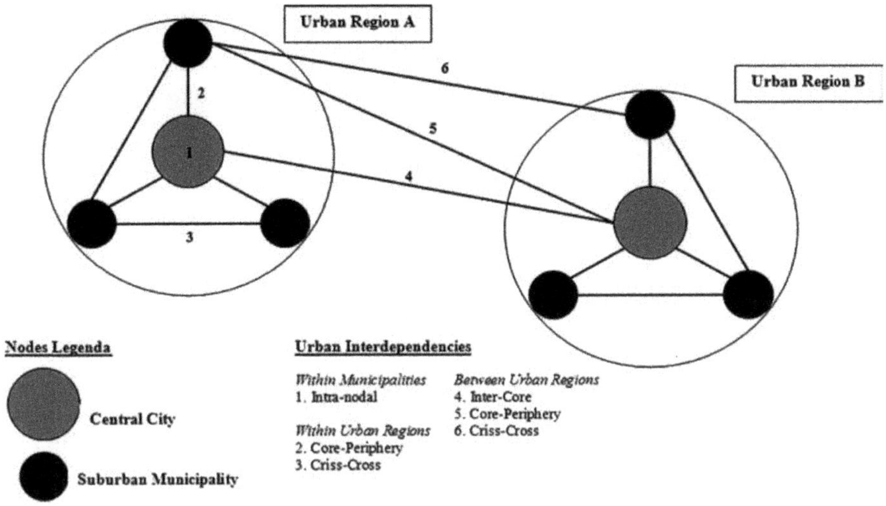

Fig. 2 Example of different spatial interdependencies within a region *Source* Van Oort et al. (2010)

more elaborate account of the estimation of gravity models can be found in Burger et al. (2009) and Broekel et al. (2014).

5 Concluding Remarks

Although the gravity model provides a framework to simultaneously assess different aspects of functional spatial structure, a difficulty in the assessment of both functional polycentricity and spatial interdependencies still constitutes the multiplexity of urban networks (Burger et al. 2014b) and integration (Meijers et al. 2018) as well as individual-level heterogeneity (Burger et al. 2014a). First, the spatial structure of different types of functional relationships is not necessarily identical and a region can, therefore, appear to be polycentric or spatially integrated based on the analysis of one type of functional linkage but loosely connected based on the analysis of another type of functional linkage (Burger et al. 2014a). Second, spatial interdependencies between different centres could have institutional and cultural dimensions besides functional dimensions (Meijers et al. 2018). Third, even when a single type of flow is taken into account, there may be a wide variety in spatial interaction patterns that can be attributed to differences among people or firms (Burger et al. 2014a). Addressing this heterogeneity is important, as networks of flows are built up by heterogeneous individual and group behaviour, and structural changes and policies that benefit one group may harm another.

References

Alonso, W. (1956), *Location and land-use: towards a general theory of land-use*. Harvard: University Press.

Baigent, E. (2004) Patrick Geddes, Lewis Mumford and Jean Gottmann: divisions over 'megalopolis'. *Progress in Human Geography*, 28: 687–700.

Broekel, T., Balland, P. A., Burger, M., & van Oort, F. (2014). Modeling knowledge networks in economic geography: a discussion of four methods. *The Annals of Regional Science*, 53(2), 423–452.

Burger, M. and Meijers, E. (2012) Form follows function? Linking functional and morphological polycentricity. *Urban Studies*, 49: 1127–1149.

Burger, M., Van Oort, F., & Linders, G. J. (2009). On the specification of the gravity model of trade: zeros, excess zeros and zero-inflated estimation. Spatial Economic Analysis, 4(2), 167–190.

Burger, M. J., de Goei, B., Van der Laan, L., & Huisman, F. J. (2011). Heterogeneous development of metropolitan spatial structure: Evidence from commuting patterns in English and Welsh city-regions, 1981–2001, *Cities*, 28(2), 160–170.

Burger, M. J., van der Knaap, B., & Wall, R. S. (2013). Revealed competition for greenfield investments between European regions. *Journal of Economic Geography*, 13(4), 619–648.

Burger, M. J., Meijers, E. J., & Van Oort, F. G. (2014a). Multiple perspectives on functional coherence: Heterogeneity and multiplexity in the Randstad. *Tijdschrift voor Economische en Sociale Geografie*, 105: 444–464.

Burger, M. J., Van Der Knaap, B., & Wall, R. S. (2014b) Polycentricity and the multiplexity of urban networks. *European Planning Studies*, 22(4), 816–840.

Burger, M. J., Meijers, E. J., & Van Oort, F. G. (2014c). The Development and Functioning of Regional Urban Systems. *Regional Studies*, 48(12), 1921–1925.

Champion, T., & Coombes, M. (2016). Is Pennine England becoming More Polycentric or More Centripetal? An Analysis of Commuting Flows in a Transforming Industrial Region, 1981–2001. Urban Transformations: Centres, Peripheries and Systems, 73.

Florida, R., T. Gulden & C. Mellander (2008) The rise of the mega-region. *Cambridge Journal of Regions, Economy and Society* 1: 459–476.

Flowerdew, R., & Aitkin, M. (1982). A method of fitting the gravity model based on the Poisson distribution. *Journal of Regional Science*, 22(2), 191–202.

De Goei, B., Burger, M.J., Van Oort, F.G. and Kitson, M. (2010) Functional polycentrism and urban network development in the Greater South East, United Kingdom: evidence from commuting patterns. *Regional Studies*, 44: 1149–1170.

Green, N. (2007) Functional polycentricity: a formal definition in terms of social network analysis. *Urban Studies*, 44: 2077–2103.

Gottmann, J. (1957) Megalopolis or the urbanization of the Northeastern Seaboard. *Economic Geography*, 33: 189–200.

Gottmann, J. (1961) *The Urbanized Northeastern Seaboard of the United States*. New York, NY: The Twentieth Century Fund.

Hall, P. (1997). The future of the metropolis and its form. *Regional Studies*, 31(3), 211–220.

Hanssens, H., Derudder, B., Van Aelst, S., & Witlox, F. (2014). Assessing the functional polycentricity of the mega-city-region of Central Belgium based on advanced producer service transaction links. *Regional Studies*, 48(12), 1939–1953.

Harris, C. D., & Ullman, E. L. (1945). The nature of cities. *The Annals of the American Academy of Political and Social Science*, 242(1), 7–17.

Howard, E. (1902) Garden Cities of Tomorrow. London: Swan Sonnenschein & Co.

Hoyler, M., Kloosterman, R. C., & Sokol, M. (2008). Polycentric puzzles–emerging mega-city regions seen through the lens of advanced producer services. *Regional Studies*, 42(8), 1055–1064.

Kloosterman, R.C. and Lambregts, B. (2001) Clustering of economic activities in polycentric urban regions: The case of the Randstad. *Urban Studies*, 38: 717–732.

Li, Y., and Phelps, N. (2018). Megalopolis unbound: Knowledge collaboration and functional polycentricity within and beyond the yangtze river delta region in china, 2014. *Urban Studies*, 55(2), 443–460.
Meijers, E.J. (2008) Measuring polycentricity and its promises. *European Planning Studies*, 16: 1313–1323.
Meijers, E. and Burger, M.J. (2010) Spatial structure and productivity in US metropolitan areas. *Environment and Planning A*, 42: 1383–1402.
Meijers, E., Hoogerbrugge, M., & Cardoso, R. (2018). Beyond polycentricity: Does stronger integration between cities in polycentric urban regions improve performance?. *Tijdschrift voor Economische en Sociale Geografie*, 109: 1–21.
Parr, J. (2004). The polycentric urban region: a closer inspection. *Regional Studies*, 38(3), 231–240.
Short, J.R. (2007). Liquid City: Megalopolis and the Contemporary Northeast. Washington, D.C.: Resouces for the Future.
Stein, C. (1964) Regional City. Baltimore, MD: Johns Hopkins University Press.
Van der Laan, L. (1998). Changing urban systems: an empirical analysis at two spatial levels. *Regional Studies*, 32(3), 235–247.
Van Oort, F., Burger, M., & Raspe, O. (2010). On the economic foundation of the urban network paradigm: Spatial integration, functional integration and economic complementarities within the Dutch Randstad. *Urban Studies*, 47(4), 725–748.
Vasanen, A. (2013). Spatial integration and functional balance in polycentric urban systems: A multi-scalar approach. *Tijdschrift voor Economische en Sociale Geografie*, 104(4), 410–425.
Veneri, P., & Burgalassi, D. (2012). Questioning polycentric development and its effects. Issues of definition and measurement for the Italian NUTS-2 regions. *European Planning Studies*, 20(6), 1017–1037.
Wright, F.L (1935) Broadacre city: a new community plan. *Architectural Record*, 77: 243–254.

Part VI
Humanistic and Multidisciplinary Commentaries

Urban Morphogenesis: Putting Mathematics in Its Place

Roger White

Abstract Mathematics is essential in a formal treatment of urban morphology because the complexity of urban form is best represented as a mathematical object, a fractal. Mathematics is equally important in models of morphogenesis. Nevertheless, mathematics in itself is incapable of representing morphogenetic processes and must be embedded in algorithms in order to capture the temporality of morphogenesis. In the case of straightforward processes of urban self-organisation, conventional simulation models are sufficient. However to capture the creativity of cities—their ability to create new agents and new types of agents with new rules of behaviour—unconventional algorithms are required, algorithms that alter themselves during execution. Ultimately algorithms, not mathematics, provide the natural language of morphogenesis.

1 Introduction

Cities are beautiful creations, whether from the street, the air or in our imagination. And morphology is the surface that presents that beauty to us. But why do we study cities? Often in order simply to understand these wonderful creations. But we also study them for more practical reasons—for example, to make planning more effective. For this we need to go deeper, to understand the processes that generate the city and its morphology: we need to understand urban morphogenesis.

Traditionally urban morphology was the subject of the mapmaker and the artist, with artists producing bird's eye views of entire towns, panoramic views from strategic vantage points, or street views; some artists, notably the *vedutisti*, specialised in representative street scenes. Morphogenesis was the province of the historian, who described the appearance of notable new buildings, significant changes to streets and public spaces, and town extensions. While this descriptive, humanistic approach is appealing in its realism and in the richness of its detail, it doesn't open the door

R. White (✉)
Memorial University of Newfoundland, St. John's, Canada
e-mail: roger@mun.ca

L. D'Acci (ed.), *The Mathematics of Urban Morphology*,
Modeling and Simulation in Science, Engineering and Technology,
https://doi.org/10.1007/978-3-030-12381-9_22

to an understanding of a generic morphogenetic process. Yet the very fact that we are comfortable applying the word "city" to settlements on five continents over a span of four millennia suggests that we believe there is a common theme, with variations of course, in the processes that generate these structures. This belief has led to the development of a range of scientific approaches to understanding the process of morphogenesis, and these approaches inevitably involve mathematics in one way or another.

2 Morphogenesis: Conventional Approaches

From a mathematical perspective, the morphology of both cities and urban systems is essentially fractal (e.g. Batty and Longley 1994; Frankhauser 1994; Thomas et al. 2008), whether in terms of the edge of the urbanised area, the land use pattern, the form of infrastructure networks or various scaling relationships. As a mathematical object, a fractal is the result of an iterated rule. In the world of natural and human systems, a fractal structure is the signature of a complex system—that is, a system that is the product of a *process* of self-organisation. In other words, time is involved, either implicitly or explicitly. This is reflected in the contributions in this volume: most are focussed either on the fractal nature of urban morphology, or on processes of urban morphogenesis, most of which generate fractal morphologies.

Scaling laws (e.g. those discussed by Bettencourt (2013) or West (2017)) are also an expression of the fractal nature of cities, though most of them do not describe spatial forms. These are typically derived statistically by regressing the characteristic of interest, such as energy use, against city size. They are not, of course, laws in the sense of the laws of physics, but rather robust statistical regularities that emerge as a result of a number of underlying processes interacting with each other. In this case, the mathematics, typically in the form of a power function, emerges as a descriptor of an empirical regularity. In the case of models developed to explain fractal morphologies, the mathematics is embedded in a simulation model—e.g. a negative exponential distance decay equation may be included in a dynamic transportation model—and the model then generates the fractal spatial structure, such as a fractal land use pattern. What is notable is that in both cases the mathematics, while essential, is contained in a non-mathematical structure—the data in the case of the scaling laws, and the algorithm in the case of the simulation models.

In contrast, approaches which are exclusively mathematical, where the mathematics is not subordinated to either data or an algorithm, essentially fail as models of morphogenesis. Key examples are the land use models derived from microeconomic theory—e.g. those of Alonso (1964), Muth (1969), Angel and Hyman (1976), Fujita (1989) or Papageorgiou (1990). In each case, in order to write equations which can be solved, radical simplifying assumptions must be made about the actors in the system—the individuals, businesses, and organisations that need land; and even then the equations can only be solved for a one-dimensional space. Thus, in spite of the fact that the mathematics is in some cases quite elaborate (e.g. triple integrals), the results are always simple: concentric bands of homogeneous land use with perfect radial symmetry. These models produce no fractal land use patterns. Furthermore, there is no representation of the *process* by which the patterns could be generated, and so the models cannot represent the growth of a city. In other words, they are exercises in mathematics that add little to our understanding of either morphology or morphogenesis. It is perhaps significant that no such models are represented among the contributions to this volume. On the other hand, mathematical models such as these can be valuable when embedded in simulation models (e.g. Caruso et al. 2007; White 1977; White et al. 2015), because they then become part of a *process*. Inside the simulation algorithm, the equations do not have to be solved because they are iterated instead, thus generating change from one time period to the next; and because they do not have to be solved, the restriction to one-dimensional space can be dropped. The result is that, as part of an algorithm, they can generate realistic land use patterns, and patterns that evolve over time. However, there is a deeper reason that mathematics as a stand-alone modelling technique is not always sufficient, and this limitation is not overcome by embedding the mathematics in a conventional simulation model. The reason is strong emergence.

3 Complexity: From Weak to Strong Emergence

Most urban simulation models, whether agent based (ABM) or based on cellular automata (CA) or other techniques, represent applications of the theory of complex self-organising systems. In this approach, order emerges in a bottom-up process from the interaction of the individual agents in an ABM or cells in a CA. The approach originates in physics, and it is entirely appropriate for many physical systems. For example, in the paradigmatic case of the self-organisation of a convection cell in a pot of water heated from below, when the temperature differential between the bottom and the top of the water column exceeds a threshold value, the random motions of the molecules are no longer sufficient to move the heat from the bottom to the top by conduction, and the trajectories of the molecules become biased in an upward direction in some regions and downward in others. This bias is self-reinforcing, so that convection cells develop, which are much more efficient at moving the heat up and out of the water. These convection cells are macro-scale patterns of movement in the water, which previously was homogeneous in that it exhibited only micro-scale, random motions of the water molecules. The threshold temperature where the convection cells begin to appear is a bifurcation point. This process was simulated at the level of individual molecules by Prigogine and Stengers (1984). Atmospheric circulation models are similar in principle, and generate large scale patterns such as the jet stream or hurricanes. In both cases, the large-scale circulation patterns that appear are instances of *weak emergence*. And in both cases, we know that the only agents involved are the relevant molecules or other particles, which behave in accordance with the laws of physics and chemistry, although in the case of the atmospheric circulation models we must aggregate to the level of cells in order to make the computations feasible.

Biological systems, however—and in particular human systems—are fundamentally different. They are not simply collections of particles interacting with each other in accordance with the laws of physics and chemistry, although they depend on those laws. As the biologist Rosen (1991) pointed out, all living systems contain models of themselves and the relevant parts of their environment, and act on the basis of those models. DNA is the medium of the most basic of these models, with neural systems providing another substrate, one that permits more flexible models. Human systems also include external models embodied in works of art, literature, scientific publications, and, of particular interest to us, computer programs. The fact that the models are part of the systems being modelled makes these systems self-referential, and self-reference generates problems such as paradox and contradiction in logic and mathematics. This is the fundamental reason that the biological and social sciences have always seemed somewhat problematic as science in the strict sense of the word, and the reason that mathematical models in these fields often seem somewhat inappropriate or inadequate. Furthermore, a structure that contains a model of itself can act on itself and its environment: it is an *agent*. Agency introduces top-down causation. The order in systems containing agents is not simply the result of bottom-up processes directed by physical laws; it is also influenced by the agents' models. This introduces intentionality into the ordering process.

In urban modelling, it is standard practice to treat human systems as if they were essentially physical ones. The behaviour generated by the intentions of individuals typically has strong statistical regularities, so behavioural rules can be used as if they were physical laws. This works because we are normally trying to model the emergence of spatial patterns, and in this context, human agents can reasonably be treated as particles. With this approach our models generate the patterns in a process of bottom-up self-organisation, a process strictly analogous to the appearance of convection cells in the pot of water.

Both urban morphology and convection cells are instances of *weak emergence*. What emerges is a spatial pattern in the particles or agents that were already there. No new agents or types of agents are generated in the model. Since in the case of urban morphology we are modelling the emergence of patterns, it would seem that models producing weak emergence are appropriate, and this is usually the case—especially when the modelling horizon is relatively short or the agents are aggregated and thus generalised, as when various retail agents, whether supermarket or art gallery, are collected into a category called "commercial" and treated as a single type. But in general, the most important cases of self-organisation in human systems are not instances of weak emergence.

Human systems generate not only new patterns, but new agents, and new *types* of agents with new types of rules. For example, a number of families and businesses individually deciding to settle in the same area will eventually produce a dense settlement which may have many of the characteristics of a city—economically and socially it will function as one. It will have a distinctive urban morphology, and we can model its emergence with a CA or an ABM. But the settlement as a whole is not an agent, because its behaviour is simply the aggregate of the interactions of the individual agents that compose it. This is again a case of weak emergence by self-organisation. However, at some point, the individual agents may decide to incorporate the settlement as a city. The city, as a legal entity with a number of powers (e.g. to issue regulations, pass laws, sign contracts) is now an *agent*, and a new kind of agent; it is different from the original person and business agents. This is a case of *strong emergence*. But what is the process by which it occurred? We can, of course, describe the individual events involved, as well as the enabling circumstances, but this would be a return to the descriptive approach of the historian. In other words, we are forced to step outside the modelling framework: in the framework of conventional modelling, the appearance of a new type of agent must be treated as an exogenous event.

Modelling strong emergence in a system is difficult, because the phenomenon represents the structural evolution of the system: the system ceases to be the system it was and becomes a new one. To model strong emergence requires that the simulation model re-program itself as it executes, so that the program itself evolves as it runs. In some cases, it is relatively straightforward to design such a model. For example, an agent-based model of the evolution of an economic system (Straatman et al. 2008) consists initially of a relatively small number of firms (firm agents) producing several products, some of which are used as inputs by other agents to produce other goods, while others are final consumer goods, purchased by people (person agents) who provide labour inputs. In this initial form, the dynamics quickly stabilises so that a stable pattern of production, consumption and prices emerges—a case of weak emergence. However, the model also includes a module that occasionally adds new agents, and some of these are different from existing agents: they use different inputs to produce an existing good (this represents a new process) or they produce a new product, which can subsequently be used by other new agents as an input to other novel processes or products. The system thus grows in size and complexity. However, because of constraints representing limitations on resources and the necessity for profitability, this growth is occasionally interrupted by crashes where firms, products, and processes are eliminated, with the result that over the long run the system is restructured and becomes larger as it becomes more efficient.

This is a very simple example, where strong emergence is always of a single kind (a firm using a different set of inputs to produce a new type of product). Nevertheless, the example makes clear the reason that strong emergence cannot be handled by a purely mathematical model, even though *at any particular iteration* the model looks mathematical. In essence, each agent is represented by an equation that shows the output of a product as a function of the inputs, and since the output of some agents is the input of others, the economic system as a whole is represented by the set of simultaneous equations. This is very similar to the input–output model widely used in economics. At this point, the system is purely mathematical. But from one iteration to the next, the number of equations in the system changes, as does the set of variables. Yet an equation, or a set of equations, cannot spontaneously generate a new variable, or generate a new equation to relate the new variable to existing variables, or alter existing equations to include the new variable. These actions can only be accomplished by an algorithm as it is executed step by step in a computer. Including a variable "t" to represent time, as is done in a conventional dynamical model, does not alter this situation. It allows us to model the dynamics of a particular system, but it does not enable us to model the evolution of that system as it continually transforms itself to become another system, thus generating a sequence of systems, each the product of the previous one. The deep reason that an algorithm can be creative by continually transforming itself into novel algorithms is that because it must be executed step by step on a computer, it embodies actual, natural time rather than a representation of time, and the future, in natural time, is open (White and Banzhaf 2019).

This is the fundamental limit of mathematics: a mathematical system cannot be creative because it is ultimately static and tautological. Mathematicians can be creative and invent new mathematics, and computers executing certain types of algorithms can be creative, as the executing algorithm alters itself to become another, unexpected algorithm; but mathematics is not and cannot be creative. Algorithms are the proper language of morphogenesis because they can be creative and thus capture the creative nature of cities.

References

Alonso, W. 1964. *Location and Land Use: Toward a General Theory of Land Rent*. Cambridge, MA: Harvard University Press.

Angel, S. and Hyman, G. 1976. *Urban Fields: A Geometry of Movement for Regional Science.* London: Pion Press.

Batty, M. and Longley, P. 1994. *Fractal Cities: A Geometry of Form and Function*. Academic Press.

Bettencourt, L. 2013. "The Origins of Scaling in Cities." *Science*, vol. 340, pp. 1438–1441.

Caruso, G., Peeters, D., Cavailhès, J. and Rounsevell, M. 2007. "Spatial Configurations in a Peri-urban City: A Cellular Automata-Based Microeconomic Model". *Regional Science and Urban Economics*, vol. 37, pp. 542–567.

Frankhauser, P. 1994. *La Fractilité des structures urbaines*. Paris: Anthropos.

Fujita, M. 1989. *Urban Economic Theory: Land Use and City Size*. Cambridge: Cambridge University Press.

Muth, R. 1969. *Cities and Housing: The Spatial Pattern of Urban Residential Land Use*. Chicago: University of Chicago Press.

Papageorgiou, Y. 1990. *The Isolated City State: An Economic Geography of Urban Spatial Structure*. London & New York: Routledge. 466 pp.

Prigogine, I. and Stengers, I. 1984. *Order out of Chaos: Man's New Dialogue with Nature*. New York: Bantam Books.

Rosen, R. 1991. *Life Itself: A Comprehensive Inquiry into the Nature, Origin, and Fabrication of Life*. New York NY: Columbia University Press.

Straatman, B., White, R. and Banzhaf, W. 2008. "An Artificial Chemistry-based Model of Economies". In Bullock, S., Noble, J., Watson, R. and Bedau, M., eds.: *Artificial Life XI: Proceedings of the Eleventh International Conference on the Simulation and Synthesis of Living Systems*. Cambridge, MA: MIT Press, pp. 592–599.

Thomas, I., Frankhauser, P. and Biernacki, C. 2008. " The Morphology of Built-Up Landscapes in Wallonia (Belgium): A Classification Using Fractal Indices". *Landscape and Urban Planning*, vol. 84, pp. 99–115.

West, G. 2017. Scale: *The Universal Laws of Growth, Innovation, Sustainability, and the Pace of Life in Organisms, Cities, Economies, and Companies*. Penguin Press.

White, R. 1977. "Dynamic Central Place Theory: Results of a Simulation Approach". *Geographical Analysis*, vol. 9, pp. 226–243.

White, R. and Banzhaf, W. 2019. "Putting Natural Time into Science". In A. Adamatzky and V. Kendon (eds) *From Astrophysics to Unconventional Computing: Essays presented to Susan Stepney on the occasion of her 60th birthday*. Springer (in press).

White, R., Engelen, G. and Uljee, I. 2015. *Modelling Cities and Regions as Complex Systems: From Theory to Planning Applications*. Cambridge, MA: MIT Press.

Not Only … But Also: Urban Mathematical Models *and* Urban Social Theory

Trevor Barnes

Abstract The chapter makes an argument that mathematical urban models should be joined with urban social theory. The two were separated historically was not because of any philosophical incommensurable difference but a result of internal sociological factors within the discipline of geography during the early 1970s. A model for integrating mathematical and social theoretical urban approaches is offered that draws on the work of the historian and philosopher of science, Peter Galison, and his idea of a trading zone.

1 Introduction

As an undergraduate studying geography at University College, London (UCL), 1975–1978, I received a schizophrenic education as it related to the urban. I took two courses in quantitative methods and modelling. I learned about the Garin-Lowry model (Garin 1966); applied the power law distribution to verify Zipf's Law; derived Alonso's (1964) concentric rings of urban land use; and read the relevant parts of August Lösch's (1954, 118) *Economics of Location* such that I could answer a question on my final exam about the number of whole settlements within any given market area. Even more extreme, I became so obsessed by the Box–Jenkins family of auto-correlation models that early on a Saturday morning I would leave my North London bed-sit to go to the Library at UCL's Bartlett School of Planning. There I would stare until my brain hurt at the seeming never-ending number of papers contained in the journal *Environment and Planning A*, and found only at that library, which used those models to represent mathematically contemporary urban and regional economies (see for example the series of papers by Bennett 1975a, b, c, d). I would continue reading those papers into late afternoon preferring to be with those winning ARMA equations than to be with my beloved winning Arsenal football team that

T. Barnes (✉)
Department of Geography, University of British Columbia, Vancouver, Canada
e-mail: tbarnes@geog.ubc.ca

L. D'Acci (ed.), *The Mathematics of Urban Morphology*,
Modeling and Simulation in Science, Engineering and Technology,
https://doi.org/10.1007/978-3-030-12381-9_23

played also on a Saturday afternoon not that far up the road at Highbury. Such was my devotion to mathematical modelling.

The other half of my urban geographical education was something completely different. It was non-mathematical urban social theory, mostly from the political left. It was Harvey (1973) on Marx, along with works by Ambrose and Colenutt (1975) on the economic injustices of *The Property Machine*, Castells (1977) on ideological contradictions embedded within *The Urban Question*, and Bunge (1971) on the many racial inequalities found in *Fitzgerald* (including a dark mirror version of Alonso's celebratory urban model but illustrating only deep-suboptimality for Detroit's black inner-city residents; Bunge 1971, 133–34). Along with these substantive urban materials, we were also taught what was bad about mathematical modelling, learning instead about the positive possibilities of other methods such as hermeneutics, phenomenology and especially the dialectic. Our touchstone here was not *Environment and Planning A*, but luridly coloured copies of *Antipode: A Radical Journal of Geography*. That journal was thought so seditious it was locked away behind the Geography Library counter in a special sturdy wooden cabinet. It could be read only at designated tables under the watchful eye of the beadle. Equally seditious, but even harder to access, was Olsson's (1975) Michigan Geographical Publication #15, *Birds in Eggs*. A post-graduate student once came to one of our classes bearing his dog-eared copy. He said he was there 'to disturb us' by reading from it. He did, but nowhere near as much as Olsson himself. He visited the UCL Geography Department in spring 1977, during my second year as an undergraduate. In his departmental seminar, he gave a brilliant triumphal performative rendering of the dialectic that he set against a regressive, status quo, power-laden mathematical equals sign (Barnes 2012). There was no doubt which side he was on.

Missing from my undergraduate curriculum, however, was how these two intellectual bodies of work related to one another. It appeared as if they lived in two separate solitudes. One could practice *either* urban mathematical modelling *or* urban qualitative social theory. Any middle ground was seemingly excluded. I believed that until I took a seminar at the end of my last year when we discussed the dyspeptic debate between Harvey (1972a) and Berry (1972) on how to do urban geography. On the surface that debate appeared as another illustration of the excluded middle: Berry maintained that doing urban geography meant using only rigorous mathematical models, while Harvey averred that the sole way forward was through deploying urban Marxist social theory along with its dialectical method. As we were picking sides to support someone said, drawing on the name of a well-known Peter Cook and Dudley Moore BBC tv comedy show running at the time, 'Why can't it be 'Not only … But also'?' Exactly, I thought, why can't it?

In the rest of this short essay, I would first like to explain how the supposed dualism between urban mathematical modelling and urban social theory initially emerged (and which I believe was for sociological rather than philosophical reasons); and second to make a case for collapsing that dualism and instead adhering to an approach of 'Not Only … But Also'. That is, to suggest that it is possible, and indeed desirable, to utilize both urban mathematical modelling *and* urban social theory.

2 Mathematics Modelling Versus Social Theory

To understand why the dualism between mathematical and non-mathematical approaches to urban geography emerged in the first place one must go back further in the story to the arrival of geography's quantitative revolution during the 1950s. That revolution began at selected sites in the United States, initially the University of Washington at Seattle, and the University of Iowa, Iowa City. Later it spread to the University of Chicago, University of Michigan, Northwestern University, and Ohio State, along with European centres at the Universities of Cambridge, Bristol, and earlier than both, Lund (Barnes 2004). Much of the focus of that early quantitative work was urban-economic and represented by writings, for example on central place theory, hierarchical urban diffusion and urban morphology. In this context, Brian Berry's work was especially notable. After he gained his Ph.D. from the University of Washington, Seattle, in 1958, he was appointed to the University of Chicago. In 1961, Berry became Director of The Center of Urban Studies that specialized in mathematical modelling of city form, particularly Chicago's.

Geography's quantitative revolution was part of a larger movement that occurred within US social sciences and humanities during the post-War period and generated by the Cold War and the rise of the military-industrial-academic complex (Barnes 2008). From early on, in that process the discipline of economics was front and centre. More than any other social science, it emphasized the importance of mathematical modelling. As an end, it had been pioneered in economics particularly by Samuelson (1947) in his signal text, *Foundations of Economic Analysis*. The volume was jam-packed with mathematical modelling techniques, making economics increasingly like a branch of physics. Samuelson's text was not politically innocent, though, but rested on neoclassical economic assumptions that acclaimed among other things individual rationality, the benefits of market competition, stable equilibrium, and dexterous movements of Adam Smith's invisible hand in realizing optimality. Furthermore, Samuelson was scornful about both those who failed to use mathematics, as well as those who sympathized with Marxist social theory along with its dialectical method. In his 1961 Presidential Address to the American Economics Association, Samuelson (1962, 12) snubbed Marx by calling him 'a minor post-Ricardian'.

When geographers first drew on mathematical models within the quantitative revolution they often drew on neoclassical economics (King 1979). Consequently, when social theory entered urban geography from the early 1970s, and represented by such works as Harvey's, Castells' and Bunge's, all of them Marxist-inspired, it was not surprising that it disparaged neoclassical economics. Furthermore, the criticism made of neoclassicism by social theorists was extended to any user of mathematics. All quantitative work found in urban geography was written off. This was perhaps clearest in David Harvey's writings. In 1972, he made a powerful statement against both neoclassicism *and* quantitative methods in an essay that appeared in *Antipode*, 'Revolutionary and counter-revolutionary theory in geography and the problem of ghetto formation' (Harvey 1972b).

His paper specifically attacked the economic basis of William Alonso's and Richard Muth's mathematical urban models 'built on neoclassical marginalist principles' (Harvey 1972b, 8). But Harvey also stretched his critique to apply to any user of quantitative methods in urban geography. As Harvey (1972b, 6) wrote:

> [Geography's] quantitative revolution has run its course and diminishing marginal returns are apparently setting in as … [it] serve[s] to tell us less and less about anything of great relevance…. There is a clear disparity between the sophisticated theoretical and methodological framework which we are using and our ability to say anything really meaningful about events as they unfold around us. … In short, our paradigm is not coping well. It is ripe for overthrow.

He further twisted the knife of criticism by also contending that formal theory was 'counter-revolutionary', which he defined as:

> A theory which may or may not appear grounded in the reality it seeks to portray, but which obscures, be-clouds, and generally obfuscates … our understanding of that reality. Such a theory is usually attractive (and hence gains currency) because it is logically coherent, easily manipulable, esthetically appealing, etc., but it is in fact divorced from the reality it purports to describe (Harvey 1972a, 41).

Here the die was cast creating an excluded middle. Harvey linked the 'logically coherent', the 'easily manipulable', and the 'esthetically appealing', that is, formal modelling, with irrelevance. In fact, worse than irrelevance because 'attention [was diverted] from fundamental issues to superficial or non-existent issues' (Harvey 1972b, 41). This is what made mathematical modelling a form of 'counter-revolutionary theory', and anathema for any self-respecting user of social theory.

As a logical argument, Harvey's contention goes as follows: neoclassicism is model-based and mathematical; social theory and particularly radical geography are against neoclassicism; therefore, social theory and radical geography are against mathematical models. Such a syllogism is logically flawed, however. It is an example of the fallacy of affirming the consequent. While the premises are right in that neoclassical economics is quantitative and mathematical, and social theory and radical geography are against neoclassicism, the conclusion does not follow. This is because mathematical modelling can be undertaken without necessarily invoking neoclassical economics.

On logical grounds, then, Harvey (and radical geography) was wrong to reject mathematical models, producing a false binary between urban mathematical modelling and urban social theory. Historically it was wrong too. If we go all the way back, Marx himself was keen to mathematize his work, learning differential calculus and undertaking complex arithmetical calculations to figure numerical values for prices and profit rates (Smolinski 1973). More recently, there have been also formal mathematical presentations of Marx's social theory including its application to the urban both as a system and as a single city entity. Indeed, after I left UCL to attend graduate school at the University of Minnesota I tried in one of my doctoral thesis chapters to join social theory *and* mathematical modelling. I brought together the Garin-Lowry model with Marx's theory of labour value and land rent with Sraffa's theory of joint production (Barnes 1983, ch. 3).

A pivotal moment in urban studies was when Harvey rejected the quantitative revolution and accepted 'revolutionary theory'. At that juncture, 'either/or' emerged in the study of the city: either urban mathematical modelling or urban social theory. On logical grounds, however, Harvey's argument was shaky, and later was contradicted by counter-examples including in my own work. Not that Harvey did anything underhand. In many ways, it was a brilliant manoeuvre. But it had larger consequences, producing a binary that was hard to shake, in fact stubbornly holding to the present.

3 From Either/or to not Only/but also

Consequently, while urban mathematical modelling and urban social theory are not intrinsically in opposition, they are rarely presented together. As Kwan (2004, 756) says, and who writes about the relationship of mathematical modelling and social theory more generally within human geography, the typical relation between the two sides of this apparent binary is 'mutual indifference and the absence of dialogue'. So, how can people be encouraged to speak to one another across this divide which though as I've argued is artificial acts no less as a real barrier?

My suggestion derives from science studies, the interdisciplinary body of literature concerned in part with how scientific knowledge is forged from divergent elements that enter it. Galison's (1998) work on 'trading zones' is especially germane. Set within the history of twentieth-century physics, it is based on his study of the interaction among different types of scientists concerned with the problems of detecting, recording and theorizing microparticles. In spite of these differences, however, those scientists managed to get along and work together, combining their insights as part of a single project.

The idea of the trading zone came from economic anthropology. It was used in that field to understand how exchange occurred between societies with very different values and meanings, objects of importance, and languages. Despite these dissimilarities, exchange still was able to proceed, fostered by the development of a pidgin trading language, an intermediary between the trading parties' parent languages. Neither fully dependent nor independent of those languages, pidgin laid betwixt and between both native languages (it was a hybrid). The development of a trading language was a first step in dialogue and gaining mutual interest between the different trading parties.

Galison (1998) works through the idea of a trading zone by studying three different groups of physicists: theoreticians, experimentalists, and instrumentalists. Each of them during the twentieth-century deal with a variety of issues—technological, experimental, metrological, mathematical and conceptual—as they endeavour to represent and explain microparticles. Furthermore, each hail from a different intellectual tradition, speaking a different language, mobilizing different methods, and holding different objects in esteem. Nonetheless, at specific moments in the research on microphysics, as Galison demonstrates, each group is willing to engage in trading, to improvise using pidgin, and to set aside its differences in order to work together with others on a common problem. Galison's (1998) prime example is the serendipitously established RadLab at MIT during the Second World War that engendered co-operation and trading among these three very different scientific partners. No superordinate principle emerged among the three groups of scientists as they worked together at the RadLab, no grand theory of everything. But as Galison shows, in the end, common purposes were achieved, and dazzling accomplishments realized.

My suggestion is for trading zones to be established between urban mathematical modellers, who are found in abundance within this book, and urban social theorists, who are barely represented. Of course, this not surprising. For almost the last half century, these two tribes of social scientists have led separate lives, inhabiting quite different camps, speaking different languages, although each studying the same entity, the city. Admittedly co-operation will be very difficult to foster. Much water has now passed under the bridge separating them. There is at best mutual suspicion, if not mutual disdain. But as Galison's study demonstrated, it is possible for quite different intellectual communities, with sometimes opposing epistemologies and ontologies, to work together, to produce results that are novel and creative, that literally change the world. To do so means setting aside prejudices and identifying particular issues and sites where conversation and trading might occur. Afterwards, the parties may go back to their respective corners, but the hope is that the very interaction makes a difference to both, producing excitement, inspiration and imagination, resulting in something new, although as yet undefined. Next time it might even produce a book called, *The Mathematics of Urban Morphology and Urban Social Theory*.

References

Alonso, W. 1964. *Location and Land Use*. Cambridge, MA: Harvard Press.

Ambrose, P. and Colenutt, B. 1975. *The Property Machine*. Harmondsworth: Penguin.

Barnes, T. J. 1983. *The Geography of Value, Production, and Distribution: Theoretical Economic Geography After Sraffa*. Unpublished Ph.D. dissertation, Department of Geography, University of Minnesota.

Barnes, T. J. 2004. Placing ideas: *Genius loci*, heterotopia, and geography's quantitative revolution. *Progress in Human Geography* 29, 565–595.

Barnes, T. J. 2008. Geography's *Underworld*: The military-industrial complex, mathematical modelling and the quantitative revolution. *Geoforum* 39, 3–16.

Barnes, T. J. 2012. Gunnar Olsson and me. In *GO: On the Geographies of Gunnar Olsson*, eds., C. Abrahamson and M. Gren. Aldershot: Ashgate, pp. 245–258.

Bennett, R. J. 1975a. Dynamic systems modelling of the North West region: 1. Spatio-temporal representation and identification. *Environment and Planning A* 7, 525–538.

Bennett, R. J. 1975b. Dynamic systems modelling of the North West region: 2. Estimation of the spatio-temporal policy model, *Environment and Planning A* 7, 539–566.

Bennett, R. J. 1975c. Dynamic systems modelling of the North West region: 3. Adaptive-parameter policy model, *Environment and Planning A* 7, 617–636.

Bennett, R. J. 1975d. Dynamic systems modelling of the North West region: 4. Adaptive spatio-temporal forecasts, *Environment and Planning A* 7, 887–898.

Berry, B. J. L. 1972. Revolutionary and counter-revolutionary theory in urban geography – a ghetto reply. *Antipode* 4 (2), 31–33.
Bunge, W. 1971. *Fitzgerald: Geography of a Revolution*. Chicago: Schenckman.
Castells, M. 1977. *The Urban Question: A Marxist Approach*. Translated from the French by Alan Sheridan. London: Edward Arnold.
Galison, P. 1998. *Image and Logic: A Material Culture of Microphysics*. Cambridge, MA: Harvard University Press.
Garin, R. A. 1966. A matrix formulation of the Lowry model for intrametropolitan activity allocation. *Journal of the Institute of American Planners* 32, 361–364.
Harvey, D, 1972a. A commentary on the comments. *Antipode* 4 (2), 36–41.
Harvey, D. 1972b. Revolutionary and counter-revolutionary theory in geography and the problem of ghetto formation. *Antipode* 4 (2), 1–13.
Harvey, D. 1973. *Social Justice and the City*. London: Edward Arnold.
King, L. J. 1979. On neoclassicism in economic geography theory. In *Retrospective 1951–76*, eds. L.-E. Hamelin, and L. Bauregard. Montreal: Canadian Association of Geographers, pp. 31–44.
Kwan, M.-P. 2004. Beyond difference: From canonical geography to hybrid geographies. *Annals of the Association of American Geographers* 94, 756–63.
Lösch, A. 1954. *Economics of Location*. Translated from the German second revised edition by W. H. Woglom with the assistance of W. F. Stolper. New Haven and London: Yale University Press.
Olsson, G. 1975. *Birds in Egg*. Michigan Geographical Publication #15. Ann Arbor: Department of Geography, University of Michigan.
Samuelson, P. A. 1947. *Foundations of Economic Analysis*. Cambridge, MA: Harvard University Press.
Samuelson, P. A. 1962. Economists and the history of ideas. *American Economic Review* 52, 1–18.
Smolinski, L. 1973. Karl Marx and mathematical economics. *The Journal of Political Economy* 81, 1189–1204.

Urban Morphology or Townscape? Wholes Made of Many Parts

Ron Johnston

Abstract Early fieldwork in a number of cities showed that townscapes—street patterns, plot sizes and building styles—rarely developed as planned wholes but rather as a congeries of mini-regions with separate characteristics inserted within the pre-urban cadaster, such as filed boundaries. Research led by Conzen, Ward, Whitehand and Larkham has developed this argument in considerable detail but there remain many lacunae for studies of emerging urban forms that could be undertaken using modern tchnology.

In my student days I had homes in Leeds and Manchester, two nineteenth-century industrial cities very different from the village in southern England where I had spent my first 18 years. As I moved through those two cities—on foot and by bus—I was intrigued by the micro-details of the townscape, how the streets didn't quite join up in many places and the building styles suddenly changed—things I could have noted in the town where I went to school (Swindon), but which were totally ignored. Then I moved to Australia and a very different city—Melbourne—but the same features stood out again: the townscape wasn't a whole but a series of independent parts, bolted together with no apparent attempt at unity.

All was revealed when I came across a paper by Ward (1962) which provided the answers I had been seeking. The key, in his terms, was the 'pre-urban cadaster': new pieces of urban fabric were added to the built-up area, not necessarily on its fringe, within the pre-existing pattern of landholdings. That pattern varied spatially in its nature, between the smallholdings into whose boundaries much of southern Leeds had been extended and the larger holdings—where I lived—to the north of the city. In many parts on the outskirts of south Leeds, almost all holdings in the mid-nineteenth century were of ten acres or less; in parts of north Leeds, on the other hand, as much as half of all holdings were of ten acres or more. The rows of terraced housing—many of them of the back-to-back variety—and the streets onto which they faced were inserted into many of those holdings. In some cases they were contained

R. Johnston (✉)
School of Geographical Sciences, University of Bristol, Bristol, England
e-mail: r.johnston@bristol.ac.uk

L. D'Acci (ed.), *The Mathematics of Urban Morphology*,
Modeling and Simulation in Science, Engineering and Technology,
https://doi.org/10.1007/978-3-030-12381-9_24

within the boundaries of individual fields, hence Ward's conclusion (Ward 1962, p. 166) that 'the location, the size and the shape of land holdings have been shown to have had a significant effect on the characteristics of the urban plan'—a situation that he claimed was representative of northern England's other industrial towns, and which my unstructured field observations in Manchester appeared to confirm (on which see Sutton 2013) as did a later paper by Mortimore (1969) on Bradford.

In Melbourne, my Ph.D. thesis was designed to bring together an appreciation of the city's social geography—based on the then emerging work on social area analysis—and elements of its townscape, notably the street pattern, building types and the distribution of shopping centres of varying size and composition (on building types, see Johnston 1969; on shopping centres, Johnston 1966a, 1968a). Much of the city's expansion occurred in the twentieth rather than the nineteenth century, but the same relationship between the pre-urban (albeit only relatively recently established) pattern of landholdings and the city's street pattern was clearly displayed. My paper on the city's streets (Johnston 1968b) clearly identified breaks in their alignment that coincided with one or both of parish and landholding boundaries, both in their orientation and the fact that, for example, north–south streets often didn't quite meet at those boundaries. Aerial photographs quite clearly identified where those boundaries had been. Furthermore, differing types of street plan—rectangular grids as against less regular patterns—were clearly associated with different types of social area; the city's social geography was apparent in those elements of its townscape.

By then I had also encountered Conzen's (1960) classic work on town-plan analysis, and seen how in some parts of a town areas that had initially been developed at relatively low densities were filled-in—what he termed repletion (something that Whitehand and Larkham 1991, later brought up-to-date in the British context). In a very different context, this is what I observed in parts of suburban Melbourne, notably those areas initially developed at very low densities for the city's upper-middle classes. By the mid-twentieth century, many of these individual properties in their large suburban estates were being abandoned and replaced by new developments at higher densities. In many cases, a new street (a cul-de-sac in a lot of them) was inserted to serve the smaller, but still relatively high status, homes and, especially, associated gardens in areas that changed their appearance but not their social standing. Mapping those redivisions of suburban space provided complementary evidence of the location of Melbourne's high-status residential areas (Johnston 1966b).

That was where my work on townscapes ended, and although I observed similar characteristics in my next home city—Christchurch, New Zealand—I made no formal study of them, though my arguments were confirmed when 'Con' Conzen visited for a year and I took him on field walks through various parts of the city's townscape. And although my next home—Sheffield—bore all of the characteristics I had observed in Leeds and Manchester a decade earlier, plus some contemporary developments charted elsewhere by Whitehand (2001), they did not attract my attention, apart from noting the uniformity that large public housing estates had imposed on the city, something that was not the case in New Zealand cities where, prior to the introduction of multi-storey flats a dominant feature was that no pair of adjacent dwellings should look the same from the street, even if the only variation was in the orientation of a basic house plan on the plot.

Townscape analysis, building on the foundations laid, in particular, by Conzen was the chosen subfield of a small group-based on the University of Birmingham, led by Jeremy Whitehand with, later, Peter Larkham (who has summarised much of the group's work—Larkham 2006). In a wide-ranging literature, they and their associates have identified many of the features of evolving townscapes, both historical and contemporary, and the agents involved in creating the parts which combined to make what in some cases was a rather chaotic whole. Modern planning has to some extent improved on that: new settlements have been planned as wholes—although most of them have had later extensions—and the layouts of new subdivisions have had to conform to planning policies with regard, for example, to street layouts and building styles: nevertheless, the separate parts are usually readily identified on maps and plans and in the field.

Whitehand's research has identified many of the separate components of these emerging townscapes. His early work on fringe belts, for example (Whitehand 1967), illustrated how certain land uses were placed at the edge of a built-up area during periods when expansion was slow—and such residential development that did take place was at relatively lower densities than at periods of rapid expansion (Whitehand 1975). When urban growth resumed, they often remained as somewhat anomalous elements within the burgeoning residential and commercial areas—cemeteries and public parks, for example, as well as allotments and playing fields. Some were later translated into other uses with, for example, residential areas inserted into the fringe belts just as had been the case on the edge of Leeds or in the infilling of low-density housing estates in Melbourne.

Townscapes—British in the case of Whitehand's research but certainly not confined to there—are thus composed of what he terms morphological periods (Whitehand and Carr 1999, 2001), characterised by their street patterns, their built forms and architectural styles (reflecting the different architects and designers involved), and the size and use of the plots on which the buildings are placed: most of them are residential areas, but commercial and industrial areas (including shopping and service centres) add further diversity to the scene. Townscapes, to use a term less popular these days than a few decades ago, are a congeries of regions—they look and feel different, and different types of people live in them.

Townscape description and analysis are now a very minor interest among urban and historical geographers, but the separate units identified in such work provide the frameworks within which the lives of modern city residents are structured: they both create and constrain their action and activity spaces.

And therein lies the challenge for new generations of researchers attracted by the wealth of big data now available and the software packages within which they can analyse, even visualise, them and undertake their work in what is now termed urban analytics. While much of their expertise might be expended using the available technology to evaluate a city's liveability—how easy is it to navigate its street system?—to understand how that structure has emerged requires analysis of how the street pattern and the built form within it have been created. Alongside research into the urban whole, attention could be focused on its many parts, that congeries of regions—some big, many tiny—that formed the building blocks within which agents created the myriad parts of complex, complicated, frequently chaotic wholes. And the construction of geocoded databases could appreciably assist in the development of what Whitehand (1975, 1977, 1992) frequently called for and made steps towards—theory building: as he noted early on (Whitehand 1975, p. 211) most research on urban form has been 'idiographic, providing little basis for either replication or further development'. The possibilities are surely endless, for comparative studies providing foundations to a wider appreciation of how urban forms have been formed.

References

Conzen, M. R. G. (1960) Alnwick, Northumberland: a study in town-plan analysis. *Transactions and Papers of the Institute of British Geographers*, 27.

Johnston, R. J. (1966a) The distribution of an intra-metropolitan central place hierarchy. *Australian Geographical Studies*, 4, 19–33.

Johnston, R. J. (1966b) The location of high status residential areas. *Geografiska Annaler*, 48B, 23–25.

Johnston, R. J. (1968a) Railways, urban growth, and central place patterns. *Tijdschrift voor Economische en Sociale Geografie*, 59, 33–41.

Johnston, R. J. (1968b) An outline of the development of Melbourne's street pattern. *The Australian Geographer*, 10, 453–465.

Johnston, R. J. (1969) Towards an analytical study of the townscape: the residential building fabric. *Geografiska Annaler*, 50B, 2–32.

Larkham, P. J. (2006) The study of urban form in Great Britain. *Urban Morphology*, 10 (2), 117–141.

Mortimore, M. J. (1969) Landownership and urban growth in Bradford and its environs in the West Riding conurbation, 1850–1950. *Transactions and Papers of the Institute of British Geographers*, 46, 105-119.

Sutton, K. (2013) Vestiges of the pre-urban landscape in the suburban geography of south Manchester. *Manchester Geographies*, 38–46 - https://www.mangeogsoc.org.uk/manchestergeographies.htm.

Ward, D. (1962) The pre-urban cadaster and the urban pattern of Leeds. *Annals of the Association of American Geographers*, 52, 152–166.

Whitehand, J. W. R. (1967) Fringe belts: a neglected aspect of urban geography. *Transactions and Papers, Institute of British Geographers*, 41, 223–233.

Whitehand, J. W. R. (1975) Building activity and intensity of development at the urban fringe: the case of a London suburb in the nineteenth century. *Journal of Historical Geography*, 1, 211–224.

Whitehand, J. W. R. (1977) The basis for an historico-geographical theory of urban form. *Transactions and Papers, Institute of British Geographers*, 56, 39–55.

Whitehand, J. W. R. (1992) Recent advances in urban morphology. *Urban Studies*, 29, 617–634.

Whitehand, J. W. R. (2001) Changing suburban landscapes at the microscale. *Tijdschrift voor Economische en Sociale Geografie*, 92, 164–184.

Whitehand, J. W. R. and Carr, C. M. (1999) Morphological periods, planning and reality: the case of England's inter-war suburbs. *Urban History*, 26, 230–248.

Whitehand, J. W. R. and Carr, C. M. (2001) The creators of England's inter-war suburbs. *Urban History*, 28, 218–234.

Whitehand, J. W. R. and Larkham, P. J. (1991) Housebuilding in the back garden: reshaping suburban townscapes in the Midlands and southeast England. *Area*, 22, 57–65.

Extending Urban Morphology: Drawing Together Quantitative and Qualitative Approaches

Peter J. Larkham

Abstract This chapter explores the complex and varied nature of 'urban morphology' and how a traditional, largely qualitative approach could be extended by the range of current and potential quantitative techniques to form a 'new urban morphology'. The problems and advantages of innovative and interdisciplinary research are explored. The potential of such a 'new urban morphology' is in developing more robust and comprehensive approaches to understanding urban form, incorporating the multiple dimensions of scale, dynamics, production, and consumption.

1 The Nature of Urban Morphology

Cities are increasingly large and complex artifacts. It seems fundamental and was recognised by Bobek as early as 1927,[1] that urban morphological research should recognise the essential trinity of function, form and change through time: cities are artifacts in both space and time, and they inevitably change. Venerandi et al. (2017, p. 1057) note that, amongst built environment disciplines, urban morphology has 'peculiarly' made change a central focus. Skeates (1997, p. 8) writes of 'the formless mess that is produced by urbanisation', and Crang (2000, p. 304) that 'the city is becoming an all-encompassing infinite space … the urban arena is made of fragmented spaces and pathic objects without a coherent totality'. If we consider Shanghai, one of the world's largest cities (24.15 million, area 6.340 km^2 and density 3,854 persons/km^2: http://www.worldpopulationreview.com) then the sheer scale and complexity of the urban object are made unmissably clear (Fig. 1). The world's population, and the rate of urbanisation, continue to grow; the problem will continue to become more intractable. Yet there is a great range of tools to help deal with urban scale and complexity, particularly of urban form. These have been developed

[1] This was Bobek's first methodological paper, written at the age of 23.

P. J. Larkham (✉)
Birmingham City University, Birmingham, England
e-mail: peter.larkham@bcu.ac.uk

L. D'Acci (ed.), *The Mathematics of Urban Morphology*,
Modeling and Simulation in Science, Engineering and Technology,
https://doi.org/10.1007/978-3-030-12381-9_25

Fig. 1 Aerial view of Shanghai (radics.geza, Creative Commons)

piecemeal over several decades and draw from a wide range of disciplines, research traditions and philosophies. At a time when interdisciplinarity is being heavily promoted as the new dominant research paradigm, it is timely to explore how some of these traditions could work better together, and some of the issues arising for urban morphology.

There are, as might be expected, confusions in 'urban morphology'. If 'morphology' is 'the study of form' as its Greek suffix implies and as is used in disciplines such as biology (Thompson 1917) and medicine,[2] then 'urban morphology' is 'the study of urban form'. However, 'morphology' is commonly used in the urban context to mean 'form': although technically misleading (Larkham 2002), this has now become generally accepted—including amongst urban morphologists. There are many other examples that could be interpreted as misuse of terms. Does this lack of precision hinder our understanding or facilitate discourse? Some have long sought precision in terms (Conzen 1969), though there is a danger of producing 'profuse nomenclature with little meaning', for which some early work was criticised (Whitehand 1981, p. 4). Amongst two disciplinary users of urban morphology, Samuels (1990, pp. 433–434) noted,

> if architects seem to use morphological terms loosely, without precision, then geographers must be tolerant. It is one of the attractions of the nexus of concepts, ideas and approaches that occupy the field of urban morphology that they are capable of being appropriated for use by different professions in different contexts who seek to use them for their own purpose

[2]Note the volume of medical citations when searching Google Scholar for 'morphology'.

> … [Morphology] is open to approach by various disciplines with their own methods and any attempt to restrict or straight-jacket the discourse could stifle it.

Much early work on urban form was essentially descriptive and classificatory—morphography (Fritz 1894; Dickinson 1934), even sometimes a spatial art history (*Kunstgeographie*) (Shaefer 1928). Topography was recognised as important in early German work, leading to its use especially where traditional documentation was absent or incomplete—an approach called 'urban constitutional topography' (*städtische Verfassungstopographie*) (Frölich 1938). A developmental perspective was also incorporated—a morphogenetic approach, recognising the detailed town plan as, in essence, a historical document in its own right (cf. Strahm 1950). Work in recent decades has substantially extended this from the more descriptive aspects of form to a critical exploration of the processes (agents and agency, cf. Larkham and Conzen 2014) that create, shape and reshape the physical form itself. For, as Pesaresi and Bianchin (2001, p. 56) state, urban form is 'the physical appearance of social reality'. In fact, much of the work in the tradition that has promoted urban morphological research, founded the International Seminar on Urban Form and its journal, *Urban Morphology*, has been broadly historico-geographical (cf. Whitehand 1977) and historico-architectural (the Italian process-typological tradition, cf. Cataldi 2003). An important strand of French work, again largely historical, has proceeded albeit with less direct contribution to ISUF in recent years (Panerai et al. 1997; Moudon 1997 identified these three national 'schools'). Although there have been suggestions that historical work has dominated the activities and output of ISUF, this historical dimension is now seen as significant in informing a range of public policy from urban design to planning for urban resilience: 'a rigorous approach towards analysing urban form from an *evolutionary* perspective is now, more than ever, relevant in interpreting its future trajectories' (Verenadi et al. 2017, p. 1057, citing Batty 2009). And a range of other work has emerged, from disciplines and researchers that may well not consider themselves as 'urban morphologists', including mathematics, computer science, remote sensing, climatology, and others. The relationship between urban form and design has become evident in the professions of urban planning and urban design—for example being identified as one of the significant areas of knowledge contributing to 'what an urban designer should know' (Moudon 1992).

Given the variations in 'morphology' and its constituent national and disciplinary parts, it is also pertinent to examine 'what is an urban morphologist' (Ley 2012). This is not as obvious as it may seem. Those who study and publish on urban form come from so many distinct disciplines, and publish in so many journals, that it may indeed be perceived as too disparate, not a unified field of knowledge at all. Certainly, some papers do not explore urban form per se, but rather the influence of urban form on variables such as urban microclimate, energy use, behaviour, and the long list goes on. Yet these works can shed welcome light, even if indirectly, on urban form, and the complex interrelationships between such factors, the physical form itself, and the means of production and consumption. 'Urban morphology gains much, especially methodologically, by encompassing so many kinds of researchers. Indeed it may be seen as a vanguard scientific field, in which interdisciplinary and transnational work was characteristic long before it became fashionable more widely' (Ley 2012, p. 78).

But is urban morphology a science? Much of its work seems rooted in the qualitative approaches of the arts and humanities. Yet there seems to have been a widespread move since the mid-twentieth century at least for such disciplines to seek to adopt the mantle of science, hence the rise of 'social science'.[3] Developments in science and technology are boosted by war (see the rise in patents immediately after 1945: Dienner 1963) and their perceived success and predictability[4] leads to these more quantitative approaches, based on measurement and replicability, being used for resource allocation. The *Oxford English Dictionary* definition of science is 'the intellectual and practical activity encompassing the systematic study of the structure and behaviour of the physical and natural world through observation and experiment'. If one broadens the definition of 'observation' to the systematic collection of data,[5] most morphology would be included. Experimenting, however, is a different matter: the expense of urban experimentation would largely relegate this to the virtual sphere, or to the ultimate in blue-sky thinking, space exploration (Millward 1979, see pp. 119–120; Greason 2011). A further suggested definition of 'science' is 'a process of constructing predictive conceptual models' (Gilbert 1991, p. 73). Much urban morphological work is concept-rich (see Whitehand's discussion of M. R. G. Conzen's historico-geographical study of Alnwick: Whitehand 1981, pp. 15–16). Some of these concepts have been of enduring and international relevance; for example, the fringe-belt concept (the *stadtrandzonen* of the geomorphologist Louis (1936) was developed by Conzen in his Alnwick study (1960), its economic grounding explored by Whitehand (1972) and their international relevance discussed by Conzen (2009), and described by Openshaw (1974, p. 10)[6] as perhaps 'the most important development in urban morphology so far'). Yet few of the concepts have resulted in 'models', fewer still have been 'predictive'. It is in the recent, more quantitative, applications that an extended urban morphology would seem to fit this definition.

2 Contributions to a 'New Urban Morphology'?

In exploring a possible 'new urban morphology' I am thinking of the parallel with geography half a century ago, and the impact of the 'quantitative revolution' (cf. Haggett 1967)—though, ideally, without the negative connotations of an artificial 'old/new' dichotomy, for there is still much of value on traditional, qualitative, critical

[3]By which I mean more than sociology; though settlements are social constructs hence sociology has a place, probably under-recognised, in urban morphology.

[4]Though military history shows their frequent fallibility.

[5]Though 'observation', physical engagement in the field, is a valuable constituent of much morphological research (Larkham 2018).

[6]Interestingly given the focus of this chapter, Stan Openshaw moved from explicitly urban morphological research to computation, automated geographical analysis tools, GIS, artificial intelligence and fuzzy logic applications.

urban morphology. To paraphrase Kwan and Schwanen (2009, p. 283), 'antagonism between critical and quantitative [morphologies] is not beneficial to the discipline … quantitative [morphology], when integrated with a critical sensibility and used appropriately, can be a powerful tool'. Some have suggested that 'there is still no established agreement on a method for the analysis of urban form' (Venerandi et al. 2017, p. 1057) and that there is an evident lack of a quantitatively rigorous, comprehensive and systematic framework for the analysis of urban form (Dibble 2016). The range of quantitative approaches now being developed and used in morphological contexts is likely to move us closer to this goal, simultaneously facilitating complex and meaningful comparative study. Yet it is the *interpretation* of such findings that will tell us more about the interrelationship between cities, form and function; so the following discussion explores a (very selective) range of applications rather than the techniques themselves.

First, it should be acknowledged that quantification, in terms of measurement, has been central to 'traditional' morphological approaches such as metrology at the plot (Sheppard 1974; Slater 1981) and block (Siksna 1997) scales. But surely we need to move beyond 'quantifying and describing' urban landscapes (Civco et al. 2002), although that is a necessary starting point. Understanding the geometric properties of urban forms can contribute to understanding—even if, in the absence of documentation, this is by inference—the processes of shaping the urban form. Hence we can differentiate areas of distinct form characteristics, and infer that these plan units were laid out at different times (Baker and Slater 1992). Measurement remains a popular activity (cf. Fleischmann 2018), and there are multiple-measure approaches that seek to provide a better understanding of urban physical form (for example, the seven-variable '*Morpho*' approach of Oliveira and Medeiros 2016).

However, one aspect that quantification and computational approaches facilitate is modelling. 'Computers act as the laboratory for experimentation on phenomena which is represented digitally with its manipulation being virtual' (Batty 1976). Yet the gap between urban morphology and urban modelling identified several years ago (Stanilov 2010) remains. At the larger scale Maske et al. (1998) model urban growth patterns, and their model offers the possibility of predicting the global properties (such as scaling behaviour) of urban forms. Many others have explored urban growth patterns using various modelling approaches such as cellular automata (He et al. 2008; Barredo and Demicheli 2003) and similar approaches can be applied to urban decline (Banzhaf et al. 2007). Likewise, land-use change can be modelled (Al-shalabi et al. 2013). With a focus on just one component of urban form, Barthélemy and Flammini (2008) model urban street patterns. Thompson's 'laws of growth' and 'theory of transformations' (1917), a fundamental approach to morphology, is of enduring value[7] wherein computer modelling is held to be 'growing today, essential tomorrow': Sharpe 2017). It seems likely that this is equally true for *urban* morphology.

[7]See the special issue of the journal *Development* (2017, vol. 144, no. 23).

Ideas of emergence and applications of complexity theory, the latter often expressed through fractal analysis, have been applied to the complex, adaptive systems of cities and their varied and changing forms. But

> complexity implies unpredictability, relativism of a kind that classical science and indeed social science find hard to deal with. It is further compounded by the fact that most complex systems span the physical – human divide, uniting the two cultures, thus opening up a veritable Pandora's box involving free will and self-determination. Cast within an evolutionary framework in which systems like cities, economies, and societies evolve from the bottom up, this implies that their future is unknowable, hence unpredictable (Batty 2009, p. 955).

This does rather contrast with the many modelling studies which have claimed to be predictive.

At the largest spatial scale, computational approaches and 'big data' allow representation and modelling of urban patterns at the city, region, national and even international scales. Much of the modelling work discussed below has been at this larger spatial scale. This raises the issue of the scale of 'traditional' urban morphology, suggested as the small scale of 'ordinary urban components (streets, blocks, plots, buildings): 'it is the interest on this *scale* which distinguishes the tradition of Urban Morphology from others which … have predominantly observed cities at a much larger scale' (Venerandi et al. 2017, p. 1057, their emphasis). The terms 'macromorphology' and 'micromorphology' that have been used in studies of urban form (e.g. Cui et al. 2013; Chen 2012) seem to have been used rather loosely, without specific delimitation of scale. Kropf (2014) also emphasises the smaller scale of spaces within buildings and elements of architectural form, especially in the context of parts and wholes. But urban morphology per se has scarcely looked at wider scales, the urban as a whole, or 'town and country' or 'urban and regional' planning (or, more recently in Europe, spatial planning).

At a smaller scale, Venerandi et al. (2017) develop a systematic and quantitative (though, as they say, 'not yet comprehensive') method of analysis of the form of places (five areas in London) which have undergone significant change in both physical form and social composition at various times since *c.* 1945. This allows them to identify links between form and social change at the neighbourhood scale—and scale emerges as a significant aspect. This, therefore, becomes a study of gentrification, and its physical consequences. The physical variables measured are: street edge, building footprint, density, street width, coverage ratio, centrality, front height and built front ratio. They demonstrate that 'features of '*traditional*', fine-grained, perimeter block-based urban form' are clearly correlated with gentrification, quantitative evidence for qualitative discussions of theorists such as Ruth Glass and Jane Jacobs. Interestingly they use street width as part of a measure of 'centrality' ('gentrified areas are found to sit *between* urban main streets, which constitute their boundaries', p. 1070). Street widths in built-up areas are difficult to change, although the nature of the surface and its use could be changed (the Dutch *woonerf* concept, for example); and other smaller scale physical changes could equally link with social change (for example the journalist Jonathan Raban characterised gentrification in the 1960s as 'the knockers-through are here'—creating larger living spaces by knocking out partition walls: Moran 2007). So Venerandi et al. do provide a useful morphometric approach, but are careful to state that their results do not imply 'any causal or universal relationship between morphological and social dynamics' (p. 1956). A qualitative approach could unpack causality.

The variability/homogeneity of an urban fabric is often an issue for planning decision-making, especially in the context of conservation. How much change will a building alteration or insertion make to an urban fabric, particularly if it is felt desirable to preserve certain characteristics of the area? Quantitative techniques can produce precise responses, replicable in different areas, potentially allowing replicability (better known as consistency) in decision-making. Hijazi et al. (2017) use a GIS-based method to analyse data extracted from 2D building footprints obtained from open source data (OpenStreetMap), which they applied in Zurich in areas 'with obvious discrepancies in their spatial configurations' (p. 1107). Measuring building edge angles and footprint areas, they can identify patterns that could characterise 'historical quarters with denser and smaller buildings [which] tend to have more organic urban fabric with winding street systems' (p. 1117). Different cities could, of course, have more rectilinear historic quarters. They suggest that their system 'could be trained to analyze historical quarters and learn from past examples about how to achieve or preserve the organic arrangement of the urban fabric. This measure can also be used as a goal function for a computational synthesis method … to automatically create new spatial configurations with a defined level of homogeneity' (p. 1118). Other methods of calculating homogeneity have been developed; Haghani (2009), for example, uses a fractal analysis approach, calculating a 'fingerprint' for specific areas from air photograph data. With a sequence of air photographs spanning several decades he can measure the extent of change over time. The procedure is a little complex, though, and the measure of fractal complexity is affected by, for example, tree foliage cover and is therefore seasonal. Rashed et al. (2006), in contrast, use 'a 'soft' approach … to identify and measure the composition of changing morphology from multi-temporal, multi-spectral satellite images … capable of deriving spatially continuous variables quantified at the sub-pixel level'. Hijazi et al. recognise that their approach could be used 'to analyze correlations between people's emotional responses to urban environments with the measured homogeneity levels of the corresponding spatial configuration' (p. 1118).

A wide range of studies attempts to make links between urban form and other variables. Xiao et al. (2016) relate urban form (in terms of urban configuration, specifically 'network accessibility matrices' to property prices in Cardiff. León and March (2016) use urban configuration models and GIS to explore evacuation routes and sheltering for tsunami evacuation in Iquique. Rose et al. (2014) relate urban form ('dominant residential building typologies') to residential heat-energy demand in Paris, London, Berlin and Istanbul. Rybarczyk and Wu (2014) use a discrete choice model to explore how micro-morphological features affect bicycle mode choice decisions. Much research of this type demonstrates correlations between form and other variables, but are these causative links?

Typological approaches have been extended by more quantitative, multi-variable approaches. The architectural building typology of Markus (1993) and the typomorphological studies of the Italian school, both of which related the historical production of spaces in buildings to people and decision-making, have been expended in computational studies, for example, to archetypal buildings, streets and block layouts (Steadman and Marshall 2005) and to urban-scale typomorphological models (Shayesteh and Steadman 2015). Shayesteh and Steadman's model of Tehran 'approximates reality' and, as with other quantitative approaches, is held to be able 'to generate various options and test what the overall built form would look like and how it would perform', thus providing 'a powerful planning tool' (p. 1145).

Space syntax has developed into a popular analytical approach for a variety of purposes, incidentally generating lively international conferences and an online journal. It seeks to explain 'how cities work—how space, movement, land uses, human activity and psychology combine to create the complex forms we occupy and experience' (Hillier 2016, p. 199). For example, Stöger (2015) generates 'novel insights ... regarding the physical environment in which Roman city dwellers lived their daily lives' (p. 61), thus demonstrating the relevance of space syntax analysis to historical urban forms where archaeological data allows. Peponis et al. (2015) explore supergrids in a range of cities. Both of these papers explore concerns central to qualitative or 'traditional' morphology in new ways. Amongst other things, space syntax depicts streets not as spaces but as networks, a useful additional perspective to the 'traditional' morphology. Di Bella et al. (2017) relate spatial configuration to urban crime environments using space syntax, typical of a large number of examples of 'applied' morphology using this approach. More broadly, Hillier (2016) challenges space syntax to explore interrelationships between spatial and social networks, posing the fundamental question of 'what are cities *for*?'.

All these are indeed new perspectives on urban form that qualitative research could not begin to explore. However, science and technology (i.e. quantification) are not normally concerned with how physical phenomena interact to create the varied urban environments in which people live (Hägerstrand 1991) and, over much of the post-war period, fields of study have tended to fragment and specialise rather than interact and integrate.

3 Conclusion: Unhelpful Dichotomies of Old and New, Quantitative and Qualitative—Moving Forward

> ... the importance of knowledge is increasingly being recognized in society today. We are living in a society which is as much characterized by the production of knowledge as anything else. Never before have there been so many systems of meaning requiring institutionalized cultures of experts and their professional discourses (Delanty 2005, p. 5).

Overall, despite the range of approaches discussed here (and there are more that could have been included), there is a tension between 'old' and 'new' urban morphology, and/or between disciplines, that still needs to be overcome. Ley (2012, p. 79), for example, states that 'there is not a satisfactory simple categorization or evaluation of cities. To attempt to render them in such ways is bound to end in shortcomings. It is necessary to set out clearly the scientific aim of the endeavour and make clear that the categories and criteria involved are inherent in the study rather than the study object … urban form cannot be satisfactorily reduced to numbers. Even a combination of various arithmetic or statistical parameters will not reflect its complexity'. I suggest that they *are* a reflection of complexity, albeit simplified; and these approaches are sufficient for the purposes for which they are designed. It is the *nature* of the approximation of reality (cf. Shayesteh and Steadman 2015) that causes concern—we need to be wary of over-claiming and over-worrying. We might also think of how *parts* can be used to reconstruct *wholes* (for a commentary, see Batty 2017). In particular, in terms of further developing the study of urban form, any either/or depictions (old/new, quantitative/qualitative) are unhelpful, even if used for rhetorical impact: they tend to fossilise established institutionalised cultures and diminish opportunities for innovation.

Interdisciplinarity has increased in many research fields especially in the past couple of decades, sometimes deliberately promoted by the allocation policies and priorities of funding bodies. The argument is that this is 'a means to address complex problems that cannot be dealt with from a single disciplinary perspective alone' (Bridle et al. 2013) and the research results are more innovative and have higher impact. An interesting aspect is whether the collaborating disciplines are distant (e.g. natural and social sciences; perhaps even quantitative and qualitative) or close (Morillo et al. 2003). However, the benefits of interdisciplinary research have been questionable, perhaps because the concept is ambiguous, because of the range of perspectives and of potential results (Huutoniemi et al. 2010). It does appear that a degree of interdisciplinarity can produce higher citation impact, while distant disciplinary collaboration may be perceived as risky and more likely to fail. There is a risk that 'scientific audiences are reluctant to cite heterodox papers that mix highly disparate bodies of knowledge—thus giving less credit to publications that are too groundbreaking or challenging' (Yegros-Yegros et al. 2015, p. 1). Looking at the institutional affiliations of authors of recent morphological papers (broadly based, including those cited here), interdisciplinarity seems relatively limited and close (e.g. between built environment disciplines and geography or history). Clearly, wider collaborations could be explored although the rationales for collaborations need to be clear from the outset. Morphologists might also seek to read more widely from other disciplines. We should be seeking to transgress disciplinary boundaries, rendering them more permeable, synthesising disciplinary knowledge in new ways rather than simply 'bolting on' new perspectives (see Friman 2010, p. 6).

One of the problems of multiple disciplines working in isolation on the same topic is that of communication. In interwar Germany, for example, the related work by architects and planning historians remained largely unknown to geographers although they were all working on aspects of urban form in the same language; this 'had a stultifying effect on plan morphology' (Whitehand 1981, p. 6). The problem of academic fashion and paradigm shift may be equally serious: the wide publication (especially, though belatedly, in English) of Christaller's central place theory 'presented urban geography with perhaps its most potent functional research model … that was ultimately to bring about a major change of course in the mainstream of urban geography during and at the end of the Second World War, involving the relegation of urban morphology to a comparatively minor role' (Whitehand 1981, p. 7). Today, the very range of disciplines involved, with their own terminologies and approaches, suggests that we should beware of mis- and lack of communication (see the example in Ley 2012, p. 79). At the very least, and to reiterate the point, morphologists may need to read more widely, out of the 'comfort zone' of their home discipline's journals.

There has been some positive evidence of comparing disciplinary approaches, or at least 'schools' (of which Kropf (1993) is probably the most detailed), combining/contrasting them (see Cataldi 2013) and combining quantitative tools (Jiang and Claramunt 2002; Ye and van Nes 2014). All appear to have been constructive and productive, but much more could be done in terms of creative interdisciplinary work.

A key, perhaps *the* most important, aspect of combining quantitative and qualitative approaches in urban morphology would be the ability to develop more robust, more comprehensive approaches to understanding urban form; encompassing its multiple dimensions, dynamics, means of production and consumption. We need to better integrate the components of the urban landscape which have often been treated in isolation by different, specialist, approaches.

Finally, quantitative and qualitative urban morphology has developed a wide range of concepts, models and tools that researchers have suggested could be useful in urban decision-making at a range of scales. Yet it is clear that few decision-makers on the ground engage closely with such research products (cf. Whitehand and Morton 2004). Urban morphology *as a whole*, though scarcely a unified discipline, would do well to explore the barriers to wider practical application of the results of urban morphological research.

References

Al-shalabi M, Billa L, Pradhan B, Mansor S, Al-Sharif AA (2013) Modelling urban growth evolution and land-use changes using GIS based cellular automata and SLEUTH models: the case of Sana'a metropolitan city, Yemen. Environmental Earth Sciences 70: 425–437

Baker NJ, Slater TR (1992) Morphological regions in English medieval towns. In: Whitehand, JWR, Larkham PJ (eds) Urban landscapes: international perspectives. Routledge, London

Banzhaf E, Kindler A, Haase D (2007) Monitoring, mapping and modelling urban decline: a multi-scale approach for Leipzig, Germany. EARSeL eProceedings 6: 101–114

Barredo JI, Demicheli L (2003) Urban sustainability in developing countries' megacities: modelling and predicting future urban growth in Lagos. Cities 20: 297–310

Barthélemy M, Flammini A (2008) Modeling urban street patterns. Physical Review Letters 100: 138702

Batty M (1976) Urban modelling. Cambridge University Press, Cambridge

Batty M (2009) Darwin at 200 and the evolution of planning. Environment and Planning: Planning and Design 36: 954–955

Batty M (2017) Benedikt's challenge: reconstructing the whole from the parts. Environment and Planning B: Urban Analytics and City Science 44: 395–397

Bella E di, Leporatti L, Persico L (2017) The spatial configuration of urban crime environments and statistical modelling. Environment and Planning B: Urban Analytics and City Science 44: 647–667

Bobek H (1927) Grundfragen der Stadtgeographie. Geographische Anzeiger 28: 213–224.

Bridle H, Vrieling A, Cardillo M, Araya Y, Hinojosa L (2013) Preparing for an interdisciplinary future: a perspective from early-career researchers. Futures 53: 22–32

Cataldi G (2003) From Muratori to Caniggia: the origins and development of the Italian school of design typology. Urban Morphology 7: 19–34

Cataldi G (2013) Thinking about Alnwick's origins. Urban Morphology 17: 125–128

Chen F (2012) Interpreting urban micromorphology in China: case studies from Suzhou. Urban Morphology 16: 133–148

Civco DL, Hurd JD, Wilson EH, Arnold CL, Prisloe M (2002) Quantifying and describing urbanizing landscapes in the Northeast United States. Photogrammetric Engineering and Remote Sensing 68: 1083–1090

Conzen MP (2009) How cities internalize their former urban fringes: a cross-cultural comparison. Urban Morphology 13: 29–54

Conzen MRG (1960) Alnwick, Northumberland: a study in town-plan analysis. Institute of British Geographers Publication 27. George Philip, London

Conzen MRG (1969) Glossary. In: Conzen MRG Alnwick, Northumberland: a study in town-plan analysis. Publication 27 (2nd edn) Institute of British Geographers, London

Crang M (2000) Urban morphology and the shaping of the transmissible city. *City* 4: 303–315

Cui J, Cui H, Duan Y (2013) The macro morphology evolution analysis of Jiangcheng town in Yunnan Province. Advanced Materials Research 671: 2463–2466

Delanty G (2005) Social science. McGraw-Hill Education, Maidenhead.

Dickinson RE (1934) The town plans of East Anglia: a study in urban morphology. Geography 19: 37–50

Dibble JL (2016) Urban morphometrics: towards a quantitative science of urban form. PhD thesis, University of Strathclyde

Dienner JA (1963) Simplifying the examination of US patent applications. Journal of the Patent Officers' Society 45: 79

Fleischmann M (2018) Measuring urban form. Urban Design 146: 6–7

Friman M (2010) Understanding boundary work through discourse theory: inter/disciplines and interdisciplinarity. Science Studies 23(2): 5–19

Fritz J (1894) Deutsche Stadtanlangen. In: Beilage zum Programm 520 edes Lyzeums Strassburg. Heitz, Strassburg

Frölich K (1938) Zur Verfassungstopographie der deutschen Städte des Mittelalters. Zeitschrift der Savigny-Stiftung für Rechtsgeschichte. Germanistische Abteilung 58: 275–310

Gilbert SW (1991) Model building and a definition of science. Journal of Research in Science Teaching 28: 73–79

Greason J (2011) A settlement strategy for NASA. Keynote address to the International Space Development Conference, Chicago

Hägerstrand T (1991) The landscape as overlapping neighbourhoods. In: Carlestam G, Sollbe B (Eds) Om tidens vidd och tingens ordning: texter av Torstein Hägerstrand. Byggforskningsrådet, Stockholm

Haggett P (1967) Models, paradigms and the new geography. In Haggett P, Models in geography. Methuen, London

Haghani T (2009) Fractal geometry, complexity, and the nature of urban morphological evolution: developing a fractal analysis tool to assess urban morphological change at neighbourhood level. PhD thesis, Birmingham City University

He C, Okada N, Zhang Q, Shi P, Li J (2008) Modelling dynamic urban expansion processes incorporating a potential model with cellular automata. Landscape and Urban Planning 86: 79–91

Hijazi I, Li X, Koenig R, Schmit G, El Meouche R, Lv Z, Abune'meh M (2017) Measuring the homogeneity of urban fabric using 2D geometry data. Environment and Planning B: Urban Analytics and City Science 44: 1097–1121

Hillier B (2016) What are cities *for*? And how does this relate to their spatial form? Journal of Space Syntax 6: 199–212

Huutoniemi K, Klein JT, Bruun H, Hukkinen J (2010) Analyzing interdisciplinarity: typology and indicators. Research Policy 39: 79–88
Jiang B, Claramunt C (2002) Integration of space syntax into GIS: new perspectives for urban morphology. Transactions in GIS 6: 295–309
Kropf KS (1993) The definition of built form in urban morphology. PhD thesis, University of Birmingham
Kropf KS (2014) Ambiguity in the definition of built form. Urban Morphology 18: 41–57
Kwan MP, Schwanen T (2009) Quantitative revolution 2: The critical (re) turn. The Professional Geographer 61: 283–291
Larkham PJ (2002) Misusing morphology. Urban Morphology 6: 95–97
Larkham PJ (2018) The importance of observation: urban morphology in the field. In: Oliveira V (ed) Teaching urban form. Springer, Dordrecht
Larkham PJ and Conzen MP (eds) (2014) The shapers of urban form. Routledge, New York
León J, March A (2016) An urban form response to disaster vulnerability: improving tsunami evacuation in Iquique, Chile. Environment and Planning B: Planning and Design 43: 826–847
Ley K (2012) What is an urban morphologist? Urban Morphology 16: 78–80
Louis H (1936) Die geographische Gliederung von Gross-Berlin. In: Louis H and Lanzer W (eds) Länderkundliche Forschung: Krebs-festschrift. Engelhorn, Stuttgart
Markus TA (1993) Buildings and power. Routledge, London
Maske HA, Andrade JS Jr, Batty M, Havlin S, Stanley HE (1998) Modeling urban growth patterns with correlated percolation. Physical Review E, 58: 7054–7062
Millward HA (1979) Geographical aspects of the 'High Frontier' concept. Geografiska Annaler B61: 113–121
Moran J (2007) Early cultures of gentrification in London, 1955–1980. Journal of Urban History 34: 101-121
Morillo F, Bordons M, Gomez I (2003) Interdisciplinarity in science: a tentative typology of disciplines and research areas. Journal of the American Society for Information Science and Technology 54: 1237–1249
Moudon AV (1997) Urban morphology as an emerging interdisciplinary field. Urban Morphology 1: 3–10
Moudon AV (1992) A catholic approach to organizing what urban designers should know. Journal of Planning Literature 6: 331–349
Oliveira V, Medeiros V (2016) *Morpho*: combining morphological measures. Environment and Planning B: Planning and Design 43: 805–825
Openshaw S (1974) Processes in urban morphology with special reference to South Shields. PhD thesis, University of Newcastle upon Tyne
Panerai P, Castex J, Depaule JC (1997) Formes urbaines: de l'îlot à la barre. Editions Parentheses, Marseille
Peponia J, Feng C, Green D, Haynie D, Kim SH, Sheng Q, Vialard A, Wang H (2015) Syntax and parametric analysis of superblock patterns. Journal of Space Syntax 5: 109–141
Pesaresi M, Bianchin A (2001) Recognizing settlement structure using mathematical morphology and image texture. In: Donnay J-P, Barnsley MJ, Longley PA (eds) Remote sensing and urban analysis. Taylor & Francis, London
Rashed T, Weeks JR, Stow D, Fugate D (2006) Measuring temporal compositions of urban morphology through spectral mixture analysis: toward a soft approach to change analysis in crowded cities. International Journal of Remote Sensing 26: 699–718
Rose P, Keim C, Robazza G, Viejo P, Schofield J (2014) Cities and energy: urban morphology and residential heat-energy demand. Environment and Planning B: Planning and Design 41: 138–162
Rybarczyk G, Wu C (2014) Examining the impact of urban morphology on bicycle mode choice. Environment and Planning B: Planning and Design 41: 272–288
Samuels I (1990) Architectural practice and urban morphology. In Slater TR (ed) The built form of western cities. Leicester University Press, Leicester

Schaefer G (1928) Kunstgeographische Siedlungslandshaften und Städtebilder. Buchdrukerei Zbinden und Hügin, Basel
Sharpe J (2017) Computer modelling in developmental biology: growing today, essential tomorrow. Development 144: 4214–4225
Shayesteh H, Steadman P (2015) Coevolution of urban form and built form: a new typomorphological model for Tehran. Environment and Planning B: Planning and Design 42: 1124–1147
Sheppard J (1974) Metrological analysis of regular village plans in Yorkshire. Agricultural History Review 22: 118–135
Siksna A (1997) The evolution of block size and form in North American and Australian city centres. Urban Morphology 1: 19–34
Skeates R (1997) The infinite city. City 2(8): 6–20
Slater TR (1981) The analysis of burgage patterns in medieval towns. Area 13: 211–216
Stöger H (2015) Roman neighbourhoods by the numbers: a space syntax view on ancient city quarters and their social life. Journal of Space Syntax 5: 61–80
Strahm H (1950) Zur Verfassungstopographie der mittelalterlichen Städte. Zeitschrift für Schweizer Geschichte 30: 406–7, 409-10
Stanilov K (2010) Bridging the gap between urban morphology and urban modelling. Urban Morphology 17: 63–64
Steadman P, Marshall S (2005) Archetypal layout: extending the concept of the archetypal building to streets and block layouts. paper presented at Solutions Conference, London
Thompson, DW (1917) On growth and form. Cambridge University Press, Cambridge
Venerandi A, Zanella M, Romice O, Dibble J, Porta S (2017) Form and urban change - an urban morphometric study of five gentrified neighbourhoods in London. Environment and Planning B: Urban Analytics and City Science 44: 1056–1076
Whitehand, JWR (1972) Urban-rent theory, time series and morphogenesis: an example of eclecticism in geographical research. Area 4: 215–222
Whitehand JWR (1977) The basis for an historico-geographical theory of urban form. Transactions of the Institute of British Geographers NS2: 400–416
Whitehand JWR (1981) Background to the urban morphogenetic tradition. In: Whitehand JWR (ed) The urban landscape: historical development and management. Academic Press, London.
Whitehand JWR, Morton NJ (2004) Urban morphology and planning: the case of fringe belts. Cities 21: 275–289
worldpopulationreview.com/world-cities/shanghai-population/ (accessed 10 April 2018)
Xiao Y, Orford S, Webster CJ (2016) Urban configuration, accessibility, and property prices: a case study of Cardiff, Wales. Environment and Planning B: Planning and Design 43: 108–129
Ye Y, van Nes A (2014) Quantitative tools in urban morphology: combining space syntax, spacematrix and mixed-use index in a GIS framework. Urban Morphology 18: 97–118
Yegros-Yegros A, Rafois I, D'Este P (2015) Does interdisciplinary research lead to higher citation impact? The different effect of proximal and distal interdisciplinarity. PLOS One 10(8): 1–21

Mathematics and Cities: A Long-Standing Relationship Fit for the Future?

Christopher D. F. Rogers

Abstract Mathematics helps us to deal with time and space, an important truism for those individuals who work on behalf of society to impose order on nature. Civil engineers in particular have to seek to harmonise their creations with the natural environment while working within temporal and special constraints. For example, engineers have to synthesise natural and man-made rhythms, meaning that timings need to be synchronised, often with a very high degree of precision. Their creations need to be harmonised with the natural landscape, and while some can hug the contours of the landscape, canals, railways, pipelines, tunnels and major roads need to cut through it such that they are linear or curved in plan and uniform in gradient. This is simple geometry made complicated by myriad engineering calculations. As we transition to a world of autonomous vehicles and robotics to mimic human thinking and actions, multiple data feeds and near instantaneous conversions into intelligent movements stretch our capacity to make calculations. Harmonising this with our infrastructure and urban systems to yield a safe, efficient and effective world today, and one that is sustainable and resilient as we look to the far future, is one of the most important mathematical challenges of the modern age. Mathematics pure and simple, or applied and complicated, permeates everything we do, therefore.

Cities respond to rhythms that range from decades or longer to heartbeats or shorter. As a civil engineer, one of my primary roles is to support civilised life in cities—cities and their citizens are effectively my clients. Therefore, my creations, as a civil engineer, likewise need to respond to these timescales, and it is largely as a result of the need to respond 'in time' that mathematics becomes a vital part of my role. Perhaps counter-intuitively, this does influence the urban form in cities as I will attempt to demonstrate.

The other primary influence of mathematics on my role as a civil engineer underpins the civil engineer's charter, which states that 'civil engineers direct the great forces of power in nature for the use and convenience of humankind'. This has con-

---

C. D. F. Rogers (✉)
Department of Civil Engineering, University of Birmingham, Birmingham, UK
e-mail: c.d.f.rogers@bham.ac.uk

L. D'Acci (ed.), *The Mathematics of Urban Morphology*,
Modeling and Simulation in Science, Engineering and Technology,
https://doi.org/10.1007/978-3-030-12381-9_26

notations of people taming nature and imposing their own patterns onto the landscape of the natural environment. These two themes are necessarily interdependent and are brought together in this article, which attempts to the (long) past, the present and the future. The two themes will weave in and out of the narrative that follows.

Looking back hundreds of years to a period in history when the UK was more of an agrarian society, time as a concept was unimportant. The working day was largely controlled by daylight hours, and synchronisation of time was largely unnecessary. A call to a gathering, for example, a church, could be achieved by ringing of bells, while early clocks had a single hand and struck on the hour. A degree of precision beyond ~5 min was unnecessary. Indeed towns and cities operated to their own timings, and these varied across the country, perhaps naturally because the hours of dawn and dusk likewise varied across the country. Our shaping of time came in the form of organising work, religion and leisure—imposing our social rhythms onto the natural diurnal pattern. It was only with the advent of the railways in the UK that a standard UK time needed to be established to synchronise cross-country journeys. Transport was thus the catalyst for this development.

As with time, our settlement patterns reflected only local considerations: the needs of safety—town and city walls, yet taking advantage of natural features such as rivers, hills and cliffs—and provision of resources. As such, towns grew upon rivers (trade) and at crossing points, or where a day's travel on foot (two or four) would bring goods to market. Not much reflection of mathematics here, then. Societal order, for example, the structuring of the urban form into grid patterns or other geometric forms, which appeared in some cultures, provided a nod to the influence of mathematics perhaps.

The Romans imposed artificial patterns on the UK landscape, for example, organising their settlements in grid formation and their transport using straight roads, created with remarkable mathematical precision, that hugged the ground. For this reason, roads can be traced over mountains in the Lake District (High Street is a spectacular walk) yet there was little attempt to change the contours, even though in Italy they had mastered the art of dealing with contours in aqueducts and viaducts—stated to be remarkable feats of engineering. One question associated with these feats is whether there was conscious use of mathematics, or simply engineering understanding.

The original dominant mode of mass transport in the UK until around three hundred years ago was by water. Many of the original city settlements occurred on navigable rivers, which afforded the means of trade with distant partners. Water transport provided one means of accessing resources, while others had to be obtained locally. In terms of water supply, this could be drawn from wells tapping into the groundwater below or, as cities grew, piped in from reservoirs. In terms of food, this would often be supplied via a network of local roads. These influences shaped the original urban form in cities, and mathematics will most certainly have played a part in all this, albeit that it largely remains hidden.

Artificial waterways developed to assist in the mass transport of goods, and to a far lesser extent people, across the countryside—thus the canal system emerged. Both the space available to receive the waterway and, in particular, engineering in the local topography that would allow for canals to be constructed as sub-horizontal transport routes with a minimum of changes in height (i.e. locks), will have shaped this aspect of the urban form. It can be seen that some cities radically changed their point of focus when the canals arrived, and therefore the local road system was adjusted to suit. Mathematics will have dictated both the route of the waterway and the points of crossing, whether above or below, and while it will have been influenced by local features of topography and geology, mathematics was being used in the design of tunnels, cuttings, embankments and bridges to form viaducts and aqueducts, as well as in precise levelling techniques.

As new mechanised forms of transport developed, railways became the dominant form of mass transport of goods and people, and these too will have responded to the local topography and geology, the requirement here being gentle curves and modest gradients. Moreover, it is the advent of the railways that brought the issue of time into play. Once a network of railway lines had been developed, then timetables and scheduling became important, and mathematics evidently had a role to play in this. Once more the point of focus of cities will have been adjusted to accommodate the railways, and hence mathematics will have been used, even if only tacitly, in this shaping. Perhaps the main point in all this, however, is that mathematics will have been employed alongside engineering knowledge and technological development to engineer the railway system and its crossing points. It is for this reason that the urban form around the railway system will have been shaped by the mathematics associated with the engineering of the system.

In the same vein, road and air transport has developed with a progressively greater emphasis on the application of mathematics to facilitate it. As the progression took place, what we see are evermore complex and interrelated systems that now need to synthesise many modes of transport in order for cities to work and their citizens to move around. The mathematics associated with this synthesis of movement —whether it be pedestrian, bicycle, car, taxi, tram, light rail, bus or a longer distance mode of transport—has become ever more sophisticated. What we now see are systems of operation that are nimble, or perhaps smart, in that where possible flows are adjusted to maximum effect by regulating the speeds across the system, while adjusting for multiple points of ingress and egress from these systems, and real-time information on the transport system (the location of the 'vehicles', estimated times of arrival, quickest routes)—split-minute timings allied to almost instantaneous information feeds.

Highly sophisticated sensor systems are now being applied to vehicles to adjust their operation in response to all of the movements around them, all of the movements associated with the city systems with which they are interdependent, leading ultimately to autonomous vehicles—multiple split-second responses to vast numbers of data feeds, all controlled by sophisticated algorithms. The transport routes of the future, and the structure and form of the built environment, will be shaped by these developments as assuredly as canals, railways and roads did before them.

The obvious question that arises from this latter discussion is 'what will the future hold in this regard?', and this is where the research base of the author provides some answers. Systems are becoming far more sophisticated, interconnected and autonomous—'smart' if you will, although 'smart' is only 'truly smart' if it delivers on all of the future agendas (Rogers 2018; Cavada et al. 2016; Rogers et al. 2014) that make our systems, and city living more generally, sustainable (Rogers 2018; Bruntland 1987), resilient (Lombardi et al. 2012; Rogers et al. 2012a, b), adaptable and liveable (Leach et al. 2016, 2017a, b). The power of distributed sensors creating the Internet of Things, where things talk to other things without humans being involved, is part of this future. Some of these developments we can foresee, especially where they simply enable what we do now to be done more efficiently, whereas some cannot. Human interaction with these new 'smart' systems also comes into question, and this raises a more complex question: what will a citizen of the future be like? A current doctoral study at the University of Birmingham is addressing precisely this question, though there are many informed guesses that we can make on the basis of the way that younger generations have radically changed their behaviour … much, unbeknown to them, being the result of advanced mathematics. As for the systems, their operation to deliver 'just in time' with a minimum of waste will reach new peaks—lights coming on and going off, doors opening automatically, taps opening and closing, indoor environmental conditions being controlled, for example, perhaps via thought processes. Such developments will introduce only minor local changes to the shape of our environment, though when summed their effect might be noticeable.

The physical infrastructures that deliver these system services will be transformed from a steadily deteriorating and progressively failing state (the proliferation of potholes in roads became a topical concern in the UK in 2018) via better mathematical models of material degradation combined with comprehensive sensing systems, new and more resilient materials (self-healing concrete), far more intelligent forms of maintenance, and better operational control. The resources used to build and operate city systems will be marshalled more effectively, acknowledging 'one planet' principles so that resource scarcity and resource security do not cripple our future way of life (Palafox et al. 2017; Rogers et al. 2017; Rogers 2017). A mathematically constructed digital twin of our cities and city systems will be created to assist in this, and once we have a mathematical model, we can explore structural changes in cities with ease, leading to potentially far-reaching changes in the sure knowledge that they will make our ways of life easier, better and friendlier to the planet. Could there be a more profound way of mathematics shaping cities of the future?

Trusting to this way of doing things provides some comfort, but the alignment of future cities with the aspirations of cities and their citizens is a missing, yet vital, feature. Methods have been developed, both via the UK Government Foresight Future of Cities project (Hunt and Rogers 2015; Government Office for Science 2017a, b, c, d) and many allied research programmes (Hunt et al. 2012, 2013), to explore the future using a variety of 'what if?' questions (Rogers et al. 2012b; Dunn et al. 2014). All have value as they prompt the user into thinking well beyond the present. The scenario techniques that I prefer are those that focus far into the future, beyond a time when predictions and projections have a relevance (Rogers 2017), and those that describe extremes (Lombardi et al. 2012; Rogers et al. 2012b). Such scenarios have been used in the design of resilience in cities and to shape cities to meet citizen aspirations, crucially in a space that is freed from current constraints.

Robustly developed scenarios, analysis of city systems (Bouch and Rogers 2017; Hall et al. 2013), mathematical models that model city and extend current trends into the future (Hall et al. 2014), and digital twins of city systems that are then combined into a single, synthesised digital twin of the city therefore provide a means of engineering our future cities. As societies have developed, with exponential degrees of sophistication and diversity, so has the mathematics that has shaped them. There is no reason to suppose that this paradigm will not continue into the future, with mathematically derived safeguards to ensure that our sophisticated 'just in time' systems are resilient to unexpected changes or shocks. Is this, then, a call to trust in the power of mathematics to deliver a better future? It parallels, yet puts into context, the idea that we need to trust in technology to deliver future sustainability and resilience, but it brings this perspective more directly onto the shape of our future cities.

Acknowledgements The author gratefully acknowledges the financial support of the UK Engineering and Physical Sciences Research Council (EPSRC) under grant numbers GR/S20482, EP/C513177 and EP/E021603 (Birmingham Eastside Regeneration—Sustainable Urban Redevelopment), EP/F007426 (Urban Futures), EP/I016133 (Resilience of Critical local Transport and Utility Infrastructure), EP/J017698 (Liveable Cities - Transforming the Engineering of Cities to Deliver Societal and Planetary Wellbeing), EP/K012398 (iBUILD: Infrastructure BUsiness models, valuation and Innovation for Local Delivery), EP/N010523 (Balancing the Impact of City Infrastructure Engineering on Natural Systems using Robots) and EP/P002021 (Urban Living Birmingham). He also gratefully acknowledges the funding from the University of Birmingham for the Policy Commission on Future Urban Living and both the project staff and his colleagues on the Lead Expert Group of the Foresight Future of Cities project.

References

Bouch CJ and Rogers CDF (2017). A Systems Thinking Approach to the Development of Alternative Infrastructure Business Models. *Proceedings of International Symposium on Next generation Infrastructure (ISNGI 2017)*, London, UK.

Bruntland GH (1987). *Our Common Future*. Oxford: Oxford University Press.

Cavada M, Hunt DVL and Rogers CDF (2016). Do Smart Cities Realise Their Potential for Lower CO_2 Emissions? *Engineering Sustainability, Proceedings of the Institution of Civil Engineers*, **169** (6), 243–252. doi: https://doi.org/10.1680/jensu.15.00032.

Dunn N, Cureton P and Pollastri S (2014). *A Visual History of the Future*. www.gov.uk/government/collections/future-of-cities.

Government Office for Science (2017a). 'Future of Cities: science of cities'. Available at www.gov.uk/government/publications/future-of-cities-science-of-cities.

Government Office for Science (2017b). 'Future of Cities: foresight for cities'. Available at www.gov.uk/government/publications/future-of-cities-foresight-for-cities.

Government Office for Science (2017c). Future of Cities: an overview of the evidence. Available at www.gov.uk/government/publications/future-of-cities-overview-of-evidence.

Government Office for Science (2017d). Foresight Future of Cities website. Available at www.gov.uk/government/collections/future-of-cities.

Hall JW, Henriques JJ, Hickford AJ and Nicholls RJ (2013). Systems-of-Systems Analysis of National Infrastructure. *Proceedings of the Institution of Civil Engineers, Engineering Sustainability*, **166** (5), 249–257.

Hall JW, Henriques J, Hickford A *et al.* (2014). Assessing the Long-Term Performance of Cross-Sectoral Strategies for National Infrastructure. *Journal of Infrastructure Systems*, **20** (3).

Hunt DVL and Rogers CDF (2015). *Aspirational City Futures*. www.gov.uk/government/collections/future-of-cities.

Hunt DVL, Lombardi DR, Atkinson S *et al.* (2012). Scenario Archetypes: Converging rather than Diverging Themes. *Sustainability*, **4** (4), 740–772.

Hunt DVL, Jefferson I and Rogers CDF (2013). Scenarios Analysis through a Futures Performance Framework. *Engineering Sustainability, Proceedings of the Institution of Civil Engineers*, **165** (5), 258–271.

Leach JM, Braithwaite PA, Lee SE, Bouch CJ, Hunt DVL and Rogers CDF (2016). Measuring Urban Sustainability and Liveability Performance: the City Analysis Methodology (CAM). *International Journal of Complexity in Applied Science and Technology (IJCAST)*, **1** (1), 86–106.

Leach JM, Lee SE, Hunt DVL and Rogers CDF (2017a). Improving City-Scale Measures of Livable Sustainability: A study of urban measurement and assessment through application to the city of Birmingham, UK. *Cities*, **71**, 80–87.

Leach JM, Lee SE, Boyko CT *et al.* (2017b). Dataset of the Livability Performance of the City of Birmingham, UK, as measured by its Citizen Wellbeing, Resource Security, Resource Efficiency and Carbon Emissions. *Data in Brief*, **15**, 691–695. http://www.sciencedirect.com/science/article/pii/S2352340917305218.

Lombardi DR, Leach JM and Rogers CDF *et al.* (2012). *Designing Resilient Cities: a Guide to Good Practice*. IHS BRE Press, Bracknell, UK, 164 pp.

Palafox G, Lee SE, Bouch CJ, Hunt DVL and Rogers CDF (2017). *The Little Book of Circular Economy in Cities – A short guide to urban metabolism and resources flows*. ISBN – 978-0-70442-950-5.

Rogers CDF (2017). *Chapter 11: Cities*. Report of the Government Chief Scientific Advisor: From Waste to Resources Productivity – Evidence and Case Studies. p. 149–164. Government Office for Science. www.gov.uk/government/publications/from-waste-to-resource-productivity.

Rogers CDF (2017). *The Value of Foresight and Scenarios in Engineering Liveable Future Cities*. Chapter in Retrofitting Cities for Tomorrow's World, Wiley Blackwell, Chichester, West Sussex, UK (Eds. Eames M, Dixon T, Hunt M & Lannon S), p. 139–152. ISBN: 978-1-119-00721-0.

Rogers CDF (2018). Engineering Future Liveable, Resilient, Sustainable Cities Using Foresight. *Civil Engineering, Proceedings of the Institution of Civil Engineers*, **171** (6). (https://doi.org/10.1680/jcien.17.00031). www.icevirtuallibrary.com/doi/pdf/10.1680/jcien.17.00031.

Rogers CDF, Bouch CJ, Williams S *et al.* (2012a). Resistance and Resilience – Paradigms for Critical Local Infrastructure. *Municipal Engineer, Proceedings of the Institution of Civil Engineers*, **165** (2), 73–84.

Rogers CDF, Lombardi DR, Cooper RFD and Leach JM (2012b). The Urban Futures Methodology Applied to Urban Regeneration, special issue on Future Proofing Urban Regeneration. *Engineering Sustainability, Proceedings of the Institution of Civil Engineers*, **165** (1), 5–20.

Rogers CDF, Shipley J, Blythe P *et al.* (2014). Future Urban Living – A Policy Commission Investigating the Most Appropriate Means for Accommodating Changing Populations and Their Needs in the Cities of the Future. University of Birmingham, UK, 60 pp. ISBN 978-0-7044-2843-0.

Rogers CDF, Hunt DVL, Leach JM, Purnell P and Roelich KE (2017). Resource Scarcity and Resource Security – A Suppressed Civil Engineering Challenge. Briefing Note, *Waste & Resources Management, Proceedings of the Institution of Civil Engineers*, 166 (2), 49–51.

Mathematics and/as Humanities–Linking Humanistic Historical to Quantitative Approaches

Tino Mager and Carola Hein

Abstract The article reflects the state of mathematics between the natural sciences and the humanities. By arguing that mathematics is a humanities subject, it suggests a close connection between mathematics and urban morphology studies. This also applies to the discrepancy between quantitative and qualitative methodological approaches. New types of research based on quantitative methods reveal previously unknown aspects of urban phenomena. They will play an increasingly important role in future research, and it is a challenge for the humanities to effectively integrate mathematical perspectives on the human habitat.

1 Mathematics as Humanities

There is something delightful in mathematics that the humanities are lacking and that has inspired other scientists: The beauty of truth emerging from itself. This dynamic is mostly abstract but serves as a basis for describing the world. Both the objectivity of mathematics and its ability to prove an infinite number of individual cases by a finite logical procedure are enviable. Unfortunately, this form of derivation of validity is virtually impossible in the humanities, where correlations can hardly be represented by elegant theories or formulas.

Mathematics and the humanities have a common starting point, largely hidden from view due to the contemporary relevance of applied mathematics and its intertwining with the natural sciences, technology, and engineering. The foundations of what we call mathematics today were present at the beginning of history. The fundamental importance of counting objects or forming numerical contexts has proven itself in everyday life over thousands of years. This knowledge was further developed and used by the ancient Egyptians and Babylonians to build the pyramids and create

T. Mager (✉) · C. Hein
Delft University of Technology, Delft, The Netherlands
e-mail: B.T.Mager@tudelft.nl

C. Hein
e-mail: C.M.Hein@tudelft.nl

L. D'Acci (ed.), *The Mathematics of Urban Morphology*,
Modeling and Simulation in Science, Engineering and Technology,
https://doi.org/10.1007/978-3-030-12381-9_27

calendars. Eventually, knowledge about numbers and their calculation itself became the object of reflection in the heyday of Greek antiquity. The Pythagoreans began to develop a philosophical interest in mathematics and set up logical evidence as a means of finding an objective truth. The study of mathematics became a theoretical matter, a purely intellectual examination with no necessary reference to the application; the mathematician Helmut Hasse concluded that "mathematics of the classical Greek epoch is pure humanities in content and methods."[1]

Today, mathematical philosophies soften the distinction between pure and applied mathematics to the extent that they trace all humanities and natural sciences back to algorithmic roots.[2] The emergence of the so-called digital humanities exemplifies this reconnection of mathematics and humanities. It signals a shift from qualitative to quantitative approaches in the humanities. Thereby, the latter is often regarded as tools for data management only. This not only reduces their potential unnecessarily, but also largely ignores a whole range of possible contributions to the humanities. How can research in the humanities and natural sciences be integrated in a way that is equally challenging in and enriching to both fields?

2 Quantitative Approaches in Urban Morphology

Examining the history of architectural and urban design, we can see many moments when scholars integrated mathematics and humanities. Over many centuries, humans have tried to bring an understandable order to the diversity of urban and architectural forms, often using mathematical formulas not only to implement a reliable system of structure, but also to provide a supernatural meaning based on numerology. The use of particular numbers has played an important role in both Chinese architecture and in European design traditions. The designation of architectural styles and the countless setups of architectural order by various antiquarians, archeologists, and architectural historians were and are means of intellectual organization. From proportions of columns to counting window axes or applying the golden ratio, architectural historians have long worked with numerical and geometrical assessments. In some cases, their analysis became the starting point for new architectural and urban design theories, including the principles of the Leonardo da Vinci's Vitruvian man, and Leon Baptista Alberti's theories on the esthetics of proportionality.[3] The German architect Ernst Neufert's standardization of interior objects and architectural elements, a much reprinted standard handbook for architects, is a more recent example of both the power of numbers in architectural design and the role such sys-

[1]Hasse, Helmut. "Mathematik als Geisteswissenschaft und Denkmittel der exakten Naturwissenschaften". Newly edited by Gabriele Dörflinger. Heidelberg: Universitätsbibliothek, 2008, p 3. URL: http://archiv.ub.uni-heidelberg.de/volltextserver/12976/1/StudGen.pdf [access: 13 March 2018].

[2]Kanitschneider, Bernulf. *Kleine Philosophie der Mathematik*. Stuttgart: Hirzel, 2017.

[3]Leon Battista Alberti, Sebastiano Serlio, Andrea Palladio, Leonhard Christoph Sturm and Claude Perrault are but a few famous writers that refined and established the classical order.

tematization as played in shaping design. The analytical mapping of urban spaces and their usage is yet another example of mathematical formulas that humanities scholars use in writing the history of the built environment.

Over centuries, such standardizations have provided an overview of existing forms, but they have also distorted the understanding of the built environment of the past. These systematizations and typologies are often only a partial representation of the diversity of the built environment. Vernacular architecture and urban morphologies that are beyond scientific focus—including workers' housing, slums, suburban urbanization, and settlement forms resulting from flight and migration—are often excluded from architectural histories that largely stick to the canon of "capital A" architecture.[4]

The desire to construct an architectural or urban history that is comprehensive (not selective or based on temporal and spatial preferences), is an attractive starting point for humanists turning to quantitative approaches. In a recent paper, for example, the urban planner Geoffrey Boeing extended several quantitative methods to the analysis of the complexity of urban form; for example, he used network-based spatial clustering to identify agglomerations of jobs or amenities.[5] The result is a multilayered typology of temporal, spatial, visual, fractal, and network measures that advance our understanding of urban complexity. Mathematical approaches, algorithms, and artificial intelligence can help us analyze and compare forms and structures that are virtually unmanageable for human brains in terms of quantity and complexity. The following two examples illustrate some novel possibilities: Vahid Moosavi, a researcher in Machine Learning, a field applying data to train computers for specific tasks without explicit programming, demonstrated that deep learning can allow a scholar to compare the complexity and shape of more than one million urban patterns worldwide.[6] Using publicly available data from Open Street Map and a specially tuned artificial neural network, he analyzes the shape of patterns in a broad variety of urban settlements across the planet, comparing them for similarities and grouping them into a large number of types based on rectangularity and complexity. The network focuses on geometric similarities without looking for man-made concepts or order structures, circumventing linguistic and conceptual conditions which bias our thinking about built forms. It "sees" no linear or dispersed settlements, no medieval town centers, no sprawl or suburbs. Rather, it discovers formal similarities that can be expressed mathematically, for example by assigning closely related structures a specific color. The result is a visualization of the current worldwide distribution of the similarities and differences of human settlements, revealing novel insights on global correlations and developments without referring to canonical categories. Quantitative methods can further deepen these insights, for example by generating

[4]E.g.,: *The Political Meaning of Informal Urbanisation* (Roberto Rocco, Jan van Ballegooijen| TU Delft), *Eine Architekturgeschichte der Armut* (Britta Hentschell| ETH Zürich).

[5]Boeing, Geoff. "Measuring the Complexity of Urban Form and Design", 2017, http://geoffboeing.com/publications/measuring-complexity-urban-form/ [access: 19 March 2018].

[6]Moosavi, Vahid. "Urban morphology meets deep learning: Exploring urban forms in one million cities, town and villages across the planet", 2017, https://arxiv.org/abs/1709.02939 [access: 20 March 2018].

connections to aspects of economy, politics, social conditions, or culture—as far as they are available digitally.

Computer vision—methods enabling computers to recognize and understand image content—can also be used for classic basic research. COMPOSITO, a research project at the University of Heidelberg, investigates the automatic recognition of architectural elements in early modern architecture.[7] The goal is to identify a structure's architectural elements—for example, column capitals, window lintels, rustic masonry—and to find other buildings with similar features. This approach aims to simplify the comparative study of buildings and to help scholars investigate the spread of architectural styles much more comprehensively than before. Furthermore, COMPOSITO experiments with the search for similarities through unsupervised learning. This means that it examines and classifies objects for similarities that do not correspond to any architectural categories, but are found by the algorithm itself. The results, which are purely formal in nature and not influenced by implicit knowledge, may, conversely, catalyze qualitative analyses of individual objects.

Another program that integrates quantitative methods into other research on the built form is the ArchiMediaL project, carried out by researchers of the Delft University of Technology and the Vrije Universiteit Amsterdam.[8] It uses publicly available data and the capabilities of a neural network to facilitate a deeper investigation of built form. In a first step, Mapillary street view images of today's Amsterdam provide a basis for image content matching, helping identify the buildings captured in more than 3,60,000 historical photographs of the city.[9] This enables the creation of a street view of Amsterdam's past. It will also allow for adding a historical layer to the neural network's training set, to make it more sensitive for recognizing the architectural content of more visual representations of the city's past. The ultimate goal is for the project to analyze imagery beyond photography—paintings, drawings, sketches—and to automatically detect buildings in these images. The outcome will be a complete set, across space and time, of the visual representations of Amsterdam's architecture. In combination with nonarchitectural data (e.g. cadastral information, election results, local income, etc.), this data set allows for automated analyses correlating architecture with urban planning and other societal conditions. The sheer amount of cases may lead to insights beyond the scope of traditional humanistic methods that are restricted by the amount of scholars working on the project.

The experiences gained from this kind of research on Amsterdam will help scholars understand how to investigate images that have less data available. That is, millions of images with architectural content lack titles, meta tags, or annotations. This matters especially if the content belongs to vernacular architecture and urban form, still underrepresented in research and not part of historical architectural and urban

---

[7] *COMPOSITO - Arthistorical Analysis of Architecture via Computer Vision* is a project conducted by Björn Ommer, Peter Bell and Michael Arnold at the Heidelberg Collaboratory for Image Processing.

[8] *ArchiMediaL* is a cooperation between historians of architecture and urban form (Carola Hein, Beate Loeffler, Tino Mager, Dirk Schubert) and computer scientists (Victor de Boer, Jan van Gemert, Seyran Khademi, Ronald Siebes). URL: http://archimedial.eu.

[9] The historical images are from the collection of Stadsarchief Amsterdam and available at Beeldbank Amsterdam. URL: http://beeldbank.amsterdam.nl.

canons. Primarily, these shortcomings apply to data of non-Western architecture. Overcoming them will enhance the further investigation and understanding of global networks of architectural ideas and techniques that lie behind the migration and evolution of form.[10]

Quantitative methods—in this case, the automatized analysis of form by computer vision—are particularly helpful in tracing developments and distributions of architectural and urban form beyond the main recognized architectural highlights. Moreover, they allow scholars to do this without having to precisely name specific features linguistically in the metadata, as they can refer to purely visual aspects. Eventually, image content matching, powered by computer vision, will help us navigate and search in visual sources as easily as we conduct research in text sources today. Then it won't matter if the object of interest is referred to as, e.g., windmill, tuulimylly, ανεμόμυλος, or 風車, as the search will concentrate on the visual representation of the linguistic term.

Linked data approaches, in this case that means the semantic interlinking of visual content and attributes, will further help to open up visual sources that are currently difficult to access due to foreign language, or deviating or even incorrect labeling.[11] Ideally, such approaches will work across public and private archives, libraries, and collections, providing scholar access to all existing sources for a specific object—or to millions of objects. The restriction is less technical than political—the willingness of archives, institutions, and private parties to allow open access to the source material—complicated by legal questions that are rather blurry for nonexperts. In the short term, however, it is clear that technology and mathematics will change the character of studies of urban morphology.

3 Imminent Changes

But quantitative methods and automatic data processing cannot replace human experience or qualitative information processing: It will hardly be possible to carry out individual studies—which rely on the deep knowledge of regional historical developments and on interdisciplinary knowledge—by using approaches that are essentially based on algorithmic approximation or even nonhuman intelligence. What these approaches can do, however, is overcome the previous focus on comparatively few objects and the reliance on types, which is only of limited validity to reality. The possibility of examining hundreds of thousands of objects, in conjunction with worldwide data acquisition, will give us a more global perspective beyond the types and

[10]E.g. Hein, Carola: *Port Cities: Dynamic Landscapes and Global Networks*. London: Routledge 2011.

[11]Löffler, Beate; Carola Hein; Tino Mager. "Searching for Meiji-Tōkyō. Heterogeneous visual media in times of global urban history, digitalization, and deep learning". In: *Global Urban History*, 20 March 2018. URL: https://globalurbanhistory.com/2018/03/20/searching-for-meiji-tokyo-heterogeneous-visual-media-and-the-turn-to-global-urban-history-digitalization-and-deep-learning/ [access: 21 March 2018].

canons limiting much of today's scientific thinking. This will also have an effect on classification, which is useful as an aid to thinking, but hitherto applies only to parts of reality. Novel classes will result from factors that have not yet been taken into account and are of greater complexity, which will make them more applicable.

Quantitative approaches can thus help us better explore urban forms and morphologies. To win this broader perspective, humanists will be obliged to deal more intensively with quantitative methods and their possibilities, as well as the possibilities of integrating them with qualitative methods. The main step, however, lies in the challenge of developing research questions in cooperation with computer scientists that enrich both fields. In this way, the potential of new technologies can be meaningfully exploited and at the same time a mutual interest can be generated. Since the human sciences are known to be difficult to monetize, they must be particularly creative in order to attract computer scientists (competing with more profitable areas such as, e.g. bioengineering). Incentives will include groundbreaking research in areas such as deep learning and computer vision. Ultimately, classical approaches in architectural and urban history will also benefit: new insights will contribute to stimulating new qualitative research, based on unprecedented results made possible only by cutting edge quantitative methods.

Urban Forms, Agents, and Processes of Change

Vítor Oliveira

Abstract Urban morphology describes and explains the physical form of cities. Adopting a dynamic perspective, it explains how different agents change that physical form over time and how diverse processes are involved in this transformation. Urban morphology may also offer insights on prescription for the design of new urban forms and for the transformation or conservation of existing forms. This chapter focuses on two dominant approaches in urban morphology, the historico-geographical and the process-typological. While briefly presenting the origins and main developments of each of these approaches, it goes one step further debating how these two can be combined with other approaches presented in the book to offer better description, explanation, and prescription of the physical form of cities.

1 Urban Morphology

Urban morphology (Oliveira 2016; Kropf 2017) is the study of the physical form of cities and of the agents and processes shaping its transformation over time. When addressing a city, an urban morphologist starts his/her analysis by the physical form of the city, eventually by a limited number of elements of urban form (for instance streets or plots). He can then add to this physical base some additional layers of information of a social, economic, or environmental nature, according to the main goals of the study. He might also have an interest on the different agents, both direct and indirect, that shape the form of those elements over time (including developers, architects, builders, local authority planning officers, and local politicians). In addition, he might have a concern on the specific processes of transformation that occur, from planned change (for instance, under the framework of a plan) to more "random" change.

The relevance of urban morphology is based on the fact that way how we build and organize our cities has an impact on different aspects of our urban life. For

V. Oliveira (✉)
CITTA Research Centre for Territory Transports and Environment, Rua Roberto Frias s/n, 4200-465 Porto, Portugal
e-mail: vitorm@fe.up.pt

L. D'Acci (ed.), *The Mathematics of Urban Morphology*,
Modeling and Simulation in Science, Engineering and Technology,
https://doi.org/10.1007/978-3-030-12381-9_28

instance, the way how we lay down our streets on the ground and on how we erect our buildings has an influence on the amount of energy that we consume in transports and buildings (Silva et al. 2017). If we think that the three main sectors responsible for our energy consumption are exactly transports and buildings, and also industry, we start to realize the importance of understanding these phenomena, especially if we consider the pressing issue of climate change.

Urban morphology had its origins at the end of the nineteenth century, in Central Europe, in a number of works on urban geography. It had a golden age in the first three decades of the twentieth century and then, it lost importance, as urban functions and urban structures become the major concerns of urban geographers. In the second half of the twentieth century, there were again innovative contributions to the study of urban form, stimulated by the activities of geographers and architects. Some of these individual contributions led to the development of schools of thought, fed by an increasing number of researchers in different parts of the world. Nowadays, we can identify four dominant schools of thought in urban morphology (Kropf 2009), with their own theories, concepts, and methods to address the physical form of cities and, as such, to impact the social, economic, and environmental aspects of life in cities. This does not mean that these four approaches encompass all studies in urban morphology. On the contrary, in the last decades, this field of knowledge has been living a stimulating moment (see the set of national reviews published in the journal "Urban Morphology"—for example Marzot et al. 2016).

These four approaches are the historico-geographical approach, the process-typological approach, space syntax, and spatial analysis (including cellular automata, agent-based models, and fractals). Due to the nature of this chapter, and the part of the book where it is included, we'll focus on the first and second.

2 The Historico-Geographical Approach

The historico-geographical approach had its origins in the mid-twentieth century in the work of the German geographer MRG Conzen, particularly in his book on the small town of Alnwick (Conzen MRG 1960)—Fig. 1. Based on the work of Conzen, JWR Whitehand, another geographer, started to build an effective school of morphological thought (Oliveira 2019). The foundation of the Urban Morphology Research Group (UMRG), in the University of Birmingham in 1974, would be crucial for this purpose.

As the name suggests, this approach addresses the city through the historico-geographical structure of the urban landscape. To do so, it focuses on three elements of the urban landscape, its tripartite division: the town plan, including streets, plots, and the block plans of buildings (see Conzen MP 2018 for a comprehensive review of town plan analysis); the building fabric, and the land and building utilization.

Most important, the school has developed a number of concepts to explain the process of urban development. We'll focus on three of these concepts that, by their nature, also illustrate the capacity of the approach of moving from the large to the

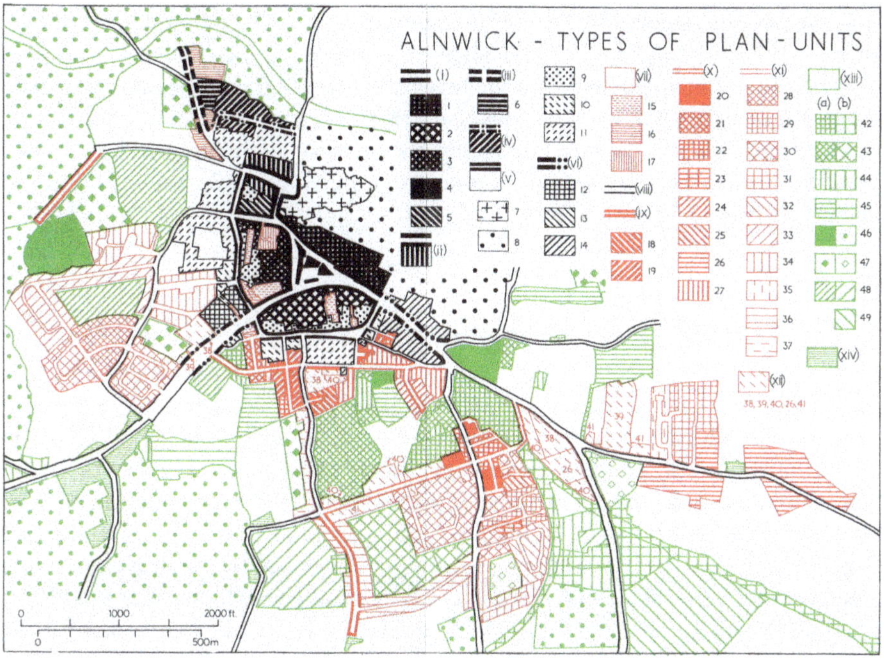

Fig. 1 Alnwick, Northumberland. A study in town plan analysis—types of plan units. *Source* Conzen (1960)

small scale of analysis. The fringe-belt concept draws on the acknowledgment that the process of growth of a city is not continuous; on the contrary, it is made of a series of outward expansions of the residential areas, and of other very different areas—relatively open and vegetated, with a sparse street network, large plots, and low building coverage (for a comprehensive reviews of the concept see Conzen MP 2009 and Ünlü 2013). Since the mid-1960s, Whitehand has been extending the limits of the concept adding it a metropolitan dimension, an economic perspective, a planning and agency perspective, and an application into different cultural contexts—for instance in China (Whitehand and Gu 2017).

The second concept is the morphological region, an area that has a unit in respect of its form, that distinguishes it from surrounding areas. For Conzen, the climax of the analysis of the physical development of an urban area was its division into morphological regions (Whitehand 2009).

The last concept, the burgage cycle, focuses on the relationship between plots and the block plans of buildings, by addressing one particular type of plot—the burgage—and its life cycle. The cycle is a particular variant of a more general phenomenon of building repletion where plots are subject to increasing pressure, often associated with changed functional requirements, in a growing urban area.

In addition, the historico-geographical approach, and particularly Whitehand, addresses in a systematic and explicit way the role of the numerous individuals, firms, and organizations involved in the ownership, design, and implementation of changes to the building fabric (Larkham and Conzen MP 2014). Since the beginning of the 1980s, Whitehand and his colleagues at the UMRG have been developing a research framework that includes: the research procedures, the types of change, the timing of the changes, the types of agents, the relations between agents, the provenance of agents, the situations of conflict, and the comparison between different functional areas (Whitehand 1992).

3 The Process-Typological Approach

The process-typological approach had its origins in the mid-twentieth century in the work of the Italian architects Saverio Muratori, particularly in his books on the operative histories of Venice and Rome (Muratori 1959; Muratori et al. 1963), and Gianfranco Caniggia, notably in his books on architectural composition and building typology written with Gian Luigi Maffei (Caniggia and Maffei 1979, 1984)—Fig. 2. The approach launched by Muratori was further developed by his Rome assistants, including Caniggia, that would leave to teach in a number of Italian cities including Reggio Calabria, Genoa, and Florence. Based in Florence, Giancarlo Cataldi had a similar role to that of JWR Whitehand in the structuring of the process-typological approach (Cataldi et al. 2002).

Muratori (1959) defines a set of fundamental concepts—type, urban tissue, organism, and operative history. According to Muratori, a certain building type could not be identified except within a particular application, in the urban tissue. The urban tissue could not be identified except in its involving context, in the urban organism. The urban organism would only become real in its historical dimension, as part of a temporal construction that is always grounded on the conditions suggested by the past. This led to the argument of a strong relation between history and planning/architecture.

Three other concepts are fundamental in this approach: route, basic building, and special building (Strappa 2018). The route is a regular course artificially obtained by consolidating a spontaneous course or constructing it according to an intentional plan. It can be classified as a matrix route, a building route, a connecting route or a restructuring route. Within the relation between each route and in the definition of the urban form, poles and nodes play a fundamental role. Node is the singular point of a continuum determined by the intersection of two continuums or by the "gemmation" of one continuum with another. Pole is the sublimation of the term node, in general, determined by the presence of more continuums, not so intersecting, as ending or starting from a point. The basic building is a housing unit that aggregates to form higher grade organisms. It corresponds to a row house, a single-family building with direct access from the route that has in common with the other unities two walls. The special building corresponds to all the nonresidential part of the built environment, also including those building types where the housing function is secondary to that which gives rise to the specialization of the type (Strappa 2018).

Fig. 2 *Studi per una operante storia urbana di Venezia – Quartieiri di S. Giovanni Crisostomo*, from the eleventh century to the 1950s. *Source* Muratori (1959)

4 Combining Different Approaches in the Study of Urban Form

Given the complex nature of cities is hard to argue that each of these morphological approaches, or others, can capture the whole complexity of their object. It might be that in one particular situation one approach could offer more knowledge on the object, while in another situation a second approach could be more useful. It can also happen that in a third situation the correct thing to do would be to combine two or more approaches. As such, we need to develop a deeper knowledge of each of these approaches. What are its main characteristics? What are the common aspects that it shares with other approaches and what distinguishes it? How does it deal with the physical form of cities? Is it only focused on description and explanation or does it lead to prescription and action?

4.1 The Historico-Geographical Approach

The historico-geographical approach shares with the process-typological approach a focus on the building fabric (the latter draws on the building fabric as a whole to reveal the fundamental building types, the main relations between them, and how they evolve over time), a comprehensive framework ranging from small to large scale analysis, and a clear statement on the importance of history offering a sense of continuity in the production of urban forms.

It shares with cellular automata, particularly with new cellular automata methods that are able to deal with irregular cells, an understanding of the plot as a minimum element of analysis (although in the historico-geographical approach the geometric characteristics of the cell are also considered). In contrast, the two approaches have different conceptions of time: in the former, history has a fundamental role in the description and explanation of the urban landscape; in the latter, the main concern is with anticipating future scenarios of urban development.

Finally, the historico-geographical approach shares with space syntax a focus on the ground plan; while space syntax is mainly concerned with streets, the historico-geographical also acknowledges the importance of plots and the block plans of buildings. As in the previous comparison, these two approaches seem to have different conceptions of time and, to a certain extent, different levels of resolution (scale).

4.2 The Process-Typological Approach

As we have seen, there is a strong relationship between the process-typological and the historico-geographical approaches. It shares with space syntax a focus on streets (although it does not address them with a configurational focus and takes into account the form and utilization of the buildings defining the streets) and on buildings (when space syntax is applied, not at the urban but, at the building scale). Despite its emphasis on history, in clear contrast with the cellular automata emphasis on the anticipation of future scenarios, both approaches stress the time dimension and the notion of process.

References

Caniggia G, Maffei GL (1979) Composizione architettonica e tipologia edilizia I: lettura dell'edilizia di base. Marsilio, Venice

Caniggia G, Maffei GL (1984) Composizione architettonica e tipologia edilizia II: il progetto nell'edilizia do base. Marsilio, Venice

Cataldi G, Maffei GL, Vaccaro P (2002) Saverio Muratori and the Italian school of planning typology. Urban Morphology 6:3–14

Conzen MP (2009) How cities internalize their former urban fringes: a cross-cultural comparison. Urban Morphology 13:29–54

Conzen MP (2018) Core concepts in town plan analysis. In Oliveira V (ed) Teaching urban morphology, Springer, Cham

Conzen MRG (1960) Alnwick Northumberland: a study in town-plan analysis, Institute of British Geographers Publication 27. George Philip, London

Kropf K (2009) Aspects of urban form, Urban Morphology 13:105–120

Kropf K (2017) The handbook of urban morphology, Wiley, Chichester

Larkham PJ, Conzen MP (eds) (2014) Shapers of urban form. New York, Routledge

Marzot M, Cavallo R, Komossa S (2016) The study of urban form in the Netherlands. Urban Morphology 20:35–43

Muratori S (1959) Studi per una operante storia urbana di Venezia I. Palladio 3–4

Muratori S, Bollati R, Bollati S, Marinucci G (1963) Studi per una operante storia urbana di Roma. Consiglio Nazionale delle Ricerche, Roma

Oliveira V (2016) Urban morphology—an introduction to the study of the physical form of cities. Springer, Dordrecht

Oliveira V (ed.) (2019) JWR Whitehand and the historico-geographical approach to urban morphology. Springer, Cham

Silva M, Oliveira V, Leal V (2017) Urban form and energy demand: a review of energy-relevant urban attributes. Journal of Planning Literature 32:346–365

Strappa G (2018) Reading the built environment as a design method. In Oliveira V (ed) Teaching urban morphology. Springer, Cham

Ünlü T (2013) Thinking about urban fringe belts: a Mediterranean perspective. Urban Morphology 17:5–20

Whitehand JWR (1992) The making of the urban landscape. Basil Blackwell, Oxford

Whitehand JWR (2009) The structure of urban landscapes: strengthening research and practice. Urban Morphology 13:5–27

Whitehand JWR, Gu K (2017) Urban fringe belts, Environment and Planning B 44:80–99

Future of Streets

Andres Sevtsuk and Diane E. Davis

Abstract Rapidly evolving mobility technologies and the associated behavioral adjustments of travelers are bringing about dramatic changes to the morphology of cities, some of which have already begun to take root. With the seemingly endless amounts of data that technology is producing about life in cities, new mathematical modeling techniques will be required to fully understand the impact these changes will have on society. As a consequence of innovation in personal mobility technologies, a combination of autonomous and electric vehicles is being seen by many as the solution to personalized and inexpensive urban transport. In this chapter, we explore some of the ways in which this version of the future of streets could reverse decades worth of efforts by cities to reduce congestion and contain sprawl unless policymakers are proactive in responding to these disruptive forces. Without active efforts to prioritize shared, public, and active mobility, the introduction of automated modes for transporting both humans and goods in an already contested public realm could mean a reversion to Modernist practices of privileging speed, efficiency, and function over human-scale interactions and serendipity. Politically contentious decisions will need to be made around questions of social, environmental, economic, and public health priorities to ensure that streets are made livable and accessible, particularly for vulnerable and marginalized groups. Ultimately, recognizing that the interests of pedestrians, bicyclists, public transit users as well as stationary street users should take priority over those of personal vehicles (autonomous ones in particular) is key to guiding changing urban morphologies in a way that places sustainability and human-focused development at the forefront.

Disruptive innovations in the field of urban transportation are changing urban morphology in ways that require new forms of analysis and action. The combined impacts of shared, electric, and automated vehicles on the nature and cost of urban travel could turn out to be as transformative as the introduction of the internal combustion engine or electric railways more than 150 years ago. And just as these nineteenth-century

A. Sevtsuk · D. E. Davis (✉)
Harvard Graduate School of Design, Cambridge, USA
e-mail: ddavis@gsd.harvard.edu

L. D'Acci (ed.), *The Mathematics of Urban Morphology*,
Modeling and Simulation in Science, Engineering and Technology,
https://doi.org/10.1007/978-3-030-12381-9_29

innovations fundamentally altered both city form and how people moved in space, technological advances in personal transportation are likewise leading cities toward a new future.

Nowhere will the transformative influence of new mobility technology play out as palpably, contentiously, and consequentially as on city streets. How city governments, technology companies, and urban planners respond to these challenges will not only shape the future form and function of streets, but quite likely, the resulting transformation of streetscapes will also decisively impact the morphology of districts, cities, and metropolitan regions. Urban planners, policy-makers, morphological analysts, and designers need to deploy the latest mathematical modeling techniques to proactively investigate these changes on the city, outlining ways in which technological changes will reshape cities and identifying strategies and tactics that would steer these transitions toward more multimodal, sustainable, equitable, and healthy outcomes.

The causes of the personal mobility revolution as manifest in the proliferation of automated, shared, and electric vehicles have been discussed at length elsewhere (Sperling 2018; Anderson et al. 2016). Here, we simply reiterate what aspects of these technologies are likely to become widespread and discuss what their adoption could mean for the design, planning, and regulation of streets.

The US transportation sector currently consumes more energy and emits more CO_2 than any other sector of the economy, including the construction or running of buildings. More efficient electric engines are thus being embraced because they promise to deliver considerable reductions in per mile energy use and per mile particle exhaust. While these benefits will only materialize if both power production and electric battery manufacturing become less dependent on fossil fuels, production technology is likely to keep improving. Carbon sequestration has lowered CO_2 emissions from coal electric plants, while total renewable energy production and consumption both reached record highs of about 10 quadrillion Btu in 2016 (EIA 2016). Given that most of US electricity is produced in-country, electric automobiles would also keep a large share of a typical household's monthly outlays in the national economy, rather than foreign oil producing nations. All major auto manufacturers are significantly investing in EV production and a number of countries, including Norway, France, UK, and China have already committed to limiting all new car sales to only electric vehicles by 2025. The trends suggest that urban transportation will become electric within the next generation.

Second, driverless vehicle technology is rapidly advancing, in no small part because it is incentivized by the large cost savings that it is expected offer to both individual owners and Transportation Network Companies (TNCs), such as Uber, Didi, and Lyft. While moving a passenger with a human driver currently costs TNCs around two dollars per mile, automated AVs are projected to initially slash these costs to less than a dollar per mile—on par with the total costs of owning a personal car. Eventually, as technology improves and adoption increases, the cost of autonomous robo-taxis is expected to decrease to even lower than the current gasoline and parking costs of a private car (Rocky Mountain Institute 2016). With Uber already providing more than 6 million rides a day, and Lyft over one million, it is not hard to see the gargantuan monetary incentive to eliminate the use of human drivers and shift passenger transport to shared vehicle fleets.

That said, AV technology is equally attractive for private automobile owners. Being able to drive door to door while working on a laptop or reading a book will make commuting a lot more attractive for a greater number of people. Most transportation policy-makers agree that both privately owned and TNC-run fleets of AVs are likely to shoot vehicle miles traveled (VMT) through the roof, shifting transit mode share to cars and creating new demand for vehicular travel through low prices. Even though TNCs could significantly reduce the number of privately owned and operated automobiles as well as the volume of parking spaces needed to accommodate a city's mobile population—by as much as 17 times by some idealistic estimates (OECD/ITF 2015)—they do so by making much more use of the vehicles they deploy. A city with fewer vehicles will paradoxically have more traffic on streets.

Some have argued that this "VMT hell" scenario can be avoided if more riders shift to car-pooling (Sperling 2018). Car-pooling has been given a boost through Lyft Line and UberPool services, which combine a number of passengers into the same vehicle by offering lower rates. But these services remain small compared to both companies' main business of serving individual clients.[1] The relatively low adoption rate of car-pooling has been unfortunate for cities, where Uber and Lyft have already been reported to increase overall traffic on streets, some of which comes from public transit ridership (MAPC 2018). There is little indication that electric and autonomous vehicles are going to change this trend. Even if a car-pooled ride on Uber would cost 10 cents as opposed to 20 cents a mile, the absolute cost savings on a typical trip might be less than a dollar—small enough for the majority of riders to opt for a shorter, more direct, and more predictable trip.

What do these combined transformations mean for streets and how will they impact a city's morphology? New spatial analysis and morphological projections are needed to explore the possible effects of shared/autonomous/electric cars on city form. Spatial accessibility models can be used to evaluate the effects of robo-taxis on other transportation mode shares—walking, biking, and transit use—and by extension, the physical patterns of development that are likely to ensue. How people will choose to travel to destinations in their city depends on the costs and benefits of different modal alternatives. If robo-taxi trips are truly cheaper, safer, faster, and more convenient than public transit, we are likely to witness a modal shift among those demographic groups who benefit. But as more people shift, new congestion will emerge as robo-taxis still take up more space to move a passenger than public transit. Congestion and slower travel speeds thus pose a limit to automotive mobility in dense city centers. A familiar trade-off emerges between a collective demand for dense, mixed-use urbanism, and comfort of personal mobility in individual vehicles. If we choose to prioritize denser forms of urbanism, then collective transit will have to play a key role in moving people far into the future. If we don't, we will

[1]Uber Pool's share is said to be around 40% in San Francisco, one of the most active ride-sharing cities in the US, but likely much lower in typical cities. Uber and Lyft don't publish exact figures. Source: Personal communication with an Uber Policy representative.

witness an even greater increase in sprawl, dispersal, and traffic than the first rate of motorization produced. A significant drop in mobility costs could also radically restructure urban land values, decreasing real estate prices in downtowns and other centers while spreading the urban land value rent gradient over a larger hinterland around a metropolitan area. But if AVs are configured to primarily support and complement public transit, the opposite land value pattern could emerge, inflating prices at already existing centers and catalyzing denser development around them. These morphological alternatives need to be modeled, simulated, disseminated, and ultimately discussed at a political level.

Which scenarios will dominate depends to a large extent on how policy-makers and planners respond to the disruptive forces of transportation technology (Davis 2018). Most experts agree that vehicle traffic on roadway infrastructure is likely to increase, and there is not much cities can do to stop that (Sperling 2018). Congestion tolls, which are in place in Singapore, London, and Stockholm, offer one remedy against traffic externalities that single-occupant vehicles create, but such restrictions on the free movement of automobiles have been politically contentious and very hard to implement. Given that AVs have to be location-aware and GPS tracked anyway, it may be that previous privacy concerns will be overcome, and that citizen pushback about the growing number of driverless vehicles operated by private transport across more firms will translate into greater political support for congestion charging. But private automobile mode share is likely to grow nevertheless, and at least in part at the expense of public transit ridership. An increase in cars and a decrease in transit ridership are bad news for cities who have been trying to shift mode share to public alternatives and active modes for decades. A combination of congestion tolls and taxes on single-occupant vehicles will be necessary to curtail traffic externalities that low-priced AVs will create, but these may be politically contentious options. And even if successfully introduced, the question is whether such revenues will go back into city infrastructure; and if so at what scale and for what purpose? Chicago's ride-share tax is expected to generate over $80 million this year, all of which the city has committed to public transport funding. There is also a movement afoot among city planners to increase investments into pedestrian and bicycle infrastructure. To the extent that new mobility solutions can generate significant opportunities for revenue that can be directed to services that have been difficult to maintain for decades, we may begin to see a repurposing of urban infrastructure. The selection of which to prioritize, however, will have a direct bearing on the future of streets, particularly as some options will enhance vehicle traffic and others will reduce it.

Perhaps even more worrisome is the potential impact of AV technology on pedestrian interactions with cars, and on the ambulatory environment of streets more generally. The technology used to locate, guide, and drive autonomous vehicles relies on real-time feature recognition from a range of different sensors. Incoming feature recognition is then processed with respect to a huge database of case information to decide which way to best operate the car in the next split second. While almost suited to navigation amidst other cars, AV technology faces significant shortcomings when it comes to interacting with pedestrians or bicyclists, who use eye contact and body language to communicate with human drivers (Ball 2018). When pedestrians or bikers step into vehicular traffic today, a rather complex communication process necessarily follows. Both look at each other and try to guess how the other will behave. Will the pedestrian step on the road or not? Will the driver stop? A pedestrian, who has confirmed eye contact with the driver might decide to step on the road deliberately, hoping that the car will stop. But it is never certain—the decision is made by a human behind a wheel and fundamentally unpredictable.

These are exactly the conundrums facing artificial intelligence experts who are deploying their skills to predict behavior in an environment where streetscapes and the culture of pedestrian-vehicle interactions are in flux. If the pedestrian would know that the driver always sees her and stops if she steps on the road, as in the case with AVs, she wouldn't think twice about the situation and could fearlessly march forward (Miller-Ball 2016). And although terrific for walkability advocates, such trolling would create frustration for AV operators, whose expensive machines would consequently move much slower in pedestrian-heavy downtown traffic than conventional, human-driven four-wheelers. Being able to drive at least as fast as traditional cars on populated city streets makes a significant incentive for AV companies to advocate for exclusive AV lanes, where pedestrians are physically separated from cars, or for clear curb limits, the crossing of which would result in hefty jaywalking fines. Such traffic interventions would decrease pedestrian options on downtown streets, dramatically reducing access to amenities and public spaces on foot or by bike (Sevtsuk 2014; Sevtsuk and Kalvo 2017). In the longer run, this could alter the location patterns of retail and service amenities along streets, moving the needle back toward car-oriented accessibility (Sevtsuk et al. 2016). We should not forget that the term "jaywalking"—inspired by the word "jay," which referred to a country bumpkin—was, in fact, socially constructed by auto-advocates of the 1910s and 20s, during the first wave of mass motorization (Norton 2007). Before private automobiles came to dominate city streets, there was no need for marked crossings or traffic lights—pedestrians could cross streets anywhere and jaywalking was really just walking (Fig. 1).

The point here is that no matter how technologically advanced automated vehicles become, their deployment on city streets competes with non-motorized activities that many cities have come to encourage as part of their social, economic, environmental, and public health priorities in the past decade. A number of mayors have implemented traffic calming, public transit priority, and "complete streets" projects that give more space and higher priority to pedestrians. Under Mayor Bloomberg, and his transportation commissioner Janette Sadik-Kahn, New York City converted more than 70 acres of the roadway to pedestrian plazas, including the world famous Times Square (Luberoff 2016). Even so, hard-earned progress on reversing the dominance of cars over pedestrians on "complete streets" is likely to face considerable opposition from AV lobbyists (NACTO 2017).

From a purely navigational standpoint too, AV feature recognition technology will require clear lane markings to situate each vehicle. If people are not required to grab the wheel when familiar road features disappear, and a vehicle gets confused—as is the case with the differently abled, under aged, or simply passengers in the back seat—then lane markings need to be uninterrupted and omnipresent, on every street,

Fig. 1 Exclusively vehicular roadway, which pedestrians must cross using elevated overpasses in Singapore. Photo by Andres Sevtsuk

intersection, and parking lot. Even a temporary road construction site will need to rearrange lane markings to support safe AV navigation around it. Similar to airport tarmacs, AVs could introduce a lot more "graffiti on the roads", as Jan Gehl likes to call it (Gehl 2010).

The widespread adoption of shared, electric, and autonomous vehicle technology will also mean that streets will need ubiquitous pickup and drop-off zones, numerous emergency parking spaces, and curbside electric charging stations. These latter features are probably preferable to current parallel parking spaces, leading to fewer idle four-wheelers on the curb and more space available for other sidewalk activities. But there could also be competition for that extra sidewalk space from a different type of AV—one that rolls along sidewalks. It is not merely passenger vehicles that will transform streetscapes and cities, infrastructure networks. Due to a rapid increase in e-commerce and home deliveries, we are already seeing a deployment of delivery robots that take packages from autonomous delivery trucks and carry them on sidewalks to designated drop-off zones, doors, and building lobbies. Like self-driving cars, they observe the world through sensors and try to dodge oncoming obstacles. San Francisco streets had three such companies in competition—Dispatch, Marble, Starship—until they became a significant nuisance to pedestrians, prompting Mayor Ed Lee and the city's Board of Supervisors to impose strict regulations on their deployment at the end of 2017 (Simon 2017). Meanwhile, other cities and states are welcoming bots on their sidewalks (Fig. 2).

Fig. 2 Starship package delivery vehicle. Photo by Starship

As we enter into a new world where technology is transforming the city and the ways that vehicles and pedestrians move through it, it is worth taking a step back and reflecting on where we are heading. Before rushing to rebuild streets according to the newest transportation technology, urban authorities and city planners should stop and consider how proposed changes fit with existing priorities of making streets more people-friendly and less machine-oriented, as well as more sustainable and less unequal. The space of a street is a highly complex economic, social, and political environment that enables communities and even democracies to work. Some of the earliest functions of streets were, in fact, not driven by mobility needs at all, but rather religious, cultural and social rituals (Rykwert 1978). Non-transient uses of streets remain equally critical today. Contemporary streets function as social spaces where face-to-face interaction occurs between passing pedestrians, where people see others while sitting on a bench or through a glazed shop window, or where a street performance or sidewalk gathering may contribute to the formation of a public sphere. Streets also function as economic space that shop owners, street vendors, and business associations cherish. They serve as play areas for children, and as spaces of livelihood for the homeless. These competing uses often come in contact and conflict with one another—street vendors fight over sidewalk space with shop owners, protesters with law enforcement and homeless folks with passersby who feel uncomfortable at their site (Loukaitou-Sideris 2011). Because of this, regulation will play a key role in determining whether and how unmanned machines enter such heated political battles over street turf.

Many city transportation departments have endorsed modal frameworks that prioritize people as they walk, bicycle, and take transit over trucks and cars on city streets (Chicago DOT 2012). Though the enforcement of these priorities is relatively new in US cities and applied only when streets require renovation, they do establish a policy framework that should be continued and extended before autonomous technology is rolled out *en masse*. Pedestrians, public transit riders, and bicyclists at the top of the pyramid should have higher priority over street space than cars and delivery vehicles. Where transit is lacking, or not viable, multi-occupant vehicles should be encouraged as a second best options. Single-occupant vehicles, and even more so zero-occupant "zombie cars", should be taxed to internalize the congestion externality they create. Package delivery robots should not be disallowed, but they should always give way to other users of sidewalks. All private vehicles on streets, either human-driven or autonomous, should give way to pedestrians and bicyclists, even when people step into vehicular lanes at unmarked locations.

Recent "complete street" design guidelines published by the National Association of City Transportation Officials (NACTO) have advocated for a clear functional separation between modes. A clear delineation of space for each type of activity—driving, walking, biking, parking—may help ensure that the less powerful and more vulnerable users of the street still command their share of street space. However, an excessive functional classification also has its problems and follows a Modernist paradigm of spatial separation, creating "a place for everything and nothing out of place". It emphasizes what Isaiah Berlin has called a negative notion of freedom—freedom *from* interference from other modes of travel or other activities that take place on a street (Berlin 1958). Berlin juxtaposes this approach with an alternative, positive concept of freedom—a freedom *to*.... In the context of street design, positive freedom might entail freedom to veer into other modal space, to cross a street at an arbitrary location, or to use a street for an activity that doesn't have its own provision—vending, sleeping, playing or gathering. An adoption of a modal hierarchy, such as that depicted in Fig. 3, could establish the rules for how the freedom to divert, and freedom to improvise applies across the different modes. A positive conception of freedom would ensure that pedestrians, onlookers, protesters, and street vendors would never be confined to the spaces they have been allocated, but can always improvise and invent, even if it interferes with the movement of vehicles or packages. The freedom of pedestrian and bicycle movement not only enhances local accessibility and contributes to higher built density and mixed-use development, but also augments multimodal accessibility in a metropolitan area as a whole in ways that automobile mode share alone can never achieve.

To the extent that building flexible streetscapes will contribute to greater public sociability, enhance the serendipity of chance encounters, and enable the cosmopolitan circumstances that define cities and produce their widespread cultural appeal, rethinking the future of streets in terms other than pure efficiency or Modernist land use logic may, in fact, be one of the most needed undertakings of the contemporary era. State-of-the-art technological change will bring positive benefits in transportation servicing while also providing data and new metrics for better strategic planning in the face of changing city form. But if these same technological changes produce

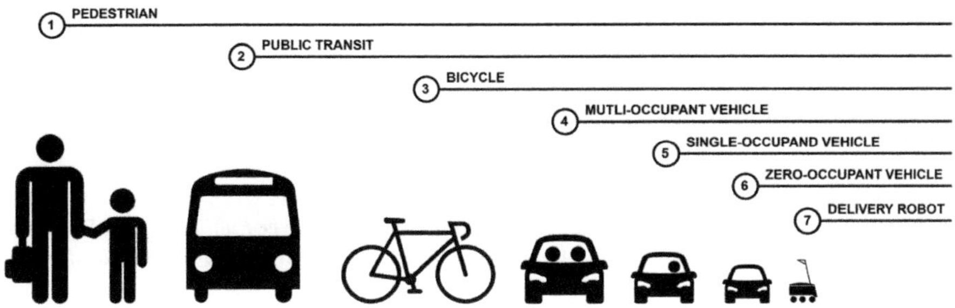

Fig. 3 Modal hierarchy in street design

street conditions that undermine the potential for human interaction, they will also introduce dystopian elements into what otherwise is being heralded as a relatively idyllic future. Let us not forget the important role that streets have played in social, political, and cultural life over the last several centuries. From Baudelaire to Benjamin, the ability to stroll has been identified as key to a vibrant public sphere and a means of connecting citizens to each other. Many major democratic transformations have emerged out of activities emanating from interaction unfolding at the scale of the street, as have significant innovations in city form and function—ranging from the introduction of urban parks and plazas to the proliferation promenades and arcades. Indeed, what some have termed "ambulatory urbanism"—or cities that are built around pedestrians more than vehicles—is precisely what has defined the most desirable cities of the world, and what differentiates European cities from their American counterparts (Davis 2015; Gehl 2010). With the automobile technology revolution, all this may be lost, unless we are prepared to rethink, reassess, and reconfigure the street's function to accommodate both past practices and future trends. It is time for urban morphology to move from retrospective analyses of historic city environments to projective depictions of changes in urban form, using spatial and mathematical models that evaluate potential outcomes of the present technological transportation revolution.

References

Anderson, J. M., Kalra, N., Stanley, K. D., Sorensen, P., Samaras, C., & Oluwatola, O. A. (2016). *Autonomous Vehicle Technology. A Guide for Policymakers.* Santa Monica, CA: Rand Corporation.

Ball, A. M. (2018). Pedestrians, Autonomous Vehicles and Cities. *Journal of Planning Education and Research, 38*(1), 6–12.

Berlin, I. (1958). *Two Concepts of Liberty.* Oxford University Press.

Chicago Department of Transportation. (2012). *Complete Streets Chicago*. Retrieved from https://www.cityofchicago.org/content/dam/city/depts/cdot/CompleteStreets/CompleteStreetsGuidelines.pdf.

Davis, D. (2018). Governance Capacity and the Smart Mobility Transition. In Greg Marsden & Louise Reardon (eds.), *Governing the Smart Mobility Transition*. Elsevier Ltd.

Davis, D. (2015). Ambulatory Urbanism. In *European Cities and Public Space*. Barcelona and Zurich: CCCB and Lars Mueller Publishers.

Energy Information Administration (EIA) (2016). The mix of U.S. energy production changes. Source: https://www.eia.gov/energyexplained/?page=us_energy_home.

Gehl, J. (2010). *Cities for People [Hardcover]*. Island Press; 1 edition.

Loukaitou-Sideris, A. (2011). *Sidewalks: Conflict and Negotiation over Public Space*. MIT Press.

Luberoff, D. (2016). Reimagining and Reconfiguring New York City's Streets. In *Transforming Urban Transportation*, Ed. Davis and Altshuler.

MAPC Research Brief. (2018). *Fare Choices. A Survey of Ride-Hailing Passengers in Metro Boston. Report #1.* Metropolitan Area Planning Commission. Boston, MA.

Miller-Ball, A. (2016). Pedestrians, Autonomous Vehicles, and Cities. *Journal of Planning Education and Research*, *1*(7), 1–7.

NACTO. (2017). *Blueprint for Autonomous Urbanism*. New York, N.Y. Retrieved from https://nacto.org/publication/bau/blueprint-for-autonomous-urbanism/.

Norton, P. D. (2007). Street Rivals: Jaywalking and the Invention of the Motor Age Street. *Technology and Culture*, *48*, 331–359.

OECD/ITF. (2015). *Urban Mobility System Upgrade*. Paris, France. Retrieved from https://www.itf-oecd.org/sites/default/files/docs/15cpb_self-drivingcars.pdf.

Rocky Mountain Institute. Charlie Johnson and Jonathan Walker. (2016). *Peak Car Ownership. The market opportunity of electric automated mobility services.* Retrieved from https://rmi.org/insights/reports/peak-car-ownership-report/.

Rykwert, J. (1978). *The Street: The Use of its History*. In Anderson, S., & Institute for Architecture and Urban Studies Ed. *On streets*. Cambridge, Mass.: MIT Press.

Sevtsuk, A. (2014). Location and Agglomeration: the Distribution of Retail and Food Businesses in Dense Urban Environments. *Journal of Planning Education and Research*, *34*(4), 374–393.

Sevtsuk, A., Kalvo, R., & Ekmekci, O. (2016). Pedestrian accessibility in grid layouts: the role of block, parcel and street dimensions. *Urban Morphology*, *20*(2), 89–106.

Sevtsuk, A., & Kalvo, R. (2017). Patronage of urban commercial clusters: A network-based extension of the Huff model for balancing location and size. *Environment and Planning B: Urban Analytics and City Science*, 1–21.

Sperling, D. (2018). *Three Revolutions. Steering Automated, Shared and Electric Vehicles to a Better Future.* Island Press.

Simon, M. (2017). San Francisco just put the brakes on delivery robots. *Wired*. December 6, 2017. Retrieved from: https://www.wired.com/story/san-francisco-just-put-the-brakes-on-delivery-robots/.

Understanding and Quantifying Urban Density Toward more Sustainable City Form

Steffen Lehmann

Abstract *Urban density* is a term used in urban planning and design to refer to the number of people inhabiting in a given urbanized area. *Density and compactness* are two closely related, but different, criteria relevant for the transformation of cities to become more resilient to climate change. While a high degree of compactness is desirable, too much density can be detrimental to liveability, health, and social well-being. More compact cities are an advantage and will help in curbing urban sprawl, but a consolidated urban form requires urban infill at densities that support compact self-reliant districts and mixed-use neighborhoods. The social dimension of such density increase is likely to be a future challenge and face community resistance. This essay argues that a quantitative approach to describe urban density helps to better understand cities; however, urban form is always more than just a mathematical formula, including nonquantifiable qualities of cities.

1 Introduction

Cities around the world are facing an ever-increasing variety of challenges to achieve sustainable urbanization. The issue of urban density is closely connected to how our cities will evolve and perform in future. *Urban density* is a term used in urban planning and architecture to refer to the mathematical number of people inhabiting a given urbanized area, and the amount of floor area built on a defined site. It is an important factor in understanding how urban development functions and architects worldwide are searching for ideal density models for tomorrow's sustainable cities. *Density* and *compactness* are two closely related, but different, criteria relevant to the transformation of cities to become more resilient to climate change. While a high degree of compactness is usually desirable, too much density can be detrimental to liveability, health, and social well-being. More compact and interconnected cities are an advantage as these will help in curbing urban sprawl, but a consolidated urban

S. Lehmann (✉)
School of Architecture, University of Nevada, Las Vegas, NV, USA
e-mail: steffen.lehmann@unlv.edu

L. D'Acci (ed.), *The Mathematics of Urban Morphology*,
Modeling and Simulation in Science, Engineering and Technology,
https://doi.org/10.1007/978-3-030-12381-9_30

form requires urban infill at densities that support compact self-reliant districts and mixed-use neighborhoods. The social dimension of such density increase is likely to be a future challenge and face increased community resistance (in Switzerland, the average realization time for a building has risen to 8 years). This essay argues that a quantitative approach to describe urban density helps to better understand cities; however, urban form is always more than just a mathematical formula, including nonquantifiable qualities of cities.

2 Defining Urban Density

In her book *The Death and Life of Great American Cities*, Jane Jacobs identified density as one of the four key ingredients for thriving and diverse cities. She wrote that in order for cities to work, "there must be a sufficiently dense concentration of people, for whatever purposes they may be there" (1961, 36). The question of urban density is closely connected to how our cities should evolve in the future. Medium- to high-density living is acceptable as long as these developments also provide for a number of requirements, such as an increase in good quality green space within walking distance in the neighborhood. What is needed is a formula and framework for high-quality urban density and more compact and interconnected cities. Consolidated urban form is achieved through urban infill at densities that support compact self-reliant and mixed-use neighborhoods. Today, density is one of the key issues in planning that can regularly create all kinds of misunderstandings and tension; it is an essential driver of our urban futures. We use the term "density" to describe the average number of people, households, floor space, or housing units on one unit of land, usually expressed in dwellings per hectare. There are different ways of mathematically measuring the density of urban areas:

- floor area ratio: the total floor area of buildings divided by the land area of the plot upon which the buildings are built (the *development plot ratio*, used as a measure of the density of the site being developed); the ratio is generated by dividing the building area by the site area;
- residential density: the number of dwelling units in any given area;
- population density: the number of persons living in any given area.

The *plot ratio* (also called floor area ratio, FAR; or floor space ratio, FSR) describes the ratio of a building's total floor area (gross floor area) divided by the size of the site (piece of land or plot) upon which it is built. The term refers to limits imposed on such a ratio—for example, to the maximum allowable ratio. For instance, a FAR of 3.0 indicates that the total floor area of a building is three times the gross area of the plot on which it is built, as would be found in a multiple-story building. The allowable plot area has a major impact on the value of the land, as higher allowable plot areas yield a higher land value.

Using such zoning regulations, municipalities have found it unnecessary to include height limitations for buildings when applying maximum floor area ratio calculations.

The FAR is commonly and successfully used in zoning regulations and planning guidelines to limit the amount of construction in a certain area. For example, if the relevant zoning ordinance permits construction on a site, and if construction must adhere to a 0.10 FAR, then the total area of all floors in all buildings constructed on the parcel must be no more than one-tenth the area of the parcel itself. An architect can plan for either a single-story building consuming the entire allowable area in one floor, or a multistory building which must consequently result in a smaller footprint than would a single-story building of the same total floor area. However, numerous urban planners have criticized the use of FAR regulation and argued that abdicating purely to floor area ratios (i.e., market forces) is the opposite of aiming for enhancing a community or neighborhood and for the diversity of ownership, as it is a poor predictor of physical urban form. Instead of pure FAR one could use the traditional design standards: building height, setbacks, or build to lines rules; these would also enable planners to make reasonably accurate predictions, recognize violations, feel secure in investment decisions and are more likely to deliver a better urban form outcome.

Urban population densities can vary widely from city to city. Asian cities have some of the highest densities (frequently reaching over 10,000 people per square kilometer, and sometimes even over 20,000 people, such as in Mumbai or Hong Kong, where a large proportion of the buildings are high-rise apartment towers). The historical European cities have lower densities and are based on the European compact perimeter block model, with densities in the range of 3,000–6,000 people per square kilometer. In the United States, Canada, and Australia, urban population densities are usually much lower, at around 1,000–2,500 people per square kilometer (Lehmann 2010; Density Atlas 2011). Hence, three clearly identifiable city typologies have their own characteristics, density profiles, and historical evolution:

- the European compact and polycentric mid-rise city with the traditional perimeter block (examples are Barcelona, Paris, Berlin, or Athens): 3,000–6,000 people per square kilometer;
- the Asian high-rise city with the distribution of individual towers (such as Shanghai, Beijing, Tokyo, and Bangkok): often around 10,000 people per square kilometer;
- the North American and Australian low-rise and low-density city typology with an urban downtown core surrounded by extensive urban sprawl (for example, Los Angeles, Phoenix, Melbourne, and Perth): only 1,000–2,500 people per square kilometer.

A high-density city is, therefore, a city that has a high average population density, high density of mixed-use built form, high-density subcenters, and high-density forms of housing. Many researchers have argued that a denser, more compact city is a more sustainable city (Hall 1988; Jenks et al. 1996; Hall and Pfeiffer 2000). Susan Roaf noted, "High density (not high rise) is probably the inevitable urban future" (Roaf 2008, 33). Today, most experts agree that compact living is sustainable living. As an important benchmark for minimum densities of new sustainable developments, the literature gives the figure of a minimum of 70 dwellings (homes) per hectare, but densities should preferably be closer to 100–120 dwellings per hectare, especially along transport corridors, to support the integration of public transport, walking and cycling to key facilities, and on-site energy generation. While a more compact city is more sustainable, expanding the city footprint farther and farther out into critical habitat areas, the loss of precious agricultural land and green space is now seen as environmentally unacceptable. Cities like Portland (Oregon, USA) have achieved good experiences by establishing an effective urban growth boundary that curbed its sprawl since the 1970s. Recent research shows that Portland's compact city design has reduced the average car use by as much as 2,000 km per person per annum.

3 Defining Density with Regard to the Compact City

There has been plenty of evidence that more compact cities with higher densities encourage the use of public transport and bicycles. It supports closer amenities, increases efficiencies of infrastructure and land use, conserves valuable land resources at the fringe, and is likely to reduce the carbon emissions of the urban dwellers (Hall 1988; Jenks et al. 1996; Lehmann 2005; Farr 2007; Toderian 2012). We should not be confused by the different denominations that have emerged to describe the "compact city", which is sometimes also called *green urbanism*, *sustainable urbanism*, *ecological urbanism*, among others—it all means the same thing.

So, what exactly is a *compact city*? A compact city is a mixed-use spatial urban form characterized by "compactness", which defines a relatively dense urban area linked by easy access to public transport systems and designed to have a minimal environmental impact by supporting walking and cycling (while low-density suburbs are incapable of supporting walking, cycling, or public transport). The compact city with four- to eight-story urban perimeter blocks and shorter block length represents the optimum use of space. However, the compact city concept is still controversial, there is no single model or formula that can be replicated as all cities, and sites are different. Development patterns with traditional urban characteristics usually have shorter block lengths with a network system of highly connected thoroughfares, local tree-lined streets, and residential alleys. Vehicle-dominated contexts have larger, longer blocks, less street connectivity, and usually no alleys; this pattern of longer blocks makes walking distances longer and, therefore, it is likely that fewer people will walk between destinations (Lehmann 2014).

The dense and compact city also increases efficiencies in urban infrastructure and services through shorter distribution networks. Higher density cities encourage reduced transit through shorter trip lengths since most amenities and public transport are more closely located. Today, architects can easily visualize scenarios, simulate the benefits of various density types, and inform policies and decision-making. Churchman (1999) defines compact city policies as policies that aim to intensify urban land use through a combination of higher residential density and centralization, mixed land use, and limits on development outside of a clearly designated area (e.g., an urban growth boundary). Jenks et al. (1996) outline three aspects of the compact city: it is high density, mixed use, and intensified. Urbanist Peter Rowe (2015) defines "urban intensity" through a formula with four factors: *Urban intensity = Density + Diversity + Connectedness + Compactness*.

However, urban density is more than just a mathematical figure. Making neighborhoods more compact and dense always needs careful consideration and a process of optimization to balance potential adverse effects and conflicting interests (Howard 1902). Higher density is beneficial at appropriate locations, but not always in every case. All urban areas have their particular social, cultural, environmental, and climatic conditions resulting in a complex urban microclimate, and density increases can affect urban wind speeds. The interplay between higher density and the increased risk of the urban heat island effect (which increases cooling energy needs) must be properly observed and taken into consideration. Density directly influences the urban microclimate. Negative effects on the urban climate can be improved by increasing greenery and vegetation and choosing materials and surfaces that minimize solar heat gain and increase the albedo effect (surface reflection).

To minimize adverse negative effects from increased densities, densification strategies should be coupled with high-quality urban design strategies, scenario testing, and real community participation, to avoid such unwanted effects as increased traffic congestion, overshading, and loss of daylight or privacy. Planners and architects need a better understanding of the impact of their design decisions on the overall performance of the urban precinct system and should introduce urban greenery in the densified areas.

Living in apartments is often the more sustainable solution, as urban perimeter blocks and residential towers share circulation systems, separating walls and roofs, therefore requiring less construction materials and heating. In the United States and Australia, researchers have now collected and analyzed the actual energy use data for a large number of residential units, and there is emerging evidence that living in inner-city high-rise buildings is a less energy-intensive lifestyle—all other things being equal—than in equivalent low rise buildings in suburbs, despite the need for elevators. This is mainly due to two aspects: the suburban house is most of the time larger and very energy intensive because of the air conditioning and other consuming devices; and the other reason is the need to commute by car to the workplace.

Nevertheless, when is a city getting too dense, and at what point is a quarter overdeveloped? Each time, higher densities require an optimization process since higher densities can create challenges for planners and designers; for instance, avoiding overshading, overlooking, loss of daylight, and loss of privacy demands clever design solutions. There are a number of other arguments against high density, which include the risk of increasing traffic congestion in the area and a potential increase in noise disturbance. We can also point to districts where densities were developed too high and the developments failed, because the lack of natural air ventilation or daylighting created unhealthy and unhygienic conditions. A well-known example of such over densification was the extremely dense Kowloon Walled City in Hong Kong, which was demolished in 1992–1993 because of the many issues that arose out of its extreme hyperdensity. It is estimated that over 50,000 people lived in squeezed conditions in dark cramped flats, on a small parcel of land of only 2 ha. While Kowloon

Walled City was a functioning urban community, it was not a sustainable and healthy place to live.

4 The Ongoing Density Debate

As urban populations and economies are expanding, and with increasing numbers of people joining the middle class (earning and spending more), consumption, energy demand, and waste generation are all rising. Due to our obsession with economic growth, the GDP-driven growth model, and excessive use of finite resources, global greenhouse gas emissions keep rising—despite all the efforts of the past 20 years to reduce them. It appears that there is a growing gap between current urbanization patterns and what would really be needed to shift to more sustainable urban futures.

Numerous governments are now actively pursuing planning policies that encourage greater residential density and support a set of new densification measures to increase housing densities (e.g., in the UK, Government is giving local councils increased powers to refuse development on the grounds of insufficient density). As more and more people live in cities, the cities have taken center stage as key players in the future of human populations. City management, governance, urban mobility, liveability, and density have all become key themes of focus for politicians and decision-makers to succeed in managing urbanization, but in conditions of rapid urbanization (especially with the dynamic exploding urbanism of Asian and African cities), controlled sustainable development has not always been achieved.

One core challenge for cities in the future will be the tension between urban form, compactness, and liveability. The modern city is always about diversity, which includes varying urban densities for different neighborhoods in different parts of the city (Howard 1902; Lynch 1960; Jacobs 1961; Rowe and Koetter 1978; Koolhaas and Mau 1995; Kostof 1999; Pont and Haupt 2010). This diversity of density types allows different demographic groups to choose how they would like to live at varying stages of their lives; for example, young professionals are now streaming back into the city and do not opt to live isolated in suburbs or far away from amenities and their workplace, expecting a more cosmopolitan lifestyle. Cities where residents do not need to drive much and efficient public transport is available have many advantages. It appears that the conflicting demands have always to be balanced through good design solutions.

Hence, *urban density* and *mixed use* are key factors in determining the sustainability of a neighborhood and its urban liveability. Urban districts have a significant complexity about them and, clearly, there is still a need for more research, comparative data, and an evidence base on the benefits and disadvantages of more dense and compact cities, which has frequently been noted by different scholars (Register 1987; Hall 1988; Breheney 1992; Jenks et al. 1996; Cuthbert 2006; Farr 2007; Girardet 2008; Lehmann 2010; Mostafavi and Doherty 2010; Beatley 2014).

There is now a strong emphasis on brownfield development and the introduction of minimum density targets for housing developments around transport hubs and along

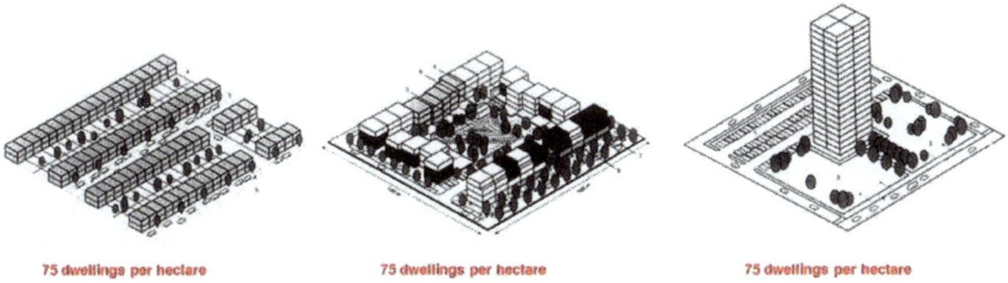

Fig. 1 Comparative diagram illustrating the different ways how 75 dwellings per hectare can be realized, from rows of terrace houses (left), to a perimeter block, or a single high-rise tower (right). Planners and architects need a better understanding of the impact of their design decisions on the overall performance of the urban quarter (Provided by the author)

transport corridors (what is commonly known as transit-oriented development). However, the link between urban density and sustainability remains a contested and often misunderstood subject of planning theory. One challenge cities face in their densification strategies is fierce resistance from residents and community groups to higher densities, as illustrated by the case of Vancouver's protesting neighborhood groups. However, if done well, higher density does not necessarily decrease liveability (as can be seen in cities such as Singapore, Barcelona, or London). Higher density living is acceptable as long as these developments also provide for new parks, gardens, and urban greenery within walking distance in the neighborhood; but a well-used park, a drive away cannot be a substitute for new green space. High-quality urban design can alleviate negative perceptions of density at the metropolitan scale. (Lehmann 2014, 708–719)

Density brings people together, and doing density well is as much about providing privacy as it is about civic life. Higher densities require new better housing typologies, a wider range of compact housing models, and innovative design solutions that integrate urban greenery and high-quality public space (Fig. 1).

5 A Proposed Framework for "Quality Density"

Cases of recent higher density housing developments have enabled residents to live closer to their workplaces. Such intensification through urban infill at appropriate density is a sustainable design strategy, as it avoids and counteracts the further dispersion and fragmentation of activity centers and helps to reduce car dependency. Urban densities must remain within a sustainable range. If density is too low, it must be allowed to increase, and if it is too high, it must be allowed to decline, to arrive at the most appropriate "quality density." However, most of the time densities are too low in cities around the world and are declining, and there is now a concern among urban experts worldwide about declining urban densities in cities globally—a situation that is exacerbating urban challenges such as sprawl and traffic congestion.

Quality density cannot always be planned or predicted; it depends on the socioeconomic context. While every city is different, some guiding principles of ideal development with quality density have been identified (Bay and Lehmann 2017), which include increasing compactness and the integration of public transport, greenery, and mixed usage. Policy makers have now to take decisive, forward-looking steps in urban planning and decision-making on density to create room for long-term physical development toward a sustainable city. As a framework for quality density, the following components have been identified as prerequisites for the appropriate quality density of a compact city:

- a strong alignment of land use and mobility: the *efficient public transport city*;
- proximity and "nearness" to amenities: the *walkable city*;
- to keep cities cool, the integration of urban greenery and green roofs need to go hand in hand with densification: the *green city*;
- high-quality architectural and urban design with more diversity and better examples of residential infill through four- to eight-story projects: the *mixed-use city*;
- more innovative design solutions need to be developed to ensure there is no negative impact on neighboring sites from densification, such as loss of privacy (overlooking) or loss of daylight (over shading): the *city of innovative housing solutions*.

Groups of buildings that form quarters are of great interest, as these can support the public realm, perform and interact, with a density of mixed functions, and connect with each other. Very dense high-rise cities are not necessarily the best option. Medium density, compact infill developments of four- to eight-story perimeter blocks are the much-preferred option and a very useful model, as the perimeter block combines a number of benefits, such as:

- smaller building envelopes (good ratio between the area of the facade and the enclosed volume), using less land and reducing heat gain in summer and heat loss in winter;
- less material used, therefore lower construction impact and reduced embodied energy;
- reduced energy consumption due to shared walls, circulation, and roofs.

6 Learning to Live in More Compact and Denser Communities

While urban density is extremely relevant, it is of course not the only determining factor for urban form. In fact, urban developments are rarely the pure result of design considerations; rather economic forces, the evolution of policies, and a range of invisible forces such as land use regulations, codes for floor space ratios, and economic power structures, shape them. There is always a multiplicity of complex forces and flows that form a city and the forces that are shaping the city are not limited to the

physical spectrum only. Long-term trends in economies, energy supply and demand, geopolitical shifts, and social change are all additional drivers of urban development.

How dense we plan our cities to be determines how efficiently we use vital resources and directly influences the quality of life of urban citizens (OECD 2012). Growth boundaries are an effective tool to contain the footprint of cities, in filling in already built-up areas to avoid urban sprawl, as cities cannot continue to expand their boundaries due to population increases. For a long time, the high infrastructure costs and inefficiencies caused by urban sprawl have somehow been accepted on the wrong assumption that sprawl would provide affordable housing. Now planners will need to increase the suburbs' densities and transform outdated urban values into acceptance of higher densities and public transport.

There is definitely a limit to urban density. Arguing for a new ethics of the urban, we can say that the traditional urbanism of the European city (such as in Barcelona, Paris, Berlin, or Athens) is also ecological urbanism (Lehmann 2010). Urban density should be embraced more strategically as the answer to a number of things wrong with today's urban developments. It is time to begin thinking about our cities in a completely new way.

References

Bay, J.H.P., and Lehmann, S., eds. 2017. *Growing Compact: Urban Form, Density and Sustainability*. London: Routledge.

Beatley, T. 2014. "Imagining Biophilic Cities", in: *Low Carbon Cities*, S. Lehmann (Ed.). London: Routledge.

Breheney, M., ed. 1992. *Sustainable Development and Urban Form*. London: Pion.

Churchman, A. 1999. "Disentangling the Concept of Density." *Journal of Planning Literature*, 13(4): 389–411. London: SAGE.

Cuthbert, A. R. 2006. *The Form of Cities: Political Economy and Urban Design*. Oxford: Blackwell.

Density Atlas. 2011. Planning resource developed by MIT faculty, available online.

Farr, D. 2007. *Sustainable Urbanism*, New York, NY: McGraw-Hill Publishers.

Girardet, H. 2008. *Cities, People, Planet: Urban Development and Climate Change*, London: Wiley.

Hall, P. 1988. *Cities of Tomorrow: An Intellectual History of Urban Planning and Design in the Twentieth Century*. Oxford and New York: Blackwell.

Hall, P., and Pfeiffer, U. 2000. *Urban Future 21: A Global Agenda for 21st-Century Cities*. New York, NY: Spon.

Howard, E. 1998/1902. *Garden Cities of Tomorrow*. London: Faber & Faber.

Jacobs, J. 1961. *The Death and Life of Great American Cities*. London: Cape; New York: Random House.

Jenks, M., Burton, E. and Williams, K., eds. 1996. *The Compact City: A Sustainable Urban Form?* London: Spon.

Koolhaas, R., and Mau, B. 1995. *S, M, L, XL*. New York, NY: Monacelli Press.

Kostof, S. 1999. *The City Shaped: Urban Patterns and Meaning Through History*, 2nd ed. New York, NY: Thames & Hudson.

Lehmann, S. 2005. "Towards a Sustainable City Centre: Integrating Ecologically Sustainable Development Principles into Urban Sprawl." *Journal of Green Building*, 1(3): 83–104.

Lehmann, S. 2010. *The Principles of Green Urbanism*. London: Routledge.

Lehmann, S. 2014. *Low Carbon Cities: Transforming Urban Systems*. London: Routledge.

Lynch, K. 1960. *Image of the City*. Cambridge, MA: MIT Press.
Mostafavi, M., and Doherty, G. (Eds.). (2010). *Ecological Urbanism*. Baden, Switzerland: Lars Mueller.
OECD. 2012. *Compact City Policies: A Comparative Assessment*. OECD Green Growth Studies. Paris: OECD Publishing.
Pont, M. Y., and Haupt, P. 2010. *Spacematrix: Space, Density and Urban Form*. Rotterdam: 001NAi Publishers.
Roaf, S. 2008. "The Sustainability of High Density. Population and the People", in: Designing High-Density Cities for Social and Environmental Sustainability, E. Ng (Ed.), 2009, London: Earthscan: 27–33.
Rowe, C., and Koetter, F. 1978. *Collage City*. Cambridge, MA: MIT Press.
Rowe, P. G. 2015. "Urban Density as a Function of Four Factors." YouTube podcast of Rowe's presentation at the Centre for Liveable Cities, Singapore, September 2015. Accessed October 25, 2015, at www.youtube.com/watch?v=IXahgHQEuMI.
Toderian, B. 2012. "Density Done Well." YouTube podcast of Toderian's presentation at the Vancouver Urban Forum 2012, June 2012. Accessed October 25, 2015, at www.youtube.com/watch?v=eRk93Wgdv1g.

To Not Talk Past Each Other: An Immodest Proposal for Cross-Conceptual Research in Urban Morphology

Michael P. Conzen

Abstract The field of urban morphology has seen welcome conceptual expansion in recent years, particularly in the spheres of mathematical modeling and topological analysis. This has posed a challenge to conceptually integrate such inherently quantitative and frequently abstract thinking with that of fundamentally historical approaches to the field in which time and place are essential culturally dependent explanatory factors. The issue is not one of the conflicts between quantitative and qualitative explanations, but of reconciling highly variable sources of evidence and definitions of relevant context. Cross-conceptual studies in urban morphology aimed at bridging perceived gaps in communication have been on the rise, yet a carefully calibrated test of all the multiple analytical approaches that have emerged applied to the urban morphology of a common geographical case remains untried. Issues of spatial scale in structuring an investigation, the relevancy of different measures of urban form to be examined, and the appropriate depth of societal context to be factored in could then be adequately addressed in direct juxtaposition. It is presumed a significant enhancement of both conceptual cross-fertilization and enriched understanding of the urban morphology of the particular case involved could result. This essay concludes with a call for such a collaborative test.

Within a relatively short span of time, the field of urban morphology has shown a remarkable tendency to sprawl across an increasing number of intellectual disciplines. As a particularly systematic and focused approach to the study of what is more generally referred to as urban form, the term "urban morphology" is creeping into more scientific literature than ever before. As cities have acquired mammoth physical dimensions and spiraling organizational complexity, it has been less and less possible to ignore the implications of the configuration of built environments in the lives and functioning of urban centers. This does not necessarily make a "morphologist" out of every scientist and scholar invoking urban morphology in the titles of their publications to signal their interest in the physical construction of cities,

M. P. Conzen (✉)
Committee on Geographical Sciences, The University of Chicago, Chicago, IL, USA
e-mail: m-conzen@uchicago.edu

L. D'Acci (ed.), *The Mathematics of Urban Morphology*,
Modeling and Simulation in Science, Engineering and Technology,
https://doi.org/10.1007/978-3-030-12381-9_31

but it does show a growing awareness of its significance in the mix of issues and factors under investigation, and, by their citations, to their knowledge of urban morphology as reflected in its accumulated specialist literature. At one extreme, urban morphology is for some writers a fashionable term that can be used *sensu lato* to invoke patterns of urban existence of almost any nature and configuration. At the other, urban morphology is a discrete field with a specific history, purpose, and set of conceptual approaches and methods.

While some historians of the field may claim the origins of urban morphology as a distinct sphere of thought should be found in the writings of the ancient Greeks (Cataldi 2018), its expression as an organized body of knowledge is far more recent. Many commentators link the emergence of a scientific consciousness of the built environment as a comprehensive spatial complex to the appearance of Leonardo da Vinci's 1502 map of Imola and the refined cartography of Nolli's 1748 map of Rome (Pinto 1976; Verstegen and Ceen 2013). And it is unsurprising that surveyors and architects, with their predisposition toward achieving areal accuracy, would be the first professionals to develop cartography as the medium for representing the detailed physical structure of whole cities. Nevertheless, it was left to geographers at the turn of the twentieth century to develop the first theoretical formulations of city structure based on urban ground plans and townscape characteristics (Schlüter 1899; Geisler 1924). By the middle of the twentieth century, the disciplines of architecture and geography would again produce theoretical advances through the foundational work of Saverio Muratori and Gianfranco Caniggia (architects), and M.R.G. Conzen and J.W.R. Whitehand (geographers) (Cataldi et al. 2002; Moudon 1997; Whitehand 1981). Both approaches, while differing significantly in their modes of specific conceptualization, draw heavily for evidence from the historical cartographic records of cities and all the difficulties that imply. Their approaches in turn led during the following decades to sustained work in what have since been characterized as the typo-morphological and the morphogenetic traditions of urban morphology.

In the 70 years, since Muratori and Conzen published their signal studies, urban morphology has greatly matured, not only through vigorous addition of concepts clearly identifiable with these two traditions but also through the emergence of two more, equally distinct but very different, conceptual approaches: spatial modeling and space syntax. Associated with the extensive work of Michael Batty and Bill Hillier, these newer fields rode the crest of the so-called quantitative revolution that began to redefine the social sciences in general during the 1960s, and have arguably had a more immediate impact on urban planning and related management policies, perhaps because these two research clusters are more directly amenable to anticipating urban futures (Batty 1976, 2005; Hillier and Hanson 1984; Hillier 1996; D'Acci 2013). Both specialties rest heavily on various types of mathematical formulation and have benefitted extensively from the rise of topological thinking and automated computation methods.

The first two approaches are closely allied with historical reconstruction of evolved townscape features (layout, building designs), in which the historicity of urban landscapes is critical as context and process, whereas the latter two have their origins in urban planning and rest more immediately on abstract mathematical principles

and generative models of projection (through simulation) and prediction. These four foundational approaches to contemporary urban morphology have been ably summarized in Oliviera (2016, 118–130) and Kropf (2017, 17–18).

In the last few decades, further spheres of substantive research interest in urban morphology have opened up, as much through investigations in disciplines beyond the established ones in the field, and owing their impetus to work in the physical sciences, additional social sciences, and the humanities. This is not the place for a detailed review of the contributions emerging from allied disciplines, but they may be summarized as follows: (a) ecological approaches that examine the urban biome and how it contributes to and responds to the morphology of places (including in particular climatological phenomena both micro and macro in scale, vegetation patterns both humanly planned and wild, and geological processes in cities, including catastrophic events); (b) the perceptual–emotional "effect" of the city's morphological content and composition on the individual psyche and group attitudes; (c) the role of modern urban design paradigms; (d) urban political and planning policy considerations; and (e) humanistic–artistic–creative engagements with the built environment. References to literature in the relevant sections of these fields lie beyond the remit of this chapter, but as hints of its breadth, see Ng et al. (2011), Batty (2001), and Foote (1997). The argument advanced here is that the expanding interdisciplinary reach of work in urban morphology, together with its contemporary societal relevance and the mostly channelized and often fragmented mode of research that supports them, lead to a single realization: that the field is in need of as-yet underdeveloped strands of cross-conceptual cooperation.

1 Desire for Integration

Jeremy Whitehand, in his panoptic position as founder and longtime editor of *Urban Morphology*, has sought through his opening editorial comments in the journal to foster not only interdisciplinary communication but also conceptual diversity. At the same time, he has long argued for integrating and creating coherence among the substantive contributions to the field, so that its conceptual richness might be more apparent to both those working in the field and those outside it (Whitehand 2015, 2016). More recently, Whitehand has drawn renewed attention to the barriers that resist the integration of the field: linguistic hurdles for many international scholars, the paucity of cross-disciplinary digestion of research findings, and the near chasm between research and practice. He cites as an example of noncommunication the gap between studies undertaken by self-described urban morphologists and those of researchers in "architectural geography" (Whitehand 2018). In speculating on the significant distance between the outlook of the latter and that of Caniggian architectural morphologists, for example, he is in essence commenting on cross-conceptual silences.

2 Cross-Conceptual Initiatives

There have, of course, been concerted attempts to overcome such barriers. As noted above, the recent appearance of Vitor Oliviera's succinct text surveying the field of urban morphology, and Karl Kropf's handbook, designed to bridge the gap between researchers and practitioners, both represent didactic landmarks in sorting out the substantive and conceptual topography of the field. And well before this, there have been isolated forays into the challenging terrain of cross-conceptual linkage. For example, Griffiths et al. sought to combine space syntax methods with historico-geographical thinking in a study of the streets of suburban centers in Greater London, demonstrating that the two approaches can prove complementary, but that the former could not solve questions posed by the latter (Griffiths et al. 2010). Oliviera et al. applied the concepts of all four of the major conceptual approaches discussed earlier to the Portuguese city of Porto, and concluded that, out of the mix, "morphological regions" might in the future offer the most promising means to integrate the separate results of each approach (Oliviera et al. 2015).

In a different vein, Stephen Marshall applied what he calls area structure and architectural morphology methods in examining in an abstract way the potential content of city blocks, claiming to unite space syntax with other approaches to buildings in space (Marshall 2015). These examples of cross-conceptual comparison studies, the result of local decisions to test differing precepts and analytical methods in a commonplace, suggest the possibilities of integration but without coming to startling conclusions. Where space syntax has offered the most potential for cross-pollination is the now-common reconstruction of evolving connectivity measures of street systems in conjunction with traditional studies of building types and land division measures. Streets as public circulation spaces are the simplest of morphological features to measure over time, but they form only a small part of a city's morphological composition. Nevertheless, extensive studies are now appearing that offer a remarkable array of comparative cases (Mahbub 2017).

A fourth cross-conceptual comparison is exemplified by Gökce and Chen (2016), who seek to unite the typological process with studies of sense of place. They suggest that quality of life (and by implication human happiness) is related to sense of place, which in turn is closely related to the layout of house, street, and neighborhood.

A more personalized challenge was posted at an International Seminar on Urban Form conference in Newcastle upon Tyne, England in 2004, when it was proposed that the historico-geographical approach be tested in Como, Italy, where Gianfranco Caniggia developed his system of typo-morphological analysis, and, to balance it, that the Italian typo-morphological approach be tested in Alnwick, in northern England, where M.R.G. Conzen worked out his historico-geographical system (Marzot 2005). In time, the challenge was addressed in part: a study of the internal morphology of Como according to Conzenian methods was published (Conzen 2010), and a hypothetical Caniggian-inspired interpretation of the region surrounding Alnwick appeared, although unfortunately it lacked any examination of the town's internal morphology (Cataldi 2013). So, the idealistic aspiration that the direct application of contrasting investigative schemes could be applied in reverse to these two iconic research sites remains for the moment unfulfilled. The internal morphological character of Alnwick, Northumberland, awaits its truly Caniggian interpretation.

3 Evidence and Theory

Regarding theoretical integration, the simplest conclusion is that distinctive analytical approaches to the immensely complex morphological character of cities each possess their specialized value, and the lack of a metatheory that forces them together is hardly to be deplored. Reviewed from the developmental history of the field, and just considering the "big four" approaches already discussed, the ontological progressions are clear both in the realm of evidence and of theory.

Data for urban morphology have progressed over time from being derived from, stored in, and analytically deployed almost wholly within the realm of mapping, historical and otherwise, to that of the present situation in which not only has remote sensing of many kinds vastly increased our ability to access spatial information about the details of the built environment, but now we also can collect, store, and examine huge amounts of data about the city's physical content in statistical form. The field has transformed from a dependence on maps as our data silos to a modern dependence on computerized data banks as our tabular silos—from which maps of all kinds can be generated with ease.

On the theoretical front, the largest impediments to integration appear to lie in the ways in which time as an existential dimension and as a methodological factor is handled, and the degree to which abstraction and concrete reality drive reasoning. In typo-morphological and morphogenetic work, historical and geographical contexts are close to paramount in shaping questions about process and outcome. In spatial modeling of any sort, reduction to essential formative and generative processes in the abstract is key to uncovering regularities that may underlie many aspects of city functioning and development regardless of local circumstances—until, that is, they are figured into the equations at whatever scale and in whatever way is deemed appropriate. One can reconstruct morphological evolution from the highly variegated evidence of history, or one can model it abstractly from first principles until it approximates observed reality, and then, theoretically, one can proceed to prognostication on the basis of trend lines from either analysis. Hypothetically, these conceptual positions as found in urban morphology represent opposite poles that should permit oscillation back and forth between them; practically, they are often hard to reconcile, more so on grounds of the choice of fundamentally different methods and evidence than from any ultimate obstacle.

Given the specialisms we acknowledge in urban morphology, including the newer interdisciplinary links considered earlier, we face an increasing compartmentalization of effort in the compilation, analysis, and display of data, the growing preoccupation with statistic-based "metrics", and the ideographic languages often used to express theoretical ideas. The result is we risk ending up essentially talking past each other. We are often so busy " staying in our lanes," to use a traffic analogy, that attempts at synthesis remain perennially delayed in the interest of ever more elaborate compartmentalized knowledge about ever smaller slivers of the whole. The arrival of "big data," including crowdsourcing of perceptions and behavior, and the almost endless opportunities thus opened up to investigate conditions in places large and small in most corners of the globe which has only added to this methodological segmentation, and in some cases sharpened its edges.

4 The Challenge

The question of conceptual integration into an all-encompassing framework remains deferred. In an effort to jolt the issue forward, however, it seems worth proposing a radically organized experiment.

As a novel step, I now issue a challenge for a grand test of all major conceptual and analytical approaches in urban morphology to be applied in a flexibly designed project focused on a SINGLE CITY—whichever is considered the most suitable testing site—in order that conditions of culture, topography, historical past, and any other structural variables may be held constant, and the special discoveries of each approach can be squarely placed alongside those of the others for comparison and interpretation.

Clearly, this project will require voluntary participation, collective leadership, compatible scheduling, and a curiosity to reveal how each dedicated conceptual system yields findings that are interesting in relation to the other parallel investigations. I make no recommendation here as to which city in the world should be chosen for this treatment, except to suggest that it must have the population size, spatial extent, and historical depth to meet the essential analytical requirements of all the investigative approaches included. Questions of suitable spatial scales of analysis will be paramount, and without question a steering committee needs to be established, preferably with various types of institutional backing and finance assured, to coordinate progress and interpret the results.

The obvious purpose and value of such an undertaking is that the constituent investigations, each following its own rules of data assembly, analytical procedures, scales of aggregation and disaggregation, and presentation—with all studies tied to the same physical urban object, and all needing to take into account its specific parameters, situational factors, and quirks—would by definition set each conceptual analytical system in the sharpest light vis-à-vis the others. The results would provide the most systematic evidence of how each system performs in a controlled test, and what findings each might be able to contribute to a heightened understanding of that test place. Oliviera and colleagues may have attempted just a side-by-side experiment, but it is not clear that a northern Portuguese city is necessarily the ideal testing ground, and other analytical approaches need adding to the experiment. Beyond that, the evidence could well illuminate the means by which, and the degree to which, compatibility and integration of different approaches might be achieved.

In the end, there may not be a path to holistic integration, yet the intellectual gains from understanding why would be significant.

A remarkable precedent for such an undertaking is the international "Historic Towns Atlas" Programme, which published over the course of more than four decades over 450 individual historic urban atlases in 17 countries. Collectively, they now provide a spectacular resource for the comparative study of medieval and early modern town and city morphology across Europe and a few cases elsewhere (reviewed in Conzen 2008). By contrast, it might be possible for the project proposed here to run for perhaps 2 or 3 years, or maybe 5 years at most, based on careful planning and a will to complete it.

If this were a "modest" proposal, it would offer its suggestions in the most tentative terms, aware that the cooperation envisioned would be hard to procure, and anticipating that its chances of adoption were small. But it is an immodest proposal, made with the view that such a scheme would be beneficial to interdisciplinary cooperation and cross-conceptual understanding in urban morphology, regardless of current preoccupations. It would be one way to counteract the all-to-common inclination of morphologists wedded to contrasting approaches, when vying for attention, to talk past each other.

References

Batty M (1976) *Urban modelling: algorithms, calibrations, predictions*. Cambridge University Press, Cambridge, England

Batty M (2001) Exploring Isovist fields: space and shape in architectural and urban morphology,' *Environment & Planning B* 28:123–150

Batty M (2005) *Cities and complexity: understanding cities with cellular automata, agent-based models, and fractals*. MIT Press, Cambridge, Mass

Cataldi G (2013) Thinking about Alnwick's origins. *Urban Morphology* 17:125–128

Cataldi G (2018) Towards a general theory of urban morphology: the typo-morphological theory. In: Oliviera V (ed.) *Teaching Urban Morphology*. Springer, Cham

Cataldi G, Maffei GL, Vaccaro P (2002) Saverio Muratori and the Italian School of planning typology. *Urban Morphology* 6:3–14

Conzen MP (2008) Retrieving the preindustrial built environments of Europe: the Historic Towns Atlas programme and comparative morphological study. *Urban Morphology* 12:143–156

Conzen MP (2010) A cartographic analysis of Como's urban morphology. In: Cerreti C, Federzoni L, Salgaro S (eds.) Cartografia di paesaggi, paesaggi nella cartografia. Pàtron Editore, Bologna, pp. 149–166

D'Acci L (2013) Simulating future societies in isobenefit cities: social isobenefit scenarios, *Futures* 54:3–18

Foote K (1997) Shadowed ground: America's landscapes of violence and tragedy. University of Texas Press, Austin

Geisler W (1924) *Die Deutsche Stadt: ein Beitrag zur zur Morphologie der Kulturlandschaft*. J Engelhorns Nachfolger, Stuttgart

Gökce D, Chen F (2016) Does the typological process help to build a sense of place? *Urban Morphjology* 20:66–69

Griffiths S, Jones CE, Vaughan L, Haklay M (2010) The persistence of suburban centres in Greater London: combining Conzenian and space syntax approaches. *Urban Morphology* 14:85–99

Hillier B (1996) *Space is the machine: a configural theory of architecture.* Cambridge University Press, Cambridge, England
Hillier B, Hanson J (1984) *The social logic of space.* Cambridge University Press, Cambridge, England
Kropf K (2017) The handbook of urban morphology. John Wiley & Sons, Chichester, England
Mahbub R (2017) The geometry of urban layouts: a global comparative study. Springer, Cham
Marshall S (2015) An area structure approach to morphological representation and analysis. *Urban Morphology* 19:117–134
Marzot N (2005) Typological analysis and hermeneutics in the Conzenian and Caniggian schools,' *Urban Morphology* 5:48–50
Moudon AV (1997) Urban morphology as an emerging interdisciplinary field. *Urban Morphology* 1:3–10
Ng E, Yuan C, Ren C, Fung JCH (2011) Improving the wind environment in high-density cities by understanding urban morphology and surface roughness: a study in Hong Kong. *Landscape and Urban Planning* 101:59–74
Oliviera V (2016) Urban morphology: an introduction to the study of the physical form of cities. Springer, Cham
Oliviera V, Monteiro C, and Partanen J (2015) A comparative study of urban form. *Urban Morphology* 19:73–92
Pinto JA (1976) Origins and development of the ichnographic city plan. Journal of the Society of Architectural Historians 25:35–50
Schlüter O (1899) Uber den Grundriss der Städte, *Zeitschrift der Gesellschaft für Erdkunde zu Berlin* 34:446–462
Verstegen I, Ceen A (2013) Giambattista Nolli: mapping the city before and after the Pianta Grande, Studium Urbis Rome Center, Rome
Whitehand JWR (1981) Conzenian ideas: extension and development. In: Whitehand JWR (ed.) *The ruan landscape: historical development and management. Papers by M. R. G. Conzen.* Academic Press, London
Whitehand JWR (2015) Seeking an integrated approach to urban form: tasks ahead. *Urban Morphology* 19:2–3
Whitehand JWR (2016) Bringing order to urban morphology? *Urban Morphology* 20:2 (2016), 87–88
Whitehand JWR (2018) Crossing boundaries: towards a more integrated approach? *Urban Morphology* 22:99–100

GPSR Compliance
The European Union's (EU) General Product Safety Regulation (GPSR) is a set of rules that requires consumer products to be safe and our obligations to ensure this.

If you have any concerns about our products, you can contact us on

ProductSafety@springernature.com

In case Publisher is established outside the EU, the EU authorized representative is:

Springer Nature Customer Service Center GmbH
Europaplatz 3
69115 Heidelberg, Germany

www.ingramcontent.com/pod-product-compliance
Ingram Content Group UK Ltd.
Pitfield, Milton Keynes, MK11 3LW, UK
UKHW022217051026
15624UKWH00001B/211
* 9 7 8 3 0 3 0 1 2 3 8 0 2 *